GUIDE TO SYSTEMS INTEGRATION

Guide to Systems Integration

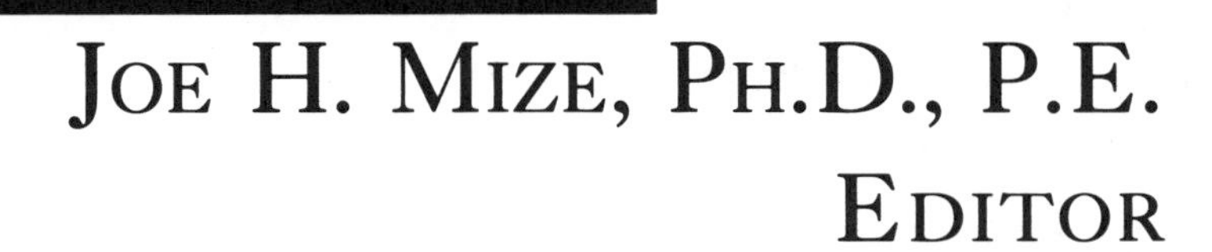

Joe H. Mize, Ph.D., P.E.
Editor

Industrial Engineering and Management Press
Institute of Industrial Engineers
Norcross, Georgia

Library of Congress Cataloging-in-Publication Data

Guide to systems integration / Joe H. Mize, editor
p. cm.
Includes bibliographical references.
ISBN 0-89806-111-3
1. System analysis. 2. Industrial engineering. I. Mize, Joe H.
T576..G85 1991
620'.001'1'--dc20 91-34120
CIP

95 94 93 92 91 5 4 3 2 1

Published in 1991
Printed in the United States of America.

Additional copies may be obtained by contacting:
Customer Service
Institute of Industrial Engineers
25 Technology Park/Atlanta
Norcross, Georgia 30092
USA
(404) 449-0460
(404) 263-8532 (FAX)

CONTENTS

PREFACE

Systems Integration has become a strategic goal of most progressive organizations as they prepare to enter the 21st century. To remain competitive in a global market, companies now realize that their entire organization must be analyzed, basic functions critically reviewed, and conventional wisdom challenged. The resulting generic components must then be integrated into a highly efficient, effective, and responsive organism.

Many companies are undergoing the most extensive and traumatic re-structuring in their history. And they are having to do this with few guideposts or role models available. Methodology for system integration is relatively embryonic and fragmented. There is no commonly accepted definition of systems integration.

This volume was prepared with two fundamental purposes in mind. First, it is an attempt to pull together into one document some of the important writings related to systems integration. The material has been organized so as to provide a logical structure relative to the planning, designing, implementation, and control (continuous improvement) of a system integration initiative. The second purpose is to provide foundation material for those who wish to prepare themselves for the Systems Integration Certification Examination administered by the Institute of Industrial Engineers.

More than fifty articles from a broad spectrum of relevant sources have been selected for inclusion in this volume. In addition, many more references are cited in the Additional Readings section of each chapter.

There is no claim that the articles presented or the references cited represent a comprehensive compilation of the material available on the subject of systems integration. Literally thousands of articles have appeared on the subject during the past ten years. An attempt has been made to select those articles that contain meaningful substance for industrial engineers who have responsibility for working in the area of systems integration.

The serious professional working in systems integration must continuously peruse a wide variety of journals and magazines in order to stay current. Attendance at a variety of professional education conferences is also critical.

We hope that the material in this volume will be useful to those industrial engineers who have a major role to play in preparing organizations for the 21st century.

Joe H. Mize
Oklahoma State University

I. INTRODUCTION

There are two primary audiences for the material in this manual. Many people involved in systems integration activities have wanted a single source of readings that cover the major aspects of systems integration. In addition, those who intend to take the Certification Examination in Systems Integration have needed a source of materials that will assist them in preparing to sit for the examination. Many people find themselves in both categories described above.

The first article in this section describes the Certification Program in Systems Integration administered by the Institute of Industrial Engineers. It develops the rationale upon which the program is founded. For those who wish to take the Certification Examination in Systems Integration, this article will provide very valuable insight into how the examination is structured, the specific knowledge and skill being tested for, and the nature of the questions that are included on the Examination. It even contains some sample questions!

Following the overview article on the Certification Program is a brief description of the organization of this study guide.

IIE CERTIFICATION PROGRAM IN SYSTEMS INTEGRATION

David W. Hess
IIE Director of Certification
Mgr., Manufacturing Services,
Distribution and Production
OKIDATA
Mt. Laurel, NJ

I. INTRODUCTION

The concept of developing certification programs for specific areas of industrial engineering application has been considered for many years and now has come to fruition with the development and administration of testing programs for systems integration. The systems integration certification program has been developed to be an adjunct to formal education, ongoing professional education, and practical work experience. To meet today's workplace demands, each industrial engineer must strive to pursue ongoing professional development and education. The rapid changes in technology, particularly in information processing, demand constant attention by those Industrial Engineers that seek to stay equipped with the latest tools to solve tomorrow's problems. Certification and ongoing professional education programs designed to help candidates prepare for certification examinations, represent a vital investment opportunity for all who participate.

This paper will seek to fully explore the role delineation process as it was pursued by the *Institute of Industrial Engineers* for systems integration. The key elements of the role delineation matrix will be reviewed and, where appropriate, examples from the author's experiences will be provided to embellish the explanation of the identified skill and/or knowledge. To further provide insights, a brief explanation of some key background historical data will be included as appropriate. Finally, this paper will seek to leave the reader with a summary of current activities and needs to ensure increasing success with system-integration certification activities. Clearly, the ongoing success of system integration should not rest with IIE staff members but rather with the general membership.

Before proceeding with an explanation of the role delineation process, however, a description of the Systems Integrator as it pertains to the *IIE Systems Integration Examination* is in order. This explanation is taken directly from the IIE handout, "*Systems Integration Certification Role Delineation Document*". (Exhibit I)

Reprinted from *1991 IIE International Industrial Engineering Conference Proceedings*

EXHIBIT I

Description of Systems Integrator
***as it pertains to the IIE Systems Integration Examination*[1]**

A Systems Integrator is the individual with *overall* responsibility for the complete integration of information, material and resources into an efficient effective system. The *functional* responsibility for complete integration may belong to a team, committee or group; however, the individual leading this group should possess the knowledge and skills representative of a Systems Integrator. The typical Systems Integrator with interest in IIE certification has a basic knowledge of accepted industrial engineering principles normally obtained from a four-year academic program, and at least five to seven years of experience in the Industrial Engineering field, focusing in the planning, design, implementation and control of systems. The particular industrial/business area (e.g., manufacturing, health care, financial services, etc.) is not as crucial as the particular knowledge and skills involved.

The *system*, as used here, can refer to one isolated system, or an integration of two or more sub-systems. Regardless of system type or size, there are particular functions and processes that should be completed to ensure successful implementation and operation. The Systems Integrator must possess the knowledge and skills necessary to ensure completion of system planning, design, implementation and control activities. This individual may not personally complete every activity himself, but rather must manage the entire project with the knowledge of the activities to be completed, the information, personnel and materials required for completion, and the interface of all involved functional areas to ensure total communication and integration. Along with the functional areas of the organization, the Systems Integrator must be aware of major issues including technical and physical issues, managerial issues, strategic/financial issues, and operational and functional issues.

The Systems Integrator generally leads a systems integration project from initial idea conception through planning design and implementation, to control, monitoring and continuous improvement of the system. As shown, the individual must perform many functions and fill many roles. But most importantly, the Systems Integrator must be an *effective* manager and leader of *people*.

---

[1] "*Systems Integration Certification Role Delineation Document*," Institute of Industrial Engineers, Circa September 1989, Norcross, Georgia

II. SOME BENEFITS OF CERTIFICATION

Consider the project engineer that becomes deeply involved in a major company project that spans a 1-3 year period. The intensity of major projects often leaves little time for ongoing professional education and sometimes the project engineer emerges from a successful project only to find that some of his skills need to be updated. In other circumstances, the project engineer emerges from a major project realizing that the project has somewhat utilized a unique application of new technology and the engineer seeks to share his knowledge within the professional engineering community. In both situations, the administration of a systems integration program has something vital to offer practitioners.

Now consider the task of developing ongoing professional education programs that will encourage local and regional chapter operations to participate for the benefit of Industrial Engineers working in an immediate geographic area. Consider that the employers of our local chapter members want to encourage professional development but they also want to show tangible benefits for their investment. Once again, the certification program has a great deal to offer. First, members can demonstrate tangibly their enrollment and successful passage of specific exam material. Second, members can provide examples to study materials with their employers to, once again, provide tangible evidence of the benefits derived from certification. Third, members can participate in leading study sessions and thus gain skill in both technical knowledge as well as presentation skills, again, providing employers with tangible evidence of the benefits of certification.

With the tangible benefits of certification come several intangible benefits. First is the definition of the skills and knowledge associated with a particular field of concentration. In the case adopted by the *Institute of Industrial Engineers*, the process of "role delineation" has been selected to define the skills and knowledge required for successful practitioners of system integration. Although this concept will be further discussed later in this paper, accept the simplistic view that "role delineation" is a process that combines brainstorming with nominal group techniques to identify necessary skills and knowledge with statistical survey validation to ensure accurate representation of the required knowledge. Once again, certification provides a tangible document that helps specifically define necessary skills and knowledge for specific fields of concentration. This document can help the *Institute of Industrial Engineers* as well as the professional to provide focus for ongoing professional education.

Consider, also, the intangible benefit that employers will realize as the number of certified System Integrators increases. Employers will gain confidence in approving project assignments and employment for those that have demonstrated successful completion of examinations. This confidence could lead to greater opportunities for Industrial Engineers as employers realize that the training for industrial engineering is both technical and broad enough to provide significant leadership growth opportunities.

A good certification program must be dynamic rather than static. New technologies must be periodically incorporated; outdated techniques retired. In addition, the relative weight assigned to specific skills and knowledge must be periodically reviewed and adjusted to reflect corresponding changes reported by integration experts. This iterative approach reflects the constant changes encountered in real-world experiences and serves as a constant service to IIE, its membership, and to employers.

III. THE ROLE DELINEATION METHODOLOGY

After conducting a feasibility study regarding cost and program methodology, the *Institute of Industrial Engineers* chose in March 1989 to retain the services of the *Profession Examination Service* (PES) of New York City, New York. PES is the oldest, nonprofit testing organization in the United States and demonstrated extensive experience in developing certification programs for many other organizations. To this end, the PES role delineation process is generally described in this section so that the reader can better understand its significance to the development of examination questions as well as other related certification program administration issues.

Before describing the role delineation process, a note of recognition and appreciation is in order. On May 18 and 19, 1989, a committee was assembled for two long days of meetings and countless hours of homework spanning the summer months of 1989. Each person added dimension to the project and all have agreed that the learning experience was intrinsically rewarding. Nonetheless, I would like to take this opportunity to, once again, thank the *Role Delineation Committee* for their unselfish contribution of time and knowledge. (Exhibit II)

EXHIBIT II

ROLE DELINEATION COMMITTEE

Allen Bensen
IIE Headquarters
25 Technology Park/Atlanta
Norcross, GA 30092

The Lehigh Press, Inc.
7001 N. Park Drive
Pennsauken, NJ 08109

Janet Fath
Dept. R16B
IBM Corporation
1500 River Edge Parkway
Atlanta, GA 30358

David W. Hess
OKIDATA
532 Fellowship Road
Mt. Laurel, NJ 08054

John E. Hughes
137 Crooked Backloop
Southern Shores, NC 27949

Joe Mize
Industrial Engineering Dept
Oklahoma State University
322 Engineering N.
Stillwater, OK 74078

James Robbins
Allied Signal Aerospace
P. O. Box 5217
Phoenix, AX 85010

Myron F. Wilson*
Rockwell Collins
MS 106-193
400 Collins Road, N. E.
Cedar Rapids, IA 52498

*Chairman

To best understand the role delineation process, we must begin with a review of terminology. The essence of this process is the application of the *Nominal Group Technique* to identify the minimum knowledge and skills that a candidate must proficiently demonstrate understanding in order to become certified. Once the role delineation process is complete, exam questions can be developed to match specific knowledge and skills considered to be representative of the entry-level professional. A major consideration of this process is that test specifications reflect *actual practice* so that the resulting examination is job related and content valid. The content validity of the examination refers to the representativeness of practice of the profession and the knowledge necessary for competent practice.[2]

A glossary of terms provided by PES is summarized in Exhibit III.[3]

---

[2] "*The Role Delineation for the Development of Test Specifications for a Certification Examination,*" Professional Examination Service, Circa March, 1989, New York, New York.

[3] IBID.

EXHIBIT III

GLOSSARY

ROLE DELINEATION: The systematic process which embraces the collection, compilation, verification, study, and application of all pertinent information about the nature of a specific job for such purposes as personnel management, job evaluation, examination development and validation, recruitment, and employee appraisal.

JOB DOMAINS: The major responsibilities or duties that make up a job. Domains are analyzed for component tasks that identify specific job-related behaviors necessary for the successful performance of a particular job.

TASK STATEMENT: A standard format used in the compilation of distinct, identifiable work activities which seeks to answer the following specific questions:

WHAT activity did you perform?
TO WHOM or **TO WHAT** was your activity directed?
WHY did you perform that activity?
HOW did you accomplish the activity?

KNOWLEDGE: An organized body of information, usually of a factual or procedural nature, which, when applied, makes successful performance of the job possible.

SKILLS: The proficient manual, verbal or mental manipulation of data, people or things. Skill embodies observable, quantifiable, and measurable performance parameters.

The role delineation process seeks to capture all of the essential elements of a specific job in order that they can be validated as they apply to personnel management, job evaluation, examination development and validation, recruitment and employee appraisal. There are several levels or layers of this process that summarize major responsibilities, job domains, and "drill down" through task statements to more specific knowledge and skills. Task statements seek to answer "*what*" activity is performed, "*to whom*" or "*to what*" the activity is directed, "*why*" the activity is performed, and "*how*" the activity is accomplished.

Although this process may seem to be initially confusing and complex, it really is nothing more than a disciplined systematic approach to a detailed job-description. Since many Industrial Engineers are familiar with the job description process, this is really an application of many familiar principles. The key principles in this process are that the resulting job description must accurately represent the minimum level of competence required by the practicing professional and that it be detailed enough to facilitate test validation and question writing.

Understanding that role delineation is actually a form of detailed job description writing will allow the Industrial Engineer to fully appreciate the following process. The first step is to use brainstorming and any other technique that a group of qualified practitioners feel is necessary to identify major competency or performance domains. These domains are principal areas of responsibility or activity comprising the job or occupation under consideration and ideally will include a brief behavioral description.

After all of the basic domains have been identified, the next step is to sub-divide the domains into associated issues. In the systems integration process, the result of defining domains and issues created a 4 X 4 matrix. See Exhibit IV.

EXHIBIT IV

Descriptions of Systems Integration Domains and Issues

Matrix design of certification module:

DOMAINS	ISSUES: Technical/ Physical	Managerial	Financial	Operational/ Functional
Planning	10.10	10.20	10.30	10.40
Design	20.10	20.20	20.30	20.40
Implementation	30.10	30.20	30.30	30.40
Control	40.10	40.20	40.30	40.40

Planning Domain: (10.XX)
This domain includes the functions and processes performed during initial planning of a system, beginning with idea conception and initial concept development. It includes investigation of requirements and problems for possible solution ideas, formulation of project scope, selection of project personnel, and other functions and processes involved with developing an idea for solution into a tangible project.

Design Domain: (20.XX)
This domain includes the functions and processes performed during the design phase of systems integration projects. It includes not only the process of designing the actual system to be implemented, but also the design of supporting systems and programs including project management systems, training programs, financial tracking systems, contract management programs, standards and procedures, and other applicable functions.

Implementation Domain: (30.XX)
This domain includes the functions and processes performed during the implementation phase of systems integration projects. It includes system component procurement and implementation, creation of appropriate system documentation, personnel training, progress and budget tracking, installation of revised cost management and performance measurement systems, changes to organization structure, compliance monitoring, system testing, evaluation and modification, and other applicable functions.

Control Domain: (40.XX)
This domain includes the functions and processes performed during the control phase of systems integration projects. It includes system evaluation, final documentation, evaluation of project schedule and financial performance to goals, communication of operational capabilities, contract close-outs, establishment of ongoing evaluation systems, maintenance processes, updates of policies and procedures, and other applicable functions.

Technical/Physical Issues: (XX.10)	These are issues pertaining to the technical and physical aspects of systems integration projects including system equipment and any necessary hardware and software, physical system layout, product, material and/or information flow, ergonomic considerations, technical information considerations, technical knowledge and skills of personnel, facilities considerations and environmental standards, and other applicable issues.
Managerial Issues: (XX.20)	These are issues pertaining to the managerial aspects of systems integration projects including strategic planning, organizational structure, goals and objectives, project management, resource utilization, training, team-building and facilitation, contract management, management presentations, personnel policies and labor relations, and other applicable issues.
Financial Issues: (XX.30)	These are issues pertaining to the financial aspects of systems integration projects including budget development and control, system costs and benefits, capital equipment and other economic justifications, risk analysis, trend analysis and forecasting, and other applicable issues.
Operational/ Functional Issues: (XX.40)	These are issues pertaining to the operational and functional aspects of systems integration projects including procedures, policies and standards, operational data requirements, compliance requirements and procedures, product/service functions and features, functional personnel considerations, continuous improvement concepts, system maintenance, and other applicable issues.

Once all of the basic domains and issues were identified, the next step was to assign component task statements to each matrix cell. Each task statement was defined as a specific, goal-directed set of activities with a common objective or type of output. Tasks were written with a verb so that they could later be associated with specific knowledges and skills.

The next step was to identify and associate learnable knowledge and skills with each task statement. Linking specific knowledges and skills to specific tasks later became the basis for test specifications and item (question) writing.

Prior to item writing, however, there was one more very important step in the process. This step was the validation of the role delineation document. The complete document was mailed to a statistically large group of recognized practicing professionals; in this case IIE chose to mail the document to more than 400 experts within the field of systems integration. Each expert was asked to review all categories and assign relative importance to each matrix combination of domains and issues. When the validation was complete, the role delineation document was fine-tuned to become the specification for exam construction and specific item (question) writing.

IV. DEVELOPMENT OF THE EXAM

In order to develop exam questions, many groups and individuals were asked to volunteer some of their time and knowledge. Each received a copy of the role delineation document with very specific instructions on how to best write multiple-choice exam questions. In a document provided by IIE and reviewed by PES, each item writer received examples of well-written items and 22 tips on how to best construct the wording of good questions. These tips were provided to reduce the subsequent editing effort that would eventually seek to correct grammar, the mis-use of pronouns and time-dependent background data, etc.

In addition to writing questions, each item writer was asked to denote the specific matrix rubric that a question was intended to test for competency. Each item writer was also asked to provide specific references (book, periodical, journal, etc.) wherever possible. The goal of IIE has been to provide solid documentation for all questions.

As questions were received by IIE, they were reviewed by a small panel of experts to confirm their rubric assignment and accuracy. Once this task was complete, all questions were sent to PES in New York City where they were screened for grammar and format. Finally, each question was coded and entered into a secure item "bank" for later reference by the *Exam Construction Committee*. Exhibit V provides an example of the materials provided to Item Writers by IIE.

EXHIBIT V

Example of Scenario Item-Writing Based on Role Delineation Task Statements

The Role Delineation Document is the basis for all items. In using the document, a Rubric category is selected, then specific task statements are selected. Knowledge and Skill statements provided should be used as guidelines; you may be aware of more relevant or up-to-date knowledge and skills to apply to the task statement. Consider the following example:

Rubric selected:	20.10.00	Design:	Technical/Physical
Task Statement selected:	20.10.03		

"Evaluate system design alternatives by choosing an evaluation method, identifying specific performance criteria (e.g. human factors, ergonomics, waste control, management criteria, cost data, etc.), collecting and analyzing data, and applying decision-making methods to select the best alternative system design."

Knowledge of:

- K-1 Evaluation methods, including factorials, randomized blocks, latin squares, confounding and factional replication, multiple comparisons, and responses surfaces
- K-2 Performance criteria identification methods
- K-3 Data collection methods, including time and motion study
- K-4 Data analysis methods including regression analysis and analysis of variance
- K-5 Decision making methods

Skill at:

- S-1 Designing experiments
- S-2 Collecting data
- S-3 Data analysis
- S-4 Translating specific performance criteria into measurable objectives
- S-5 Good decision making

Item Development:

A commonly used method of system design evaluation, although not specifically mentioned in the Knowledge and Skill statements, is simulation. Construct a scenario using simulation as an evaluation method.

Scenario:

A simulation model must be developed to evaluate whether the existing physical layout of equipment in a particular department will allow production of required quantities of a new product at the specified quality levels.

In a given period, the required quantities of the new product should agree with the following distribution:

Quantity Required	Probability of the Quantity
100	.15
200	.30
300	.25
400	.20
500	.10

Items: (Note: In item-writing, always use "A" for the correct answer.)

1. The simulation model most appropriate for this type of evaluation is a/an:

   A. discrete-event simulation model
   B. pseudo-number simulation model
   C. analog simulation model
   D. physical simulation model

2. If the random number of 44 had been generated by the simulation, the desired quantity of the new product would be:

   A. 200
   B. 100
   C. 300
   D. 400

3. If time is incremented on the basis of occurrences within the simulated production line, the model would be referred to as a:

   A. next-event simulation model
   B. continuous-occurrence simulation model
   C. random-time simulation model
   D. continuous-clock simulation model

4. Which of the following is a widely used method for testing the significance of a hypothesized distribution in a simulation model?

   A. Chi-square
   B. Exponential distribution
   C. Number of degrees of freedom
   D. Parameter estimation

The information required by the Institute for use in the examination would include the specific task statement number used, the scenario statement, the items (with "A" as the correct answer), and a reference (book, periodical, journal) with author, date of publication and publisher.

V. THE EXAM CONSTRUCTION

A small group, known as the *Exam Construction Committee*, first met in New York City at PES headquarters in February, 1989. The objective of this group was to select 150 multiple-choice questions that would match the role delineation specification. Questions were selected by rubric to match the weighted distribution developed by the statistical validation study. Questions not selected for the initial examination were returned to the "item bank" for future exam applications.

The need to generate new exam questions is an ongoing requirement. PES prefers to "retire" approximately 25% of the exam questions after each exam administration. In addition, sample questions are needed for study materials and to provide applicants with an idea of what to expect prior to the real exam.

After the first test administration, PES tabulated an extensive table of statistics regarding responses to each question. When the Exam Construction Committee met for the second time, the statistical analysis was carefully reviewed for each question. Where appropriate, adjustments were made to increase the discrimination between answers and to ensure that questions were neither too hard nor too easy. The exam committee also reviewed new questions to confirm their rubric assignment and accuracy before adding them to the "item bank." Finally, the committee selectively replaced 25% of the questions from the first exam to meet the PES objective for retiring questions.

VI. DISCUSSIONS CONCERNING JOB-RELATED ACTIVITIES

At the time of this writing, it is not the intention to provide examples of personal job-related activities that can be associated with the specific rubrics. However, as time permits during the presentation of this paper, it is the author's hope that time can be devoted to interaction with attendees to use on-the-job experience to best demonstrate the role delineation process.

The essence of the certification program and the current examination process is that the knowledge base is dynamic and the exam constructive is iterative by design. Periodically, IIE expects to update the basic role delineation document and changes will be made to keep pace with changes within the profession.

On-the-job experiences that the author may choose to reference during the presentation of this paper will most likely be associated with barcode applications or inventory system enhancements. The author recently participated in a major system-conversion from a PRIME computer base to a UNISYS computer base with a UNIX operating system. This new corporate information system emphasizes distributed processing. In addition, the author is currently part of an ABCD industry task force, representing more than 2,100 member companies, that recently developed new barcode packaging standards.

VII. SOME THOUGHTS CONCERNING MEMBER PARTICIPATION

The certification program is now better than two years old. In less than one year, the role delineation process was completed for systems integration and the first test was administered. During the second year, improved study materials have been developed by IIE. Applicants can now receive a recommended bibliography, a summarized re-print of selected articles, sample questions and recommendations for specific IIE publications that could help them study.

While this is impressive, there is much more that could be done. The *Certification Board* hopes that the enthusiasm for this program continues to grow and that additional study materials can be developed for member utilization. The goal for this third year is to assemble within the membership, a group that will continue to develop better study materials as well as to become the ongoing Exam Construction Committee.

The process has been initiated and a new product has been brought to the marketplace. The exam is a quality product and, make no mistake, it is challenging. But the exam is fair and the process of development has been disciplined. Now the product must move toward maturity.

The "*ownership*" for this program cannot rest with the IIE staff. To continue to provide quality, the membership must be the driving force. The membership knows the "state of the art" and should maintain a high interest for sharing this knowledge with its profession. Involvement is very rewarding and will help those participating to stay current with their skills. Involvement can also provide members with tangible justification for their efforts that should help their career. Credentials can be enhanced with certification and employers will likely find this activity easier to support than generalized membership fees. Think of the opportunities and then . . . get involved!

EXHIBIT VII

CERTIFICATION: REQUIRED ACTIONS vs OPPORTUNITIES

CANDIDATE REQUESTS, APPLICATION INFORMATION REGARDING CERTIFICATION	
PACKAGE IS SENT FROM INSTITUTE HEADQUARTERS WITH: 1. SUGGESTED READING LIST; 2. SAMPLE QUESTIONS FOR SELF-ADMINISTRATION; AND 3. APPLICATION FOR EXAMINATION	
CANDIDATE REVIEWS THE "APPLICATION PACKAGE" AND DECIDES WHETHER OR NOT TO SUBMIT APPLICATION FOR EXAMINATION. CANDIDATE MAY ELECT TO READ RECOMMENDED MATERIALS AND/OR ADMINISTER SELF-TEST OF SAMPLE QUESTIONS	REQUIRED ACTION
CANDIDATE SEEKS ADDITIONAL STUDY/PREPARATORY ALTERNATIVES	ELECTIVE ACTION OPS
IIE-SPONSORED SEMINARS (TO BE DEVELOPED)	ELECTIVE
LOCAL/AREA CHAPTER STUDY COURSES (TO BE DEVELOPED)	ELECTIVE
IIE SPONSORED PUBLICATIONS (MEMBERSHIP SUBMITTALS)	ELECTIVE
CANDIDATE SUCCESSFULLY COMPLETES EXAM ADMINISTRATION	REQUIRED ACTION
CANDIDATE BECOMES MEMBER OF EXAM REVIEW BOARD; ITEM WRITING EFFORT, ETC.	ELECTIVE
CANDIDATE HELPS CONDUCT REVIEW COURSES, WRITE STUDY MATERIAL, ETC.	ELECTIVE

BIOGRAPHICAL SKETCH

David Hess is a senior member of IIE and currently is the Institute Director of Certification. David has been involved in Institute Certification activities since 1988 when he participated in a task force study concerning certification.

David has been employed for 2-1/2 years by OKIDATA, a rapidly growing distributor of products associated with the personal computer industry, most notably, impact and non-impact printers. At OKIDATA, David is responsible for Industrial Engineering, Test Engineering and the East Coast Warehouse and Packaging functions. Prior to joining OKIDATA, David had more than 12 years of experience in positions of Industrial Engineering and Material-Control Management at companies including GE, Wheaton Glass Co., and Oscar Mayer and Co.

REFERENCES
(ADDITIONAL INFORMATION ON ITEM DEVELOPMENT)

Brown, F. G. (1983) **Principles of Educational and Psychological Testing.** Third Edition. New York: Holt, Rinehart.

Lindeman, R. H., & Merenda, R. F. (1979) **Education Measurement.** Second Edition. Glenville, IL: Scott, Foresman.

Wesman, A. G. (1971) Writing the Test Item. In R. L. Thorndike (Ed.), **Educational Measurement.** Washington: American Council on Education, Pages 81 - 119.

ORGANIZATION OF GUIDE TO SYSTEMS INTEGRATION

In the preceding article on the IIE Certification Program, the "Job Domains" of Systems Integrators were defined as Planning, Designing, Implementing, and Controlling. We will use these and other categories to organize the material for inclusion in the *Guide to Systems Integration.*

There have been so many different meanings associated with the term "integration," we feel that it is important to establish some common understanding of its usage relative to industrial engineering activity. Consequently, Section II of the *Guide* is entitled "Fundamentals of Systems Integration."

Having a foundation upon which to build, four consecutive sections are devoted to the major job domains mentioned above. Section III, "Planning for Systems Integration" deals with the critical area of tying systems integration initiatives to the strategic business goals of the organization. The selected articles also describe various processes for developing comprehensive plans for systems integration.

Section IV deals with "Design Issues in Systems Integration." All major systems in an organization are subject to a critical design review. This includes not only *physical systems* (such as equipment selection/layout, material handling, etc.), but also *information systems* (manual and automated information systems, networks, databases, software, etc.) and *decision/control systems* (master production scheduling, inventory management, quality assurance, performance measurement, etc.). In many cases, organizations are essentially re-created. The end result of the Design stage is a detailed specification of the "To-Be" situation.

Section V, "Implementation Issues in Systems Integration," is concerned with carrying out all the plans and designs created in the previous stages. There are many barriers to successful implementation which must be overcome. Experience shows that the more critical barriers are behavioral and organizational. Usually, large numbers of people are being asked to change rapidly, in fundamental ways. The successful systems integrator must have skill in dealing with this type of problem.

Section VI, "Control Issues in Systems Integration," is concerned with maintaining and continuously improving the system once the system integration initiative has been implemented. It is proper to visualize a feedback control loop, in which system performance is measured, continuously re-assessed and continuously upgraded.

While Sections III through VI cover the four job domains of systems integration, there are other topics that are also important. Two sections are included to cover these. Section VII, "Enabling Technologies," deals with some of the particular technologies that are critical for successful integration of most industrial systems. Specifically, data communications, networks, software, and information system design are covered in the articles included in this section.

Section VIII, "Tools and Techniques for the Systems Integrator," describes particular methodologies that are useful in system integration. Tools for systems analysis and design include modeling approaches, computer simulation, graphical diagramming tools, and data flow diagrams. Other tools deal with means for justifying investments in advanced manufacturing technologies, including both tangible and intangible factors. Project management tools are described, explicitly as they may be applied in systems integration initiatives. Finally, group dynamics tools are described, for the purpose of providing help to the systems integrator in the critical area of group problem solving and decision making.

The final section in the book is Section IX, "Systems Integration Case Studies." Three papers are included to illustrate how certain organization approached the systems integration challenge.

In each section, several articles are presented from a wide variety of sources. Each was carefully scrutinized and selected as being particularly germane to the topic highlighted in that section. At the conclusion of each section is a list of "Additional Readings" that will provide further insight into the topics.

II. FUNDAMENTALS OF SYSTEMS INTEGRATION

The term "systems integration" means different things to different groups of people. To an aircraft design team, system integration deals with the problem of assuring that all major sub-systems (airframe, propulsion, hydraulics, electronics, etc.) are properly inter-related, such that the overall aircraft performance is optimized. To the electronics specialty firm that designs and manufactures the avionics package for the aircraft, system integration deals with assuring that all the sub-components of the avionics equipment (computer chips, diodes, DRAMs, printed circuits, etc.) are properly inter-related, such that the avionics function performs as required. The geometry of the "box" into which the sub-components reside must also be consistent with the space available for this equipment. Interfaces to other systems are also critical.

The electrical engineer or computer scientist responsible for the various data communication networks views system integration in terms of the flow of "bits and bytes." They, too, are concerned with the various sub-components working together in a harmonious whole. There are many critical interdependencies among the sub-components.

The industrial or manufacturing engineer engaged in system integration may at first seem to be engaged in a totally different type of activity from those described above for other professionals. Upon further reflection, however, the industrial engineer doing system integration work is responsible for performing the same *basic* functions as are the other people. The major difference is that the industrial engineer must deal with very large and ill-structured systems. Indeed, the entire organization is the "system" being studied, analyzed, rationalized, and hopefully optimized. Rather than combining electronic components into a functioning system, the industrial engineer must combine functional organizations, equipment, material handlers, policies and procedures, workers at all levels, computer/information systems, suppliers, etc., into a harmoniously working total system.

The papers in this section have been selected to provide a conceptual foundation for systems integration. An attempt has been made to select writings that avoid flowery rhetoric and faddish phraseology.

FUNDAMENTALS OF INTEGRATED MANUFACTURING

Joe H. Mize
Regents Professor of Industrial Engineering and Management
Director, Center for Computer Integrated Manufacturing
Oklahoma State University

INTRODUCTION

In 1973, two of the most respected scholars in the area of operations management, W. L. Berry and D. Clay Whybark, wrote the following: "The MRP crusade has been officially closed and pronounced successful....Successful MRP system applications have been reported by APICS members from a wide variety of industries and production organizations." (3)

Another significant event that occurred in 1973 was the publication of the landmark book by Joseph Harrington, *Computer Integrated Manufacturing*. (10) Ah — 1973 was a very good year! The MRP struggle finally over, and Computer Integrated Manufacturing (CIM) poised to burst upon the scene. How are these two events related? Primarily in the irony of the perceived achievements and relative expectations of the two "movements", MRP and CIM. We now know that neither the perception of MRP being a perfected methodology nor the expectation that CIM would quickly become the wave of the future would prove to be accurate.

Actually, in the research paper by Berry and Whybark, the authors warn the reader that much remained to be accomplished in MRP. This has certainly proven to be the case. Even today, 20 years later, less than 25% of the attempts to implement MRP in U.S. companies are successful. (16)

Similarly, the excellent book by Dr. Harrington received essentially no attention at the time. It was destined to lie in relative obscurity for a full decade before the concepts of CIM would begin to be discussed on a wide-scale basis. After ten years of intensive exposure, there is still much confusion over the meaning of integrated manufacturing.

Unfortunately, CIM became exactly what Dr. Harrington did not want it to become—a buzzword; another fad; this week's answer to all problems. Since 1984, literally thousands of articles and conference presentations were published proclaiming to explain the meaning of "CIM". Many of these articles have caused more confusion than they have helped in terms of clarifying the underlying concepts of integrated manufacturing.

In the following sections, we will attempt to "sort out" some of the confusion regarding integrated manufacturing and to provide a conceptual basis for understanding the elements involved in the design of integrated systems. We will also discuss what role may be played by industrial engineers in systems integration.

THE CURRENT STATUS OF CIM

Depending on who you listen to, CIM is either "right on track" or "dead in the water." Laura Conigliaro, first vice president for equity at Prudential-Bache Capital Funding (New York City) was quoted at a Design Automation Seminar as saying, "There are a lot of jaded people right now hearing empty promises of integration." (7) Another revealing statement was given by Ronald Reimink; "Integration is one of the most overused, abused, and misunderstood words in the technical world, and yet it is a uniquely important factor in system success!" (11)

We see in the last quotation the dilemma we all face: CIM has clearly become another shallow, meaningless, overworked term having a myriad of meanings to different people and companies, and yet, the concept of system integration is so fundamentally critical, we must continue our struggle to give it meaning and substance.

CIM has been portrayed by countless zealots as the ultimate perfection in system design, in which everything connects to everything else, and the system works perfectly every time even with the lights turned out. Robert Malone, Editor-in-Chief of *Managing Automation*, said it well: "We do not need more and more vagaries, more levels of control, more abstract profundities filled with spacious reasoning and elegant surfaces.... We must not think of CIM as some nice but unobtainable perfection." (11)

How have we allowed ourselves to get into such a confusing situation? There are probably many reasons, but the predominant factor is the insatiable-appetite of U.S. Managers for the "instant solution," the "quick fix"—in truth, the fad. This insatiable appetite is fed by a large and growing "fad industry." Computer graphics have enhanced the fad perpetrators' ability to add more and more glitter to their "solutions."

Consider, for example, the hypothetical advertisement by a "systems integration house" shown in Figure 1. The author put these words together by taking phrases from several advertisements. Is it any wonder that managers are being turned off to the concept of integration?

FOR YOUR TOTAL CIM SOLUTION,
CALL TODAY

We will design for you and help you implement a Computer Integrated Manufacturing System organized around a central data base that bridges your existing islands of automation, capitalizes on the power of distributed processing, provides maximum flexibility in manufacturing, eliminates inventory, assures full machine utilization in a fault tolerant environment.

Figure 1. Hypothetical, but Typical Advertisement by a "Systems Integration House."

There are no short cuts to CIM. There are no easy answers. CIM cannot be purchased. The sooner we realize this, the sooner we can get our act together for a long-term program of continuous improvement. The simple truth of the matter is that we do not know how to actually achieve a fully integrated production environment. Our basic concepts regarding integration are still in the formative stage. We must make significant progress at the conceptual level before we can hope to develop sound methodology for achieving CIM.

As usually happens at the beginning phases of a major fad, several self-proclaimed CIM experts have stepped forward claiming to have omnipotence in system integration. In hundreds of journal articles and conference presentations, CIM is loudly proclaimed to be "the answer" to all the woes facing a company, but the articles do not make clear *what questions* CIM is an answer to. The six or eight discrete steps commonly listed for achieving CIM are conceptually shallow and provide little new insight into how to approach such an endeavor in an actual situation.

A SAMPLING OF CIM DEFINITIONS

While there is no commonly accepted definition of CIM, it is instructive to look at how a variety of authors and experts have defined the term.

> "A collection of machines tied together by a material handling system and controlled by a single computer or hierarchy of computers." (8)
>
> "A production facility that consists of a group of process equipment units, such as machine tools, auxiliary equipment (inspection machines, washing stations, etc), linked with an automatic materials handling system that reaches every process station, the entire facility being integrated under common computer control." (2)

These first two definitions imply that CIM applies to a given set of machines, such as a flexible manufacturing cell. Both place emphasis upon the material handling system, with one saying this system should be automated and the other not making that requirement. Both of these definitions are limited to a subset of the factory floor. Computer control is common to both.

> "CIM is the integration of key product-related data in a company, where the integration of various computer-based automation activities leads to improved productivity in all business areas from marketing to product shipment." (1)
>
> "CIM is the vehicle that links the operations of the entire company together which results in a cohesive system." (9)

The preceding two definitions broaden the scope of CIM to include company functions outside manufacturing. The word "automation" appears in one of the definitions, implicitly tying automation and CIM together.

> "CIM is a rounded concept that rests on a central manufacturing database. Linked to this database will be the key functions of engineering design, manufacturing engineering, factory production, and information management." (5)

The definition above adds the concept of a central manufacturing database to our understanding of CIM.

> "Computer integrated manufacturing is the automation and integration of the business of manufacturing from product design to distribution." (6)

This definition again seems to suggest that automation is an essential element of CIM.

> "Integration is the process of interfacing subsystems such that a total system performs its prescribed functions under all defined conditions." (18)

This definition carries a lot of appeal; however, it equates "interfacing" and "integration." As we shall see shortly, these two terms have different meanings and should not be used synonomously.

Interesting enough, the term "CIM" is not defined in the first definitive book on CIM, written in 1973. (10) The author provides a thorough explanation of CIM, but he does not define it explicitly. Neither is CIM defined in the *Computer Integrated Manufacturing Glossary*. (21)

It is not necessarily bad that different people in different companies view CIM in different lights. Not all companies have the same needs or experience the same situations. However, much of the nonuniformity of CIM definitions must be attributed to the lack of a sound conceptual base for system integration methodologies.

WHAT CIM IS NOT

There is a growing awareness that CIM is not a specific technology that a company can buy. CIM is a *concept*, one which involves the intelligent combination of many technologies. There is not one generic set of CIM technologies or applications that will serve the needs of all manufacturers. Each company has its own unique set of needs. There is no generic factory; therefore, there is no generic CIM.

There is also growing awareness that CIM is not something that can be achieved quickly, painlessly, or inexpensively. Moving to a CIM concept requires a company to undergo fundamental, pervasive changes in its policies, methods, equipment, control systems, management practice, and basic values. A common misconception is that CAD/CAM is equivalent to CIM. CAD/CAM constitutes a significant portion of CIM, but it is only a portion.

A similar misconception is that CIM is characterized by a high degree of automation. This is not necessarily the case. CIM has a lot more to do with the systemization of data flows than with equipment/process automation.

CIM is not a set of software packages, although any CIM environment would most certainly include a large number of such packages. Similarly, CIM is not a data communications network, although such a network will likely be included in a CIM implementation. Perhaps the most difficult characteristic of CIM to deal with conceptually is that CIM is not some defined, perhaps idealized, state. CIM is a strategic direction, not a specific destination. It is a long-range target, and our understanding of that target continuously changes.

A GENERIC APPROACH TO CIM CLARIFICATION

One of the real disservices associated with CIM falling prey to the "fad monster" is that many people seem to view CIM as an end in itself. The solution has become more important that the problem. What is it, specifically, that we hope CIM (however we define it) will do for us? Some of the desirable outcomes to which we would expect CIM to contribute are shown in Figure 2. The fundamental reason for pursuing CIM, however, is the conviction that CIM is the most effective means conceivable for achieving our strategic business objectives year after year.

An examination of Figure 2 suggests that in a CIM environment, several traditional performance measures do not appear. For example, machine utilization, worker idle time, and labor productivity are not included. These measures are not as relevant in a CIM environment. This fact has far-reaching implications for the practicing industrial engineer.

STRATEGIC OUTCOMES

- Improved Customer Service
- Improved Quality
- Improved Competitiveness
- Improved Responsiveness, Flexibility
- Greater Profitability, Long Term
- Shorter Time to Market
- Greater Return on Assets Employed
- Greater Value Added per Square Foot
- Greater Market Share

OPERATIONAL OUTCOMES

- Greater Manufacturing Velocity
- Improved Schedule Performance
- Lower Total Cost
- Reduced Inventories, Greater Turns
- Enhanced Stability of Employment
- Lower Economic Batch Quantities
- Shorter Vendor Lead Times
- Greatly Reduced Congestion in Plant

Figure 2. Desirable Outcomes of CIM

There are uncertainties about each of the three words in CIM. What does the word "computer" in the term imply? We know what a computer is, of course, but what meaning does it have in this context? About the best explanation we can contrive is that we know the manufacturing systems of the future will be driven by enormous amounts of data, and that computers are essential in the overall processing and transmission of data throughout the manufacturing environment. We also know that many machines and processes will operate under real-time computer control.

The third word in CIM, "manufacturing," also raises certain questions. The primary question is one of boundary. Does CIM include *only* the production operations? Does it also include engineering design? What about purchasing? Some companies have decided that integration should apply to the total firm, including sales, accounting, strategic management, etc. A new term (wouldn't you know it!) is beginning to be used to characterize the total integration of a company. That term is *Computer Integrated Enterprise (CIE).*

Having dealt with the first and third words in CIM, we have saved the most difficult word for last. The word "Integrated" in CIM, while the most critical, is by far the least understood.

If we carefully and critically examine the implied meaning of the word "integrated" as used in most writings on CIM, we would find that the authors are actually using the wrong word. The manner in which these writings view the combining of system elements is really that of *interfacing*, not integration. It is useful to resort to dictionary definitions to distinguish between "integrate" and "interface."

Definition of "Integrate"

- To bring together parts into a whole.

Definition of "Integrated"

- An integrating or being integrated.
- (Psychol.) The organization of various traits, feelings, attitudes, etc, into one harmonious personality.

Definition of "Interface"

Noun: A point or means of interaction between two systems.
Verb: To interact with another system.

It is interesting to utilize a chemical analogy to illustrate further the distinction between "integrate" and "interface." Consider the difference between a mixture and a compound. Both are combinations of two or more elements. In mixtures, the ingredients do not lose their individual characteristics, and they may be separated by physical means. In compounds, however, the ingredients lose their individual characteristics, while the compound exhibits new characteristics. The constituents of compounds cannot be separated by physical means. Clearly, interfaced elements are analogous to mixtures and integrated elements are related to compounds.

Unfortunately, the analogy breaks down at this point. Whereas chemical elements behave and interact according to well defined laws, there currently exists no equivalent science base for describing and understanding the behavior of the interacting components of manufacturing systems. Glittery brochures and gaudy graphics from vendors of "CIM Solutions" are hardly the sound principles and fundamental knowledge that is so desperately needed for real progress in our ability to achieve integrated system designs.

The distinction between "interfacing" and "integration" can be illustrated with a concrete example. A division of a large company has used a particular part numbering system (which we will call Scheme A) for many years. A different division has also used its own part numbering system (which we will call Scheme B) for many years. The two divisions are being merged. Many of the parts under the two schemes are identical except for their part number.

If the merged divisions would develop one common part numbering system, say Scheme C, that would satisfy the needs of both groups, this would be an act of integration. However, if the two groups insisted on retaining two separate numbering systems, constructing a translation table for cross referencing, this would be an act of interfacing.

Are we implying that interfacing is bad? Not at all! In fact, the only feasible way of proceeding toward improved systems at this time is to interface many existing elements. But please! If we are interfacing, let us not claim to be integrating. False claims hinder our long term aspiration of learning how to achieve true integration.

A PROPOSED GENERAL DEFINITION OF CIM

Almost all the definitions of CIM proposed in the literature include the word "integrate." This is circular reasoning, and is therefore irrational. The definition of Computer Integrated Manufacturing shown in Figure 3 is proposed as a general definition of CIM. Like so many definitions of complex concepts, this definition requires further elaboration. Specifically, the word "rationalize" needs to be explained.

> **COMPUTER INTEGRATED MANUFACTURING DEFINED**
>
> Computer Integrated Manufacturing is an approach to the organization and management of a firm, in which the functions of design, manufacturing, and production management are mutually rationalized and completely coordinated, through the use of appropriate levels of computer and information/communication technologies.

Figure 3. General Definition of CIM

Definition of "Rationalization" (22)

- Making to conform to reason.
- Applying modern methods of efficiency to an industrial concern.

In the field of systems engineering, the term "rationalization" has come to mean the following: The systematic division of a system into its most fundamental elements, determining the cause-effect relationships between system elements, and characterizing the total system and its components in terms of the most important performance variables of the system. (15)

Applying this definition of rationalization to a production system, the following would be performed: The entire system, from product definition and raw material acquisition to the disposition of the final product, is carefully analyzed such that every operation and element can be designed to contribute in the most efficient and effective way to the achievement of clearly enunciated goals of the enterprise.

Realizing that these are just words, in the absence of a meaningful frame of reference, Figure 4 is included to illustrate the distinction between a traditionally designed system and a rationalized system. The reader can adapt and elaborate this figure to provide greater understanding of how to apply these concepts to his/her own organization.

DEFINING THE SCOPE OF CIM

In an earlier section, we said that CAD/CAM is not equivalent to CIM. We can go further and show the relationships between many of the "buzzwords" that are being thrown around today.

As a generic reference model for manufacturing system flow, consider the diagram in Figure 5. This diagram shows the major elements of a manufacturing system which involves product design, production planning, and manufacturing. Other functions of the firm will be added later.

TRADITIONAL SYSTEM	RATIONALIZED SYSTEM
Sales makes delivery commitment arbitrarily	Commitment cannot be made until impact on production schedule has been evaluated
Engineering designs components in isolation	Design must give proper consideration to cost, producibility, and quality impact
Engineering introduces design changes at will	Design change introductions are managed, considering schedules, inventories, and customer support
Manufacturing Planning modifies the design arbitrarily	Configuration control is exercised to assure product performance
Manufacturing Planning designs tooling "from scratch"	Design retrieval exercised; adapt existing tooling where possible
Factory Management splits orders and expedites to meet shipment quota	Scheduling discipline exercised, focusing on meeting customer delivery schedule
Factory Management deemphasizes spares production	Customer support given explicit schedule priority
Manufacturing deviates from process plan	Discipline exercised in following prescribed methods
Manufacturing fails to follow maintenance schedule	Fail-safe maintenance procedures employed
Production Control releases orders to shop early to provide work for work centers	Orders released only when needed to prevent "clogging up the shop"
Accounting focuses on historical reporting and taxation-related concerns	Accounting data used for managing the operation–integrated with performance data
Management focuses on high machine utilization and low worker idle time	Focus shifts to manufacturing velocity
Quality Control philosophy is to "cut it, measure it, rework it till it fits"	Do it right the first time–no rework stations, no material review boards
Quality Control procedures inadequate for assigning explicit responsibility	Quality problems surface immediately, and the line shuts down until problem is corrected
Administration focuses on monthly shipment quotas and quarterly P & L	Focus shifts to customer delivery schedule and long range competitiveness
Management focuses on tactical and operational issues	Focus shifts to attaining strategic objectives of the enterprise

Figure 4. Distinctions Between Traditional and Rationalized Systems

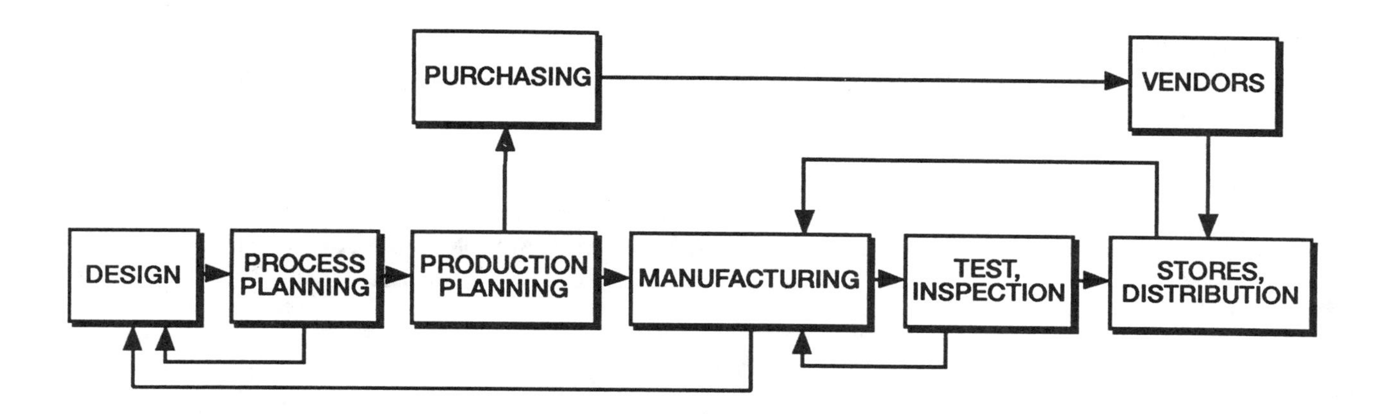

Figure 5. Generic Reference Model for Manufacturing System Flow.

We can illustrate on this generic model a large number of the manufacturing related technologies that are commonly associated with an advanced manufacturing environment. This is done in Figure 6. The relationships among the enabling technologies are portrayed in such a way that CAD/CAM is clearly not equivalent to CIM, as many articles and conference speakers imply.

Companies that have made progress toward implementing some of the elements shown in Figure 6 are discovering that they encounter additional problems when they attempt to tie this portion of the company to the remaining functions.

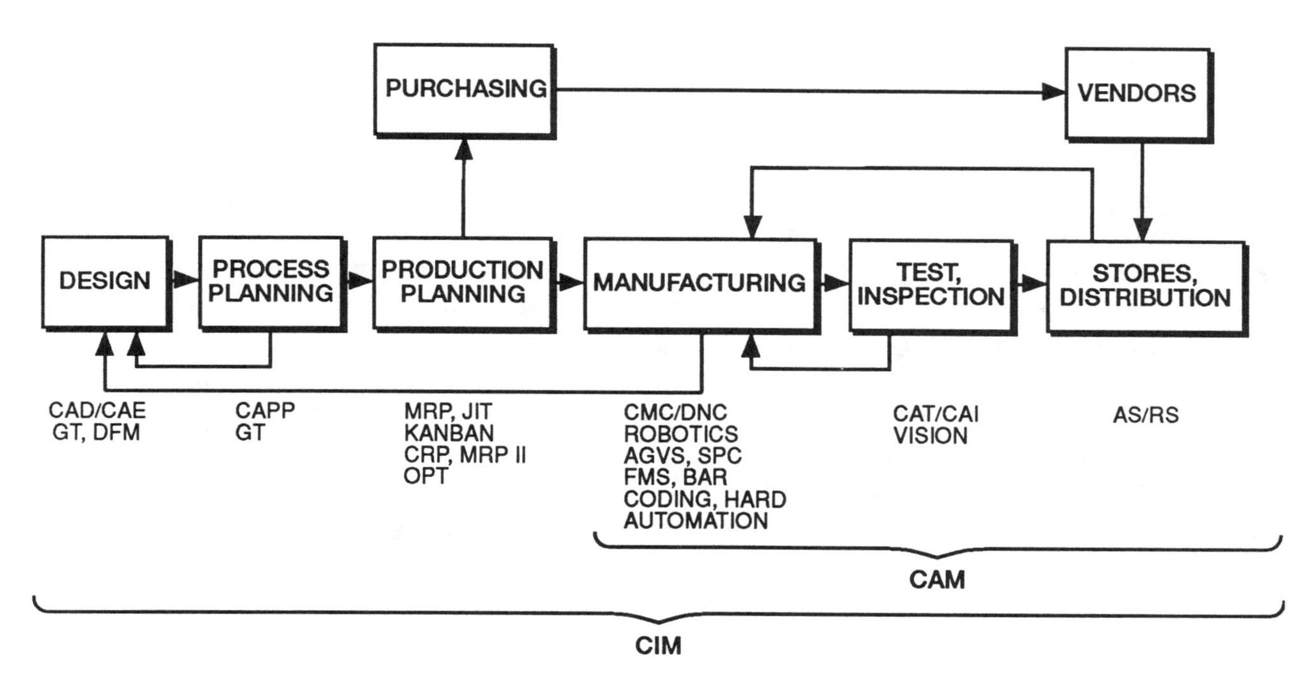

Figure 6. Relationships of Manufacturing Technologies.

In Figure 7, we portray the additional elements of the company that must be considered. (Details are omitted for the sake of clarity.) The Computer Integrated Enterprise (CIE) concept alluded to in an earlier section is portrayed here. The CIE/CIM/CAM relationships can be viewed as the

"Technology Dimension" of an integrated company. Shown on the vertical axis are system elements that cut across all company functions. We can refer to this set of elements as the "Information/ Communication Dimension".

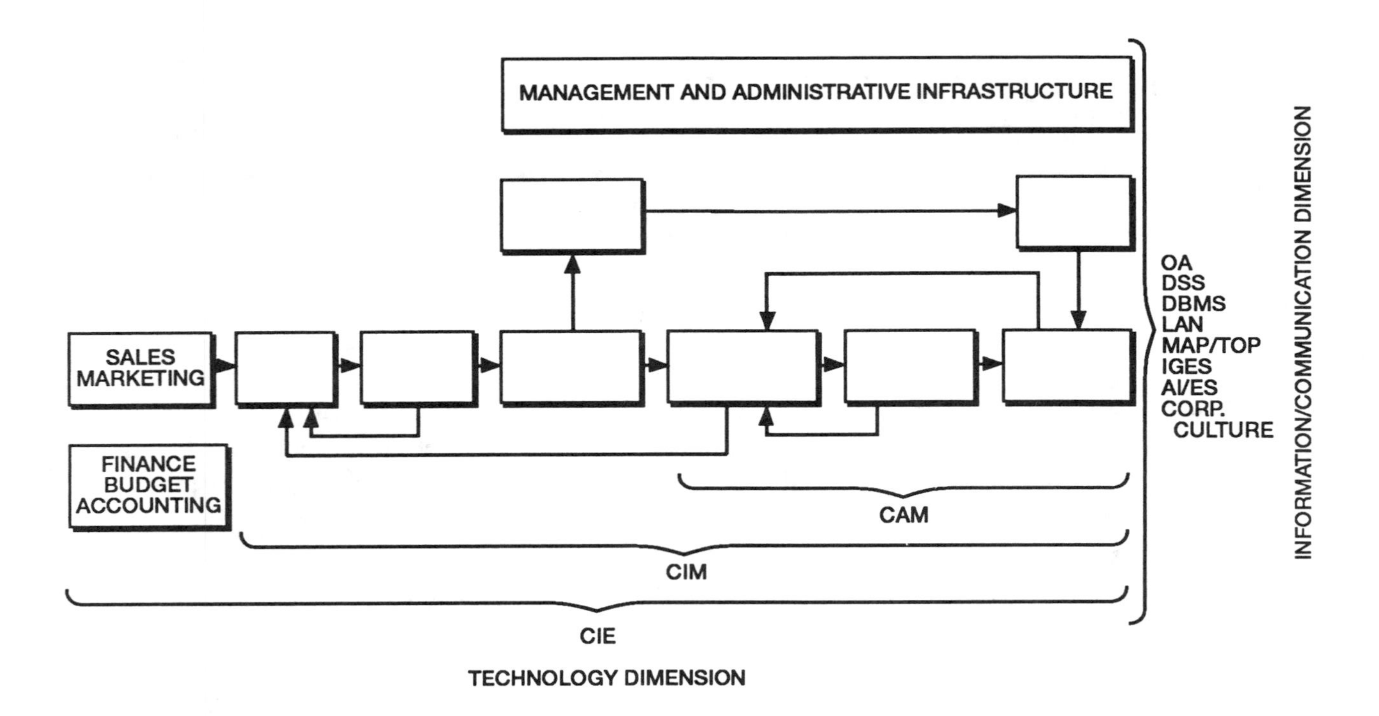

Figure 7. Relationships of All Company Functions.

We now have a rational framework within which we can consider the question, "What should be the scope of CIM?" We also have a meaningful starting point for performing an in-depth analysis of our present system. Based on this analysis we can determine if some fundamental structural changes should be made before proceeding with an integration initiative. For example, should product design and process planning be performed in series as shown? Some companies are finding that they should perform these two functions simultaneously. As we learn more about the concept of "feature-based product design," we are discovering that most of the knowledge required to create the optimal process plan is available as a consequence of executing the feature-based design rules.

In a similar manner, we should perform a high-level critique of our system structure, focusing especially at the interfaces between the several functions. This critique amounts to a company-level rationalization of our organization. The mission statement of each resulting organizational unit should be developed or refined as part of this rationalization process.

THE CRITICAL ROLE OF PEOPLE IN INTEGRATED SYSTEMS

How have companies survived without CIM over the past two hundred years of the industrial revolution? The truth is, many companies have achieved a very high degree of rationalization through the use of intelligent people. Regardless of the degree of automation, today's manufacturing systems are integrated primarily through expert human intelligence. We can call this "Knowledge-Based CIM Support," but the kind of knowledge we mean is human knowledge, not "machine knowledge," if there is such a thing.

The knowledge of human professionals constitutes the bridges between the subsystems. High-level corporate goals and programs cannot be realized without these human integrators.

Factory management systems depend significantly upon the presence of humans to interface between various elements of the factory. Neither the physical systems nor the Management/Information systems are currently capable of anticipating all the possible variations and contingencies that can occur during the operation of the factory.

Humans are required throughout the factory to perform "Physical Interfacing" (e.g., loading parts onto a machine) and "Judgemental Interfacing" (e.g., recognizing excessive vibration on a machine, and calling for maintenance action).

A third type of human interfacing can be called "Conceptual Interfacing." For example, a work center supervisor reviews a printout of the status of work orders at "Upstream" work centers and the status of queues at "Downstream" work centers, then re-sequences the jobs at his/her own work center for the maximum benefit of the total production schedule.

In a typical large discrete part factory, thousands of workers are continuously and simultaneously providing the three types of interfacing described above. Even in the best managed factories, mistakes occur, errors are made, orders are misinterpreted, parts are lost,.equipment breaks down, inventory records incur errors, and bad-quality parts go undetected and into the next assembly. While many of these adverse events are caused by the presence of humans, human thought processes are required for their resolution.

A major point to make here is that people will be *more* important in the "factory of the future," not less important.

TWO FUNDAMENTAL PATHS TO CIM

Conceptually, there are two fundamentally different approaches to achieving a CIM environment. The first approach places emphasis on getting several computerized applications in place and worrying about pulling them together at a later time. This approach has been dubbed the "islands of automation" approach. Proponents of this approach argue that a company should build realistically sized modules, then link them into an integrated system. We have seen that this actually results in interfacing, not integration.

The second fundamental approach to CIM development is to start at the top (total system), and design a comprehensive system from the very beginning. Proponents of this approach argue that this is the only way to assure that the system pieces will "fit together." Opponents of this approach argue that this is analogous to the "total systems" fad in the 1960s and 1970s, in which complete, company-wide management information systems were to be designed and built as one integral system. The magnitude of such an endeavor prevented companies from achieving their "total systems."

A third approach to CIM, which is a hybrid of the two fundamental paths includes the best features of each. At the company-wide level, a general framework for the total system can be specified. This framework would include standard communication protocols, a data dictionary, standard file structures, etc. Then, individual modules are implemented through time, each having been designed using the overall framework and standards. This approach is sometimes called, "Design from the top, implement from the bottom."

A COMPANY MUST START WHERE IT IS

The realities facing a particular company will normally prevent the company from taking a "Greenfield" approach to CIM. The fact is, the company has already made progress in several areas of computerization. It will not usually be in a position where it can afford to throw away years of software and data base systems. Figure 8 shows many of the functions that have been computerized on a local basis.

It is very important for a company to develop an on-going, comprehensive, top down design of its CIM implementation. It can then determine specifically how it can build on existing modules, which additional modules need to be added, and develop a time-phased plan for implementing the eventual system.

This is not an easy task. A company should develop its own internal capability for achieving CIM, rather than relying entirely on consultants and vendors. While outsiders can provide valuable assistance, they cannot do the company's thinking.

MODERNIZATION IS A NEVER ENDING PROCESS

The concept of an integrated manufacturing system is, itself, a moving target. A company must develop and maintain a process which continuously redefines its vision of an integrated system, based on technological trends and developments.

The manufacturing company of the future will be in a continuous state of change. At the same time it is implementing an improved segment of the factory; it will have already begun the process of conceptualizing the next improvement of that same segment. The modernization of facilities will be continuous, not something that occurs every 15 or 20 years, as has been the practice in the past.

BARRIERS TO REALIZING CIM

There are five major barriers that are inhibiting our progress toward the achievement of computer integrated manufacturing:

Non-standard Communications Protocols: It is simply not possible, at the present time, to connect directly all the components of a manufacturing system. The biggest single problem is the lack of a standard format for the electronic exchange of data between components. The initiative of General Motors with its Manufacturing Automation Protocol (MAP) standard shows promise of addressing this problem. The MAP standards are being developed to be consistent with the Initial Graphics Exchange Specification standards. This barrier should be effectively removed by 1993.

Inadequacy of Integrated Methodology: There is, as yet, no commonly understood methodology for achieving manufacturing system integration. The approaches being used are largely intuitive and ad hoc. They are also highly fragmented, with each technique (such as facility layout) applied to individual functions. In the absence of a comprehensive methodology, we have no way of even hoping to achieve an optimal factory design nor do we have a way of predicting the performance of a proposed design. Such a comprehensive methodology will not likely be available before 1997.

FUNCTION	COMPUTERIZATION
Product Design	CAD/CAE, Group Technology
Process Planning	GT, GAPP
Methods/Standards	Automated Work Measurement, Computerized Cost Estimating
Facilities Management	Computer-Aided Layout
Maintenance	Computer-Aided Preventive Maintenance Automated Machine Diagnostics
Production	NC/CNC/DNC, AS/RS, AGV, Robotics, Machine Vision, Programmable Controllers, Factory Networks
Quality Control	Automated Inspection, Statistical Process Control
Production Management	MRP/MRPII, Inventory Control On-line Inquiry

Figure 8. Locally Computerized Functions in Manufacturing.

Lack of Understanding of Change Management: Even more traumatic than the changes in our equipment and processes will be the pervasive changes that CIM will require in the social aspects of the factory. CIM will require that we re-think our traditional approach to organization design. It will also require new approaches to management practices at all levels. A very large portion of what we now consider "middle management" will either disappear or be radically modified. We do not know how all these changes will work out. Worse still, most companies are not even aware that they should be concerned about such things.

Lack of Appropriate Methodology for Justification: Traditional approaches to cost justifying the acquisition of individual pieces of equipment are inadequate for justifying a comprehensive upgrading of a manufacturing system. The synergistic benefits are not shown in the traditional approaches.

Lack of Qualified System Integrators: There is a tremendous shortage of qualified engineers and systems analysts/designers who know how to properly use the available technology. Each company that is serious about moving to a CIM environment should establish a group for this purpose. The composition of the group would include industrial engineers, electrical/computer engineers, computer scientists, mechanical engineers, and manufacturing engineers/technologists. The preceding specialties are stated in terms of typical academic disciplines rather than typical company job titles.

OTHER ISSUES RELATED TO CIM

There are many other fundamental issues related to CIM, but space does not permit a detailed discussion of them. Rather, we simply provide the following list of some of the important current issues:

- Is CIM an absolute or a relative concept?
- What will CIM do for an organization?
- How much will it cost?
- Is CIM actually attainable? Completely? Partially? In Phases? Not at all?
- How should a company proceed?
- Has any company been successful in implementing a "true" CIM system?
- Where do U.S. companies stand in relation to Japan? Western Europe? Soviet block?
- What are the barriers to more rapid progress toward CIM?
- Is CIM really new?
- Is CIM for everyone?

THE PROGNOSIS FOR CIM

So, where do we go from here? Is CIM just another buzzword that will soon give way to some new fad that the consultants and the academics try to peddle?

We have already seen that "CIE" is beginning to be used. How about "SIE," Strategically Integrated Enterprise? Labels are not particularly important, except when they *hinder* progress by creating confusion, mistrust, etc., or when they *contribute* to progress by providing a consistent rallying point for an important concept.

MRP is a case in point. For many years, it had all the characteristics of a fad. It was pushed as *the* solution to all the world's problems. Many unqualified people peddled "complete MRP solutions". A frighteningly high percentage of implementation attempts ended in total failure. And yet, MRP survived and is now, some twenty years later, reaching a level of maturity and stability that will assure its survival far into the future.

MRP *survived* because it addressed a critical, fundamental need. Even the early, crude MRP systems were improvements over the clumsy, patch-worked material control systems that they replaced. MRP *flourished* when its underlying principles were researched, tested, refined, and eventually communicated as a cohesive body of knowledge. It is heartening to see additional new concepts, such as JIT and SPC merge with MRP to form even better production management systems. Even if the label "MRP" eventually gives way to some other label few would argue that the MRP initiative has been a landmark development in the evolution of production management systems.

CIM could very well enjoy an experience similar to that of MRP. In fact, CIM implicitly includes a disciplined MRP-type control system. It is entirely conceivable that CIM will survive the fad phase, for the exact same reason that MRP survived. *CIM addresses a critical, fundamental need.* Will CIM flourish as MRP eventually did? It can if we make progress in developing a cohesive body of knowledge regarding CIM.

IMPLICATIONS FOR IEs

Who will develop the science base for integrated systems? Who will use this science base to develop methodologies for achieving integrated systems? And finally, who will apply the methodologies within the tens of thousands of organizations that could benefit?

The answers to these questions are far from clear. For anyone who responded, "IEs" to all three, our track record is not encouraging. Our traditional research strength in areas such as human performance, operations research, and quality control does not appear to be sufficient or appropriate for approaching the fundamental issues of large scale system integration.

Probably the group best qualified to perform basic research in this area is that relatively small number of people who received special training in the 1960s and early 1970s in large-scale system design and development. These are people who worked on many of our large space and defense programs. This is the closest the U.S. has ever come to producing true "systems engineers" and "systems scientists."

Can IEs be expected to contribute to the second area, that of developing methodology for designing optimal integrated systems? Yes, many IEs can contribute to this effort but they cannot do the total job alone. Specifically, the inputs of computer scientists, marketing specialists, accountants, and strategic planners will be very important in the overall development of these methodologies. Will IEs lead these efforts to develop methodologies? Some IEs are qualified to lead such efforts, and some are not.

What about the third area, that of applying the methodologies? The IE can expect to play a more central role in this area, but again, IEs will not be able to do the complete job. There are talents required, such as digital electronics, which the IE simply does not possess in sufficient strength. However, the broad background of the IE is such that in many situations an IE will be selected to lead the system integration effort.

There are many other implications for the IE profession, but space does not permit an elaboration on each one. Rather, the following list of implications are included to stimulate the reader's interest:

- Do IEs have the tools needed to be successful system integrators? If not, what additional tools do we need? How and where can we get them?

- Do IEs have the mission within their organization to be the system integrators? If not, how can we add this to our charter?

- Are IE curricula preparing students adequately to be system integrators?

- Will IE researchers contribute to the conceptual advancement of CIM?

- Should we try to do the job alone? Are we pretenders to a throne that we are not qualified to occupy?

SUMMARY

"CIM" has taken only the aura of a fad. Many people are misusing the term. Nevertheless, CIM has a good chance of surviving because it addresses a critical, fundamental need.

IEs have a role to play in the further advancement of CIM, but we are not the only players in the game. Indeed, we must learn to serve as a team player in pursuing CIM initiatives. Those IEs who are qualified will be selected to lead CIM implementation teams. Every manufacturing firm in the world, whether in Tokyo, Detroit, Taipei, London, or Bombay, is entering unchartered water. There are many unknowns, and few precedents to follow.

The correct choices are rarely evident, yet managers are compelled to make decisions and move aggressively toward "the factory of the future." Some decisions will yield bad results. Many companies will fail in their initial attempts to achieve CIM. Some will even go out of business. But, as great as the risks are, *not* moving aggressively is even more risky. Manufacturing systems will be fundamentally different in 2000 from those we have today.

Like all "trail blazers" throughout history, we are *driven* and *guided* by (1) a degree of intellectual intensity not encountered in the normal realm of things, (2) a basic survival instinct that elicits the best in our thinking and strategizing—a phenomenon that transcends traditional wisdom and rational thinking, and (3) an insatiable curiosity of the unknown—of how far our innate wits can carry us beyond the current boundaries of our knowledge base.

The industrial engineering profession is facing its greatest challenge in leading the CIM movement. Opportunities accompany challenges. The opportunities for our profession have never been greater.

** The concepts presented in this section represent an amalgamation of several papers and presentations by the author. See specifically references (12), (14), (15), (16), and (17). Used with permission of Computer Integrated Management, Inc.*

REFERENCES

1. Allen, M. A., Chairman of the Board, Computervision Corporation, *Keynote Address*, WESTEC '84, March 20, 1984.

2. Barish, N. M., "Computerized Systems in the Scheme of Things," *1979 IIE Fall Conference Proceedings.*

3. Berry, W. L. and D. C. Whybark, "Research Perspectives for Material Requirements Planning Systems", Research Paper No. 434, Purdue University, November 1973.

4. Burbidge, et. al, "Integration in Manufacturing", *Computers in Industry* 9, North Holland, 1987.

5. Clancy, John, Senior Vice President, McDonnell Douglas Automation Company, quoted in *Industry Week*/March 4, 1985, p. 52.

6. Conaway, Jack, Manager of CIM, Digital Equipment Corporation, quoted in *Industry Week*/March 4, 1985, p. 49.

7. Conigliaro, L., quoted by S. Walsh, "Integration Dilemma Addressed at Seminar", *Managing Automation*, August 1988, p. 20.

8. Davis, R. P., et al., "Manufacturing Systems Planning - the Key to Production Control," *1979 IIE Spring Annual Conference Proceedings.*

9. Greene, T. J., "CIM Goes Beyond Production and Inventory Control," *Production and Inventory Control Division Newsletter*, Institute of Industrial Engineers, Vol. XIX, No. 1, Summer 1984.

10. Harrington, J., *Computer Integrated Manufacturing*. Krieger Publishing Company, Malabar,Florida, 1975.

11. Malone, R ., "Clouds of Unknowing", *Managing Automation*, August 1988, p. 4.

12. Mize, J. H. and D. J. Seifert, "CIM - A Global View of the Factory", *1985 IIE Fall Conference Proceedings.*

13. Mize, J. H., D. J. Seifert, and F. S. Settles, "CIM From a Corporate View - The Garrett Experience," *Industrial Engineering*, November 1985, Vol. 17, No. 11.

14. Mize, J. H., "Success Factors for Advanced Manufacturing Systems", *1987 IIE Spring Conference Proceedings.*

15. Mize, J. H., "Prerequisites for CIE", *Proceedings, CAM-I Computer Integrated Enterprise Interest Group Meeting*, October 15-16, 1987.

16. Mize, J. H., "CIM - A Perspective for the Future of IEs", *1987 IIE Fall Conference Proceedings.*

17. Mize, J. H., "Planning for CIM", Kickoff Panel, Productivity and Total Quality in Manufacturing, *1987 IIE Fall Conference Proceedings.*

18. Reimink, R. L., "Integrating AGVs with Automated Manufacturing", *SAE Technical Paper Series*. International Congress and Exposition, Detroit, MI, Feb. 1986.

19. Sibbald, G. W., "Roadblocks to CIM Success", *CIM Review*, Vol. 4, No. 3, Spring 1988.

20. White, J. A., "The Industrial Engineer As the System Integrator", *Proceedings, 1986 IE Managers Seminar*. Institute of Industrial Engineers. Norcross, GA.

21. *Computer Integrated Manufacturing Glossary*, Society of Manufacturing Engineers, Dearborn, MI, 1984.

22. *Webster's New World Dictionary of the American Language*, 2nd Edition, Simon and Schuster, 1984.

FUNDAMENTALS OF SYSTEMS INTEGRATION

James A. Bontadelli, Ph.D., P.E.
Dept. of Industrial Engineering
University of Tennessee

Kenneth E. Kirby, Ph.D., P.E.
Dept. of Industrial Engineering
University of Tennessee

The word system(s) is used both as a label (e.g., a railroad system), and as an abstract concept which is applicable to most fields of endeavor. In this summary paper, both meanings of the word will be used. Hopefully, the context of each use will indicate the intended meaning.

Two terms commonly used in reference to systems are systems thinking and systems approach. Examples of the use of a systems approach, familiar to most of us, were the development (and application) of methodologies in operations research and management science, systems analysis, and systems engineering. In a broader context, we can consider systems thinking as a way of dealing with complexity. For example, it is a critical part of developing and implementing total quality or continuous performance improvement processes within organizational systems. System ideas and concepts, which formalize the framework of systems thinking, are an essential part of the planning, design, problem solving, and improvement activities that most of us are routinely engaged in. Thus, understanding them in order to achieve their effective use in these and other types of activities is very important.

We are interested today primarily in the application of systems thinking, not theory, and particularly within organizational systems. However, it is necessary to first "set the stage" with a discussion of some selected system ideas and concepts. This will include added emphasis on the concept of systems integration. Then a basic, and general, application framework for these ideas and concepts will be presented. In the later part of the paper, some specific recommendations for using systems thinking in improving the effectiveness of organizational systems will be presented and discussed.

SELECTED SYSTEM IDEAS AND CONCEPTS

In general, the 1940s can be considered the beginning of the systems age, followed by significant expansion of systems thinking in the 1950s. During this period, the idea of reductionism was being supplemented (not replaced) by the idea of expansionism. The basic doctrine of reductionism is that everything can be reduced or decomposed down to simple elements or indivisible parts (e.g., the atom in physics). Expansionism, in contrast, considers everything to be parts of larger wholes (e.g., a flexible manufacturing cell as part of a production system). As Ackoff [1] said, "It is another way of viewing things, a way that is different from, but compatible with, reductionism. It turns attention from ultimate elements to wholes with interrelated parts, to systems." It is this focus on the whole and its interrelated parts that gives systems thinking the potential for dealing better with complexity.

Based on a more extensive definition of a system given by Ackoff, a system can be defined as a set of two or more elements of any kind where the

Reprinted from *1991 IIE International Industrial Engineering Conference Proceedings*

elements (and the set of elements) have the following three properties:

1. Each element has an affect on the set of elements.
2. Each element depends on at least one other element in the set.
3. The set of elements cannot be organized into independent subsets.

Hall [8] defined a system more simply as "a set of objects with relationships between the objects and between their attributes."

Included in systems thinking, as part of its formalized framework, are several important ideas and concepts. Selected for this discussion are the following:

1. Wholeness
2. Boundary/Interface
3. Hierarchical Order
4. Environment
5. Systems Integration
6. Inputs/Tranformations/ Outputs
7. Feedback/Adaption/Stability
8. Hard versus Soft Systems

As previously mentioned, systems thinking is based on the idea of expansionism which views any defined system as part of a larger whole. In Figure 1, which is based on an arrangement of organizational systems, the concept of wholeness is implicitly illustrated. We will also use Figure 1 to discuss the concepts of boundaries and interfaces, hierarchical order, environment(s), and systems integration.

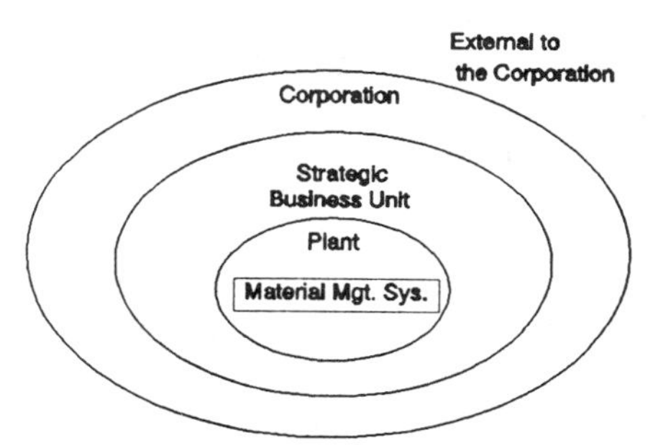

Figure 1: Hierarchical Order of Systems

Fundamental to identifying a set of elements as a system is developing a basic definition for it, as well as delineating its boundary. The basic definition of a system should begin with its purpose (mission) and principal operating objectives. Also included should be the primary transformation process(es) or activities, responsibility for (or ownership of) the system, customers (if any) and their expectations, and the principle constraints externally imposed on the system. Development of the basic definition is an essential part of delineating the system boundary (and interface) with the next system in the hierarchy.

The Material Management System (Figure 1), in hierarchical order, is shown as part of a plant (system), which at the next level, is part of a strategic business unit (system). The later, then, is a part of the total corporation (system). Everything external to the boundary of the Material Management System (represented by the rectangular box) is within its environment. At each level in the environment (plant, and so on) there are factors which affect the system, and are affected by it. For example, consider customers which may exist within each level of the environment (e.g., the production planning and scheduling function in the plant, another plant in the strategic business unit, and so on). If timely delivery of products (or information) is an expectation of each customer, then the operating characteristics of the system will determine how well it contributes to meeting these expectations. Thus, the original system design should include the necessary value adding features based on these expectations. Also, operational performance measures for the system should show explicitly how well the expectations are actually being met.

Systems integration is an important concept in the systems thinking framework, and includes two basic dimensions. The first dimension involves the factors within the environment which affect the system, or are affected by it. We have already discussed how customer expectations require certain value adding properties to be designed into a system. Similarly, goals and objectives established by higher level organizational systems in the environment have to be met. Another environmental factor, such as available technology, may place constraints on what a system can be designed to do. This illustrates what we mean by systems integration as it relates to the system environment. That is, good integration is achieved on this first dimension when all the requirements, constraints, and so on related to factors in the system environment have been efficiently and effectively met.

Systems integration on the second dimension relates to achieving synergism in the relationships between elements, and subsets of elements, within the system itself. For example, in the Materials Management System represented in Figure 1, several functions within the plant will be involved with operating the system. This means that organizational interfaces are "traversed" by the system. Achieving efficient and effective operation of the system across these functional interfaces is one important aspect of systems integration on the second dimension.

In Figure 2, a general schematic for an organizational system is shown. The boundary of the system is indicated by the rectangle, and factors in the system environment other than suppliers and customers are not shown.

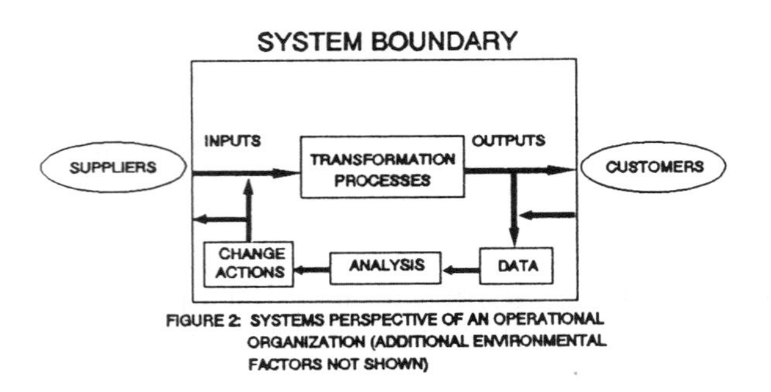

FIGURE 2: SYSTEMS PERSPECTIVE OF AN OPERATIONAL ORGANIZATION (ADDITIONAL ENVIRONMENTAL FACTORS NOT SHOWN)

The schematic illustrates the system concept of inputs, transformation processes, and outputs. Since the example illustrates the organization as a system, the inputs include all resources used by the transformation processes to produce the goods, services, and information (outputs) required by customers and other entities (e.g., government agencies) in the systems environment.

Also illustrated in Figure 2 is the system concept of feedback, which is necessary for system adaptation to changes in its environment. It is by adaptation through the feedback loop that systems integration is maintained on the first dimension. Also, adaptation to changing requirements in the environment maintains system stability, a prerequisite for long-term survival. For example, when an organizational system does not have an effective feedback (performance measurement) mechanism, adapting to changing requirements in its environment is delayed, or does not occur. Consequently, stability cannot be maintained and outcomes, such as market share, suffer. Competitors then increase their market share, and survival of the organization is in jeopardy. Or, in the case of a government type organization, its services gradually become irrelevent and the need for the organization dissipates.

An important point shown in Figure 2 is that data used in the feedback loop are from both the output side of the transformation processes, and from the system boundary (interface). In the case of organizational systems, the later data (which indicates outcomes) are often not collected adequately. The changing requirements in the environment are then "missed", and unsatisfactory outcomes (e.g., with customers) result.

Data collected in the feedback loop are analyzed, the results evaluated (in a separate step not shown), and change actions are initiated based on the evaluation. These changes occur as both an input to the transformation processes, and at the system interface with entities in the environment, such as suppliers.

The last system idea selected for this discussion is that of hard versus soft systems. We may work with both types of systems, and often the system involved is a hybrid of the two types. However, implementing total quality concepts and continuous improvement processes within organizational systems is now a pervasive effort. Since many important organizational systems are in the soft systems category, we need to be aware of some differences in characteristics with hard systems. A comparison of selected characteristics between hard and soft systems is shown in Table 1. As indicated by the comparison, the level of human activity is a major difference between hard and soft systems. This difference also "drives" the other differences in characteristics shown in Table 1.

Table 1. Some Characteristics of Hard Versus Soft Systems

HARD SYSTEMS	SOFT SYSTEMS
• Low level of human activity	• High level of human activity
• Structure, elements, and boundaries are explicit	• Structure, elements, and boundaries harder to define
• Emphasis on efficiency in achieving explicit goals	• Emphasis on learning and improvement
• System "problems" can be well defined	• System "problems" can be made explicit, but defining them is more difficult
• A system to be designed based on engineering principles	• A system to be evolved in its design with careful consideration of the human elements

GENERAL APPLICATION FRAMEWORK

A general systems application framework is shown in Figure 3. It has three major components: the set of systems ideas and concepts; a specific application area; and a basic methodology for the application area. The application may be the design of a hard system, and the steps of a basic design methodology (or, in the case of a large hard systems program, a basic systems engineering methodology) would be used. Similarly, the application area may be planning (e.g., the development of a master plan for a plant manufacturing system), and the steps of a basic planning methodology would be followed. In either of these cases, or for any application area, the system ideas and concepts would be integrated into the steps of the basic methodology employed. The underlying philosophy of this systems application framework is that there is not a unique "systems methodology" that can be generally used. Instead, a systems approach is developed for an application area by effectively incorporating the systems ideas and concepts (the formal framework of systems thinking) into an applicable basic methodology for the area.

As an example, consider a basic problem solving methodology within the context of an existing soft system operation (e.g., the material management system in Figure 1). In this case, the desired outcome is implemented improvements. The steps of a representative basic methodology for this application is shown in the first column of Table 2. The second column lists the most applicable system ideas and concepts, among those previously discussed, for incorporating into each step of the basic methodology. Even though we do not have time in our discussion to examine in more detail developing a systems approach for this application, this illustrates how to operationalize the general systems application framework (Figure 3).

RECOMMENDATIONS FOR APPLYING SYSTEMS THINKING

Although systems concepts and ideas have been growing in importance since about 1950, only recently has the "broader" significance of using this approach to address and improve organizational systems been more fully appreciated. Many managers continue to ponder the best way to implement organizational change. They recognize that they must somehow dramatically increase their rate of improvement in order to survive in an increasingly competitive global market, but are at a loss to identify the best way to make that happen.

W. Edwards Deming [5] provides some insight as to how organizational improvement might best be achieved when he defines the three primary roles of a top manager:

1. Provide the theory upon which the individual components of the system can relate to the purpose of the system.
2. Transform the basic structure of the system to one which uses more responsibly the components available to the system.
3. Keep the purpose of the company in harmony with the broader aspects of a healthy, prosperous community and society.

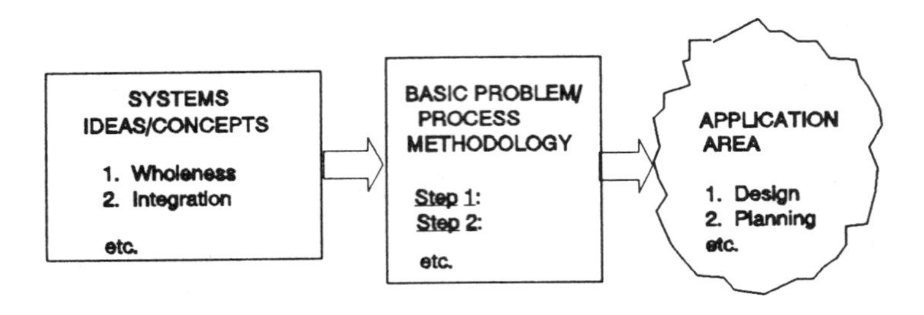

FIGURE 3: GENERAL SYSTEMS APPLICATION FRAMEWORK

In order to fulfill these roles, managers must view the entire organization as a system. That system resides in an environment that consists of governmental constraints, suppliers, competitors, potential competitors, and most importantly, customers. In order to survive, the organization must intimately understand what customers value and establish specific capabilities to provide products or services that are valued. The systems that reside within the organization are the means of achieving those desired capabilities. Ultimately, the future of an organization will depend on how well its systems can provide those capabilities that are most valued by customers. Systems will function most effectively when their component parts are well integrated (on both dimensions previously discussed), and aligned to system purpose.

There are many recommendations that could be made regarding the use of systems thinking in improving organizational systems. This discussion will focus on the following:

1. Managers must continuously seek to understand the complex relationships between the various elements, and sets of elements, of the organizational system, and how each contributes to providing some capability that is valued by customers.
2. Managers must design an effective system for understanding what customers now value and what will be valued in the future as the initial step in any successful continuous improvement effort.
3. Managers must be assigned ownership for those internal systems that provide capabilities valued by customers. These systems are extremely complex and are affected by most parts of the organization. Work to improve them must be orchestrated and coordinated in order to avoid suboptimization.
4. Improving these internal systems will require that measures be established for total system performance as well as for significantly contributing system components.

Understanding System Relationships

The decade of the 1980's has seen most U.S. manufacturing companies concentrating on the system that provides product quality. Significant drops in market share to foreign competitors for many different products (e.g., automobiles, electronics, computer chips) left no room for doubt that perceived inferior quality of U.S. products was quickly eroding our manufacturing base.

Initial work on improving the product "quality" system was focused on the manufacturing systems component. There were structural, procedural, as well as attitudinal changes that took place in the quality system in many organizations. The responsibility for product quality moved from an inspection department to the line organization in many companies. There was increased emphasis on satisfying the needs of the customer, with the customer in some cases defined as an internal customer; essentially the next step in the

Table 2. Basic Problem Solving Methodology and Systems Thinking

Steps in Basic Methodology	Most Applicable Systems Concepts
1. Problem definition	Wholeness, Boundary/Interface, Hierarchical Order, Environment
2. Review/Assessment of present operation	(Above concepts plus): Integration, Inputs/TP/O, Feedback/A/S
3. Selection of decision Criteria	Wholeness, Environment, Integration
4. Development of improvement alternatives	(Same as Step 2)
5. Analysis/Comparison of Alternatives	(Same as Step 3)
6. Selection of preferred alternative	(Same as Step 3)
7. Implementation	(Same as Step 2)

manufacturing process. Detailed analysis of process steps identified waste and rework to be eliminated. Process studies revealed cause and effect relationships between process parameter characteristics and product quality characteristics that were used to change procedures and operating practices and further improve the system. Attitudinal changes were achieved when operators were given the responsibility to shut down the line if there were any questions regarding product quality. Dramatic improvement in quality resulted. A 1988 research report of the Boston University School of Management Manufacturing Roundtable [10] revealed an average 19 percent improvement in conformance quality during the years 1985-1988. The percent improvement varied by industry with the consumer packaged goods industry generating the lowest value (13 percent) and the electronics industry achieving the highest (29 percent).

Although dramatic improvements in product quality were achieved by concentrating on the performance of manufacturing processes, many companies discovered that studying manufacturing processes exclusively ultimately placed constraints on performance. Managers with some awareness of systems concepts began to recognize that the "true" quality system was much more complex than the small portion currently being given consideration. Product quality was heavily influenced by the product design system; as much as 80 percent of product cost and quality attributes for some products are firmed at the point of product design. The maintenance system, the procurement system, the production planning and scheduling system, the order entry system, the finance/accounting system, the marketing/sales system, and the distribution system all had parts to play in achieving desired levels of capability in product quality. These so-called "independent" systems all had linkages that affected achieving a strategic capability such as quality and needed to be integrated, on both dimensions of systems integration, and aligned if competitive levels of performance were to be achieved and maintained.

The importance of understanding these complex linkages between organizational components to provide specific capabilities is further highlighted in a recent article by Serge [13]. Serge's article proposes that the leader's new work is to build "learning" organizations, an important soft systems characteristic. He discusses the Japanese versus U.S. manufacturers approach to achieve competitive advantage. U.S. managers used rigid controls on inventories, incentives against over production, and strict adherence to demand forecasts. In contrast, the Japanese focused on reducing delays in a "broader" system context, believing this to be a much higher-leverage approach to improving both cost and customer loyalty. He quotes George Stalk of the Boston Consulting Group as follows:
"...the Japanese saw the significance of delays because they saw the process of order entry, production scheduling, materials procurement, production, and distribution as an integrated system."

Although, in general, U.S. managers appear to

be a step behind their Japanese counterparts, many are beginning to appreciate that understanding interrelationships within the organization, and designing systems that make use of those known relationships to provide some strategic capability is the key to organizational improvement. The systems that are most important in an organization, those that truly address achieving a capability valued by the customer, are complex. They are comprised of many parts and stretch horizontally across the entire organization. Improving their performance should be readily apparent to the customer. Their identification will allow managers to focus resources on work activity that is most valued. Some capabilities for a manufacturing organization that could be desired in addition to quality are listed below:

1. Rapid delivery
2. Consistent delivery
3. Innovative products
4. Rapid design changes
5. Flexibility to change product mix and volume
6. Interpreting customer expectations
7. Low cost

The Initial System: Determining What Customers Value

As managers identify those internal systems within their organization that are most important, the first and foremost system that should be given consideration is the one that identifies and continuously tracks what is most valued by customers. This will likely require dramatic changes in the system that exists in most organizations. It will likely involve efforts to gather details of customer needs that currently do not exist. The output from this system provides insight as to which system(s) within the organization are valued highly so as to be allocated corresponding resources. The value of having such a system is exemplified in an article by Sirkin and Stalk entitled "Fix the Process, Not the Problem." [14]. The authors discuss the dramatic progress a floundering paper company achieved when their managers improved the system used to accumulate details of the attributes valued by customers. Insights gained allowed the company to move from dead-last among the five available suppliers in their industry to the number one supplier in two-and-one-half years.

An organization will typically have five to eight strategic capabilities that are desired. Each of these capabilities must be provided by a complex, cross-functional system. To make matters more demanding, the capabilities most valued will change over time. The external environment of an organizational system is always changing. Therefore, success will not only depend on how well it adapts to current demands by understanding and designing systems to provide what is now most-valued, it will depend on how well the organization is able to forecast what will be valued in the future and how rapidly it can respond with changes to its systems to meet the new demands (i.e,. prospective adaptation). The Boston University Report referenced earlier indicated that improving quality was the most important capability over the three years 1985-1988, specifically through the increased use of Statistical Process Control (SPC) and vendor quality approaches. The report also noted that time-based capabilities (e.g., delivery speed, after-sales service, product/volume flexibility) had grown in importance more rapidly during the past four years. Thus, it appears that as companies in a given industry improve their system for providing a selected capability (e.g., product quality), demanding customers will move to a higher level of expectations for some other capability.

Assignment of Systems Ownership

These large internal systems that provide an organization with selected capabilities valued by customers typically have no owners in present organizations. Our experience has been that when a given set of managers are asked who owns these critical systems, two typical responses are (1) no one owns them, or (2) we all do. In either case, the result is the same. No single manager has been assigned the responsibility to identify the scope of the system, its boundaries, the transactions that take place, the component parts, and other essential information that is prerequisite to systems improvement. We, as do other authors such as Rummler and Bracke [12], believe the assignment of system ownership is one of the most important factors in achieving system improvement. Bracke and Rummler provide criteria for selecting systems owners:

1. Holds a senior management position
2. Holds a position that has the most to gain if the

system is improved
3. Manages the largest number of people working in the system
4. Understands the entire system
5. Has an overall perspective of the effect of the environment on the system and the effect of system performance on the business
6. Has the personal ability to influence decisions and people outside of his/her line management responsibilities

System ownership, to be successful, will require that all functional managers be aware that they are subservient to these owners. Obviously, this is likely to generate some heated debate in some companies that are politically driven by the functional or vertical formal organization structure.

Developing System Measures

Last, in order to support continuous improvement, performance measures must be developed for each of these critical systems. We believe, very strongly, in the cliche "if you can't measure it, you can't manage it." Care should be taken in developing these measures. Performance measures can invoke different individual behaviors. Poor system performance measures can generate undesired human behaviors. Not only must good measures be established for total system performance, but the intended contribution of each system component must be defined and acceptable measures established for it. The definition of component contribution allows one to understand how system integration must occur. Also, establishing measures for system components allows for evaluating the effectiveness of the system integration, making system changes where necessary (adaptation from the feedback loop) and evaluating whether the changes had an effect (achieving and maintaining stability).

SELECTED BIBLIOGRAPHY

[1] Ackoff, R.L., "The Systems Revolution," Long Range Planning, September, 1974.

[2] Checkland, P., "From Optimizing to Learning: A Development of Systems Thinking for the 1990s," Journal of Operational Research Society, Vol. 36, No. 9, 1985

[3] Churchman, C.W., Ackoff, R.L., Arnoff, E.L., Introduction to Operations Research, John Wiley and Sons, New York, 1957.

[4] DeGarmo, E.P., Sullivan, W.G., Bontadelli, J.A., Engineering Economy (8th Ed.), Macmillan, 1988.

[5] Deming, W.E., Out of the Crisis, MIT Center for Advanced Engineering Study, 1986.

[6] Eden, C., Graham, R., "Halfway to Infinity: Systems Theorizing for the Practitioners?," Journal of Operational Research Society, Vol. 34, No. 8, 1983.

[7] Flood, R.L., Carson, E.R., "Dealing with Complexity," An Introduction to the Theory and Application of Systems Science, Plenum Press, 1988.

[8] Hall, A.D., A Methodology for Systems Engineering, Van Nostrand, Princeton, N.J., 1962.

[9] Malouin, J.L., Landry, M., "The Mirage of Universal Methods in Systems Design," Journal of Applied Systems Analysis, Vol. 10, 1983.

[10] Miller, J.G., Roth, A.V., "Manufacturing Strategies: Executive Summary of the 1988 North American Manufacturing Futures Survey," Research Report, Boston University School of Management Roundtable, 1988.

[11] Mize, J.H., Siefert, D.J., 1985, "CIM--A Global View of the Factory," Proceedings: Fall Industrial Engineering Conference.

[12] Rummler, G.A., Bracke, A.P., Improving Performance: How to Manage the White Space on the Organizational Chart, December, 1989.

[13] Senge, Peter M., "The Leader's New Work: Building Learning Organizations," Sloan Management Review, Fall 1990.

[14] Sirkin, H., Stalk, G. (Jr.), "Fix the Process, Not the Problem," Harvard Business Review, Jul.-Aug., 1990.

[15] Wright, R., Systems Thinking: A Guide to Managing in a Changing Environment, Society of Manufacturing Engineers, 1989.

BIOGRAPHICAL SKETCH

Dr. James A. Bontadelli, Professor of Industrial Engineering, University of Tennessee. Dr. Bontadelli was Director of Industrial Engineering at the Tennessee Valley Authority from 1974 to 1990 where he managed industrial and systems engineering activities in support of both power generating and nonpower operations. Also, he was an Adjunct Professor in Industrial Engineering at the University of Tennessee from 1976 - 1990. From 1962 - 1974 he served in various technical management positions at Battelle Columbus Laboratories providing systems analysis and systems engineering services to a number of companies and government agencies. Dr. Bontadelli has extensive experience in systems related activites, including continuous improvement and total quality processes, in both private sector and government organizations. He is a Fellow of IIE.

Dr. Kenneth E. Kirby, Professor of Industrial Engineering, University of Tennessee. Dr. Kirby came to the University after fifteen years of industrial experience with the Aluminum Company of America (ALCOA). His last position with ALCOA was Manager of Industrial Engineering for the Company's Tennessee Operations. Dr. Kirby's current research interests and teaching responsibilities are in the areas of systems analysis and design, applied statistics, and manufacturing systems/facilities design. His academic interests have been reinforced by establishing a broad-based consulting practice. Dr. Kirby currently teaches in the Institutes for Productivity Through Quality offered at U.T. The Institutes for Productivity Through Quality are a structured set of management development offerings that focus on enhancing competitive position through the analysis and improvement of management systems. He is currently active in the Institute of Industrial Engineers, the American Production and Inventory Control Society, and the Society of Manufacturing Engineers. He received his BS, MS, and Ph.D. degrees from the University of Tennessee.

COMPUTER INTEGRATED MANUFACTURING FROM A PRAGMATIC PERSPECTIVE

Philip M. Wolfe
Arizona State University
Tempe, Arizona

F. Stan Settles
Garrett Engine Division
Phoenix, Arizona

ABSTRACT

The CIM philosophy has not been as widely accepted as initially predicted. Some reasons are discussed. Guidelines are presented for successfully implementing a CIM project.

INTRODUCTION

This paper is concerned with the success or failure of CIM as applied to discrete part manufacturing. During the last three years, literature on the subject of CIM and related manufacturing topics has proliferated at a very rapid rate. Manufacturing for Competitive Advantage, by Gunn is one of many books that have been published. Business oriented magazines have published numerous articles, such as "High Tech to the Rescue". New journals have been launched, such as CIM Technology and CIM Review, and each month short courses on the topic of CIM are presented at major metropolitan locations.

This great interest in CIM is not unfounded. Management is looking for better ways to run their firms because of the increased difficulties that U.S. firms are having competing in international markets. Associated with CIM have been some outstanding competitive improvement stories that have received substantial publicity. An excellent summary (see Figure 1) of what can be realized from successful implementation of CIM concepts was documented in a study by the National Research Council Manufacturing Studies Board. After reviewing Figure 1, it is not surprising that CIM is receiving a lot of attention.

Although earlier CIM related articles were very supportive, recent publications have been reporting less encouraging results. Some examples of these articles are: "Detroit Stumbles on Its Way to the Future," Business Week; "Automakers Discover Factory of the Future Is Headache Just Now," Wall Street Journal; "No Easy Road Seen to Implementation of CIM," Management Information Systems Week; and "Is Management to Blame for the Unfilled Dream of CIM," Production Engineering. With a growing amount of negative publicity, people are beginning to question what can really be expected from CIM. Consequently, the growth in CIM related service and hardware industries has slowed significantly.

ACHIEVEMENT	RANGE	AVERAGE
Manufacturing Productivity Increase	20-200%	120%
Product Quality Increase	60-200%	140%
Product Design Thru Production for Sale Lead Time Decrease	30-100%	60%
Receipt of Order to Shipment Lead Time Decrease	30-50%	45%
Capital Equipment Utilization Increase	20-1500%	340%
Inventory Work In Process (WIP) Decrease	30-100%	75%

Source: National Research Council Manufacturing Studies Board

FIGURE 1

TECHNOLOGICAL POTENTIAL OF CIM

Although some very successful CIM implementations have been achieved, they may be more the exception than the rule. This realization is not untypical of how other new technological developments have been accepted. Initially, as the media describes a new emerging technology, the potential may seem almost unlimited (the media seems to have a tendency toward exaggeration). However, as the technology becomes better understood, the inherent limitations become apparent. At that point, a more accurate assessment of the potential can be made. Our understanding of CIM has increased dramatically in the last two years; consequently, we can now make a much more accurate assessment of the potential benefits of CIM.

MANAGEMENTS' UNDERSTANDING OF CIM

CIM is difficult to understand because it is difficult to define. Experts in this field are having a hard time reaching a consensus on a definition for CIM. A cursory look at the literature will validate this

Reprinted from *1989 IIE International Industrial Engineering Conference Proceedings*

observation. A sample of these definitions follows:

CIM is the use of computer and information/communication technologies to effectively integrate all of the:

* engineering/design functions
* manufacturing planning functions
* equipment/process technologies
* manufacturing control processes, and
* management functions

necessary to convert:

* raw materials
* labor
* energy, and
* information

into a high quality, profitable product, within a reasonable amount of time (Allied/Signal).

CIM is the conceptual basis for integrating the application and information flow of product design, production planning, and plant operations (D. E. Winosky).

CIM systems are flexible manufacturing systems having a blend of materials, machinery, people, and information interfacing with each other and with management through integrated material handling and storage and a distributed network of computers (Battelle).

CIM is a powerful technical concept that can be used as a framework for implementing the factory of the future. CIM is a total management approach toward making a company a more viable competitor in world markets (T. G. Gunn).

CIM is a philosophy that may be used by management to improve productivity through the simplification and integration of its operations (T. Hill).

Looking over these definitions, we can see considerable differences. In addition, each one contains nonoperational words, such as system and integration, that compound the difficulties in obtaining a consensus. After reading these definitions, one is left with the desire to find a "better" definition.

If it is difficult for experts to define CIM, it is easy to accept that management does not understand CIM. Furthermore, many experts have approached CIM as an end unto itself. Combining this observation with awareness of some recent negative articles and the substantial investment required, one can readily understand why management in most firms have not aggressively promoted CIM implementations.

The validity of this observation is reinforced when you realize that a typical CIM project may require two or more years to complete. U.S. management has often been criticized as being short term oriented versus long term. In other words, the emphasis is on current quarter profits and today's stock price. Since a typical CIM project may not be completed until after the present manager has been promoted and the project probability will not be reflected in the stock price, U.S. management is reluctant to promote CIM projects. Showing support could represent unwarranted risk to an individuals career. The U.S. industrial environment does not encourage applications of new technologies having long implementation times, especially if much risk is involved.

TECHNOLOGY LIMITATIONS

As we gain a better understanding of CIM, our ability to understand the associated technology requirements has substantially increased. One important reason why CIM concepts have not been successfully implemented in many firms is because the necessary technology does not exist.

These limitations can best be understood by considering some of the tasks that one might expect to accomplish in a CIM environment. Ideally there will only one part model (geometry, specifications, notes, etc.); consequently, the same part model will be used by all functional organizations. A part model would be used in the following tasks:

* Design assemblies and perform tolerance analyses on those assemblies.
* Prepare engineering drawings of assemblies individual parts, tooling, fixtures, and manufacturing facilities.
* Create analytical models of parts for structural and thermal analysis.
* Calculate weights, volumes, centers of gravity, and costs of manufacturing.
* Classify existing parts according to shape, function, and the processes by which they are manufactured and retrieve these parts, from the data library on demand.
* Prepare parts lists.
* Prepare process plans for individual part manufacture and assembly.
* Program NC tools for machining complete parts.
* Program tools for bending and punching sheet metal parts.
* Design robotic work cells and program the movement of robots in those cells.
* Draw isometric sketches of parts and

assemblies for use in process planning sheets and technical manuals.
* Control the effect of part design changes on assemblies and their manufacture.
* Prepare inspection programs, including programs for coordinate measuring machines.
* Analyze the effect of design changes on work-in-process and finished goods inventory.

Although this list is not exhaustive, it illustrates that many tasks utilize a part model.

It is interesting to compare this list with how most companies are functioning in the design engineering/manufacturing environments. If a company is using computer aided design, many are using some CAD, you will find that several different models of a part are made (sometimes five or more). These models have several different uses, such as stress and thermal analysis, aerodynamic studies, process planning, and tooling design. Some of these will be wire frame models and some may be solid models. After the part model is "complete" on a CAD system, it may be passed to manufacturing as a paper drawing even though the model exists in a digital form because no way exists to transfer data between dissimilar CAD systems.

After manufacturing engineering receives the drawing, they may recreate the model in a CAD system. Then in some industries where the parts may be very difficult to manufacture, such as aerospace, elaborate manufacturing process plans are created using this CAD model. These plans include graphical representations of the part in different stages of fabrication. After the process plan is completed, NC programming and verification may be done on a different system requiring that another model of the part be created. Finally, quality assurance may use some type of automation, such as a coordinate measuring machine (CMM), to validate a design. Preparing the programs for the CMM may require another version of the part model.

So, although CAD and CAM are used in many companies, very little integration of the "islands of automation" has occurred. It is very obvious that this integration needs to occur. However, with existing technology this is a difficult problem to solve, because very few standards exist for transferring a part model between dissimilar computer systems. Solving this problem is fundamental to achieving a CIM environment!

Also, standards do not exist for representing complex geometric shapes. One reason is that the associated technology is still evolving. For instance, a complex part surface might be represented (approximated) using a bezier, b-spline, or non-uniform rational b-spline function. No one of these will satisfactorily handle all types of surfaces. Some CAD systems do not provide all of these functions; and, where the same function is provided on dissimilar systems, the implementation may vary. As a result, systems utilizing apparently similar functions for modeling surfaces may be operationally incompatible. Consequently, transferring complex geometry between dissimilar CAD systems is not a trivial problem; it will remain a problem until some type of standards evolve.

Companies have tried to mandate a solution to these difficult integration problems by buying only one brand of CAD system. However, this has not provided a satisfactory solution because no one system is superior in all aspects; this may not be the case in the future as this technology rapidly evolves. In addition, since most companies purchase some parts from other companies, there is a need to transfer part models from one company to another. For the defense industry, CALS (Computer Aided Acquisition and Logistics Support) demands that in the future all contractors will be required to provide data through an electronic representation. Mandating one type of graphics system in this environment is virtually impossible.

As CIM becomes better understood, we have realized that only one part model should be maintained for each release. This is an important integration concept. This model should contain any information required by any functional group from design engineering to field support. When this philosophy is applied, the way parts are designed and manufactured changes. Some of these differences are reflected in the following ways: errors from using a wrong model are eliminated; assemblies will fit together because they have been modeled as an assembly; and concurrent engineering and manufacturing are stimulated. When paper drawings and 2-D graphics systems are used, engineering designs often are incomplete and contain errors. A 3-D part model, however, provides another perspective and permits a more accurate evaluation of assemblies; consequently, many design errors can be eliminated. Also, realizing that the design of a complex part is normally an evolutionary process, concurrent engineering and manufacturing will be facilitated because the model can be easily accessed and the access can be controlled such that the appropriate model is utilized. As a result, design and manufacturing lead times are reduced, productivity increases, and quality improves.

The concept of "one part model" has some important ramifications, one of those being data management. A part model will involve graphical data, textual and numeric data, and analytical data. There may be large amounts of each type. Most data management systems in use today were developed for business applications characterized by textual and numeric data that can be managed in similar ways. Virtually no CIM data management systems are available today except in a developmental stage. Because we are just being to understand CIM and what is required to successfully create and support such an environment.

The magnitude of the data management problem is easier to comprehend when the tasks (listed above) anticipated to be performed in a CIM environment are considered. Note the many ways that data needs to be associated if it is to be efficiently managed (stored, retrieved, modified, and reported). Also, the amount of data involved will be enormous. Compare this with the current environment in a typical manufacturing firm where it is difficult to manage data such that inventories and bills of material are accurate enough to support MRP. Consequently, we have a lot to learn from the technical as well as human aspects of data management.

This discussion on technology limitations could be continued (network standards, distributed data bases, etc.); however, the point has been adequately made that initial CIM expectations were not realistic considering the current technology. Yet it is surprising that most CIM literature does not address this issue. For instance, a recent study, conducted by the Digital Equipment Corporation, with the objective of identifying issues that are blocking the implementation of CIM, did not include technology limitations (see Figure 2).

* Lack of understanding, knowledge of CIM
* Lack of management support, commitment
* Resistance to change
* Cost justification
* Turf protection
* Job security
* Training, retraining, education
* Organizational restructuring

FIGURE 2
ISSUES BLOCKING THE
IMPLEMENTATION OF CIM

This list does contain some issues noted here (management support and understanding of CIM).

Although limitations exist, progress is being made in many areas. Extensive research is being performed on methodologies for modeling complex surfaces. Practical solid modeling systems are coming to the market place. NURBS (Non-Uniform Rational B-Splines), a recent advance for modeling complex surfaces, capability is being offered on most graphics systems. Therefore, it is safe to say that modeling techniques are evolving and are becoming mature enough that we can anticipate using one part model to represent a part throughout its life from inception to field support.

IGES (Initial Graphics Exchange Specification) was initially developed to facilitate the transfer of graphics data between dissimilar systems. This effort started in 1979 and is still evolving. Today users of most major graphics vendors utilize IGES.

More than graphics data needs to be passed between computer systems in a CIM environment. In 1981 an effort was started to develop a Product Definition Data Interface (PDDI) that would support a part model robust enough to be utilized by manufacturing engineering as well as design engineering. The product data types considered were: geometry, topology, dimensions and tolerances, features or shape data, nonshape notes, first level assembly, and administration. Most of the associated work has been funded by the Air Force and done in the aerospace industry. In 1984 this project was expanded to include data required to support product in the field. This expanded effort is entitled the Product Data Exchange Specification (PDES). The objective is do develop specifications that would be adopted as an international standard for exchanging a part model between dissimilar computer systems. Included in this model is any data required in the life cycle of a part. Since the integration of information is a fundamental basis for CIM, this philosophy can never be very successful until standards such as PDES have been adopted. However, the need for such a standard is understood and progress is being made toward defining and establishing this standard.

As noted above, no existing data base management systems are adequate for supporting a CIM environment. However, research is being done in this area; for example, some of the major vendors, such as IBM, are sponsoring and performing research in this area. Some early product offerings are on the market; one of the most advanced is a system called SIMPLEX, which is offered by Automation Technology Products (ATP). All of the available products have some limitations and some are based on unproven technologies, such as a relational data base; consequently, additional research is required before these products obtain wide acceptance. From a positive viewpoint, the need is recognized and significant progress is being made toward developing products that satisfy the need.

Data cannot be readily transferred between dissimilar computer systems without standard network protocols. During the 1980's significant progress has been made in network associated technologies and standards. Some examples are MAP, TOP, OSI, DECNET, and SAA. Without these recent advances, the CIM philosophy could not be successful. Although much research remains to be done, the early 1990's will see tremendous progress toward networking dissimilar computer systems and the application of distributed data base concepts.

CIM SUCCESSES

As was noted earlier, there have been some very successful CIM applications. Most of these success stories have some common characteristics. The production environments were of high volume, such as printers and keyboards. In these environments, a large effort could be devoted to optimizing the part design and manufacturing processes. Also, in most of these case, the companies were faced with making significant changes in how they did business or the business would cease to exist. The reluctance to change was overcome by the fear of job loss. Another interesting observation is that several of the successful applications occurred in companies that market products and/or services in the CIM industry.

Initially, in the rush to implement CIM concepts, many companies spent large amounts of money on automation and integration. We now understand that the part design and manufacturing processes must first be simplified and discipline instilled before automation and integration can be successful.

We have also noted that CIM has been successful in companies where it is an element of the firm's strategic plan. In conjunction, top management avidly supported CIM. This support is especially important because CIM applications are long term projects involving large amounts of information, new technologies, and interdisciplinary skills from many different functional organizations. Therefore, success requires strong leadership from top management.

We should note that most of the successful CIM stories involved integrating only a portion of the manufacturing environment. This is understandable when the technology limitations noted above are considered. Possibly CIM should be considered a goal, something that we continuously work toward but never fully achieve.

GUIDELINES FOR SUCCESS

CIM projects can be successfully implemented. From comments made above, we know that realistic objectives should be established. Keeping this in mind, where do we start? In today's manufacturing environment, direct labor is approximately 10 percent of product costs, and this percentage is projected to become smaller. However, indirect labor may account for 45 percent of product costs with the remainder being attributed to materials. With this insight, we can see that automation of direct labor activities should not be a major objective, but indirect labor tasks appear to offer significant potential cost savings. Although this may appear to be an obvious conclusion, many companies have spent millions of dollars on direct labor automation before realizing that the return on investment was not at a desirable level.

Indirect labor costs offer significant opportunities for product cost reduction; consequently, CIM projects comprising these functions promise the most potential benefits. Pursuing this reasoning, it is important to identify the indirect labor tasks in a typical discrete parts manufacturing firm. A new part design or an engineering change to an existing part can result in the following actions:

* Sketches and drawing retrieval
* Engineering analysis
* Testing and simulation
* Layout and checking
* Engineering review
* New drawings
* Manufacturing modeling
* Process plans
* New molds, dies, tools, and fixtures
* New programs for machining
* New inspection procedures
* Quality assurance and testing
* New process plans, routings, and time standards
* New assembly instructions
* New field and service manuals

Many people from several departments will be involved in completing these tasks and the cycle time will be weeks if not months.

Another way to look at product costs is to look at the relationship of cost versus type of decision. Figure 3 depicts such a relationship. Note that decisions made early in the life cycle of a part have a much greater impact on product costs than those made later.

At this point we have a good idea where our efforts should be concentrated. However, before a CIM strategic plan is developed, some guidelines for successfully completing a CIM project should be understood. The following guidelines have been compiled from

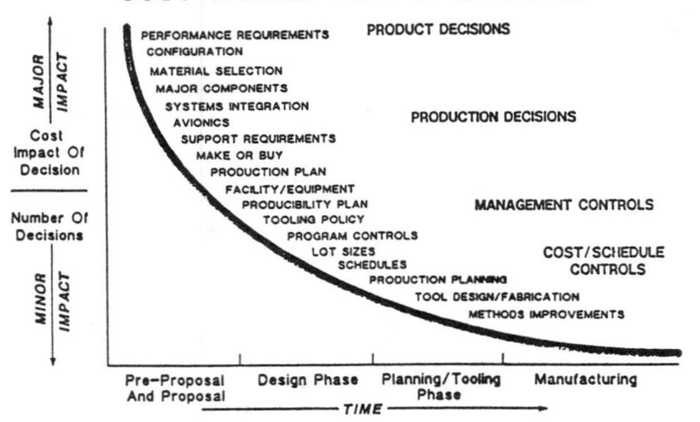

FIGURE 3
IMPACT OF COST VS DECISION

several CIM success stories:

* Define CIM
* Make CIM part of a long term strategic plan.
* Plan from the top (must have top manage support); implement from the bottom.
* There are no isolated manufacturing problems; develop for the long term.
* Simplify, automate, and then integrate.
* Emphasize basics (data integrity, discipline, quality, etc.)
* Place emphasis on indirect labor, engineering, and management functions.
* Strive to "pay as you go".
* Emphasize education (management and work force).
* Communicate, communicate, communicate!

SUMMARY

Although the CIM philosophy has not been as successful as initially projected, we should not underestimate what can be achieved. Lack of understanding of the meaning of CIM and technology limitations have been major impediments to success. However, these impediments are being overcome. Tremendous benefits can be realized with today's technology if it is judiciously applied. Anticipated technology advances promise even greater benefits. We must be realistic in our expectations and follow guidelines assure success, such as those proposed in this paper.

BIBLIOGRAPHY

Gunn, Thomas G., Manufacturing for Competitive Advantage, Ballinger Publishing Co., Cambridge, Mass., 1987.

Halbrecht, Herbert Z. and John W. Nostrand, "Is Management to Blame for the Unfilled Dream of CIM", Production Engineering, July, 1987.

Nag, Amal, "Auto Makers Discover `Factory of the Future' Is Headache Just Now", Wall Street Journal, May 13, 1986.

Port, Otis, "High Tech to the Rescue", Business Week, June 16, 1986.

Rhein, Bob, "No Easy Road Seen to Implementation of CIM", Management Information Systems Week, Jan. 26, 1987.

What's in a Name: Plain Talk About CIM

Most current models for computer-integrated manufacturing betray a natural human tendency to inflate and obscure. They usually ignore the nontechnical aspects of the manufacturing business. True integration requires an uninterrupted flow of electronic information, in carefully laid out channels, about all aspects of an enterprise.

Jack Conaway
Manager, CIM Marketing
Digital Equipment Corp.
Marlboro, Mass.

CIM, the acronym for computer-integrated manufacturing, has almost eclipsed CAD/CAM in the lexicon of high-tech jargon. As with other acronyms, there is still widespread confusion regarding the definition of CIM. Also in question are the motivation for pursuing it, the steps toward achieving it, and how to pay for it. We can put these issues in perspective by examining the paths traveled to reach the current state of automation in manufacturing and from there the directions that are likely to be taken in the future.

What is CIM?

Joseph Harrington coined the term *Computer-Integrated Manufacturing* in 1973 in his book of the same name [1]. Unfortunately, many have deviated from Harrington's remarkable vision of an entire manufacturing corporation which runs off the uninterrupted flow of electronic information. Most of today's CIM models betray a natural human tendency to inflate and obscure. They usually don't bother to consider the nontechnical aspects of a manufacturing business: marketing, sales, service, finance, and administration.

A number of vendors claim to have undergone the miraculous, overnight tranformation from turnkey CAD/CAM system house to CIM supplier. Actually, there is no such thing yet as a CIM supplier. In order to assemble all of the computers, software, and factory automation equipment required for a CIM solution, the customer must deal with many different vendors.

CIM may be defined generally as the automation and integration of a manufacturing enterprise through use of computers. An enterprise may consist of many corporate entities that cooperate from the initial conception of the product to its distribution, installation, and maintenance. In order to achieve progress toward integration, an enterprise must first consider the electronic transfer of data across organizational and geographic boundaries.

Why Do It?

The decision to pursue CIM is a strategic one that rests with top management. They are most concerned about achieving corporate business goals and critical success factors such as a short design/production cycle, flexible product mix, high quality, customer satisfaction, and ultimately the survival of the company. The automation of various tasks, while necessary to CIM, cannot by itself achieve these global goals.

The justification for CIM should be related primarily to long-term

Reprinted with permission from *Computers in Mechanical Engineering*.

Inexpensive graphics systems, coupled with electronic mail, help manufacturing and purchasing personnel become familiar with the designs well before they have to be executed.

revenue generation and market share enhancement, rather than to cutting costs in the pursuit of profit. In many industries, short-term goals must be sacrificed. New financial measures are needed to relate corporate goals directly to expanded revenues over the long period of time necessary for CIM implementations (10 years or more). Companies that are talking about two- or three-year CIM projects are not addressing the global CIM problem as defined here, but some subset of integration.

The Heterogeneous Environment

Many people think that CIM means putting all of a corporation's data on a data base management system (DBMS). This is neither a desirable nor achievable goal. The great revolution in mini- and microcomputers was largely fueled by the poor performance of large shared systems. Even logical centralization of data is a spurious goal for all data of a enterprise. One should not expect that corporate planners would be interested in stresses on a part or that an engineer would be interested in the maintenance status of a machine on the shop floor. Rather, data should be organized so that people or machines that share a set of functions have access to the information about them. A centralized data base ignores the heterogeneity of data management strategies and tools used in manufacturing today.

Islands of automation. Most companies no longer write their own programs for CAD, numerical control, finite element analysis, testing, simulation, or manufacturing planning, as they had to do in the infancy of these technologies. Instead, their internal programming efforts are now directed to newer technology areas and to tailoring the systems purchased from outside suppliers to the specifics of their operations. The islands are mostly purchased from independent suppliers who have their own data management strategies. To complicate the problem, these suppliers change the data entities and attributes in their applications independently and release these changes on an independent schedule. Any real CIM solution must deal with the variety of tools that are being used today for task automation.

But the data management heterogeneity is only one of the integration issues that is posed by the islands of automation. The modern manufacturing enterprise must also find a way to integrate computers, operating systems, communications technologies, workstations, and automation equipment from different vendors. In most cases, the original make/buy decisions for these components were based on optimizing task automation with little or no regard to integration.

Layers of integration. The opportunities for integration occur in networking/communications and data/information control, as well as the application data management layers. Integration in organization will be discussed later under CIM implementation. The networks and administrative control layers need to be globally linked through functions across the corporation or enterprise. In general, the application data need to be linked to closely associated applications that may be geographically and/or logically connected through organizational departments or projects in the office or factory.

CAD/CAM and OA. Fully three quarters of the personnel of a large modern manufacturing corporation

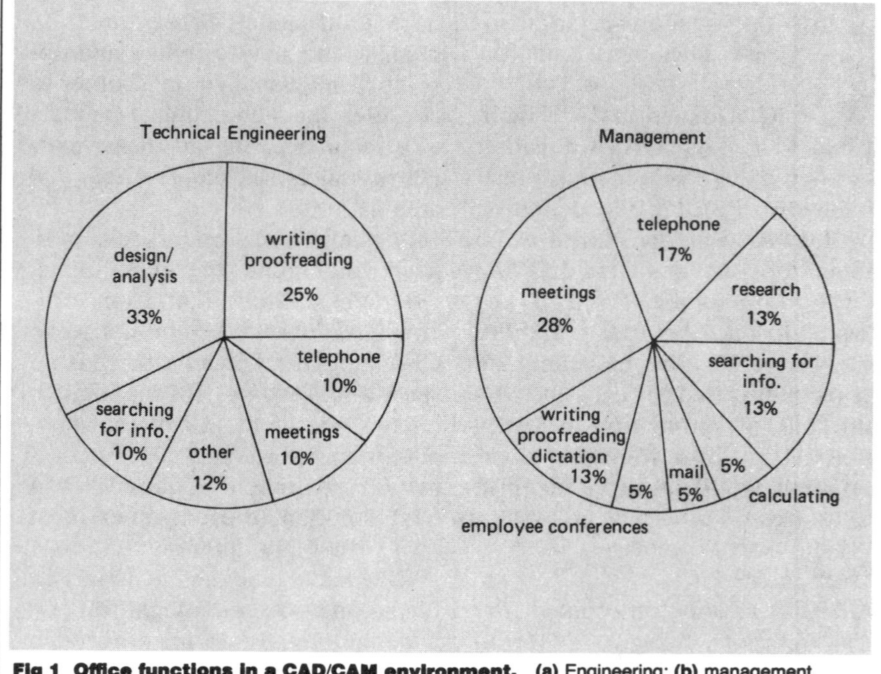

Fig 1 Office functions in a CAD/CAM environment. (a) Engineering; (b) management.

work in an office. Vendors of office automation (OA) and CAD/CAM systems market their products as if they were going to two separate markets. The major reason is that most vendors do not have both products. In reality, the products end up in the same offices and should be integrated. A study of the time spent by technical personnel and their managers (Figure 1) indicates that only a third of an engineer's time is spent using a CAD/CAM system [2]. The other two-thirds is spent on activities that could be more effectively attacked using OA tools. All of the manager's tasks within the same department are enhanced by OA tools. Fortunately, major information consultants have realized this and are busy educating end users and vendors on the requirements for integrated office environments [2, 3].

A Workable Strategy

Manufacturing companies want to tie their islands together into a unified system that still preserves special characteristics, local control, and high performance. But they also want to attain consistent data and control over larger segments of their operations, more typical of centralized systems. The variety of these islands suggests that a workable CIM strategy is a modular approach to integration, aided by the adoption of local data exchange standards that allow free selection of software applications as they become available. The alternative approach of reducing the choice in applications and machinery and building closely coupled systems based on these reduced options will only lead, in the long run, to reduced creativity, reduced competition, and stagnation.

A workable strategy for flexible automation in CIM starts with an open computer architecture (Figure 2). Customers who are interested in true integration should stick with computer suppliers that have a commitment to a single architecture across a wide range of their products. This approach allows flexible and cost effective choices for processing power and storage to suit local computing requirements. It also preserves investment in software programming and user training when changing processors. An open architecture has easily recognizable characteristics: the networking architecture and communications are tied to the International Standards Organization's (ISO) seven-layer Open System Interconnect standards.

Defacto standard operating systems such as Unix, CP/M, and MS/DOS are offered, as are standard computer languages such as Fortran-77, C, and Pascal. Standard data base management (e.g. Codasyl compliant) products can be purchased. Standard graphics-workstation interfaces such as GKS and Siggraph Core are also available.

Provisions should be made for the connection of standard local area networks, such as Ethernet, with other communication media for wide area networking, e.g. public-packet switching networks and satellites. Also, systems on the local area network should be able to communicate with other networks such as IBM SNA, PBX, and token bus in a factory environment, as the standards for these become available.

Each engineering program has its own data structure (Figure 3). The little integration existing between applications from different vendors is usually accomplished by translators operating on a one-to-one basis between the different internal formats. For each set of two applications, a new translator must be written. And if either of the appli-

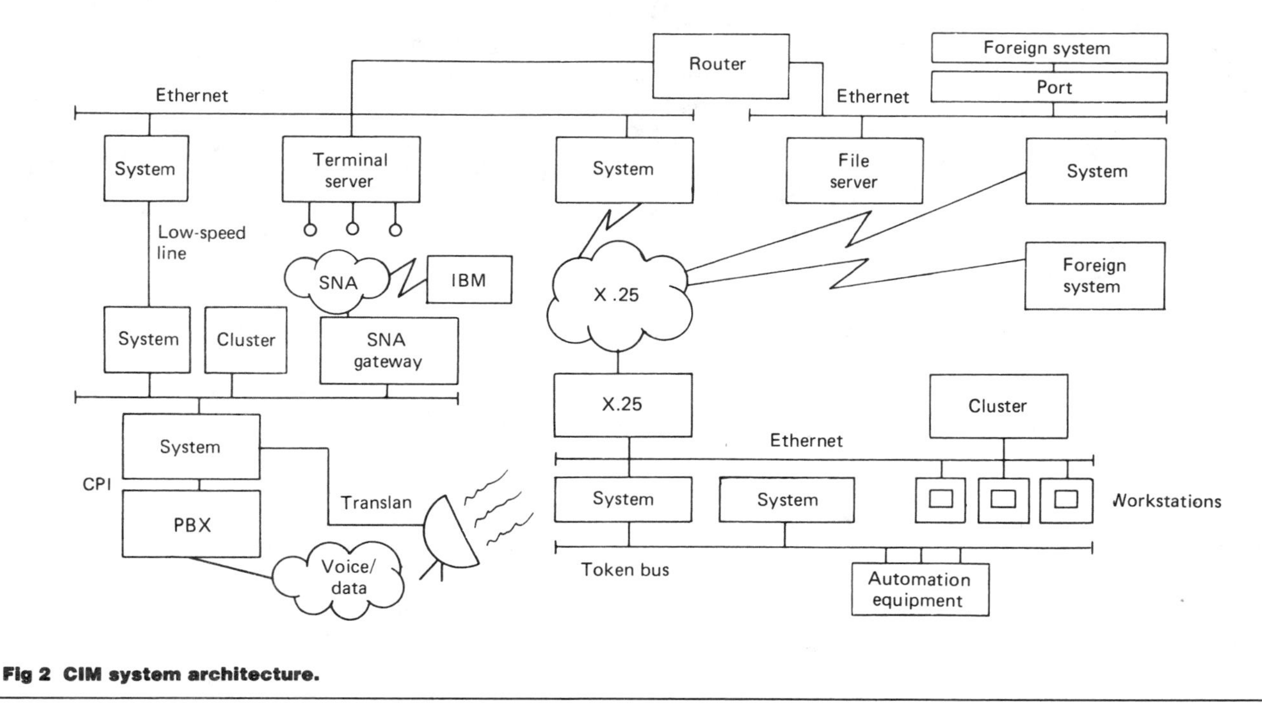

Fig 2 CIM system architecture.

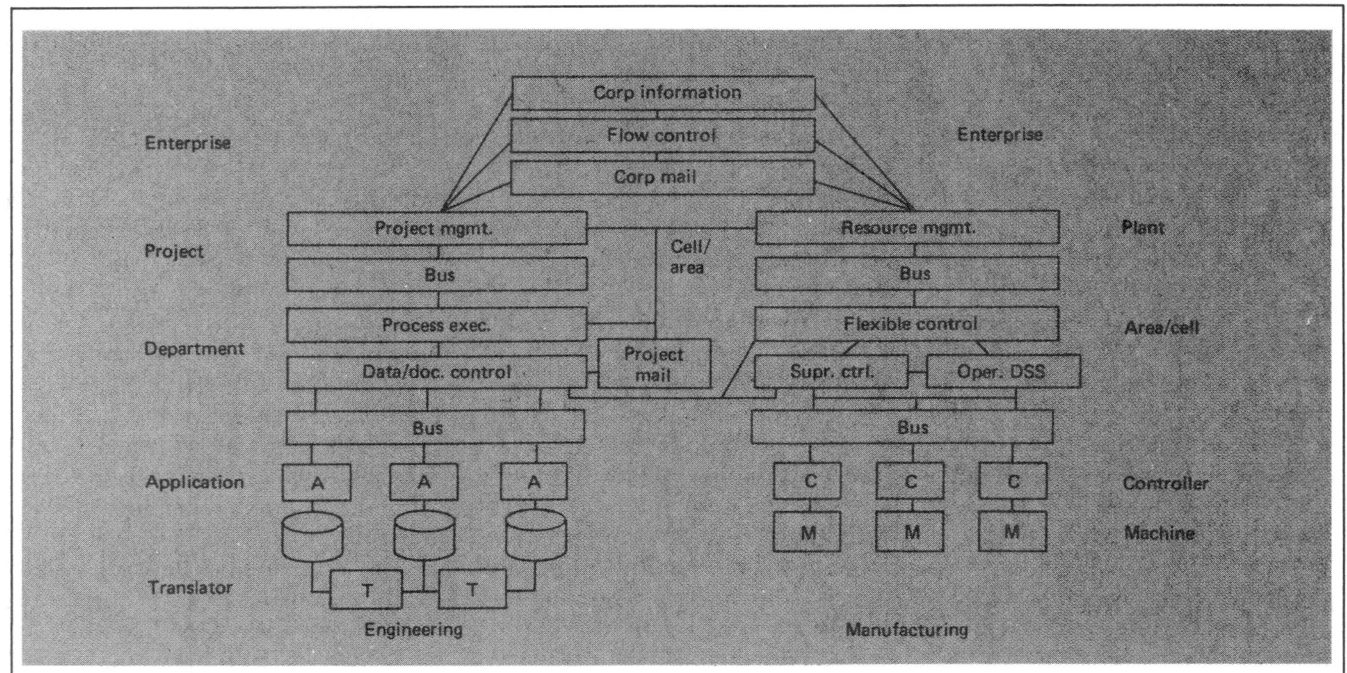

Fig 3 CIM data structure.

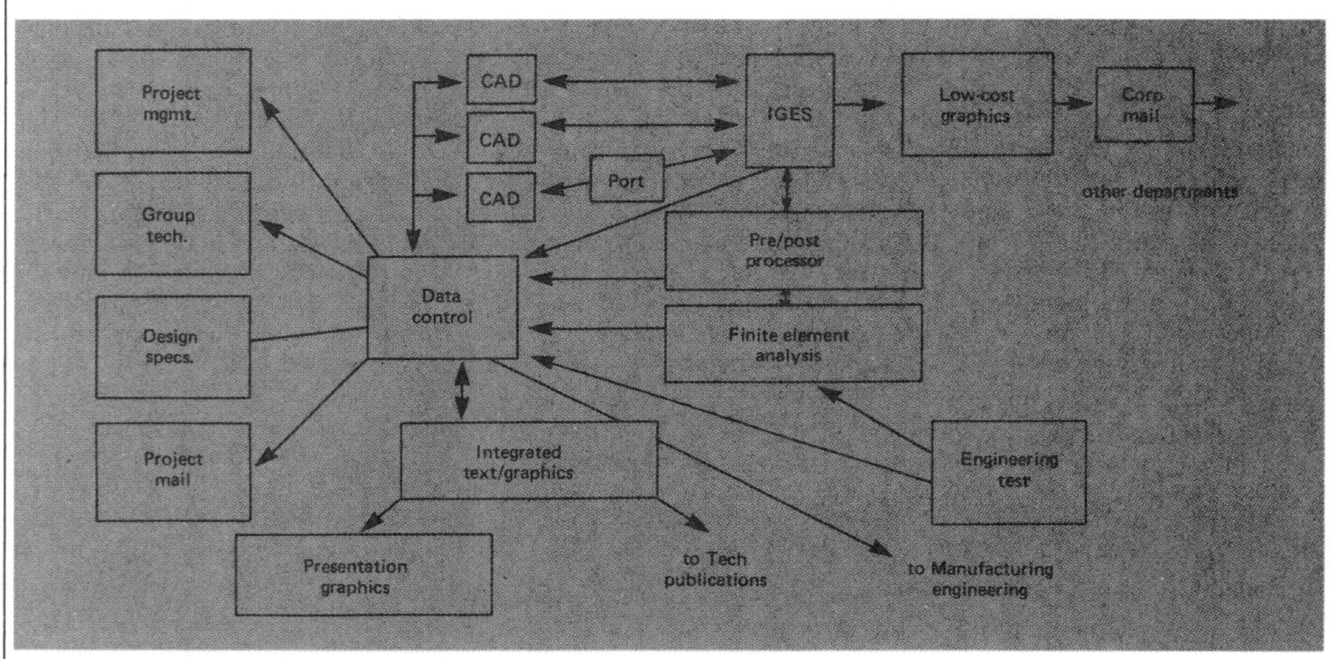

Fig 4 Mechanical engineering fuctions.

cations is modified, the translator may need to be modified as well. Translators are usually run in batch mode and operate one way, so they do not provide an intelligent interactive environment for decision making. Also, translators lead to inconsistencies when modifications are made to data downstream.

A much more effective strategy is the use of departmental software buses, or data exchange standards. As the specifications for these buses become standard, e.g., IGES, or PDES in mechanical engineering, the software and turnkey vendors are required to provide a stable interface with each new product release in order to compete in the market. Also, vendors who are developing data control, integrated text/graphics, document control, and project management applications can more easily integrate their products because they no longer have to worry about data translation problems. The applications may be backed up by data controllers which provide project security, audit trails, revision and change order control, and notification of design status. Process executives which execute and chain applications based on changes of primary design information can be employed.

The use of software buses, project mail, and process executives makes the data produced by the applications more consistent and timely across the department, removing the argument against distributed processing.

In the factory, the same principle holds true when applied to factory automation equipment: bus standards are needed for classes of factory equipment (such as numerical control machines, robots, automation test equipment, and coordinate measurement machines, etc.). Once these standards are established, it should be the responsibility of the factory automation vendors to supply interfaces to the standard every time they introduce a new product or new model of an old product. The vendors that are developing supervisory control, decision support, and flexible automation systems could then communicate instead of worrying about internal data formats for a variety of equipment. Plant resource management applications such as MRP II could be connected with generic flexible automation and control software to integrate planning and production.

There will always be a need to extract data to support corporate-wide activities such as personnel and corporate finance. These functions should be treated just the same as the engineering and manufacturing departments that are supplying project status, personnel, budget, and expense data. The corporate functions have their own local computing requirements for data from all over a corporation. Mail systems can be used to communicate between these corporate groups and other functions over the network.

Consultants who offer to conduct CIM studies in six to eight weeks probably do not have a rigorous enough planning method or may fear that companies will not engage them for a longer period of time.

The Productivity Stream

At present most discrete manufacturing companies working on CIM are concentrating on linking the product engineering, manufacturing engineering, and production functions. This activity is of particular interest to enterprises dealing with complex assemblies of parts. There is less emphasis now on linking resource management with engineering by passing bills of material and process plans. There also appears to be less activity now in connecting up the shop floor with resource management in manufacturing. Interest in these secondary productivity streams should accelerate as success is achieved with the primary integration.

The stream from product engineering through production and distribution is primarily associated with corporate goals of reduced time-to-market and increased product quality. Despite substantial task automation by many manufacturers, a five-year time-to-market is typical for complex products. Those companies that can reduce their time-to-market by a year or two while maintaining or improving quality will have a competitive advantage.

Mechanical engineering. The product engineering functions within a discrete manufacturing enterprise are dominated by mechanical and electronic engineering (Figure 4). It is not unusual for a single engineering department or large project to be utilizing software and turnkey systems from seven to 10 suppliers. Most mechanical CAD systems are able to read and write in IGES formats [4]. Although IGES is an imperfect standard, it greatly simplifies the connecting of mechanical CAD programs. It is only a matter of time before IGES or another related standard is adopted.

But this is only a small part of the picture. Many manufacturers with sophisticated CAD systems are finding out that they are not making significant progress in reducing time-to-market in spite of the dramatic productivity increases in mechanical engineering. Design data and downstream applications in manufacturing engineering are often still managed by use of pen and paper. Moreover, as CAD systems crank out more new designs and modifications, the revision control is not exercised and the supporting data (finite element models, analysis results, and test results) are no longer in step. Erroneous data slip through to manufacturing, leading to a waste of valuable time and incremental costs in retooling, reworking, and recalling of products. Also the information on a product's design often surpasses that in its documentation, even though the service documentation must accurately reflect the product changes long after the plant has geared up for new models.

Fortunately, several programs and turnkey systems either on the market now or soon to be delivered address these problems. These include: data controllers for security, back-up, data concurrency, and auditing while maintaining strict control of engineering change orders; document systems that manage complex drawing data in an integrated text/graphics system; and low-cost graphics systems for viewing and manipulating designs where expensive workstations are not required. Combining these new systems with office automation tools such as electronic mail, it is possible to send drawings for review through the mail and to notify all project members immediately of design changes.

It is very important to electroni-

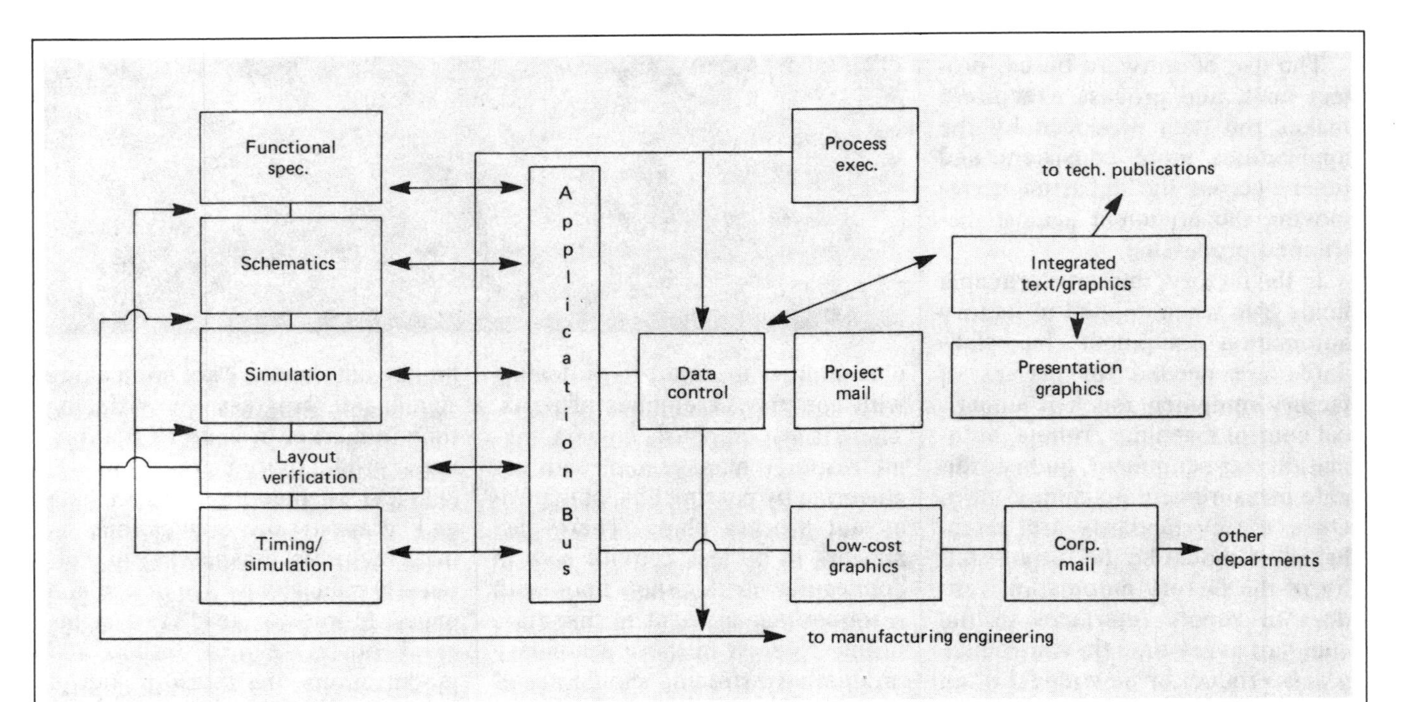

Fig 5 Electronic engineering. Steps in PCB and IC design are executed more regularly than those in mechanical design, although examination and reevaluation of earlier results is necessary.

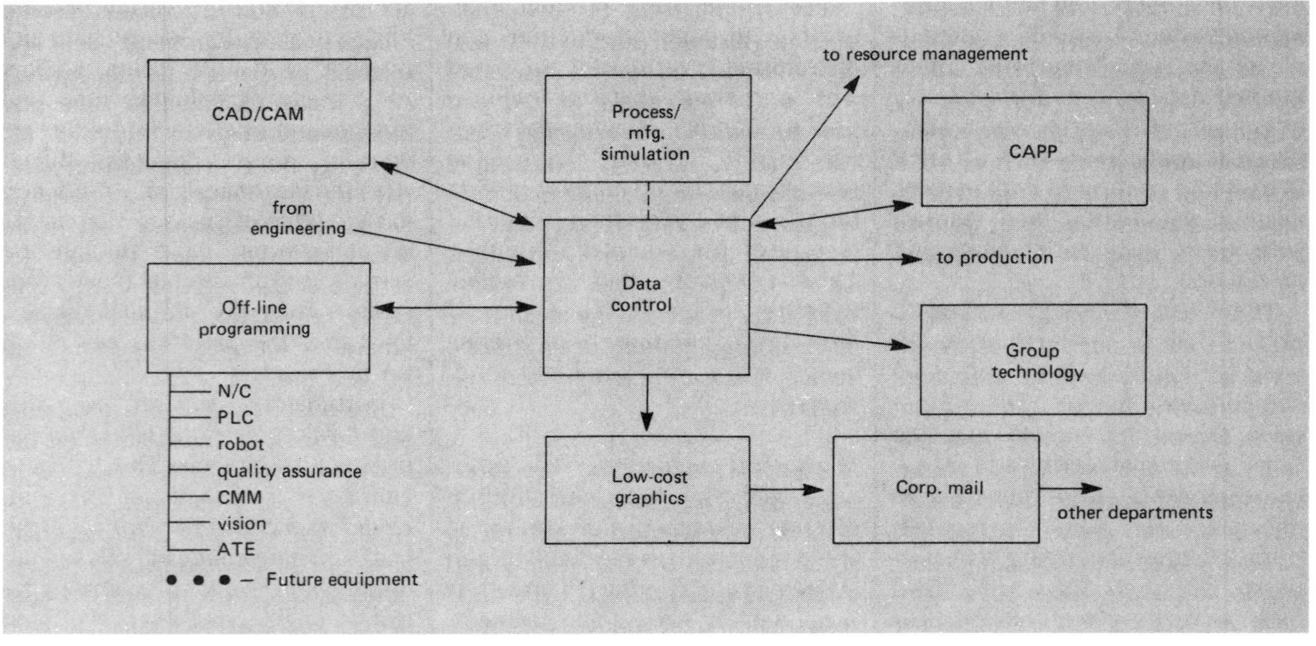

Fig 6 The role of CAD/CAM in process engineering.

cally pass the design information to other departments in order to avoid manually re-entering data. Manual operations are prone to error and tend to destroy data consistency. The most important data in the time-to-market stream are the design geometry and text. The design is needed for the tool, die, and fixture design and part programming and process planning in the process engineering stage. There is an increasing desire to use the design data from product engineering in quality assurance operations. Low-cost graphics systems, coupled with electronic mail, help manufacturing and purchasing personnel become familiar with the designs well before they have to be executed.

Electronic engineering. Most of the preceding comments also apply to the electronic engineering of printed circuit boards (PCBs) and integrated circuits (ICs). The steps required to design PCBs and ICs—functional specifications, schematics, logic simulation, layout verification, and circuit timing and simulation—are executed more regularly than those in mechanical design, although as in any real design process there is a good deal of backtracking to examine and reevaluate

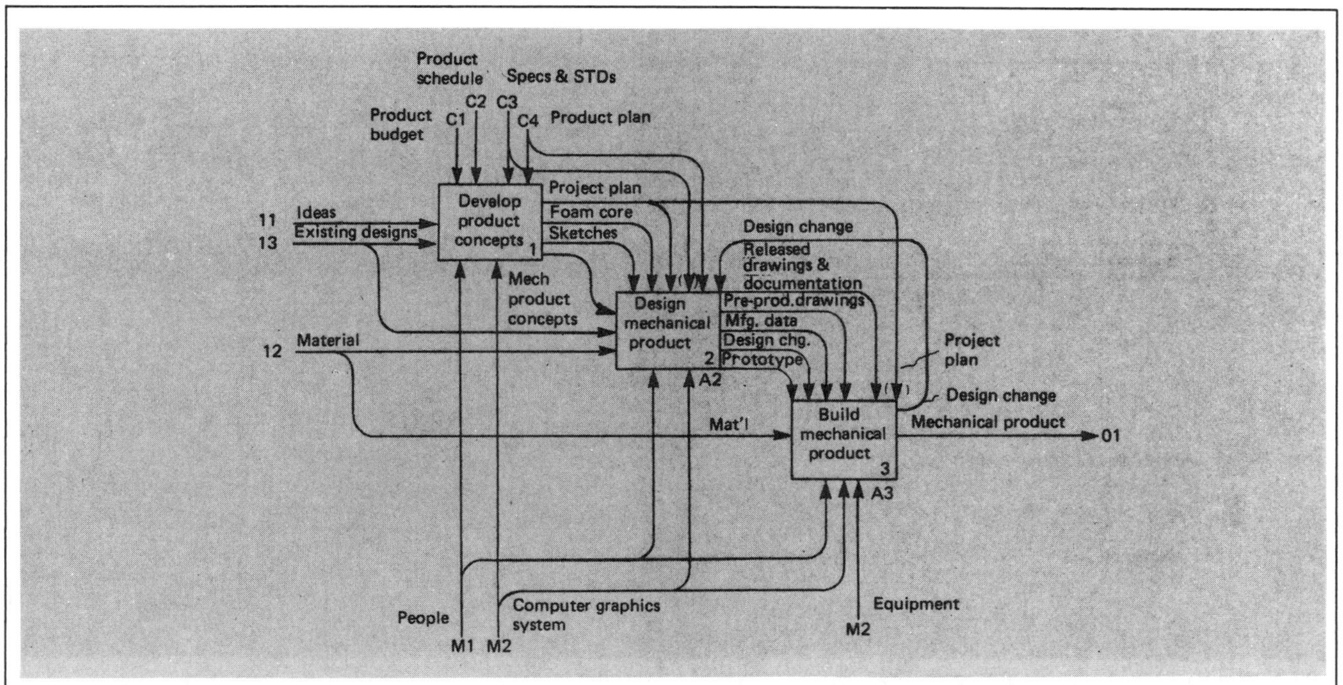

Fig 7 Modeling an enterprise's functions and their connections: Input, output, storage, and documentation requirements. The number of users, frequency of execution, and average program size must also be considered.

earlier results (Figure 5). In most companies these steps require the use of several turnkey systems, leased software, and in-house developed software. Vendors who specialize in schematic capture, for example, may not offer logic and fault simulation. Circuit simulation software usually comes from yet another source. The search for the perfect auto router can lead to other vendors, as well.

Software and turnkey systems from separate suppliers are loosely integrated with translator programs working with internal file formats from adjacent programs.

In the IC world, the electronic design interchange format (EDIF) is emerging as a possible departmental application bus, and some thought has been given to extending the standard to PCBs [5]. A proposal has also been accepted to extend the IGES specification to PCB design. It is important to resolve the question of standards to provide a stable medium of data exchange. Tying the mechanical and PCB design worlds together would also be desirable for electronic product packaging.

Because of the structured nature of electronics design, particularly using standard cell methods and gate array technology, it may be possible to control the execution of the program and translator via a preprogrammed process executive. As changes to the schematics, based on simulation and verification results, are fed into the system, downstream applications can be chained to provide automatic data consistency across the department. Proprietary process executives for electronic engineering, which also link to manufacturing programs, can be found in the internal operations of computer companies, but are not yet commercially available.

The remainder of the data control and office functions shown on the diagram are the same as required for mechanical design. By using modular software design principles, a data bus and format-independent data control, much of the same software used in mechanical design can also be employed in electronics design for these functions.

In the the time-to-market stream, numerical control and robotic control information for the drilling, device insertion, and assembly of PCBs into modules should be transferred electronically to manufacturing or process engineering and production. The same path should be followed for test pattern generation to automated test equipment and mask data to photolithography or E-beam devices for manufacturing ICs.

Process engineering. CAD/CAM systems are often used in process engineering for tooling, design of dies and fixtures, and plant layout (Figure 6). Some of these systems may also be used for simulation of numerical control and robotic cells as well as creating part programs for these devices. Often, separate stand-alone programs are used in off-line part programming for various production, material handling, and testing equipment. There are few commercially available process planning software packages at this time. Many companies are still struggling with paper process sheets or have in-house programs. Automation of process sheets can be accomplished fairly simply with inexpensive graphics by combining the design drawings with a forms-management system. In this way, electronic process sheets are created and then reviewed over the net-

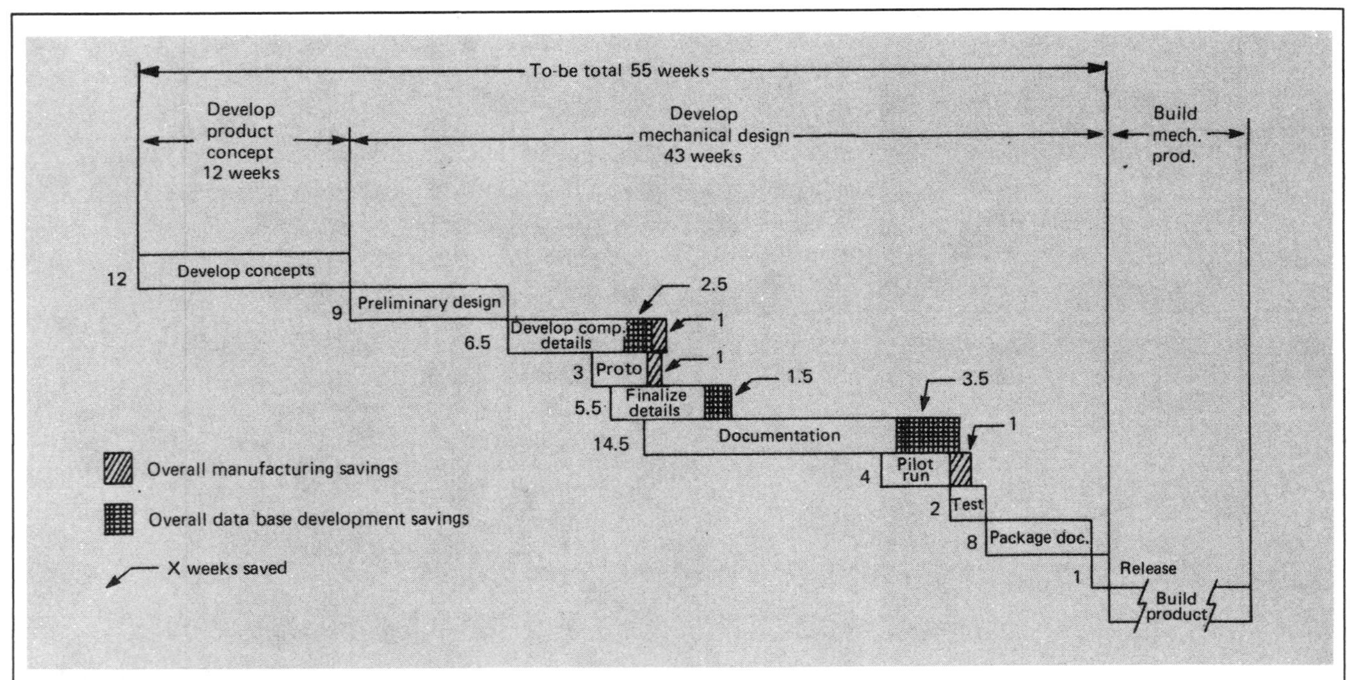

Fig 8 Time-to-market savings with CIM. Manufacturing, data base development, and time.

work. Group technology programs for process plan retrieval are available from outside vendors.

As the number of changes in product and process engineering increases due to increased use of CAD, it will be necessary to cover the tool, dies, fixtures, part programs, and process plans under the same revision control as design engineering in order to avoid errors in production.

Production/assembly. The production, assembly, and testing functions should receive released versions of part programs for management and down-line loading by the supervisory control software in the factory. Software (for example, DEC's Baseway) that collects data from devices on the shop floor and provides decision support, data management, and down-line loading capabilities is just beginning to appear commercially. The next level of flexible cell and area control software needed in an automated factory is still in development. Limited forms of this software appear in large turnkey FMS systems. As flexible control becomes available, it will also be necessary to electronically transfer process planning information, as well as device programs, from process engineering to production.

Factories use different brands of equipment for production, data acquisition, material handling, and testing. Use of standardized buses would place the burden of interfacing new products on the automation vendors. Control programs operating at a higher level could then work independently of the machines selected. The success of bar code standards illustrates the proper way of interfacing. Because of this standard, bar codes printed from a variety of devices can be read by several vendors' equipment. Much work still needs to be done to define the data exchange standards for other important classes of devices found on the shop floor.

Above the flexible control level, application buses may be used to communicate with resource management programs used in production planning. This approach will allow control and shop scheduling software to interface with a variety of manufacturing resource planning, quality assurance, and energy management programs.

CIM Planning

The planning for CIM should proceed from the top level of the corporation down, and implementation from the bottom up. Top-down planning assures that the business goals will be tied to the implementation and that there will be reference points against which to measure results. The complexity of the manufacturing process dictates that implementation be modular and phased in from the lowest level of operation.

Enterprise planning should begin with global business goals. This starts with fundamental questions such as: what business are we in, and what business do we want to be in in five or ten years? These "as is" and "to be" schemes should then be translated into specific objectives that support the goals. Because of the long implementation time for CIM, relevant trends in the market, economics, politics, technology, and competition should be considered at this stage. John Rockhart from MIT has developed a method for clarifying objectives. [7].

A functional model of the enterprise or segment under study should depict the functions and their connections. These models are an important foundation for planning. They can be used to look

The strategically important weapons of manufacturing are no longer labor or capital equipment for mass production. Better information is the means that forward-looking organizations are using to reach towards their business goals.

at the efficiency of applications and uncover missing capabilities required for integration. They can also be used to understand the physical distribution of computers and applications by location, facility, computer room, or CPU.

The functional models (Figure 7) can next be decomposed into data models by looking at the input, output, storage, and documentation requirements of each function. By examining parameters such as number of users, frequency of execution, and average program size, the planners can get a good idea of the storage and communication requirements.

A complete audit of all present hardware and software should be a part of the "as is" model. But it is also important to emphasize the "to be" model at the bottom levels. An understanding of current systems is absolutely necessary, but there is a natural tendency to spend too much time on the details of the old rather than possible changes. Functional and data flow models are necessary to determine the applications, systems, and communications that must be acquired and linked to achieve integration. Good models help the planners locate critical paths of data, resource expenditures, and time, which relate back to the basic business objectives (Figure 8).

The design for the CIM system follows from the data flow models. At this point, conventional MIS measures for computer capacity and performance on networked systems may be used to determine what configurations are needed.

Planning is normally done by a program office or task force consisting of process experts, technology experts, and management. The work of a core group of experts is usually reviewed by additional groups. The core group is often assisted by external consultants. Serious CIM studies completed in the past few years have involved interviews with 50 to 2000 people over a period of three to 18 months. These figures depend on the extent of the operation being analyzed and the rigor employed. Consultants who offer to conduct CIM studies in six to eight weeks probably do not have a rigorous enough planning method or may fear that companies will not engage them for a longer period of time. The studies that result from these short contracts may shed some light on top management's objectives and indicate possible areas of further study, but real integration will require further study.

Implementation

The planning process should produce a time table and action plan, which will probably be modified before implementation actually begins. In most companies the top-down planning for CIM is highly biased towards computer and factory automation technology. But a substantial advancement in technology affects the organization of people and definition of tasks, and these changes must be taken into account.

Organizations have "turf" issues that need to be addressed. People resist change. In many instances, the tasks performed within a function need to be completely redefined in order to take advantage of CIM technology. And as a result, several manual tasks will likely be eliminated. The CIM action plan should be widely published and include organization, personnel, and task changes as well as technology proposals.

CIM implementation is difficult because of the level of attention to detail required for real success. After a long planning period, the tendency is to get bogged down by the enormity of the challenge and the intractability of turf and personnel issues. Continued commitment and pressure from top management is vital. At the same time, a procedure for resolving conflicts should be established.

Lessons learned in the painful installation and frequent failure of early MRP systems are important. Unless the correct data are collected on a timely basis to fuel the system, the technology will be worthless. Automating large segments of a manufacturing business eliminates many chances for old-fashioned human adjustments. From the start, data must be precise and correct.

The post-industrial, or information, age is upon us. The strategically important weapons of the manufacturing enterprise are no longer labor or capital equipment for mass production. Better information is the means that forward-looking organizations are using to reach towards their business goals, face growing competition, and survive. ■

References

1 Harrington, J., *Computer Integrated Manufacturing*. Huntington, N.Y.: Robert E. Kreiger Publishing, 1979.

2 Poppel, H., "Who Needs the Office of the Future?" *Harvard Business Review*, 1983.

3 *The Consultant*, September/October, Digital Equipment Corp., 1984.

4 *Initial Graphics Exchange Specification (IGES)*, Version 2.0, National Bureau of Standards Report NBSIR 82–2631 (AF).

5 *Electronic Design Interchange Format (EDIF)*, Version 0.8, EDIF Steering Committee, April 30, 1984.

6 Burgam, P., "Marrying MRP and CIM," *CAD/CAM Technology*, Winter, 1983.

7 Rockhart, J.F., "A Primer on Critical Success Factors," Center for Information Systems Research, MIT, June 1981.

Additional Readings on Fundamentals of Systems Integration

Hall, George M. 1991. *Strategy, Systems, and Integration.* Blue Ridge Summit, PA: TAB Books.

Mitchell, F. H., Jr. 1991. *CIM Systems.* Englewood Cliffs, NJ: Prentice Hall.

Ranky, Paul G. 1986. *Computer Integrated Manufacturing.* Englewood Cliffs, NJ: Prentice Hall.

Rembold, Ulrich, Christian Blume, and Ruedicer Dillmann. 1985. *Computer Integrated Manufacturing Technology and Systems.* New York, NY: Marcel Dekker, Inc.

Savage, Charles M. 1990. *5th Generation Management.* Bedford, MA: Digital Press.

Teicholz, Eric and Joel N. Orr. 1987. *Computer Integrated Manufacturing Handbook,* New York, NY: McGraw-Hill.

III. PLANNING FOR SYSTEMS INTEGRATION

Many organizations which launched aggressive programs in automation and robotics during the middle 1980s were greatly disappointed that their large expenditures did not yield the improved performance they were expecting. In many such cases, the companies launched the programs for fundamentally unsound reasons. Many were simply enamored with the advanced technologies and assumed that the payoffs would "come naturally." Others were motivated by a desire to reduce or eliminate direct labor from their operations.

Most companies today are much wiser than before. There is a growing realization that any modernization initiative must be designed explicitly to address strategic business objectives. Acquiring a high degree of manufacturing flexibility, for example, is neither a good thing to do nor a bad thing to do, in and of itself. If, however, a critical driving force in the industry is rapid response to customer changes, then having a high degree of flexibility can contribute to achieving this strategic business objective. The same can be said for any technology, system, or management practice.

Industrial engineers engaged in system integration initiatives cannot be successful without mastering the fundamental concepts of comprehensive business planning. The articles selected for this section are focused particularly upon the application of planning for systems integration.

Vision Into Action™

An Overview

by

Dr. John Samuels,
Vice President, Continuous Quality Improvement, CONRAIL

Alexander Ogg,
President, LEADERSHIP STUDIES INTERNATIONAL

Introduction: Report from the Year 2002

In these turbulent times, the pace of change is rapidly accelerating, sweeping our organizations into chaos and turmoil. In the coming years, many currently excellent organizations will not survive. They will die -- not because they were bad businesses -- but because they lacked powerful visions of their future, or failed to transform their visions into results.

To get a sense of who will survive, and who will not, let's spend a moment in the future. It is now the year 2002. Let's look around and see how organizations appear from a leader's point of view. What do we find?

- Entrepreneurial, vision-driven firms emerging as powerhouses, suddenly dominating their industries through innovative services and technologies.

- Well-established companies hopelessly clinging to obsolete products, market share eroding, slowly bleeding to death.

- Authoritarian leaders being swept aside by tidal waves of human desire for freedom.

- Firms with established, protected markets suddenly finding no place to hide from the unforgiving reality of a global economy.

- Organizations torn apart from within because management has not learned to lead an increasingly diverse, multi-cultural work force.

A rapidly fading memory is the spectator sport of watching giant companies stomping about like clumsy elephants, trying to crush their competition with the weight of traditional business practices. Those who survived learned that agility coupled with adaptive strength and endurance is the key to success. Many organizations who cling to the old ways desperately fight to keep alive nostalgic memories of a time long gone. Like zombies, most are already dead. They just haven't realized it yet. Here in the world of the 21st century "tiger" organizations dominate the marketplace.

What caused this radical transformation?

Throughout the later decades of the 20th century, megawaves of change relentlessly transformed the business landscape. Like tidal waves dramatically altering the landscape as they crash ashore, these social, political, technological and economic megawaves are swamping "old style" companies, destroying their established ways of operating. Here in the year 2002, organizations that did not anticipate change and failed to prepare themselves proactively are being crushed. Those with the foresight to adapt and move quickly forward are benefiting with increased profits and operating efficiencies, triumphantly riding the megawaves of change into a welcome future.

What are these megawaves of change?

Megawave: *The violent and sudden rejection of non-responsive and inflexible leadership.* Many were surprised at the fall of the Berlin Wall in 1989. With the benefit of hindsight in 2002, the fall of the wall was an obvious and natural step in an inevitable process.

The rejection of non-responsive national power structures and inflexible, ego-centered leadership is also sweeping through many organizations. We see the litter of dead organizations in worthless junk bonds, decaying buildings and shattered lives. Amid this chaos, some organizations thrive and prosper because they are responsive to the needs and requirements of their customers, employees, and society.

Megawave: *The insatiable need for training and development.* Change is a dramatic and powerful force, rendering today's work methods and techniques obsolete. Change, like all agents of entropy, is a silent yet ruthless enemy of continuity. Change constantly erodes the very foundation of an organization, diminishing the value of corporate memory and skill and experience, making executives leadership-illiterate. Companies who fail to respond to change stand vulnerable, like arthritic elephants, waiting to be devoured by tiger-like competitors.

Megawave: *Nations clustering into powerful trade groups, building economic muscle and erecting barriers.*

Inspired leadership has stimulated the revolutionary changes sweeping the world. Individually and in groups, people with the courage to turn their visions into action are making important changes everywhere in our global society. The good news is that most of these changes are making a positive contribution to human well-being. The best news is that despite the potential for chaos and anarchy, the vast majority of these changes are being accomplished in a peaceful manner.

Looking back from the year 2002, it's clear that the decade of the 90's was one of turbulence, chaos, and change. The survivors all had one thing in common. Their leaders had a vision of the future that allowed their organization to navigate the unseen, but very real whitewater that lay just ahead. They had the flexibility to adapt and make painful changes that allowed their visions to become reality. Looking ahead, they were not fooled by apparent calm and the promise of unbroken prosperity. Instead, successful leaders imagined a future attained by vigorously maneuvering in a once-familiar environment suddenly transformed into permanent whitewater.

How did these leaders accomplish this?

Most leaders in the 90's realized the importance of *vision*. Vision was a popular and widely used buzzword, capturing the essence of the new competitiveness required for organizational survival. Leaders looked ahead and bravely declared they were not afraid of the future, regardless of how intimidating or impossible the situation appeared. They reported to employees, stakeholders, and the media that they had the courage to envision a vivid and successful future for their organization. And many leaders did create visions, some of them quite powerful and compelling.

But one single factor -- and one alone -- separated successful leaders from those that failed:

> *Successful leaders had the ability and willingness to transform **vision** into **action**.*

The world of 2002 has been dramatically transformed from the relatively calm and peaceful period of the 90's. Debris of broken organizations litters the landscape. Let's now return to the present, and examine how the survivors of the millennium seized the initiative and insured their survival.

Chapter 1: VIA™: A New Way of Leadership

What's the secret to transforming vision into action?

There is no secret formula for successfully implementing visions. For an experienced leader, the steps are really quite basic. First, create a powerful vision of where you want to go. It is imperative that the vision is clear and powerful. Your vision must be *vivid* -- easy for your organization to understand and share.

Next, *interpret* the vision appropriately through the entire organization. This is the hard part. Many leaders make the mistake of believing that once they have created a vision and announced it to the organization, their job is done. In fact, at this point, the work has just begun. Leaders must relentlessly keep the vision squarely in view of the organization, constantly interpreting the vision for people at all levels. The vision must grow and assume a life of its own, becoming a vital of an organization's daily operation.

Unfortunately, this is where many excellent leaders with powerful visions fail. They see their role solely as the creator of the vision. Once done, they leave the "dirty work" of implementation to everybody else.

How do you see yourself as a leader?

As we travel through the permanent whitewater decade, many leaders still think of their organizations as if they were crew shells gliding along a straight, calm waterway. They see themselves as a member of a crew team. The roles in this team/organization are very well defined. Most of your crew are strong, experienced, well trained, and obedient. The leader's role is to steer this craft pretty much in a straight line. The leader barks out orders in a constant rhythm: more! more! more! Your organization seems on a predetermined path to a predictable goal. Yet things just aren't the same. New competitors have a better crew, or more sophisticated crew shell. The once placid, predictable course has now become choppy and turbulent.

If you are leading this way in today's environment, you sense with increasing urgency that you are failing. You are failing because you are practicing yesterday's leadership. As that narrow, inflexible crew shell of an organization is swept inexorably into the turbulent environment we are now facing, it is going to snap, it is going to break. Treacherous boulders, whirlpools, waterfalls and rapids await unseen to destroy you. The environment is in turmoil; it is a churning, surging maelstrom. Many of our excellent crew-shell organizations are in fact being smashed to bits, breaking apart and sinking rapidly into oblivion.

But imagine an alternate craft -- a rubber boat -- which is designed for the kind of environment we face today. One minute I row, the next minute you row. At this moment I'm pulling on the oar, then you're pulling on the oar. I'm bailing, you're bailing. One minute I lead and the next minute you lead. This type of organization is the type of organization that is likely to be most effective as we move forward. A team pulling together to accomplish a common vision.

How can you transform your organization from a crew shell to a rubber raft? And how can you prepare your people for this change?

The Vision Into Action (VIA)™ approach is a new way of thinking about leadership in a more global and strategic manner. As the megawaves of change sweep toward you, your way of thinking about and using leadership must change to cope with new opportunities and threats. This means you must seek out and embrace new leadership concepts and techniques. Concepts and techniques that may seem daring and original -- different as crew racing is from white water rafting. Choose to use them or not: megawaves of change are a certainty and your personal and organizational survival is at risk.

The issue is simple. Will you ride the megawaves of change to a successful future or will you be crushed by them?

The main role of a leader is to organize success. To stay alive and prosper you need to see the realities of the 21st century. You need to acquire new leadership perspectives, skills, and techniques. A characteristic of megawaves with an important parallel in your leadership development efforts is that you must be moving with the wave to take full advantage of its power. You cannot remain stationary. As a leader, you must be proactive. Doing nothing means failure.

To stay competitive, you must have a head start, an advantage. To reach your personal potential, and direct your organization to reach its potential, you need to be "up to speed." You need a forward thrust, a new way of thinking about leadership in a much more global and strategic manner.

VIA™: Vision into Action, is a leadership system that helps you navigate permanent whitewater. VIA™ lets you seek the uncertain future with open arms and welcome the potential it has to offer.

Chapter 2: Vision-Driven Leadership

VIA™: Vision into Action is a global approach to organizational survival. VIA™ is a dynamic, holistic system that maximizes an organization's most powerful resources -- the brains and hearts of its people. VIA™ is an influential process that helps organizations navigate the permanent whitewater of today's chaotic environment.

VIA™ is a dynamic new approach to leadership that builds on the classic *Situational Leadership®* process to liberate the full capability of individuals, teams, and organizations. VIA™ is an influence process designed to expedite the transformation of vision into action. Like *Situational Leadership®*, the VIA™ process has a dual emphasis, first on diagnosing ability and willingness, or *readiness* for change, and then matching the appropriate leadership style to maximize effectiveness.

In addition to these vital people issues, VIA™ is unique because it also addresses business issues faced by an organization. VIA™ tightly integrates the needs of *people,* through effective leadership, and the needs of *business,* through appropriate management practices. This holistic approach has the required breadth and depth required to effectively implement vision in an organization.

Vision-driven leadership is a top-down process of transforming "what to do" to "how to do it". Visions can be transformed into business objectives. Business objectives are then transformed into strategies. Strategies are transformed into goals, and goals become specific tasks for individual people to accomplish.

Vision and Mission

The heart of VIA™ is *vision.* Vision is a concentrated, intense foresight into a future that can only be achieved through determined force of will and effort. Good visions are lucid, describing a vivid picture of the future at a specific point in time.

Before an organization can establish a focused vision, it must set a direction -- it's *mission.* Many people confuse vision and mission. Vision paints a picture of what our business will look like in the future. Mission answers the questions: what business are we in, and what direction will we take on our road to success.

Many organizations focus almost entirely on this question of where we are *today* -- most even create this year's plan by looking at last year's results and extending that into the future. This approach to navigating in today's turbulent environment is like driving down a crowded freeway looking only in the rear-view mirror. Vision-driven organizations, on the other hand, create a compelling and powerful vision of what they will be the future. Then -- and only then -- they plan forward to make the vision happen.

Vision is an expression of the will of an individual, group, or organization. Vision-driven leadership focuses on the future -- what needs to be done to achieve a desired goal. It's not enough just to express a vision to an organization -- to tell your people what you see ahead. The vision must be *interpreted* -- articulated as a tangible result you want the organization to achieve, the end result of a process with a specific and definite time frame.

What is a Mission Statement?

... an enduring statement of purpose that reveals an organization's products, services, markets, customers, and philosophy... [It] provides the foundation for priorities, strategies, plans and work assignments. It is the starting point for the design of managerial jobs and structures. It specifies the fundamental reason why an organization exists. -- John A. Pearce and Fred David

The "fundamental reason why an organization exists" should be a direct reflection of the vision. It should also embody the core *values* that will guide and define the organization as the vision is implemented.

Key Components of a Mission Statement

1. Target customers and markets
2. Describe principal products/services
3. Define geographic domain
4. Describe core technologies
5. Commitments to ... survival, growth and profitability
6. Key elements of company philosophy
7. Company self-concept identification
8. Desired public image

Proposed by John A. Pierce and Fred David

What is a Vision?

If a mission statement gives an organization direction, vision describes "an image of our business success" at a given point in time. By creating a clear vision for their organizations, leaders establish a coordinating focus, a source of motivation, and a reason to continue even when short term odds are against them.

Visions excite people. They challenge people to reach for their best effort to realize a compelling future.

Visions are reflections of value-anchored wills. Visions mirror fundamental, core beliefs. Visions come from people's perspectives of the world and the events around them. Will is a core part of a person, the inner being that reveals itself in thoughts, concepts, dreams, visions. Vision creates the potential for success and turns realized potential into profits... into reality. Vision enables people to accomplish what once seemed impossible. If there is a vision, there is a way.

The major difference between a dream and a vision is time; visions are time bound. A leader should dream regularly since dreams of things that have not happened are sources of creative visions.

A Well-Known Vision Statement

> I have a dream... that we will be able to work together, to pray together, to struggle together... to stand up for freedom together, knowing that we will be free one day... And when this happens, and when we allow freedom to ring, when we let it ring from every village and every hamlet, from every state and every city, we will be able to speed up that day when all of God's children, black men and white men, Jews and Gentiles; Protestants and Catholics, will be able to join hands and sing in the words of the old Negro spiritual, "Free at last! Free at last! Thank God Almighty, we are free at last! -- Martin Luther King, Jr.

Does an Organization Need Both a Vision and a Mission?

Absolutely. Why?

Most organizations have a mission, whether it is explicitly articulated or not. Ask any employee "what does your organization do?" and the answer you receive is the mission of the company. A formal mission statement synchronizes the answers of all employees, becoming a direct reflection of a common path and a common set of values. This is the value of a mission statement: clarifying the purpose and meaning of an organization.

A vision statement, on the other hand, is vital because it represents specific goals or objectives for the organization at a specific time in the future. Mission is ongoing and eternal. Vision, however, is much more specific. It paints a vivid and clear picture of what the future will look like, and what must be done to get there.

Organizations need both a mission statement, to keep the ongoing direction clear, and a vision statement, to keep desired objectives clearly in sight.

You as a leader must be the driving force behind setting the mission and creating the vision for your organization. It is not enough for a leader to **create** a vision; a leader has a duty to **implement** the vision. Vision is implemented only if lucidly communicated and interpreted for the organization.

Chapter 3: The VIA™ Leadership Model

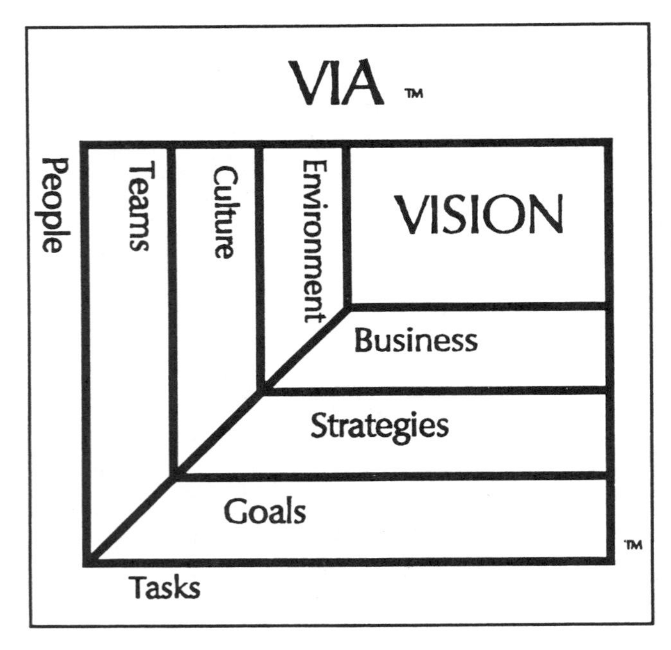

Once vision and mission have been clarified, the VIA™ Model provides a powerful framework for implementation. VIA™ is a top-down process, moving from abstract to concrete. The specific actions leaders must take move from vision in the upper right corner of the model down to the specific tasks people must do in the lower left of the model.

The integration between people issues and business issues is seen clearly in the VIA™ model. In the next few pages, we'll describe the crucial issues a leader faces when implementing a vision.

People issues are represented on the left side of the model. These are *leadership* issues, involving the business environment, organizational culture, teams, and individual people.

Business issues are represented on the right side of the model. These *management* issues -- business objectives, strategies, goals, and tasks -- are also crucial to implementing a vision.

Good leaders understand that effectiveness requires a balance between people and business issues, forging a symmetry between leadership and management. In the age of permanent whitewater, it's no longer a choice between leadership or management. Once an either/or situation, it is now a both/and situation for organizations.

The VIA™ model emphasizes the integration between people and business issues. Let's walk through the VIA™ model, starting at the top level, where visions are created, and move down the levels of the model as vision is implemented into action.

The Business/Environment Level

Business: Business Objectives are the operational tools of visions.
Environment: Environment is the complete human sphere in which an organization operates. It includes customers, competitors, regulatory agencies, shareholders.

At this level of the VIA™ model, the process is one of solidifying vision into a business idea with specific dimensions -- what we want to be, how big we want to be, and the results we want to achieve. Vision must anticipate megatrends in the environment -- customer needs, shareholder wants, and the requirements of other key players.

Visions are interpreted on the right, or management side of the VIA™ model, as business objectives:

- Our size in terms of sales volume
- Our financial outcomes in terms of specific results
- Our way of doing business in terms of qualitative principles

By defining these we can set long term aims for our organization. Visions and objectives should be set in the same time frame. If an organization has determined its "Vision 1995" or "Vision 2000," its business objectives should be parallel in time.

The Relationship Between Business and Environment

Business is conducted in an environment composed of stakeholders, both internal and external to the organization. Stakeholders are individuals or groups who either influence an organization or are influenced by the organization. They include external players including customers, shareholders, strategic partners, debt holders, government, and internal ones: employees and managers. An organization should be aligned with stakeholder expectations and attuned with their needs and values. If influential stakeholders do not "buy-into" the vision, it is not going to be implemented without substantial difficulty.

External and internal stakeholders present two different kinds of leadership challenges. External stakeholders are the key players whose values, will, and perception of purpose might be the same as or different than the internal key players. If the products or services that the organization produces do not match with customer needs, it might be easier to change the business than the environment.

In contrast, internal stakeholders, by definition, can be influenced on a cultural level, a behavioral level, to "buy into" the vision. While both levels are value driven, one must approach them differently.

A well designed mission with a clarifying vision supported by realistic business objectives when processed by key players gives an organization the supportive environment it needs to get its vision implemented. Managers as one of the most important internal stakeholder groups play a particularly important role. At the same time they are buying into the vision, they are responsible for implementing the vision. Before managers will devote their full efforts, they ask the question, "what's in it for me"? In the process of answering this question they form their own interpretation of the organizational vision and develop their own individual visions. Making sure managers are on the team is a critical role of senior leadership.

The Strategy/Culture Level

Strategy: Strategy is the general scheme for the conduct of business. Strategies tell us *what* we should do on the lower levels, and *how* we can get the vision accomplished. Strategies serve as guidelines, determining what kind of goals should be established on an operational level.

Culture: Culture is sum of the beliefs, values and behavioral patterns of an organization -- "the way we do business". Culture represents an organization's history. What personality is for an individual, culture is for an organization.

The next step for an organization to transform vision into action is to create and implement vision-driven strategies. Effective strategies focus on three critical areas: Satisfying customer needs, creating competitive advantage, and maximizing the use of people and resources within the organization.

The moment of truth for a strategy is when it confronts an organization's culture. *Culture* is an extremely powerful force that can add significant leverage to strategies. However, without support of the culture, a strategy has little or no chance of survival.

On the highest level it means organizational strategy and involves complex assessments of all resources, organizational priorities, and environmental factors that are probably global in reach.

Mission, vision, and business objectives are implemented through strategies. Business objectives tell the "what to do;" strategies are the "how to do it." Strategies must not only be accepted but supported by the organization's culture which represents people's willingness and ability to contribute to a strategy's success. Organizations do not have just one culture; they have many subcultures that, when combined, represent the whole organizational culture.

The Relationship Between Strategy and Culture

Business objectives are reflected in strategies that describe how these objectives are to be implemented in an organization whose culture consists of a system of shared beliefs, habits and symbols, and collective organizational personality.

The Goal/Team Level

Goal: A **Goal** is a mindpicture, an image of the preferred end result anchored in human mind(s) by emotions.

Team: A **Team** is a group of people with a common goal.

At the next level of the VIA™ model, strategies are transformed into *goals* that include specific roles and measurements. To achieve a strategy-driven goal, systems must be put in place to provide feedback -- because knowledge of "how we're doing" is vital -- possibly more important than the goal itself.

Teams are becoming the most important work unit for accomplishing goals. As tasks become more complex and organizations flatten, teams provide versatility and a combination of skills that may go far beyond the capabilities of individuals.

At the goal-team level of the VIA™ model, strategies are further refined into specific goals to be accomplished by teams, particularly self-managed teams which many companies are developing. Team acceptance of goals is critical. If the team does not agree with the goals, the goals are not going to be fully accomplished.

The Relationship Between Goals and Teams

The goal-team match represents the essential interplay between the goal(s) the organization has developed to help implement its strategies and operating teams. Teams are defined as two or more people interacting in which the existence of all (the team as a team) is necessary for the needs of the individual team members to be satisfied. **Effective** teams have common goals or purposes. In an organizational context, a team may be as small as two individuals or as large as a department.

Specific, timely goals are not enough. There must be feedback on the quantity and quality of goal achievement as well as on the performance process. The bottom line is that specific goals and effective teams are inexorably intertwined.

One concluding thought. The purpose of teams has changed dramatically in recent years with the increase in responsibilities shared with them. Historically, teams were a collection of individuals formed to carry out leaders' orders. Today, with the emphasis being placed on self-managed teams, the teams themselves are the leaders. This means that the roles of individual team members are far more complex than they have ever been. And with this complexity comes increased need for training and development to prepare team members for these roles. At the same time, leaders need training and development in how to build effective teams.

The Task/People Level

Task: A **Task** is the input you have to do to move towards the goal.
People: **People** are the fundamental and critical elements of an organization. At some point,

all visions, strategies, goals, and tasks must be accomplished by individuals.

The final level of the VIA™ model focuses on the tasks that make up goals, and the **people** who perform those tasks. **Empowerment** is the key here, focusing on the strategic and psychological power issues that arise when others are given the freedom to act.

If people do not want to perform the task, they just are not going to do it. A leader's ultimate objective is to influence people to accept responsibility or ownership of the task; to see the task as viable for themselves.

The Relationship Between Tasks and People

This foundation level of the VIA™ model focuses on the interrelationship between tasks and individuals. If one refines all that has been learned about the behavioral sciences throughout decades of behavioral science research, what is left in the bottom of the crucible are the two elements of tasks and people. They are the building blocks of the entire VIA™ model. The international success of Situational Leadership® over almost three decades with more than one million managers developed annually, is clear evidence of the importance forward thinking firms place on the task-individual match.

What is The Task Dimension?

Task leadership focuses on the "what to do" of a firm's responsibilities.

What is The People Dimension?

The people dimension focuses on the "who is to do it" of an organization's responsibilities. Both dimensions, interdependent and interacting in many different effective combinations, are essential for an organization's effectiveness. This is one of the most fundamental conclusions of decades of behavioral science research.

The Implementation Gauntlet

Visions live a tenuous life. On their path to implementation they must run a gauntlet of contending forces. Your organization is a sea of powerful, sometimes conflicting forces. Visions refined into business objectives must run the gauntlet of contending key players and competing business objectives; strategies must run the strategy-culture gauntlet; goals, the goal-team gauntlet; and tasks, the task-people gauntlet. As the implementation of a vision proceeds down the gauntlet of supporting and opposing forces, your leadership issue is this: How can we make these visions, business ideas, strategies, goals, and tasks strong and robust? How can we use the powerful supporting forces in our organization to help give the vision a forward thrust through the synchronization of these forces?

Knight's point is that implementation requires leadership. It is not enough to have a formal plan; the plan must capture the hearts and minds of those who must implement it.

The next chapter addresses the implementation issue, and describes specific elements of the

VIA™ process that improve the chances of a vision coming to life.

Chapter 4: Leadership vs. Management

Alignment and Attunement

VIA™ is a unique approach to vision-driven leadership. It goes beyond traditional one-dimensional approaches that emphasize either the business side or the people side of an organization. At every level, VIA™ forges a tight integration between business issues, from plans to results, and people issues, from individuals to organizations.

At all levels, VIA™ combines the managerial and leadership processes. On the right side of the VIA™ Model, the purpose of the managerial process is to create *alignment*. Alignment occurs between the levels -- tasks should be aligned with goals, goals with strategies, strategies with the business idea, and the business idea aligned with the vision.

Alignment should also occur within each level. For example, structures and policies should align with strategies ... roles and measurement systems should align with goals.

On the left side of the VIA™ Model, the purpose of the leadership process is to create *attunement*. When an organization is attuned, people are committed, they feel purposeful and empowered -- fundamentally, attunement means that people enjoy their work. Attunement should exist at all levels -- individuals are attuned with tasks, teams with goals, organizations with strategies.

Organizations are basically technological, social, and process systems. If system components are to function together, smoothly and effectively, they must work together. The process of alignment and attunement forges links between groups and systems that are characteristic of world-class organizations. In this process, the linkages must be so crafted that each link contributes to the strength of the whole.

Alignment Through Management

The minds dimension is represented by the right or management process side of the framework. It is similar to the traditional planning activities of management. Visions, business objectives, strategies, goals, tasks must be aligned if intended results are to be achieved.

Management is defined as the process of working with and through individuals and groups and other resources to accomplish organizational purposes. Leadership is a broader concept than management. Management is thought of as a special kind of leadership in which the achievement of organizational goals is primary. The key difference is in the word--organization. Leadership occurs any time one attempts to influence the behavior of an individual or team, regardless of the reason. It may be for one's own goals or for those of others, and these goals may not be congruent with organizational goals.[30]

Attunement Through Leadership

The hearts dimension is represented by the left or leadership process side of the framework. Plans do not implement themselves. They must be implemented at the appropriate environment, culture, team, and people levels through effective leadership.

Attunement means that the human resources of an organization must act with the same frequency to be "in tune." In the VIA™ model, the focus is on the most critical resource-the human resource.

This means that the vision must be shared in the hearts and minds of the human resource. The vision must be accepted by the human resource. When the frequency of a piano key does not match the technician's tuning fork, the key is "out of tune." The same concept is true in organizations. When people are not on the same frequency as the organization's vision, they are "out of tune" with what is happening. Just like an "out of tune" piano, the organizational result is discord and disharmony.

Integrating the leadership and management sides of the model requires constant attention. It requires effort, flexibility, and a willingness to make leading a full time job.

Conclusion: VIA™: The Way to the Future

The key elements of VIA™ involve creating a vision and implementing that vision through a development process that achieves sustainable, profitable competitive advantage in all business areas through providing customer value. This world, ravaged by war and unsound leadership and management practices, needs a renewed vision. A vision that will achieve not only a world of peace, but a world dedicated to developing each individual's potential. The skilled and informed use of research based leadership concepts, tools, and techniques can unleash this potential. VIA™ offers a highly promising "way to go" in creating this world of opportunity for all peoples.

What an interesting and intriguing future awaits you, as intriguing as when you will look backward in time on January 1, 2002, and say, "I feel proud of our leadership accomplishments."

Leadership Studies International
230 West Third Avenue
Escondido, CA 92025
619-741-6595

CIM Success Requires Integrated Planning Techniques

CARL L. JOHNSON
Austin Consulting

The CIM movement is stalled. Negative experiences by early CIM implementors have caused corporate decision makers to go slow in authorizing CIM projects. What is the problem and what can be done? This paper analyzes the planning process used in failed and successful CIM projects. We find that "islands of specialization" are not adequately integrated into the planning process. Computer-integrated manufacturing fails because planning is not integrated. Techniques for successfully integrating users, engineers, specialists and decision makers into CIM planning are presented. the target audience is individuals who realize high-level and broad-based support increases the likelihood of CIM success.

When you see an article on Computer Integrated Manufacturing (CIM) in a magazine's table of contents what do you do? Read it first? Make a point to read it later? Or, make a point to avoid it?

Many of us who used to read any material we could get our hands on concerning CIM are much less avid readers than we used to be. Why? Because we wonder if the reality of CIM will ever compare to the promise it originally held.

We've heard too many horror stories of projects with overruns and delays. CIM plans are being scaled back, postponed, or cancelled. Management continues to think and talk about CIM, but is reluctant to act. In short, the momentum toward significant use of CIM by U.S. manufacturers seems to have greatly slowed. Many people are now wondering if we will ever see real, significant benefits from CIM for more than a handful of companies.

This situation is caused by two problems which need to be corrected. The first problem is that we have a narrow definition of what CIM actually is. The second is that we don't effectively plan for CIM. This paper suggests one potential definition for CIM and then presents a detailed discussion of how to plan for it. The paper is intended for individuals who realize that CIM planning must be addressed at the proper levels within an organization if it is to be successful.

NARROW DEFINITION OF CIM

Most CIM definitions stress the objective of widespread access to manufacturing information via a corporate data base. The idea is to gather, store, and provide access to manufacturing data through computer and communication systems so that decisions can be based on timely, accurate information. The notion of a corporate data base has led to an expanded definition of CIM which includes nonmanufacturing information as well. In fact, terms such as "computer integrated business" and "enterprise networking" have been suggested to replace "CIM."

Although these definitions are helpful, they overlook an important aspect which should be stressed in all planning for new information systems, "Do proposed systems actually help improve the quality of work performed by the organization and its individuals?" That is, although CIM must ensure that the right information is gathered and shared, any serious look at computer and communication systems in today's organizations must consider whether employees have access to the tools and resources necessary to do the best job possible.

For example, much concern is expressed for the integration of Computer Aided Design (CAD) and Computer Aided Manufacturing (CAM). This interface across functions is an obvious candidate for an integrated data base and can improve lead time and manufacturing quality. However, integrated CAD/CAM does not of itself enhance the quality of the items designed nor assist the flow of new products to the marketplace. To do this, the organization must look at other tools within each function (design and marketing, in this case) to remain competitive.

Thus, a broader definition of CIM is required which focuses management attention on technologies which lead to the benefits most required for the company to become more competitive. A suggested definition which attempts to accomplish this is:

> **"CIM is a management orientation which assumes that competitive advantage can be achieved through creative application of information technologies which facilitate decision making and communications within and across corporate functions."**

The scope of planning under this definition is represented in Exhibit 1. It indicates that CIM planning should encompass each functional area and the exchange of information between functions.

Note that this definition does not focus on manufacturing but instead leaves management free to determine the areas where computer and communication systems would have the most impact. It obviously raises planning issues which corporate management must address, such as which functions should be linked together first, and which functions require more internal attention before they can be externally linked. This leads to the second topic, how to plan for CIM.

DIFFICULTIES IN CIM PLANNING

Most corporate planning is "product-oriented" versus "process-oriented." Product-oriented planning creates a document defining goals, objectives, tasks, schedule, and budget. Process-oriented planning also creates a formal plan, but it explicitly considers corporate mission and objectives, organization, and strategy. It emphasizes that the approach or process used to create the plan must address these broader corporate issues.

CIM planning using a product-oriented approach produces plans which can not be implemented, or can not be effectively implemented. CIM planning using a process-oriented approach produces plans with a high probability of successful implementation. Following are examples of product and process-oriented approaches as applied to CIM.

Product-Oriented CIM Planning for a Tool Manufacturer

The Vice President of Manufacturing for a tool manufacturer requested permission from the President to develop a CIM plan to be implemented throughout several plants. Upon approval by the President, the Vice President met with his staff to develop goals and objectives they wished to achieve through CIM. They then delegated further work to systems analysts, product designers, manufacturing planners, and engineers who researched equipment needs, compatibility, and costs. The Vice President's staff collected all information and developed a document including goals, objectives, tasks, schedule, and budget. The document was submitted to the President for approval.

Process-Oriented CIM Planning for a Wood Products Company

The Vice President of Manufacturing for a wood products company attacked the same issue differently. He approached the President with a recommendation to involve all departments and levels to some degree in the planning process. He also stressed that not just the President's verbal commitment but his active participation throughout the process would be required.

The President created and convened a CIM Policy Board for an off-site, three-day, planning workshop. The Board, consisting of the current Executive Committee (Vice President of Marketing, Manufacturing, Finance, and Administration) and key staff from each functional area, numbered approximately fifteen people. The workshop first focused not on CIM but on company growth strategy, goals, and objectives. The participants, led by an experienced facilitator, then drafted a high-level, functional description of how CIM should support the corporate direction. Based on this document, additional task forces were created to develop concepts and detailed plans. Cross-functional task forces were used throughout to ensure that CIM would meet all user needs. Members of the CIM Policy Board met frequently with task forces to assure overall company objectives were honored.

Comparison of Results

The high-level and broad-based input to the wood product company's CIM planning effort resulted in a seamless transition from CIM planning to implementation. No formal declaration was made that the CIM plan had been approved; in fact, some CIM implementation tasks, such as the "As-Is" analysis, were initiated before the planning was completed. By contrast, ten months after the tool company's CIM plan was first created, no decision had been made yet about its implementation.

COMPARISON OF THE TWO PLANNING APPROACHES

The experience of the tool manufacturer highlights the basic problems with the traditional, product-oriented approach. First, top-level commitment is not achieved early or is evident only to a few individuals. For example, individuals assume their CIM planning responsibilities do not have high priority because the importance of CIM has not been stressed. Also, top-level input on key issues is not provided. If these concerns are not addressed early, it may be too late to incorporate them later.

Secondly, planning does not cross functional areas and no overall integration occurs. For example, if the CIM effort is perceived as manufacturing's responsibility, marketing's information needs may not be recognized, resulting in no improvement in customer service.

Thirdly, the process focuses on the use of experts working in their own specialized areas. They may develop technically correct plans, but the plans may not achieve the overall objectives. For example, a data base may be recommended by systems analysts because it is efficient and fast, but the security needs of users may not have been considered. Also, cross-fertilization--finding a new solution to a problem through synthesis of different opinions--is not achieved.

In short, product-oriented planning depends on "islands of specialization" which can not generate integrated CIM plans, much as islands of automation do not constitute an integrated plant.

Major Benefits of Process-Oriented Planning

These problems are overcome with the approach used by the wood products manufacturer. One of the most obvious and important characteristics of the process-oriented approach is that high-level management commitment was not simply expressed, but was evident in the active involvement of policymakers in planning. There is a tremendous difference in saying that management is behind a project and in having management involved in planning sessions.

Another difference is the broad-based, cross-functional input that went into the initial and ongoing stages of planning. This accomplishes two objectives. First, it provides input concerning each functions' CIM needs, as previsouly discussed. Second, it creates ownership of the CIM effort. Planners have known for years that quality is improved and the probability of implementation is increased when people have ownership of plans. The best way to encourage ownership is to involve people in the nitty-gritty elements of establishing goals, analyzing trade-offs, and recommending solutions. Without ownership, it is too easy for individuals to fall back on the not-invented-here excuse for inaction.

By contrast with product-oriented planning, which depends on islands of specialization, process-oriented planning integrates these islands just as CIM integrates an entire organization.

IMPLEMENTING A SUCCESSFUL CIM PLANNING SYSTEM

Although the process-oriented approach has obvious benefits, it is difficult for many organizations to implement because of old habits. Management must commit to five specific efforts to break with the past. These commitments are:

1. Create a CIM Policy Board to provide high-level guidance.

2. Designate a Core Team and Task Forces to ensure leadership and broad-based input.

3. Conduct a CIM Planning Workshop to launch the planning process and to establish the basic planning context.

4. Link corporate vision and CIM strategy to enhance competitive position.

5. Estimate and assign resources in advance to remove roadblocks.

While these commitments do not guarantee success, success will be elusive without them. Each is discussed in detail in the following sections.

CIM Policy Board

The function of the CIM Policy Board is to provide overall guidance and direction to the CIM project. It consists of policymakers plus select individuals with specific knowledge or skills. Individuals with leadership, technology, planning, and implementation experience are necessary. Consideration of the mix of personalities is also valuable.

The Board's primary concern is that the CIM plan support company objectives. No single individual can ensure this will occur, it takes watchful eyes from different perspectives. The Board helps establish the link between objectives and CIM during the workshop (as discussed later), but it also monitors subsequent planning and implementation.

Core Team and Task Forces

Outside systems integrators or consultants should not have a major role in the early stages of CIM planning. Ninety percent of the issues can be resolved by resources already available to the company--its own people. They know what the problems are and, given the right environment and leadership, they can find the solutions.

Part of the right environment is emphasis on teamwork. CIM projects are too complex for one or a few people to plan. Simply stating the functional requirements requires inputs from many people; determining how to meet them takes even more.

The best approach is to designate a Core Team with full-time responsibility for CIM. The Core Team coordinates planning, arranges resources, disseminates information, and acts as a collective "champion" for CIM. The Core Team is supplemented by ad hoc task forces which undertake specific tasks as assigned, e.g. financial justification.

The relationships between the Core Team, Task Forces, and Policy Board is represented by Exhibit 2. All three are critical. With the Policy Board providing overall guidance, the Core Team taking day-to-day responsibility, and task forces tackling specific issues, a balanced integrated CIM plan is possible.

CIM Planning Workshop

The purpose of the CIM Planning Workshop is to kick off the planning effort and to set the planning context in a highly participative, group problem-solving setting. It should be off site to avoid interruptions and distractions.

Ideally, the workshop will be led by an outside, objective facilitator experienced in CIM planning; this will ensure the most effective use of participants' time. Alternatively, an internal facilitator may be used if the individual remains unbiased and keeps the process flowing.

The workshop requires anywhere from two to three days depending on the input data available and the level of detail desired. A three-day format might be structured as shown in Exhibit 3; although less time may be necessary depending on the amount of information available in advance. Note from Exhibit 3 that each day has a specific focus: Day One - Understanding where the company is and where it wants to be; Day Two - Identifying the challenges facing manufacturing and setting objectives; Day Three - Creating the actual CIM strategy. The schedule itself reinforces the notion that the CIM strategy should flow from broader corporate objectives; the plan must not be created independently.

Note that more than three days could be dedicated to several of the topics shown in Exhibit 3. The workshop is not intended to resolve all issues and make all decisions. Rather, it must highlight issues and paint in broad brush strokes the overall CIM strategy. Refinement and expansion of the strategy and plans will be undertaken as follow-up assignments to task forces. Experience indicates that a well orchestrated CIM Planning Workshop produces credible results and enthusiastic support.

Linking Corporate Vision and CIM Strategy

Two of the most important topics of the workshop are corporate vision and CIM strategy. The CIM strategy should flow from corporate vision and support it by focusing on those CIM technologies which produce the greatest improvements. Since no company can afford to invest equally in all technologies (CAE, robotics, etc.), company objectives should dictate which ones to pursue.

One approach which pictorially shows the results of the workshop is to create a "Corporate Vision Chart" and a "CIM Strategy Chart." These charts are created by participants in a structured group exercise. Exhibits 4 and 5 show hypothetical charts as they might result from a workshop for an apparel manufacturer selling to a variety of retailers.

The center column of the Vision Chart in Exhibit 4 indicates that the company's competitive position depends primarily on improved service to retailers who sell apparel under their private label. The services required were identified as frequent delivery of small orders to store doors, and better order and shipment tracking, allowing customers to hold down inventory costs. The right-hand flanking columns indicate that lowering costs by reducing work-in-process and finished goods storage are key elements. The left-hand flanking columns indicate that improving quality through training and quality control programs would also be required.

The center column of the CIM Strategy Chart in Exhibit 5 shows that remodeling and expanding the existing distribution center is critical to achieving the overall vision. Flanking columns indicate that modern business systems (right-hand columns) and 100% inventory tracking (left-hand columns) are also required. Note there is not a one-to-one relationship between elements of the CIM strategy and the vision. Collectively, however, the strategy elements achieve the vision.

Such charts can serve as useful management tools. One Vice President of Manufacturing presented his staff with framed copies of the strategy chart and carries his copy to planning meetings; it helps keep discussions on track.

Approaches other than charts could be used to portray company vision and CIM strategy. The important issue is to ensure that the CIM strategy is not created in a vacuum and does not attempt to use state-of-the-art technologies simply because they are available. Although additional research might be needed to finalize the CIM strategy, the workshop participants need to spend enough time discussing options so that they see the connection between company objectives and CIM strategy.

Assign Resources in Advance

Many projects which fail do so because the resources required for success were not available when needed. Or, optimistic estimates were made and management withdrew support when these were exceeded. Management should realistically estimate up front the required resources, and then be sure they are made available.

SUMMARY AND CONCLUSION

The CIM movement is slowed because we have a narrow definition of CIM, and because we don't effectively plan for it. CIM should be considered a management orientation which seeks to achieve competitive advantage through creative application of information technologies which facilitate decision making and communication within and across corporate functions.

Effective CIM planning uses process-oriented techniques which ensure that high-level and broad-based input is available whenever needed. This is accomplished by using a dynamic team consisting of a CIM Policy Board, Core Team, and Task Forces. A participatory, structured CIM Planning Workshop is used to link corporate vision and CIM strategy.

The CIM movement can be revitalized if U.S. manufacturers take an integrated approach to new CIM projects.

Exhibit 1 - Scope Of CIM Planning

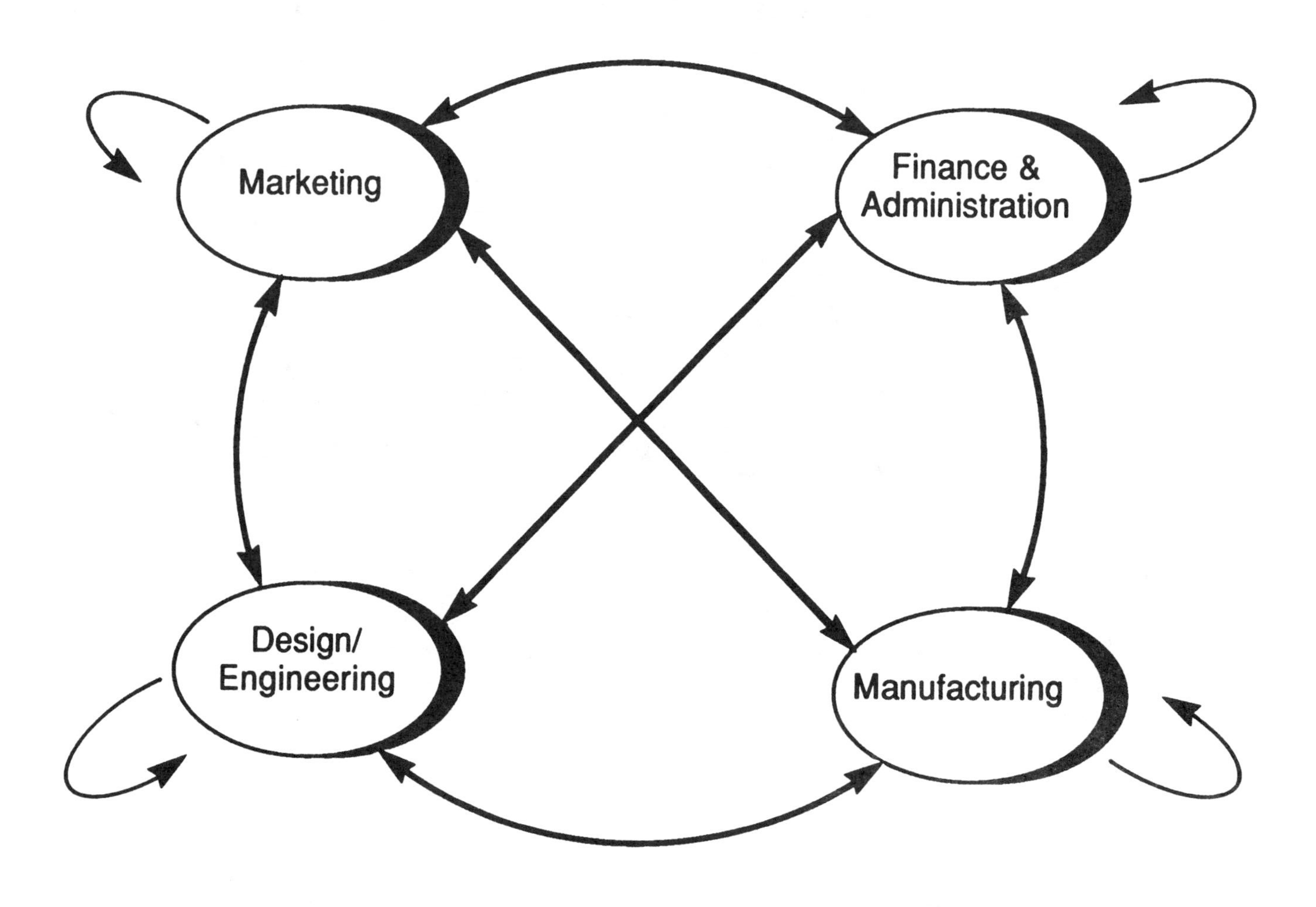

Exhibit 2 - Relationships Within CIM Planning Team

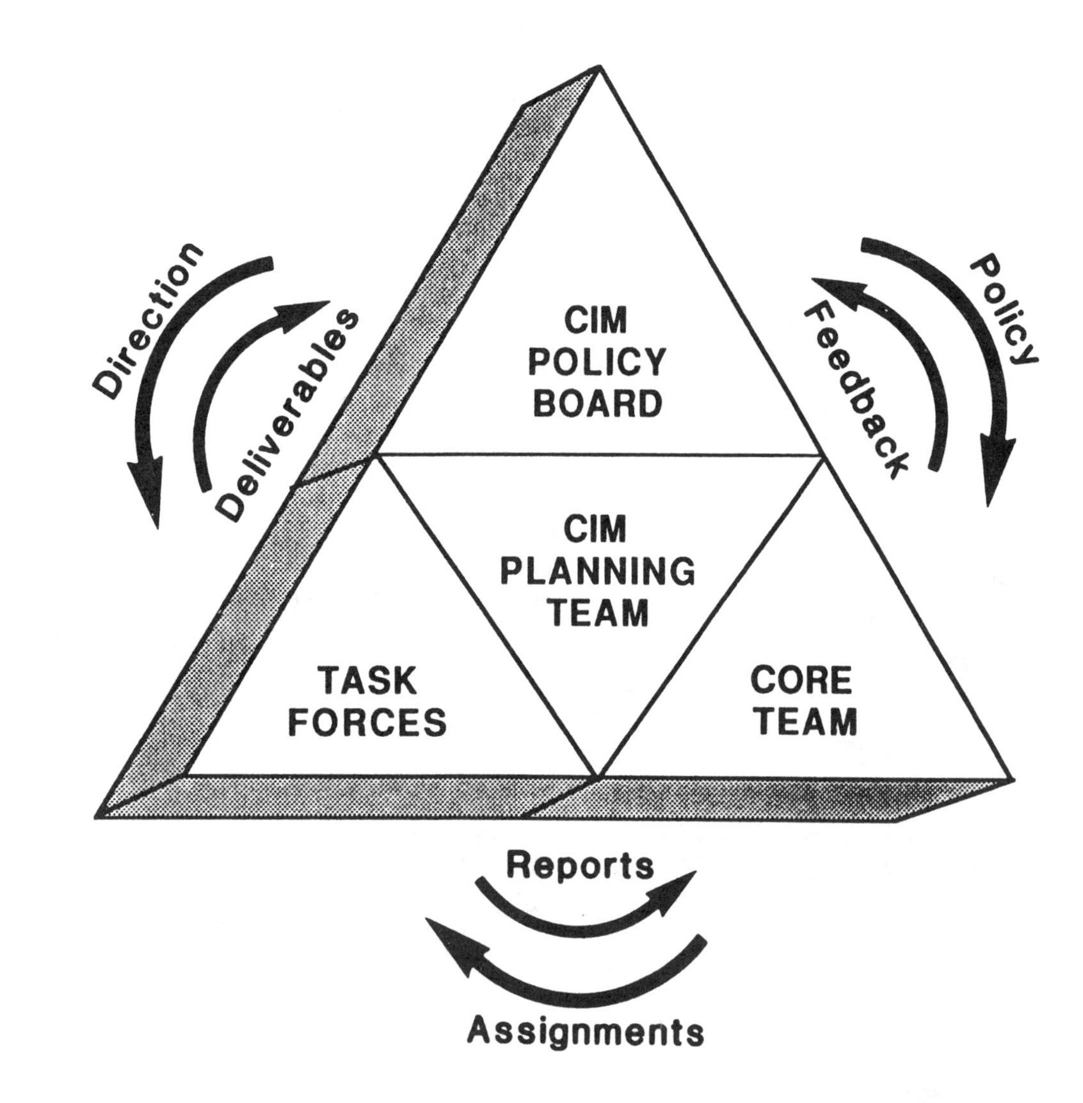

Exhibit 3 - Possible Three Day CIM Planning Workshop

DAY ONE

COMPANY BASELINE INFORMATION
- Historical Information
- Customer/Marketplace Trends
- Corporate Strengths And Weaknesses

COMPANY MISSION AND OBJECTIVES
- Five-Year Company Vision
- Company Mission Statement
- Competitive Advantage
- Company Objectives
- Issues Facing Manufacturing

DAY TWO

MANUFACTURING'S CRITICAL CHALLENGES
- Manufacturing's Interfaces
- Traditional Stengths
- Areas Of Improvement

MANUFACTURING'S OBJECTIVES
- Historical Performance
- New Performance Objectives

DAY THREE

CIM STRATEGY
- Information Needs
- Candidate CIM Technologies
- Five-Year CIM Vision

CIM SCHEDULE
- Critical Activities
- Project Milestones
- Resources Required And Available
- Follow-On Responsibilities

Exhibit 4 - Corporate Vision Chart

<table>
<tr><th colspan="5">CORPORATE VISION CHART</th></tr>
<tr><td colspan="5">Increase Market Share Through Better Service And Higher Quality</td></tr>
<tr><td colspan="2">Higher Product Quality</td><td rowspan="2">Improved Service To Retailers</td><td colspan="2">Lower Inventory Costs</td></tr>
<tr><td>Human Resources</td><td>Quality Control</td><td>Work In Process</td><td>Finished Goods Inventory</td></tr>
<tr><td rowspan="3">Frequent Training Programs</td><td>Vendor Inspection Agreement</td><td>Smaller, More Frequent Deliveries</td><td>Efficient Material Handling</td><td rowspan="3">Double Inventory Turns</td></tr>
<tr><td>In Process Inspections</td><td rowspan="2">Better Order And Shipment Tracking</td><td>Just In Time Scheduling</td></tr>
<tr><td>Quality Mindset</td><td>Uncongested Work Areas</td></tr>
</table>

Exhibit 5 - CIM Strategy Chart

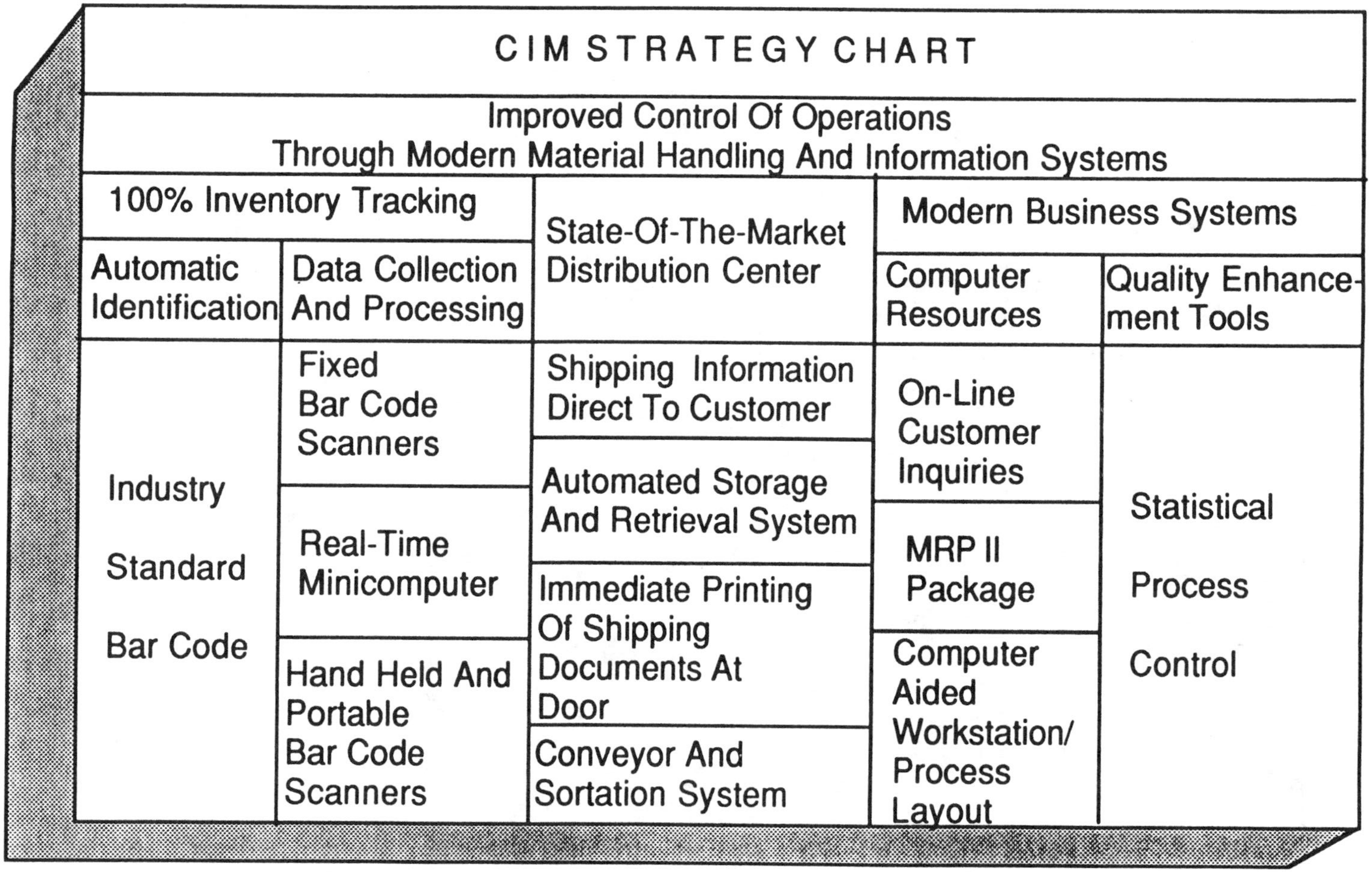

<table>
<tr><th colspan="5">CIM STRATEGY CHART</th></tr>
<tr><td colspan="5">Improved Control Of Operations
Through Modern Material Handling And Information Systems</td></tr>
<tr><td colspan="2">100% Inventory Tracking</td><td rowspan="2">State-Of-The-Market
Distribution Center</td><td colspan="2">Modern Business Systems</td></tr>
<tr><td>Automatic
Identification</td><td>Data Collection
And Processing</td><td>Computer
Resources</td><td>Quality Enhance-
ment Tools</td></tr>
<tr><td rowspan="3">Industry
Standard
Bar Code</td><td>Fixed
Bar Code
Scanners</td><td>Shipping Information
Direct To Customer

Automated Storage
And Retrieval System</td><td>On-Line
Customer
Inquiries</td><td rowspan="3">Statistical
Process
Control</td></tr>
<tr><td>Real-Time
Minicomputer</td><td>Immediate Printing
Of Shipping
Documents At
Door</td><td>MRP II
Package</td></tr>
<tr><td>Hand Held And
Portable
Bar Code
Scanners</td><td>Conveyor And
Sortation System</td><td>Computer
Aided
Workstation/
Process
Layout</td></tr>
</table>

ORGANIZING FOR SYSTEMS INTEGRATION

BY

Johnson A. Edosomwan
JOHNSON & JOHNSON ASSOCIATES

ABSTRACT

This paper communicates a clear vision of the nature of the systems integration discipline. The role and skills of a systems integrator are presented. Guidelines are provided for planning, designing and implementing integrated systems.

INTRODUCTION

The last two decades can be described as the era of technological exlposion. Increasing competitive pressures have stimulated the development of new systems, tools, technologies and methods to help organizations improve performance. While the increased development, introduction and use of the various forms of systems and tools in the work environment offers great promise to productivity and quality improvement, it is doubtful if adequate infrastructure are in place to address the problems created by islands of automation that do not communicate with each other. The need exits for an integrated systems approach to factory automation. Systems integration has evolved as an inevitable new discipline to address the problems associated with running an enterprise with islands of automation and to provide integrated systems that enables organizations to produce competitive and acceptable goods and services.

DEFINITION OF SYSTEMS INTEGRATION

Systems Integration is concerned with the successful completion of system planning, design, improvement and control activities for any industry or business. It facilitates the integration of the activities of an organization's network of systems and subsystems to respond effectively to environmental changes and customer needs.

THE ROLE OF A SYSTEMS INTEGRATOR

A systems integrator combines information, tools, materials and resources to plan, design, implement and control systems for an enterprise or business unit. Systems integrators are linking pins which help facilitate vertical and horizontal communications on continuing basis to enable the business units to operate in an adaptive fashion and deliver acceptable goods and services to the customer. The systems integrator's role varies from industry to industry. The responsibility of a systems integrator is dependent upon the organizational circumstances. In most organization, the systems integrator is viewed as a leader with the knowledge and skills to ensure the successful completion of system planning, desin, improvement and control activities. The systems integrator work as a team leader to enhance the traditional organizational functional roles in a manner which enables the organization to adapt to changing environmental circumstances and operate effectively.

SYSTEMS INTEGRATION SKILLS

The skills listed below are those which help the systems integrator facilitate the successful development and integration of systems, structures, tools and methods:

1. Knowledge of existing and emerging systems and technologies.

2. Understanding of system engineering and knowledge of the important technical aspects of the functional area which influence customer value and needs.

3. Understanding the key interface issues that affect organizational work units that use various types of technologies.

4. Knowledge of data bases, process control tools, expert systems and management information system for improving decision making and total systems integration.

5. Ability to serve as a change agent, clarifying system requirements and customer needs. Managing resistance from work groups and preparing employees to meet the challenges of the new technologies.

6. Understanding the key technology requirements to satisfy both short term and long term goals.

7. Understanding the continuous process improvement principles, initiatives, and the critical interface that influence total systems integration for quality and productivity improvement.

8. Ability to justify, sell, and implement integrated systems across multi-functional disciplines within which the systems integrator operates.

GUIDELINES FOR PLANNING, DESIGNING AND IMPLEMENTING INTEGRATED SYSTEMS

1. Prepare and define a systems integration plan that include technologies, tools, methods and activities for production and service processes. The plan should also incude long term automation strategy.

2. **Assess the total feasibility of the systems integration plan and how it fits into the overall business strategy for the enterprise.**

3. Evaluate the production and service processes for improvement opportunities, screen all new technologies for relevance and appropriateness. Simplify the process, automate and then integrate.

4. Identify all the functional specifications of existing and new systems.

5. Develop a phased implementation plan that addresses key events and activities.

6. Involve all potential users of the system during implementation. Sell the entire systems integration plan to management, unions, and employees.

7. Emphasize preventitive solutions, data integrity, quality, reliability, availability, discipline, productivity and cost reduction.

8. Arrange for adequate training of work-force on the commissioning, calibtrating, and maintenance of the systems and technologies. Select and train system intrgrators.

9. Perform ongoing characterization of exixting and new systems and technologies. Allow for system flexibility, modification of tools specifications, controls and process parameters.

10. Emphasize total systems solutions through pay for performance, competitive advantage and team work.

CONCLUDING REMARKS

The success of a systems integrator depends on the organizational culture, the abilities of the individual and the context in which systems integration occurs. Other important success factors for systems integration work include the content , quality, and timeliness of information available to the integrator and the effectiveness of the tools and systems available to support effective decision making. Integrated systems will continue to receive increased attention because of its unique importance in allowing organizations to operate in an open system mode to respond effectively to customer needs.

REFERENCES FOR FURTHER READING

Edosomwan, J.A. (1989) Integrating Innovation and Technology Management, John Wiley and Sons, New York.

Edosomwan, J.A. (1986) " Managing Technology in the Workplace: A Challenge for Industrial Engineers", Industrial Engineering, February issue.

White , J. A. (1986) " Becoming A Systems Integrator Before the Year 2020" Fall Industrial Engineering Conference Proceedings

Edosomwan, J. A. (1990) " New Manfacturing Technologies: Managerial and Technical Obstacles " Proceedings for the annual Industrial Engineering Conference.

Edosomwan, J.A. (1990) A Ten Step Approach For Systems Integration, Monograph, JOHNSON & JOHNSON ASSOC.

BIOGRAPHICAL SKETCH

Johnson A. Edosomwan is president and principal consultant of JOHNSON & JOHNSON ASSOCIATES, international consultants in quality, productivity, innovation, technology management and organizational development. Dr. Edosomwan is author/editor of 12 books and over 100 papers. His contributions as an engineer, educator, consultant and scholar has earned him over 40 awards including "The IIE Technical Innovation Award in Industrial Engineering" and " The Outstanding Young Industrial Engineer Award" He has completed over 100 cosulting projects for manufacturing and service organizations worldwide. Dr. Edosomwan is a senior member of IIE. He is on the IIE publications policy board and also currently serving as Chairman of the IIE Taskforce on Women, Minorities and Disabled in Engineering.

Why Strategy Drives Integration

Earl D. Bennett, Barbara M. Fossums, Roy D. Harris, Sarah A. Reed, and J. Franklin Skipper

Computer-integrated manufacturing can shorten the product development life cycle and improve production cost controls. Based on responses to their 1986 survey of Fortune 500 manufacturing executives, the authors found the road to integration paved with serious technological and organizational obstacles. Quoting respondents, this article defines those obstacles in the words of integration-experienced executives and suggests integration solutions that advance manufacturers' competitive fortunes.

Earl D. Bennett, PhD, is a professor of accounting at Texas A&M University, College Station.

Barbara M. Fossums is executive vice-president at Factorial Systems Inc, a company based in Austin TX that designs software for shop floor systems.

Roy D. Harris, PhD, is Centennial Professor for Management of Innovative Technology at the University of Texas, Austin.

Sarah A. Reed is an assistant professor of accounting at Texas A&M.

J. Franklin Skipper is a senior consultant at D. Appleton Co, a management and technology consulting firm based in Manhattan Beach CA that works mainly with defense contractors.

Many US manufacturers have countered foreign competitors in traditional US markets by acquiring automated and computer-integrated production systems. Robotics, computer-aided design (CAD), computer-integrated manufacturing (CIM), just-in-time (JIT) inventory systems, and flexible manufacturing systems (FMSs) have enabled US manufacturers to reduce their time from concept to delivery of products to paying customers.

These manufacturers can boost productivity by shortening the product development cycle and improving their production cost controls. These companies are also ready to take advantage of other promising technologies as they emerge. The gap between those manufacturers who pursue and those who avoid computer integration is widening as product windows shrink. Manufacturers who don't integrate will find themselves in uncompetitive positions with limited means to catch up.

Our 1986 study, which focused on integration of new technologies in large manufacturing operations, identified the problems these new systems cause financial executives and executives responsible for system implementation. It then helped us develop guidelines for manufacturers considering integration.

Under the auspices of the Financial Executives Institute, we mailed questionnaires in summer 1986 to 450 members who worked for Fortune 500 manufacturers.[1] Financial, manufacturing, and distribution executives from 67 companies responded to their respective segments of the survey.

Responding manufacturers accounted for 13% of Fortune 500 companies in April 1986. Collectively, they accounted for $450.3 billion in sales in 1986, controlled 26% of the Fortune 500's assets, earned 37% of its profits, and employed 23% of its employees.

Integrated Manufacturing

Manufacturing executives were asked to describe the interfaces their companies use for manufacturing operations and their modes of data transfer. Interfaces were categorized as manual, offline batch transfer, online batch transfer, and real-time (see Exhibits 1 and 2).

Computerized interfaces for five functions were reported by 70% of responding manufacturing executives, who also reported cost accounting, payroll, and budgeting. But less than 20% of the accounting interfaces generated real-time data.

Engineering, operating, and quality control applications, which were found to be less dependent on computerization, were integrated in 50% to 70% of respondents' systems. Real-time (i.e., interactive) integration appeared to focus on certain manufacturing subsystems (e.g., materials requirements planning, engineering and manufacturing bill of materials [BOM], and computerized shop floor control).

Less than 60% of the respondents used any form of computerized shop floor control system. But more than 50% of respondents with computer capabilities used real-time interfaces. And 60% of companies using manual systems were planning to computerize their manufacturing control operations in the near future.

Real-time manufacturing control capabilities do not yield their full benefits, however, unless other subsystem interfaces (with shop floor control) also support real-time operations. These interactive manufacturing subsystems force financial executives to be alert to the possibility that new manufacturing data bases in their companies may communicate with their financial planning, reporting, and control systems. For example, manual time-card systems do not need maintenance when a computerized shop floor data collection system is in place.

Formal communication with suppliers is the least integrated function in manufacturing; only 26% of the respondents said they used any computer interface, and less than 10% of these interfaces are interactive. Connecting to supplier systems is, however, planned by 22% of responding manufacturing executives.

Although CIM is intended to totally integrate real-time systems, implementing CIM is complex, time consuming, and expensive, and the simultaneous integration of all functions is certainly a high-risk venture.

Obstacles

CIM's difficulties can be organized into two general but not mutually exclusive categories: technological and organizational.

Technological

Disconnections. Although small, independent integrated systems tailored to specific products or manufacturing technologies are being developed, integration can be hampered when there are a large number of business and support systems. Transition or conversion software is then needed to link the different systems. Human representatives for all subsystems should therefore be involved when systems changes and interface adjustments are implemented. For example, an executive reported the following: "Integrating a multiplicity of different business and associated operational support systems (e.g., engineering versus manufacturing, commercial versus government business) [is a major challenge]. Creative systems interfaces are emphasized as a way to minimize existing system changes."

Engineering, operating, and quality control applications were integrated in more than 50% of respondents' systems

Another said, "With each operating division managing its own systems development efforts, [the] level of sophistication depends on size of division, stages of development, resources available, etc."

Incompatibility. Incompatible hardware, software, and communications systems can foster a parochial view that hinders development of systems supporting multiple departments. Companies must not only prepare themselves to support short-term technical solutions where appropriate but be willing to abandon existing systems when more comprehensive and effective answers are

Exhibit 1. ***Survey Respondents' Integration of 14 Manufacturing Functions***

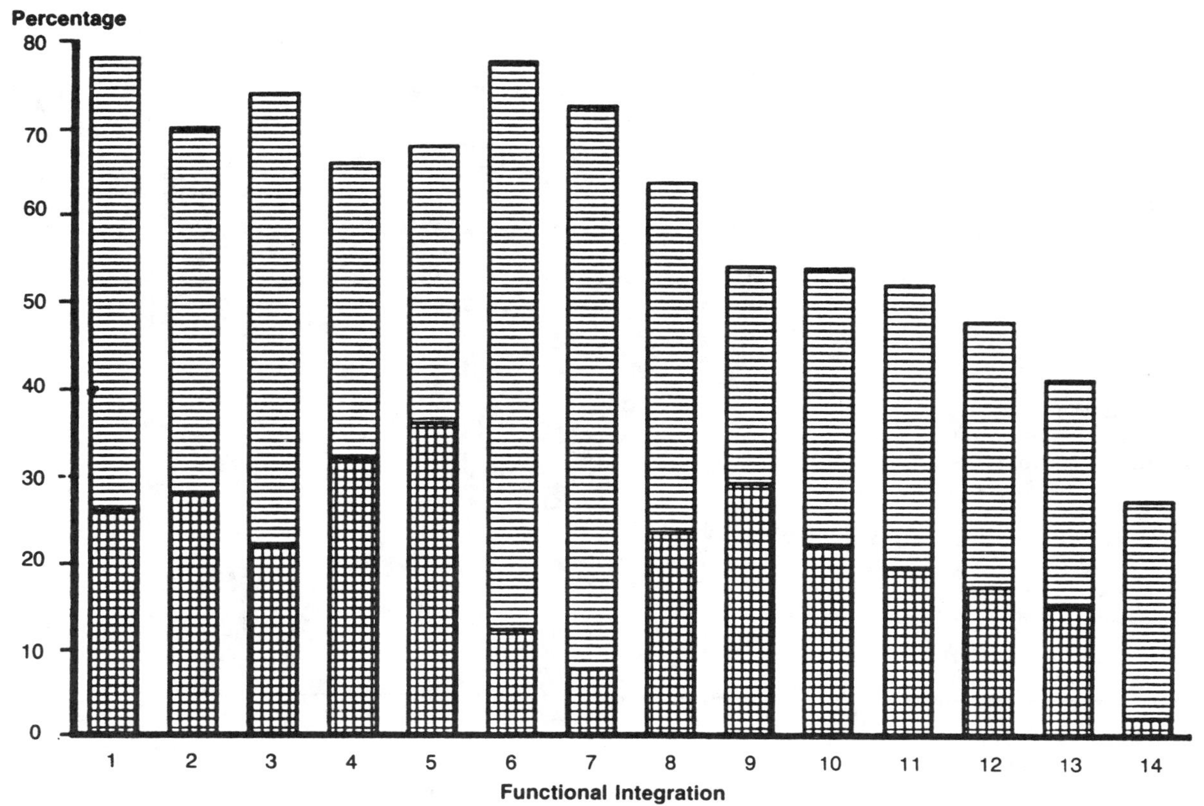

Function

1. Material requirements planning with purchasing
2. Master scheduling with sales/order entry
3. Master scheduling with forecasting
4. Material requirements planning with capacity planning
5. Engineering BOM with manufacturing BOM
6. Cost accounting with shop floor data collection
7. Payroll and budgeting with shop floor data collection
8. Design engineering with process equipment (CAD with CAM)
9. Computerized shop floor control
10. Quality control with shop floor data collection
11. Testing and inspection with fabrication or assembly
12. Work-in-process tracking with quality control
13. Customers' systems with your business unit's system
14. Suppliers' systems with your business unit's system

Key:

Computerized interfaces as a percentage of total interfaces

Real-time interfaces as a percentage of total interfaces

Exhibit 2. ***Respondents' Computerized and Manual Integration***

Function	No Interface (%)	Manual Interface (%)	Computerized Off-Line Batch Transfer (%)	On-Line Batch Transfer (%)	Real Time (%)
1. Material requirements planning with purchasing	2.0	21.6	23.5	27.5	25.5
2. Master scheduling with sales/order entry	2.0	29.4	11.8	29.4	27.5
3. Master scheduling with forecasting	2.0	25.5	17.6	33.3	21.6
4. Material requirements planning with capacity planning	5.9	29.4	19.6	13.7	31.4
5. Engineering BOM with manufacturing BOM	7.8	29.4	17.6	11.8	33.3
6. Cost accounting with shop floor data collection	3.9	21.6	27.5	35.3	11.8
7. Payroll and budgeting with shop floor data collection		27.5	27.5	37.3	7.8
8. Design engineering with process equipment (CAD with CAM)	7.8	33.4	13.7	23.5	21.6
9. Computerized shop floor control	5.9	43.1	9.8	13.7	27.5
10. Quality control with shop floor data collection	2.0	45.1	13.7	17.6	21.6
11. Testing and inspection with fabrication or assembly	9.8	43.1	13.7	15.7	17.6
12. Work-in-process tracking with quality control	9.8	47.1	9.8	17.6	15.7
13. Customers' systems with your business unit's system	9.8	52.9	7.8	15.7	13.7
14. Suppliers' systems with your business unit's system	5.9	68.7	11.8	11.8	2.0

found. For example, Thomas J. Dominise, data base manager at Honeywell Inc, said, "I see our main problems as integrating heterogeneous hardware and software solutions. High concern is the availability of dictionary and access methods to allow retrieval of data from distributed systems."

Henry D. Libera, a systems project manager at Xerox Corp, said, "Main difficulties arise when integrating over multiple computer technologies. Communications capabilities become high-cost and high-risk activities."

And another executive said, "Data and data base administration functions are key. More awareness and education [are] needed [in] integration concepts and opportunities. [Should we] build islands of automation and integrate later or do dynamic integration? [And] communications technology improvements are needed to support a multivendor environment."

Financial interfaces. Investments in computer integrated plant and equipment must be made in large chunks. After a major investment is made in a certain combination of machines, technologies, and people, plant operations cannot be upgraded significantly or changed without incurring major retrofitting costs and organizational trauma. Careful planning and process upgrade before integration can forestall some later compatibility problems.

For example, David G. Harner, controller at FMC Corp, said, "Today it is no longer practical to view cost accounting and financial control systems for the manufacturing facility as separate from manufacturing systems. The financial systems must be viewed as that part of the manufacturing system, which values and controls (in dollars) the manufacturing activity. To date, we have installed islands of technology and have done relatively little in regard to implementation of CIM. Our efforts from the financial function have been geared toward ensuring that cost accounting and other financial control systems are integrated with the manufacturing system and that financial systems are changed to keep in step with changes in the manufacturing process."

John Luranc, FMC's director of applications research, said, "A continuing concern that is foremost in our planning and recommendations of integration-type projects is to make certain that process simplification occurs before any integration is implemented."

James P. Conroy, director of MIS at Reliance Electric Co, reflected that "Manufacturing systems and minicomputers have been distributed to plant locations. [A] program to distribute CAD and CAM is currently under way with equipment and software installed at four locations. Major problems relate to equipment compatibility, local area networks, and upgrading our existing manufacturing systems to reflect the best parts of the just-in-time and CIM methodology."

According to Leon V. Level, vice-president and treasurer at Unisys Corp, "The most significant change in financial systems related to introduction of computer information systems in manufacturing and distribution focuses on the integration of those systems. Financial information generated by manufacturing and distribution systems must be accessible and usable by the existing financial applications. Even without fully integrated systems, the need exists to furnish a consistent level of information from new manufacturing and distribution systems to financial systems in place.

"The business reality is that all systems can't be replaced at once while continuing to run and support the current business operation. Therefore, the integration of new, improved applications must be done on a priority basis, with those systems providing the greatest return receiving implementation priority. Those new tools must be interfaced with existing systems and remain so as other systems are implemented in the priority queue."

And another executive said, "Given the technical aspects of daily operations [continuous processing], automation within our company has tended to evolve on a separate path from its business and financial counterparts. The distinction has been perpetuated with separate technical computing organizations and processing sites, each optimizing process operations to maximize yields, upgrade product mix, and save on energy consumption.

"While there is only limited evidence of integration of daily operations with business and financial systems today, the benefits of doing so have become increasingly apparent in the last few years. As such, several development efforts currently under way will link the technical processing systems with the financial systems. In [our] environment, this typically means that the operational systems capture the input data and automatically generate appropriate entries into the accounting and financial systems. Such links, with emphasis on source data capture, eliminate several redundant steps of data capture, validation, and controls currently required in financial systems while ensuring more timely, accurate, and consistent information."

Organizational

Boundaries. Manufacturers need both corporate policy guidelines and top management support in order to develop and install integrated systems. Because the financial function traditionally crosses departmental boundaries, financial executives' interests include many departments. They also know strategic business plans from both technical and financial sides and possess the financial tools needed to transform vague organizational commitments into specific tactical implementation plans. For example, William R. Lanyard, manager of manufacturing information resources at the Upjohn Co, said, "In the past, integration was governed by corporate policy; most applications were standalone. Currently, a major effort called the Integrated Manufacturing Project is designed to provide the basis for integrating manufacturing, control, and engineering systems."

"Given the technical aspects of daily operations, automation has tended to evolve on a separate path from its business and financial counterparts"

Another executive said, "[We had a] mixed bag of success across manufacturing sites and within the sites. In general, lack of integration [was] caused by lack of direction as to the technological architecture and the information architecture plan."

From an executive at a major aerospace company: "Approximately 10 years ago our firm achieved a high level of functional integration through [the] batch transfer of files. The resulting interdependencies make it difficult to define and execute development of comparable real-time capabilities whose scope is less than the whole system. The minimum logical project is too large to manage successfully. For example, a planned online engineering release and data distribution system incorporates requirements of more than 60 separate existing applications in multiple functional organizations."

And another executive said, "There are rarely significant changes all at once because we have been committed to the computer integrated approach for so many years—in manufacturing, design, and management information systems. Recent changes have been refinements to the overall approach."

Conflicts. Conflicting philosophies about information can lead to difficulties. If a single system is incapable of producing two types of data, compromise is necessary. Manufacturers must be aware of the danger that systems can produce inadequate information or that system controls cannot maintain the integrity of the data. For example, one executive said, "Financial and accounting problems and considerations have to be dealt with in systems design. [The] financial staff has to participate in systems design. That's what we mean by [an] integrated system—service to all functional areas. When trade-offs are necessary, they are highlighted to the level of management where both activities in conflict report."

Another executive said, "No real significant changes are required other than redesigning the interface requirements. There is, however, a difference in objectives between financial systems and manufacturing systems. On the financial side, accuracy and timeliness are much more important."

Expertise. Expertise may not be available to implement some new systems. Many companies need more trained systems designers, installers, operators, and users. For example, one executive said that "Qualified operational personnel are not available in many divisions to assist in the effective design and implementation of new systems. This results in systems that require longer time to implement and may be less efficient than desired."

John L. Larson, executive director, materials management and purchasing at Eli Lilly and Co, said, "Education relative to the formal closed-loop planning and manufacturing process for people who use the computer tools will greatly increase the success of installing integrated manufacturing systems."

And Raymond Patla, manager, finance systems, at Schering Plough Corp, said, "Principal problems are finding the most suitable software packages; installation and acceptance by [the] staff; and convincing [the] staff to drop personal, manual systems and rely exclusively on [the] new subsystem."

New Finance

Manufacturers usually develop comprehensive integrated systems in a piecemeal fashion. As a result, their financial departments have been forced by integrated systems to change both their traditional base for overhead allocations and detailed manufacturing information.

Integrated systems have reduced the direct labor production component while increasing overhead costs. When materials are a significant percentage of total product costs, materials acquisition and handling costs can be used to develop a basis for material overhead allocation. Computer use might also become part of an allocation procedure. Overhead could also be assigned directly. For example, John P. Campi, director, accounting services, at Parker Hannifin Corp, pointed out that "Traditional concepts of cost accounting and product line profitability are virtually irrelevant. Cost techniques for inventory control and valuation are not relevant in a just-in-time environment. Overhead allocation is no longer a feasible method for product costing when overhead is now the lion's share of product costs. We are moving to replace all work order control systems with process costing techniques, and looking at direct overhead cost association to replace allocation of overhead costs."

Another executive said, "Where we have introduced cell technology, we have had to forgo applying overhead to direct labor hours, and base overhead absorption on actual machine hours spent in the cell by a given work order or part. This has also been a problem in determining work center capacity and shop floor scheduling. For example, if a cell contains five machines, and a particular part requires use of only three, operators may operate other machines manually, concurrent with running the scheduled part. This can be dealt with from an actual cost viewpoint, but not from a shop scheduling viewpoint, and is in effect free capacity."

"Cost techniques for inventory control and valuation are not relevant in a just-in-time environment"

A financial executive at a major aircraft company offered: "Our firm is working on changes to accounting practices and financial systems in two areas: development of new concepts and processes for job costing and absorbing overhead where production processes are highly automated. (Since our factories are being automated incrementally, the concepts must simultaneously support machine- and process-based costing for some functions and labor-based costing for others). [And] redefinition of accounting practices for job order identification and inventory allocation ground rules in MRP-based production control systems. Because of our company's mixed military and commercial business base, we are developing two solutions to this set of problems. A major difficulty is devising a way for both forms to coexist in a plant that does both kinds of work.

"A major consideration arising out of the development of integrated systems is the large cost of adapting them to existing disparate accounting practices at our plants."

Another executive said "Integration of MRP with [our] standard cost system has enabled us to detail cost accumulation at the work order level. Variance application to define [the] whole cost of [a] product has been a problem. Deterioration of labor content has necessitated a new look at overhead. And material burdening is being considered."

G.R. Simon, vice-president and controller at the Trane Co, a division of American Standard Inc, said that "Further work needs to be done to have our manufacturing methods and manufacturing accounting interface well. The requirements have yet to be well defined. We are using a major product development project, which includes factory automation, as a prototype."

And Jayne Benish, director of finance and administration at Honeywell Inc, said, "The shrinking labor cost element as a percent of total product cost in the automated factory has resulted in the need for new performance measurements that reward groups for applying resources to the objective [and] the need to find new bases for the allocation of variable and fixed burden.

"Conversely, the large investment in capital equipment requires that greater emphasis be placed on the burden cost element of product cost. As a result, less traditional depreciation methods (e.g., units of production) will play a more important role in matching costs to revenue. This may mean that more than one allocation base will be used.

"Likewise, as the labor cost element decreases as a percent of total product cost, the material cost element will gain more attention. Just-in-time inventory methods will contribute to minimizing the material cost element."

Management Control

Computers have been used to emulate clerical functions. Integration is now shifting toward management control and productivity systems. These higher-level systems differ from clerical systems in their interrelationships with the work force. Clerical systems are viewed as aids to supervisors and workers. Management control and productivity systems monitor events and interact with employees on the factory floor. Physical units rather than cost units are being developed as performance measures.

The rapid increase of the availability of data raises the question of whether that data is used effectively. Automation allows a significant increase in specificity, but creates interfacing difficulties. For example, an executive said, "Costs directly identified with a manufacutring process and organizational entity increase 20% to 30% and allocations of overhead costs are reduced by an equal percentage. Labor and parts-claiming procedures can be simplified. New procedures to perform the cost accounting functions of identifying, counting, accumulating, or decrementing work in process, labor, supplies, or indirect costs in the production line need only be performed at the gate in and out of the target department."

James B. Gustafson, director, IS, at the Stanley Works, said, "A new MRP system was implemented with a bill of material significantly more detailed than the one used previously in a product cost system. A new product cost system was then integrated with the MRP bill of material to reconcile the differences between building and costing the product."

Another executive reported two problems: "Revisions of standard cost computations to meet materials planning needs for corporate bills of materials. [And] development of interface programs between new manufacturing/materials tracking systems and existing financial systems."

And another executive said "Automatic updating of raw material inventory balances is done through a bill-of-materials file. The BOM assumed yield losses, correct from a cost estimating perspective, but incorrect for translating actual finished goods production into assumed raw material usage. The BOM file was modified to allow both calculations."

Strategic Increments

Manufacturers, whose business strategies dictate reduced production cycle time, lower product costs, and the flexibility to respond to consumer demands, need integrated manufacturing and business systems. Although implementation of these systems is incremental, each must fit the strategies' grand designs. ▲

Notes

1. E.D. Bennett et al, *Financial Practices In a Computer Integrated Systems Environment* (Financial Executives Research Foundation, 1987).

Developing a Corporate Charter for CIM

Thomas E. Sprimont

US manufacturers are facing their greatest challenge—to survive in a fiercely competitive global market. Concentrating on engineering and manufacturing quality products and parts, meeting customer demands, and delivering on schedule becomes essential. Corporate management must understand that to accomplish this all departments should plan together to achieve a common corporate goal—CIM. This article proposes a top-down approach for planning and implementing goals and objectives leading to successful CIM, and will focus on a corporation with several divisions.

Thomas E. Sprimont has a BS in CAD and has worked at Digital Equipment Corp as a field consultant dealing with engineering and manufacturing accounts. He is currently CAE manager with Rogerson Kratos, a division of Rogerson Aircraft Corp in California, and teaches computer systems performance analysis and evaluation at UCLA.

Manufacturing departments that plan and procure hardware and software without consulting with divisional or corporate management often end up with islands of automation that are unable to perform export or import functions from one application's data base to another's. Each department is, in effect, seeking solutions to its own problems.

For example, an engineering department may purchase CAD hardware and software without requiring input from other departments such as testing, manufacturing, technical publications, resource planning, or information services. Although special engineering problems may be resolved, the overall result can be catastrophic for the rest of the company. While each department is following its own directions, the entire company is travelling farther away from achieving the preliminary foundation of CIM—the total integration of all hardware and software.

Each new purchase request typically goes through a standard cycle consisting of:

- Need generation and definition.
- Product justification, review, demonstration, and benchmarking.
- Pricing.
- Vendor negotiations.
- Product selection, recommendations, and ordering.

But each department performs these tasks independently. The more hardware and software acquired in this manner, the more difficult integration becomes. To solve this problem, CEO-supported corporate- or companywide policies (charters) must be created to institute guidelines that will direct individual departmental procurement processes.

Communicating Corporate Goals

Corporate charters must begin at the top, and be executed at the lower levels. The content of these charters is the same for both interdivisional and intradivisional situations. An overview of the process by which CIM may be accomplished is illustrated in Exhibit 1.

The charter begins with setting goals for op-

Exhibit 1. ***The Process by Which CIM Is Accomplished***

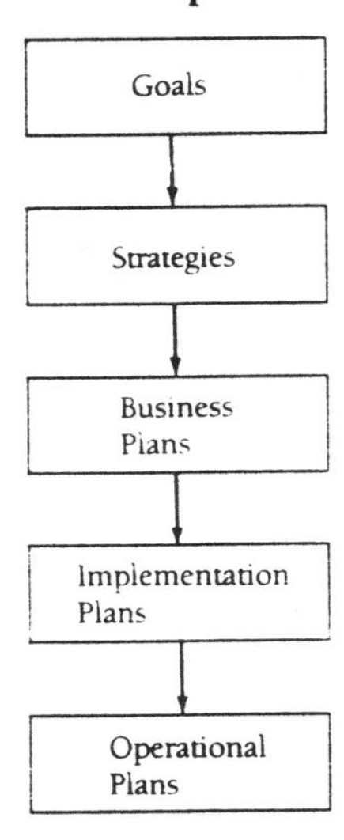

erational implementation. For example, as an entity, a corporate board must include CIM as a long-term goal, but because CIM is a concept, it can only be mapped through strategies.

Strategies give meaning to goals (goals being a vision of the CEO and board of directors) and are the methodologies to attain those goals. For example, factory of the future is part of CIM. It consists of several key areas: around-the-clock operations, easy and quick changeovers and setups, and highly versatile machines that must be integrated into the factory environment.

These strategies must be elaborated upon in the business plan, which acts as a medium-range road map or progress report for strategic implementation.

The next step is creating an implementation plan to accomplish the objectives stated in the business plan. Timetables, equipment, delivery schedules, and installations are part of this plan.

The final phase is the operational plan, which details the daily activities and resource allocation necessary to accomplish the implementation plan.

Creating Corporate Charters

Setting corporate and divisional policy regarding the procurement of hardware and software is essential to the accomplishment of CIM. The corporate staff is responsible for understanding the concepts as well as the technical issues of achieving CIM. It is at this level that interdepartmental needs are recognized and solutions planned.

For example, knowing precisely what departmental data is needed where and when includes passing:

- CAD data to the CAE process.
- CAE data to automatic test equipment.
- CAE data to automatic insertion equipment.
- CAD drawings to technical publications.
- CAD models to the CAM software data base.
- CAM processed output to numerical controllers.

Once the charters are accepted as policy, they should be adhered to by the various departments. This prevents each division from diffusing the integration needed for CIM. When a hardware or software need is recognized, the corporate charter should be referenced. The investigation, analysis, benchmark, evaluation, and selection stages should also be made within the guidelines of the charter.

Charter goals

A corporatewide strategy for CAD/CAM/CAE/CIM supports interdivisional and intradivisional engineering and manufacturing, including the design, development, testing, documentation, and manufacture (including MRP II) of product lines. This strategy can only be accomplished with:

- The ability to pass data electronically.
- The ability to export or import information between data bases.
- Standards support from hardware and software vendors.
- Software integration.

Charter benefits

The charter guidelines should enable each division to remain flexible enough to satisfy its individual needs. However, mechanical and electrical drawings, parts lists, bills of materials, and inventory are targets for interdepartmental transfer. Although each department has unique hardware and software requirements, a companywide charter will protect all the investments of the company.

Charter scope

The scope of a corporate charter treats each department as a generator and evaluator of information. These are the areas that identify, justify, and fulfill computing needs. Communication needs as well as high-level information transfer must be addressed, and paths must be made for file exchange between departmental computers.

The ability to automate information exchanges for areas such as parts lists, bills of materials, CAD and MRP II, engineering drawings or specifications, and technical publications eliminates costly and time-consuming reworking.

Another challenge of the charter is to set guidelines for a wide area network (WAN)—this is important to the overall structure of a corporation. Corporations consisting of groups of companies in vertical markets must be able to support useful information exchange between their distant locations, and the charter must specify how to establish the network. If different divisions purchase equipment, the charter will define requirements for remote communications.

The authors of the charter must be well qualified—the long-range goal of CIM depends on the technical validity of the charter's contents—strategy fundamentals and the engineering and manufacturing processes must be understood.

Standards

The charter should define standards. Standards specifications are beyond the scope of this article, but the basis for a corporate computing architecture must be investigated and addressed. Without computer communications between departments and divisions, other integrations become moot. Beyond the communications aspect, other standardization issues are data integration between application software data bases, graphics standards, and hardware and software standards. (For a further discussion on standards, please see the article, "A Standards Framework for the Computer-Integrated Enterprise" by Albert J. Gibbons, in this issue of CIM REVIEW.)

The network

Standards are based on a network that represents information sharing and is the first step toward planning. Standards that are widely accepted and supported by hardware vendors should be targeted. This ensures a homogeneous communication environment, even in a heterogeneous computer environment. Two types of networks will be targeted in the charter: the local area network (LAN) and the WAN.

The LAN. The LAN involves a single geographic site. It is best, but not necessary, to support a single type of LAN. Large companies must support more than one type of LAN for business purposes. The evaluation of LANs should involve the review of national and international standards. Organizations such as the IEEE (Institute of Electrical and Electronic Engineers) and CCITT (Consultative Committee on Telephone and Telegraph) set many communications standards.

Commonly supported LANs are IEEE 802.3 (Ethernet standard), IEEE 802.4 (Manufacturing Automation Protocol), the token bus standard, and IEEE 802.5 (token ring passing protocol). With Ethernet, the CAD, manufacturing, MIS, test, technical publications, software development, and MRP departments could procure different computer vendor's products, provided each supports Ethernet.

Next level. The next level of support is communications software (Ethernet is the medium of file transfer, but does not perform this function). The level of information exchange will dictate the type of communication software specified in the charter.

The method of establishing a WAN must also be included in the charter. A software package supporting X.25 packeting and communication over PSDNs (Packet Switching Data Networks) may be the solution.

Larger companies may need to support more than one type of LAN. A large manufacturer may run MAP on the factory floor, and Ethernet for departmental communication such as electronic mail, file transfer, and remote services. Therefore, specifications for bridging the two LANs must be detailed in the charter, including hardware devices, cabling, protocols, and high-level communication software.

Exhibit 2. *Divisional Method for Planning and Implementing a Charter*

<table>
<tr><td colspan="12">Corporate Goals</td></tr>
<tr><td colspan="12">Corporate Strategies</td></tr>
<tr><td colspan="12">Corporate Business Plan</td></tr>
<tr><td colspan="12">Corporate CAD/CAM/CAE/CIM Charter</td></tr>
<tr><td colspan="4">Divisional Goals and Corporate Strategy Support</td><td colspan="4">Divisional Goals and Corporate Strategy Support</td><td colspan="4">Divisional Goals and Corporate Strategy Support</td></tr>
<tr><td colspan="4">Divisional Dept Strategy Plan</td><td colspan="4">Divisional Dept Strategy Plan</td><td colspan="4">Divisional Dept Strategy Plan</td></tr>
<tr><td>Dept Plan</td><td>Dept Plan</td><td>Dept Plan</td><td>Dept Plan</td><td>Dept Plan</td><td>Dept Plan</td><td>Dept Plan</td><td>Dept Plan</td><td>Dept Plan</td><td>Dept Plan</td><td>Dept Plan</td><td>Dept Plan</td></tr>
<tr><td>MRP</td><td>CAD/CAM</td><td>Tech Pubs</td><td>ATE</td><td>MRP</td><td>CAD/CAM</td><td>Tech Pubs</td><td>ATE</td><td>MRP</td><td>CAD/CAM</td><td>Tech Pubs</td><td>ATE</td></tr>
<tr><td colspan="4">Local Area Network Division 1</td><td colspan="4">Local Area Network Division 2</td><td colspan="4">Local Area Network Division 3</td></tr>
<tr><td colspan="12">Wide Area Network</td></tr>
</table>

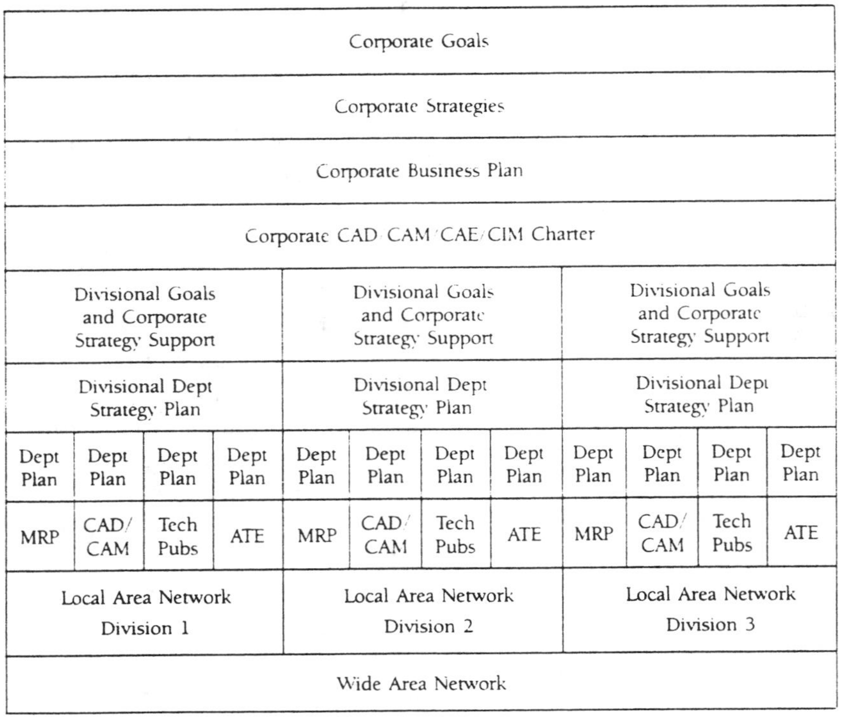

Notes:
Dept Departmental
Tech Pubs Technical Publications

Communications guidelines

The communications guidelines must incorporate all computers and related devices (such as automated test equipment, numerical controllers, and data collection) into the same network, yet remain flexible. New technologies and advancements are being made in networking. As network applications such as NFS (Network File System) and committees such as NCF (Network Computing Forum) mature, they will begin to play a larger role in software integration.

The objectives of any standardization group are to achieve supportable hardware and software networks for all current and projected needs and to maintain an open architecture for growth and the incorporation of future technologies.

Execution and Support

Exhibit 2 illustrates a divisional method for planning and implementing a charter within an organization. These plans must include capital equipment projections. The corporate guidelines should be consulted before equipment is purchased; however, the final recommendation should still be sent to a corporate approval board to ensure that all recommendations are within the established guidelines. If there is an inconsistency, the recommendation should be refuted and sent back to the division with a statement which details the technical reasons for noncompliance. If there are no inconsistencies, the recommendations are endorsed.

Summary

The charter, when it is established, must maintain and support interdepartmental integrations that are not apparent to the individual departments. The individuals may only be concerned with addressing their local needs, but by adhering to a companywide charter, they automatically build the support necessary to achieve interdivisional integrations. Thus, a company has the foundation on which to accomplish CIM. ▲

Additional Readings on Planning for Systems Integration

Betts, Jerry. 1986. Integration of existing systems. *Systems I Conference Proceedings*, Dearborn, MI: Society of Manufacturng Engineers. March 24-26.

Drucker, Peter F. 1990. The emerging theory of manufacturing. *Harvard Business Review*. Boston, MA: Harvard Business School Press, May-June.

Gerelle, Eric G. R., and John Stark. 1988. *Integrated Manufacturing: Strategy, Planning, and Implementation*. New York, NY: McGraw-Hill, Inc.

Vesey, Joseph T. 1991. The new competitors: Thinking in terms of 'speed-to-market'. *Manufacturing Systems*. June.

IV. DESIGN ISSUES IN SYSTEMS INTEGRATION

Design activity among mechanical engineers and electrical engineers is easily comprehended, because the results are items that one can touch and feel; a brace for an aircraft wing, or a printed circuit for a control panel. Design methodology for these cases is relatively mature.

Industrial engineers engaged in systems integration also perform a design function, but is more nebulous than those mentioned above. Also, design methodology for systems integration is more embryonic than for other fields of engineering. Design is a creative activity. One must create something that was not there before, or at least create an improvement in something that previously existed.

There are many facets of design, and no one engineer can possibly master them all. It is probably true that engineers of all types spend much of their careers embellishing their design skills.

Much of the work of systems integrators involves the *redesign* of current systems. Consequently, among the articles included in this section are two that deal explicitly with system redesign.

Issues in the design and implementation of a system architecture for computer integrated manufacturing

ALBERT JONES, EDWARD BARKMEYER and WAYNE DAVIS

Abstract. The advent of sophisticated automation equipment and computer hardware and software is changing the way manufacturing is carried out. To be competitive, manufacturing companies must integrate these new technologies into their existing and future factories. In addition, they must integrate the planning, control and data management methodologies needed to make effective use of these technologies. This paper discusses several issues related to the design and implementation of a system architecture which can serve as the basis for integration. The architecture includes separate architectures for production planning and control, data management and data communications.

1. Introduction

The recent worldwide metamorphosis in manufacturing has centred around the introduction of new computer and production technologies. Efforts are under way at many companies to use these technologies to increase their productivity and profits. Increases are expected to come from using computers to automate and integrate manufacturing processes and related functions. The hope is that automated, integrated factories will produce parts which are of higher quality, cheaper and on time.

In the recent past, many companies have been successful in automating physical manufacturing processes. They have been less successful at automating many of the human-intensive functions such as part design, process planning and scheduling. The integration of manufacturing processes with engineering and production functions into a computer integrated manufacturing (CIM) system has progressed at an even slower pace. Our view is that the basis for achieving this integration lies in the design and implementation of a system architecture for CIM.

Authors: Albert Jones and Edward Barkmeyer, National Institute of Standards & Technology, Graithersburg, MD 20899; Wayne Davis, University of Illinois, Urbana, Illinois 61801, USA.

We believe that a system architecture must include three separate but related architectures for production management, data management and communications. Production management includes all of the functions related to the order, design, fabrication and inspection of parts. Data management includes all functions related to the delivery of accurate and timely information to the production management processes. Communication management includes those functions required for the reliable transmission of messages between computer programs.

Each of these architectures has been the subject of intense research, both public and private. But little has been done to incorporate the results of these research efforts into a single, integrated, real-time system architecture. This paper examines several issues related to the design, implementation and integration of each of these individual architectures into the overall system architecture. It also discusses some of the work being done to resolve those issues.

2. Issues in production management

2.1. What is it?

As noted above, production management includes all functions related to the order, design, fabrication and inspection of parts. In our view, executing these functions in a CIM environment means: (1) automating each function to the extent possible; (2) developing an architecture which can be used to integrate them; and (3) providing uniform data structures for all shared information. To understand the difficulty involved in achieving these goals, it is necessary to understand what these functions do, their inputs and their outputs.

Reprinted courtesy of Taylor & Francis from *International Journal of Computer-Integrated Manufacturing*, vol. 2, no. 2.

2.1.1. The functions. We now discuss a major subset of these functions: marketing and sales, manufacturing data preparation, aggregate and detailed production planning, production scheduling, process supervision, and quality assurance. Figure 1 shows the interactions and data flows among these functions.

Marketing and sales provide the primary interfaces between a manufacturing facility and its customers. They inform customers of available products, generate orders for selected products, price those products, negotiate delivery schedules, track shop floor performance in meeting those schedules, and ensure customer satisfaction after delivery. They assist in assessing the reliability of existing products and producing specifications for new or improved products. In addition, they often conduct a needs analysis to determine potentially profitable new products.

Manufacturing data preparation includes all of the functions required to generate the data needed to manufacture a product capable of meeting a particular customer's requirements. First, it must translate a set of requirements (product information) into products designs which include drawings, geometry data, tolerances, materials, and other required manufacturing specifications. These designs are then used to generate a process plan which is a complete list (including any possible alternatives) of raw materials, tools, machines, fixtures, and the precise machining instructions to be used during the entire fabrication process.

Aggregate production planning (APP) uses both the current and projected demands established by marketing to set production quotas and inventory requirements for each product type during the next planning period. Those inventory requirements include all raw materials, tools, etc. needed to meet the demands. The APP also establishes quotas for subassemblies that will be assembled during subsequent production periods into finished products.

Detailed production planning (DPP) uses these assigned quotas to generate production and inventory 'jobs' for each time period. Before a production job is released to the shop floor for scheduling and processing, it is assigned a priority and a due date, and a check is made to verify that the required materials are on hand. Looking at future production quotas, the DPP may issue orders to external vendors to replenish inventories for input materials. The DPP monitors the differences between the assigned and anticipated job completion dates, and if needed, changes both the due date and the specification of the criteria to be considered in the scheduling function. This information is used by the APP continually to monitor and update its production quotas and inventory policies. Production scheduling develops detailed (usually daily) schedules of the operations required to complete the jobs issued by the DPP. These operations are then assigned to the various processes together with their anticipated start and finish times. Once a production schedule has been generated, the scheduler must coordinate activities at each process to ensure that the schedule is met. This inter-process coordination function requires continuous monitoring of the feedback from process supervisors. This feedback is then used to update the existing schedule, when unexpected delays occur.

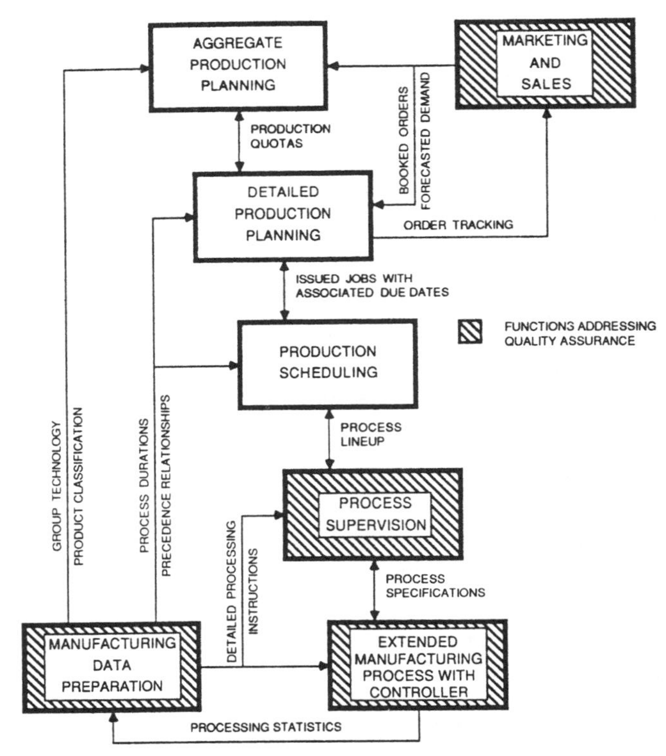

Figure 1. Manufacturing functions comprising the CIM environment.

Each process has its own supervisor who has three responsibilities. First, he executes each operation using the instructions in the process plan. He also monitors the process during its execution to verify conformance to those instructions. Finally, he may be required to perform some on-line inspection to verify quality assurance and machine performance. He can compensate for minor changes in the processing environment, such as a tool wear, as long as they do not result in substantial deviations from the original instructions. Major problems, such as tool breakage, will force him to wait for a new set of instructions (a process plan) and a new schedule before completing the assigned task.

Quality assurance (QA) is divided into two major functions. First, it verifies that the output from each manufacturing process meets the specifications contained in the original design. As noted, some of these checks are performed on-line by the process supervisor. Others are the result of off-line inspections. Detected and corrected errors are used to improve the designs, the process plans, and the processes themselves. Next, QA keeps historical records which are used to improve the quality of all phases of the manufacturing system. In some cases, these records reflect the results of statistical studies which are used to track past and predict future equipment performance. In other cases, information is archived on each product. This helps guide future decisions regarding the fabrication of those or similar products.

2.1.2. Data driven production. It is clear from the preceding discussion that parts of the same data package are used by many production management functions. This means that providing a uniform structure for these data packages will be a key to the successful integration of these functions. Two of the most important of these data packages are product information and process plans.

Electronic product information includes all of the main raw data needed to design, fabricate, test, inspect and maintain a product during its entire life cycle. A complete, neutral form for that data is critical to external relationships with both customers and suppliers. It also plays an important role in the internal integration of many production management functions. Externally, it allows for the reliable, and unambiguous transfer of product information. That transfer can take place between customer and producer, or between two or more facilities involved in the manufacture of a single complex product. Internally, it is the main guideline to the manufacturing data preparation function which provides the link to many downstream production functions. That link is the process plan.

The generalized concept of 'process plan' described above is expected to play a major role in integrating production functions in the CIM environment. Major changes will be required in existing process planning systems to fulfil this role. All of the data (resources, timings, routings and alternatives) required to transport, fabricate, inspect and ship a part must be included in the total process plan for that part. The total plan should be decomposed into subplans having the same generic structure. One or more subplans will be used by each production function to make decisions about each 'job' it executes. The process planning system must be flexible enough to allow the evolution of processing capability associated with new technological innovations. Finally, the system must be intelligent enough to produce new plans and modify existing ones quickly and accurately.

2.2. Current approaches to designing architectures

Hierachical structures have been frequently used to provide the coordination necessary to manage production activities in traditional human-based factories. The number of levels and the responsibilities of the individuals at each level can vary dramatically from one company to another. In many small companies, all decisions are made at the top, and subordinates simply implement various decisions at their own level. In most larger companies, people at every level are expected to make certain decisions, based on input form a superior, and exert the control necessary to have subordinates execute their decisions.

Recent attempts to design and implement production management (sometimes called control) architectures for CIM systems have used the same hierarchical approach (Jones and Whitt 1985). Their designs are based on three principles (Albus *et al.* 1981). First, levels are used to reduce the size and complexity of the problem and to limit the scope of responsibility and authority. For CIM, each level will consist of one or more computers which decompose commands from a supervisory computer into procedures to be executed by other computers at that level and subcommands to be issued to one or more subordinate levels. Second, decision making and control always resides at the lowest possible level. That is where the most complete, up-to-date and deterministic information is available. Third, planning horizons decrease as you go down the hierarchy. At the higher levels, the horizon can be months or years. At the lowest level, the horizon can be a few seconds.

There are, however, three attributes that distinguish one hierarchy from another. First, there are the number of levels and the assignment of functions to levels. In most designs, this 'decomposition' depends on both the complexity of a given function and the actual physical configuration of the system. The Advanced Factory Management System (Liu 1985), which is based on the approach of grouping similar machines close together, has four levels. The Automated Manufacturing Research Facility model (Jones and McLean 1986) is based on a group technology approach to shop floor layout and has five levels. The Factory Automation Model (Ottawa 1986), which attempts to accommodate both, has six.

The second characteristic involves the identification and direction of control paths. Control relationships can be assigned once and remain static or they can be assigned dynamically as the situation dictates. The question of direction is independent of assignment. Most architectures allow only vertical control flow. This means that each control module can have only one supervisor who issues commands. This supervisor always resides at the next higher level. Some structures allow vertical and peer-to-peer control paths. This means that a control module at one level can issue a command to another control module in the same level. While there is a substantial theoretical basis for a hierarchy with vertical control flows (Libosvar 1988, Saleh 1988), we are not aware of any theoretical foundations for a hierarchy with peer-to-peer control relationships.

The last distinguishing characteristic is the method of handling data. In some architectures, data management is viewed as one of the control functions at each level and is included in the existing control hierarchy. This means that the data needed to carry out a given command is either part of the command itself or tightly coupled to the control path. In other systems, there is a totally separate data management architecture which serves the control hierarchy. In this case, control modules must access the data needed to execute a command from the data management system. We believe that the latter approach is preferable for three reasons. First, it eliminates the time spent passing data packages, such as NC code, through intermediate control levels. Second, it allows control modules to be developed independent of the sources of the data they need to perform their assigned functions. Third, it allows the control architecture to be independent of the data flow paths (Barkmeyer 1989).

2.3. *Implementation versus design*

While hierarchies have worked well in manufacturing plants where people make the decisions, there is no *a priori* reason to believe that they are the best choice for CIM systems. In fact, although numerous designs have been proposed (Jones and Whitt 1985), major problems have arisen in the implementation of each of these computer-based control hierarchies.

2.3.1. Automation and integration. There have been and continues to be problems developing automated decision-making methodologies which adequately duplicates the way humans conduct the negotiations needed to make complex decisions. Even if we could automate these decisions, we do not know how to integrate the results into the distributed hierarchical computing environment discussed above.

There are three principal reasons for these difficulties. First, the best-known paradigm for distributed decision-making is decomposition of mathematical programming models of the system (Geoffrion 1970). It assumes a static, deterministic, single criterion problem. Manufacturing problems, on the other hand, are dynamic, stochastic and multi-criteria. Secondly, forcing these functions into a hierarchical structure masks the complex relationships that exist among the functions themselves. That is, hierarchical architectures make it appear that the only functions that interact are those in adjacent levels. The fact is (see Fig. 1) that all of these functions interact in very complex ways. Finally, there are several different time scales associated with these functions which affect the way they can be grouped and the way the control feedback information can be used to update decisions.

2.3.2. Choosing the best design. Another major implementation problem arises because CIM systems and their control hierarchies are typically imbedded as part of an existing factory. All the decisions related to the tasks that are performed in the CIM system are made by factory systems external to the CIM heirarchy. This includes which products will be produced, the production schedule, and the essential resources and data needed to support that production. This means that the implementation is usually limited to the control functions at the lowest levels. As a result, there is no proof that: (1) control hierarchies are the most appropriate architecture for production management in CIM systems; and (2) the entire architecture can be implemented as designed. In addition, there is no way to determine whether a given design is appropriate for a particular application before, during, or after initial implementation.

2.4. *Where do we go from here?*

Two things are clear. First, we cannot, in the near future, implement hierarchical production management architectures for CIM which incorporate automated control and decision-making at every level. We do not know: (1) whether hierarchies are the best choice; (2) how to automate many of the decision-making processes; (3) how to integrate the results into a distributed, hierarchical framework. We advocate a 'back to basics' approach in which we attempt to build dynamic models for the various functions in section 2.1. The fact is that while CIM systems may vary considerably in size, scope and complexity, they all perform these same functions. The key is to develop models which provide the insight needed to design *effective and implementable* production management architectures. We believe that the models described in (Davis and Jones 1988, 1989) provide insight into: (1) the impact of the dynamic CIM environment on these functions; (2) the data flows and relationships that exist among functions; (3) methods for automating these functions; and (4) keys to integrating them into a production management architecture which is suitable for the particular application.

Second, there is little hope that a single architecture for production management will emerge which can serve as a reference model for all CIM applications. Several different designs have already been proposed; but, none of them has been completely implemented. In addition, there is also no way to determine in advance which designs are implementable and which one is the 'best' for a given CIM system. We agree with the view in (Nof and Moodie 1989) that a combination of quantitative/qualitative performance measures and analysis tools must be developed which can be used to compare different designs and select the 'best'.

3. Issues in data management

3.1. Overview

The main purpose of a data management system in a CIM environment is to support functions in the production management architecture with timely access to all essential data. It is essential to design and implement a data management architecture which is separate from but integrated with the production management architecture. It allows the two structures to be developed independently, provided their interrelationships are well understood. There are, however, many characteristics of a CIM environment which make this approach difficult to implement. Su *et al.* (Su 1986) have discussed this at length, and we now present a brief discussion of these characteristics.

3.1.1. Heterogeneous system environment. The computers and production equipment which make up these CIM systems will be purchased from a variety of vendors over a long period of time. This implies that data is likely to be physically distributed across a network of heterogeneous computers. These local repositories will have a wide range of data access, storage, management and sharing capabilities. The CIM data management architecture must make these differences transparent to users, i.e. users simply make requests and receive data. In CIM systems, the users are production management functions. In addition, users should not be concerned about the effort required to satisfy their requests. To achieve these goals, the data system must provide users with a common method of accessing information. The data system must deal with the problem of translating requests in the common form to operations on the underlying data repositories, wherever and whatever they may be. This implies transmission and translation of component operations to the appropriate database management systems, assembly of information from multiple sources, and conversion of the information to the form the user expects.

3.1.2. Real-time operations. A variety of data is used by computer systems that control shop-floor equipment to make real-time decisions. If that data is not present when it is needed, erroneous decisions or no decision may be made, resulting in processing delays and reduced plant throughput. To complicate matters, some of that data may be shared by several users with different 'real-time' access requirements. This implies that the data system must enable asynchronous interchanges of information between production processes which are effectively communicating with one another. This in turn requires the replication of some information units on two or more systems and the frequent and timely updates of those units.

3.1.3. Data delivery and job scheduling. Highly automated systems are highly dependent on electronic information. It is important to realize that data delivery, like material delivery takes time and must be included in the planning of each job. Actual part production cannot start until all of the required information is transferred to the computer responsible for controlling the process performing that production. The notion that this transfer is effectively instantaneous is becoming obsolete as the speeds of automated systems themselves increase. This means that data is quickly becoming a critical resource which must be scheduled. Poor 'data scheduling' will lead to delays, bottlenecks and idle equipment. Therefore, the scheduling decisions made by the data manager have a direct impact on the scheduling decisions made by the production scheduler. This implies the need for coordination between the data scheduling function in the data management architecture and its counterpart in the production management architecture. To the author's knowledge, this type of integration does not happen in any existing CIM system.

3.2. What constitutes an architecture?

To meet the requirements and constraints described above, a data management architecture must address three major concerns: data modelling, database design and data administration.

3.2.1. Data modelling. Developing a 'conceptual model' of all the information involved in the entire production management spectrum is critical to the success of any integrated data management system for CIM. Because the amount of information is so large, a 'divide-and-conquer' approach to performing the analysis must be taken. Experts on individual functions in the production management architecture will perform the analysis and develop a conceptual information model for each functional area. This results in models for product data, process plans, CAD designs, knowledge bases, etc. Then the resulting 'component' models must be integrated into a single 'enterprise model'. The enterprise model is the conceptual representation of the global information base.

To be successful, this integration must be based on the identification of common real-world objects and concepts, rather than trying to identify the 'common data'. A conceptual model', therefore, must represent the relationships between information units as they apply to the real world, rather than the structured and limited relationships between these units as they are stored in a data system. Several powerful modelling techniques now available allow representation of the real-world objects themselves, as well as the information units which describe and distinguish them. Using such a technique, one can distinguish a concept from its computer representation. This avoids problems with multiple representations of the same concept or similar representations of somewhat different concepts. This capability is vital to the development of an integrated data model.

From this enterprise model, it is possible to extract subsets which represent 'views' of the global information base possessed by individual production management processes. But, the processes actually want to use data, and they want that data organized in a specific way, which we call an 'external view'. An external view is an interchange representation and is one of the elements of the interface between data management functions and the production management functions. Generating these views correctly is a matter of considerable current research interest (Mark 1987). They are extremely important because they make the abstract objects disappear and the modelled information units acquire specific physical representations.

3.2.2. Database design. Having arrived at a global enterprise model, we have the problem of mapping this model on to live databases. We must now choose systems, organizations and representations for the information units in the model. This process is called database design. It must result in databases which are consistent with the model and tuned to the timing and access requirements of the production management functions that use them. Because of the evolutionary nature of CIM, it is not possible to start this process from scratch. There are vendor-supplied databases and data systems, which are difficult to alter or augment. There are also 'legacies'—large reservoirs of previously developed information which already have an imposed organization. Moreover, even when one has total freedom of design, the design of databases to perform optimally under the multitude of views possessed by different production functions is a black art. The only solution which meets production management requirements and the legacy and autonomy constraints is to divide up the data itself into multiple databases serving specific production functions well. Since much of that data must be shared, two problems results:

(*a*) partitioning—some production functions must simultaneously access information stored in two or more databases; and/or
(*b*) replication—some data must be stored simultaneously in two or more different databases, and maintained consistently.

The available options for the placement of databases in the CIM computer system complex, and for the selection of specific data management systems to support them, are dictated to a large extent by the architecture of the 'global' data administation system.

3.2.3. Data administration. The administration portion of the data management architecture provides the data services controlling access to all data:

- 'query processing', which is concerned with the command/response interface to user programs, the data manipulation language and its interpretation, and the validation of user transaction requests;
- 'transaction management', which is concerned with identification of databases participating in a given transaction, transaction scheduling and conflict resolution; and
- 'data manipulation', which involves execution of the operations on the databases.

There are three control architectures for data administration systems which have been used with varying degrees of success in various business applications: centralized data and control, distributed data and control, and hybrid systems.

The totally centralized approach is the traditional design, the simplest, and the most workable. Whether this is a feasible architecture for CIM in the long-run is unresolved. There are currently available high-speed, internally redundant, fault-tolerant, integrated centralized sytems. But even if such systems can keep pace with the growing demands and time-constraints of automated production systems, the centralized architecture is not workable from the point of view of subsystem autonomy. Vendors of design and planning systems, for example, cannot assume that every customer will have such a facility, and will therefore develop *local* databases and data services to meet the needs of the products they provide. Consequently, the nominally centralized architecture will in fact consist of a collection of autonomous systems copying information to and from a single centralized facility, according to some externally specified plan. At best, if the external plan provides uniformly for concurrency control, security and the like, what results is centralized data with distributed control.

The canonical architecture for the totally distributed approach (Fig. 2) consists of local data management systems which process locally originated and locally satisfiable requests and negotiate with each other to process all other requests. In this case, difficult problems of concurrency control, distributed transaction sequencing and deadlock avoidance occur and must be resolved by committee. While there has been considerable research in these areas, satisfactory solutions have not been found. Moreover, the dynamic evolution of CIM leads to the problem of configuration changes in the complex, which requires informing all existing participants and modifying their distribution information.

The hybrid architecture depicted in Fig. 3 attempts to combine the best features of both centralized and distributed architectures. Subsystem autonomy and high throughput are achieved by allowing local data systems to process locally originated operations on local data. Operations which transcend the scope of a local system are sent to a centralized 'global query processor' for distribution to the appropriate sites. The global query processor acts as a central arbiter for resolving the characteristic problems of distributed transactions and for handling configuration changes in the data administration system itself. There are a number of ways of implementing such an architecture, but they are all characterized by standardized interfaces between the local data systems and the global query processor.

We note that many new commercial distributed data systems are of this hybrid variety, but in general, they have not yet resolved the control problems associated with distributed transaction management to the extent necessary to provide robust support for CIM (Thomas *et al.* 1988). Unless the local system is aware of potentially global consequences of local changes, and can propagate those correctly, the integrity of the global information bases is always in doubt.

3.3. *Our approach*

It is our view that separating the query processing and transaction management functions from the data manipulation functions, producing a layered hybrid architecture, is essential for effective distributed data management in CIM systems. Each module within the layered architecture now manages a queue of database operations needed which result from the decomposition of some complex query. Each query decomposition can be posed as an optimization problem having both static and dynamic characteristics. Techniques used to solve the static problem within a centralized system (Chu 1986) work equally well here. But, not much is known about approaches to solving the dynamic optimization aspect of

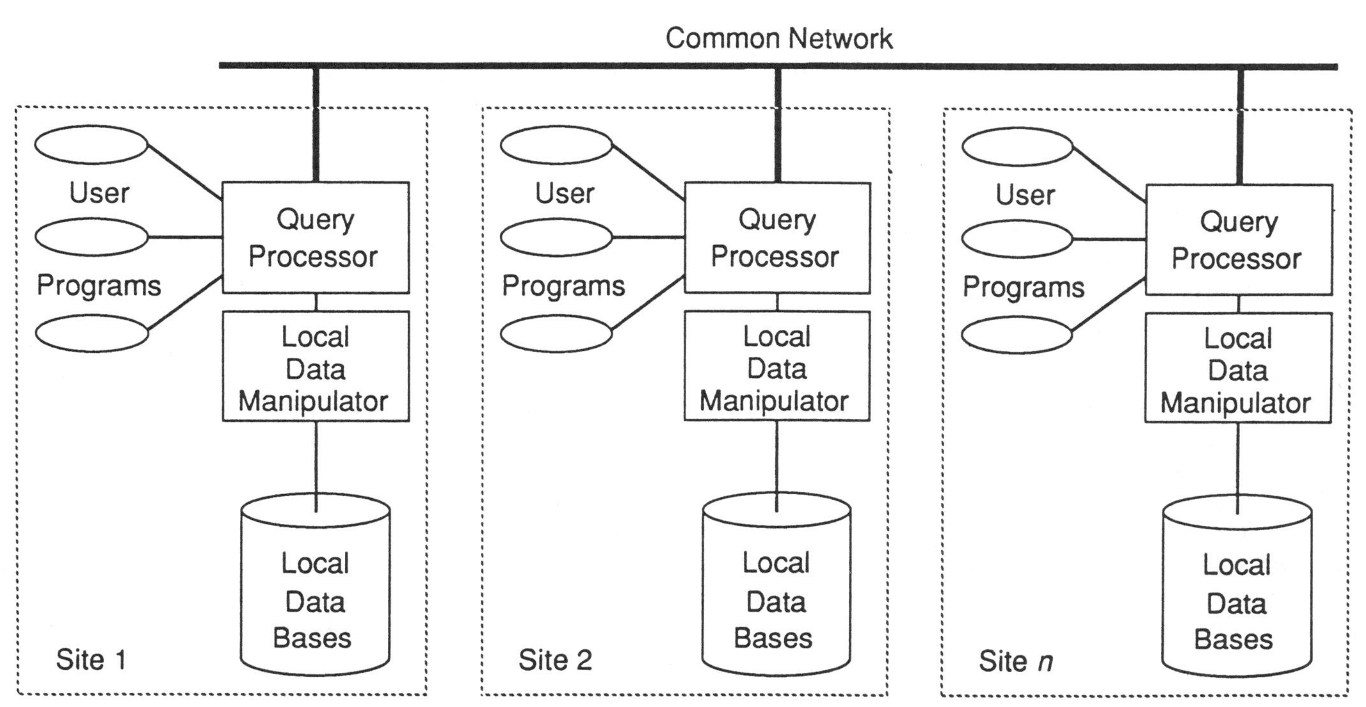

Figure 2. Distributed data system with distributed control.

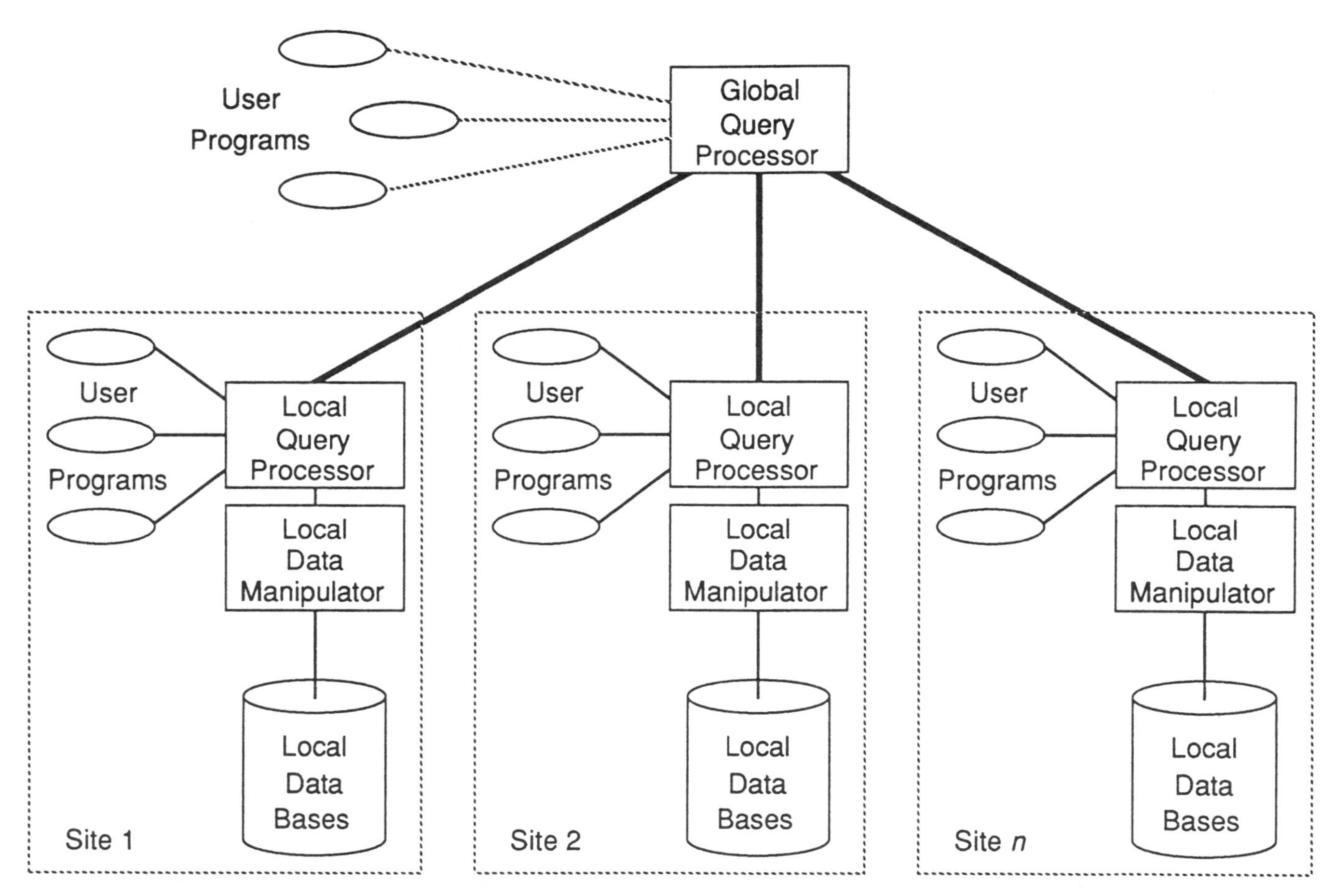

Figure 3. Distributed Data System with Hybrid Architecture.

query decomposition problems in a distributed environment.

Once this decomposition has been completed, the resulting operations must be scheduled and sequenced. These scheduling problems and job scheduling problems have similar characteristics. Scheduling these database operations is complicated by the difficulty involved in: (1) estimating the time required to complete a given operation: (2) obtaining a 'due date': and (3) coordinating the database operations across multiple layers which may be involved in the completion of a single complex query. Little is known about approaches to solving these problems.

These real-time decisions must be integrated into the data management architecture. We have already discussed some of the problems in integrating decision-making into a distributed computing architecture. Furthermore, as noted above, these scheduling decisions must be integrated with the job scheduling decisions that are made in the production management architecture. To the authors' knowledge, no research is being done to address this critical problem.

4. Issues in communications

The CIM communications system provides those functions needed to transmit messages between computer programs executing production and data management tasks. In planning an integrated CIM network, we believe that three ideas are fundamental:

(1) that the production management and data management programs themselves use one common connection service specification for communication with other programs, regardless of function or location;
(2) that the physical networks are transparently interconnected, so that any program could conceivably communicate with any related program anywhere in the CIM complex; and
(3) that the technology and topology of subnetworks are chosen to provide optimal communications responsiveness for the primary functions.

In this section we describe a CIM communications

architecture whose design is based on these ideas. This architecture ensures that *any* production management or data management architecture that is deemed to be desirable can be conveniently constructed with the CIM network as-built.

4.1. *Types of communication*

Communications can be divided into two classes: those *within* computer systems, and those *across* computer systems. The first type is often referred to as 'interprocess communication' while the second is often called 'network communication'.

Interprocess communication is dependent on features of the operating system. Many systems provide no such facility at all, or provide only for communication between a 'parent' process and 'child' subprocesses which are created by the parent. On such systems the coordination of multiple production management and data management activities is extremely difficult. On the other hand, properly implemented 'network communication' software provides for the case in which the selected correspondent process is resident on the same computer system as the process originating the connection. That is, the proper solution for the future is to make local interprocess communication a special case handled by the network software. This solution has the added advantage that all communications by a production management or data management process, regardless of the location of the correspondent process, has the *same* interface.

The accepted paradigm for network communication is the Open Systems Interconnection Reference Model (OSI) which separates the concerns of communication into seven layers (Day and Zimmerman 1983). These are listed below.

(1) Physical layer deals with cables, connectors and signals, and the protocols governing 'who talks when' on a shared medium.
(2) DataLink layer deals with packaging the signals into elementary messages (called frames), checking for errors in transmission and (perhaps) recovering lost frames. It provides the control and checking for the physical link.
(3) Network layer deals with making logical end-to-end connections out of one or more physical connections, i.e. finding a path that gets a message from station A to station B.
(4) Transport layer controls and checks the end-to-end connections so that complete messages are delivered in logical order and without losses.
(5) Session layer distinguishes separate processes or functions communicating between the same two stations, and implements rules for message flow between those processes or functions.
(6) Presentation layer converts between local data representations and interchange data representations.
(7) Application layer deals with establishing links between processes and the relationships between interprocess links and functions being performed.

Traditionally, 'network communication' has meant concentration on the lower four layers and 'exposure' of the Transport layer to production management programs. The important aspect of this model is that it formalizes and separates the logical process-to-process link (in layers 5–7) from the physical network service considerations (in layers 1–4). By exposing only the Application layer service, which implements a common program-to-program communications capability, to the production management and data management programs, we insulate them from the networking concerns. (We note that even local interprocess communication has elements of layers 1, 2, 5 and 7.) Consequently, we believe that it is meaningful and proper to build an Application layer interface which is common to *all* program-to-program communications, both 'local' and 'networked'.

4.2. *CIM network architecture*

A great deal of flexibility is created by implementing the OSI model. On one hand, a single physical medium can multiplex many separate process-to-process communications. On the other hand, a given process-to-process communications can use several separate physical connections with relays between them. This gives rise to the general CIM network architecture shown in Fig. 4. Ideally, all stations on the network implement common OSI protocol suites in the intermediate layers (3–5) and some globally common protocols for moving data sets in layers 6 and 7. In addition, other standard application layer protocols will be shared among systems performing related functions. The choices of protocol suites in the Physical and DataLink layers and the connectivity of individual stations will vary. They will depend on the physical arrangement and capabilities of the individual stations, and their functional assignments and performance requirements. There may be one physical network, or many. All of these separate physical networks, however, must be linked together by 'bridges' that implement the proper Network layer protocols. This results in a *single logical network* on which any given production management or data management process can connect to other process regardless of location. We note, that because this architecture is layered, multiple subnetworks

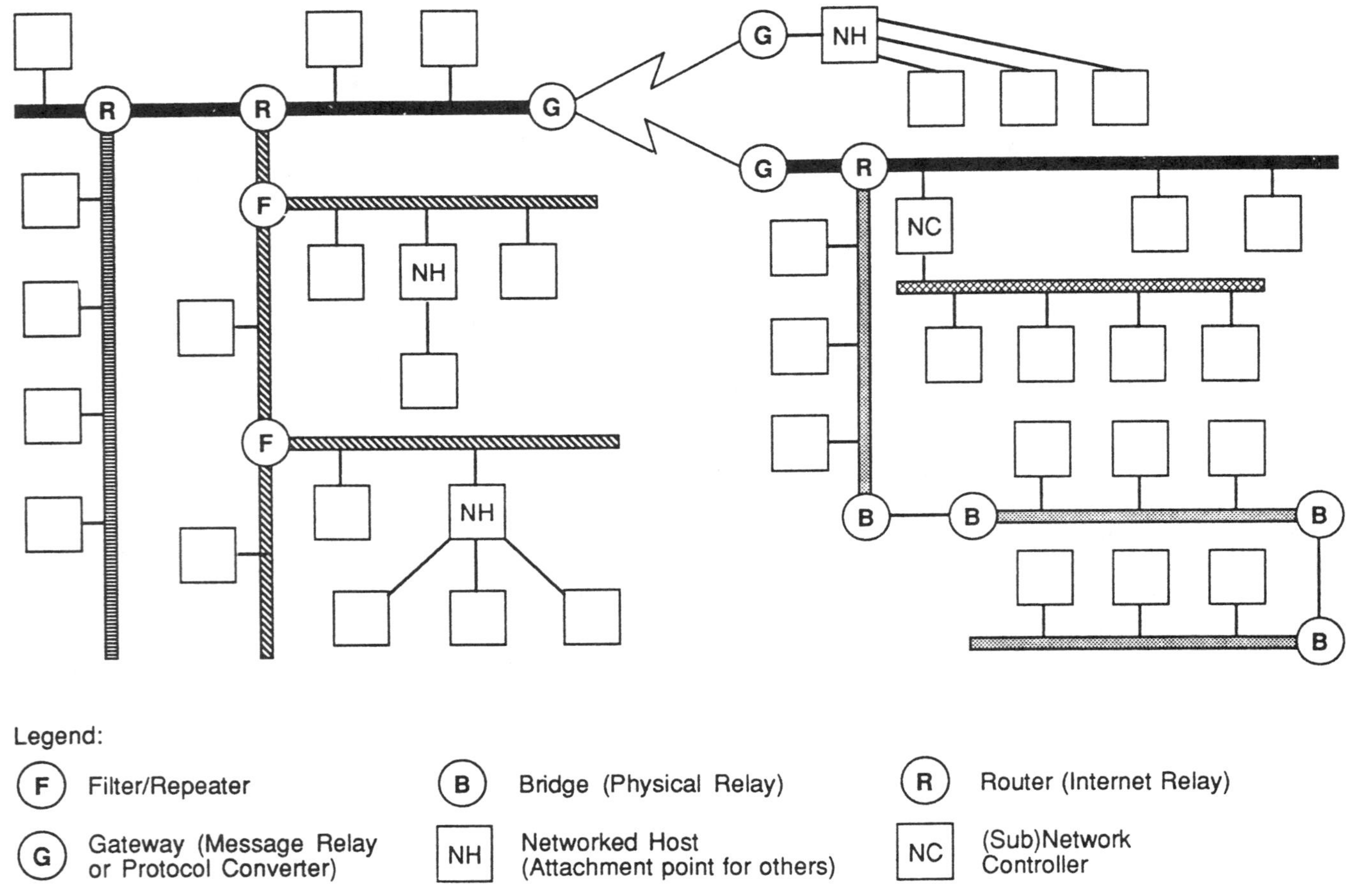

Figure 4. Generalized CIM network.

become transparent to our interprocess communication paradigm.

The Manufacturing Automation Protocols (MAP) concept of one physical bus connecting all factory-floor stations may be appropriate for some manufacturing facilities. It is not, however, general enough to meet all communications requirements of the CIM systems of tomorrow. However, the 'enterprise neworking' concept, connecting MAP control networks with Technical Office Protocols (TOP) engineering networks, demonstrates that the generalized CIM network architecture is, in fact, currently practical. We believe that this will lead customers to demand, and vendors to produce, products consistent with that architecture (MAP and TOP 1988).

It is likely that emerging physical networking technologies will, in time, make the physical layer standards selected by MAP/TOP obsolete. This will lead to the addition or substitution of subnetworks with new physical and datalink protocols to current CIM networks. Nevertheless, the transparent, multiple, subnetwork architecture we are advocating should result in little or no impact on CIM networks already in place and on process-to-process communication. At the same time, adherence to at least the layering, but preferably also the intermediate layer protocol suites, in various types of 'gateway' machines, provides for the transparent interconnection of subnetworks based on proprietary, or nonstandard protocol suites in the lower layers.

4.3. Technology

There are now many standard protocol suites for the DataLink and Physical layers, and there will soon be more. They all provide frame delivery and integrity checking; some provide for reliability and recovery, others defer those considerations to the transport layer. The real distinguishing characteristics among these standards are the signalling technologies and the sharing algorithms. Loosely speaking, the signalling technology determines the raw transmission speed, the relative immunity to electronic noise, and the cost. The sharing algorithm determines the nature of network service seen by the station. There are generally three choices:

(*a*) connection to one other station or one other

station at a time, with fixed dedicated bandwidth (point-to-point, time- and frequency-division);

(*b*) connection to multiple stations simultaneously, with variable bandwidth with fixed lower and upper bounds depending on the number of stations connected to the medium (token bus and ring); and

(*c*) connection to multiple station simultaneously, with variable bandwidth from zero to the bandwidth of the medium depending on the traffic generated by all stations connected to the medium (CSMA/CD).

In general, engineering and administrative activities, which have infrequent and variable communications requirements, can tolerate and use the type (*c*) services more effectively. The production control activities, which have frequent and regular messaging requirements, however, prefer type (*a*) or (*b*) services.

There are also several 'standard' protocol suites for the intermediate layers as well. But in this area, the differences are historical rather than functional. It is clear that the existing intermediate layer protocols will be *the* standard in the near future. In the upper layers, standards are still evolving. Here the only problem will be to determine the suite of protocols necessary to a given production management or data management function.

4.4. Topology

Topology is that aspect of network design which concerns itself with the connection of stations to subnetworks and the interconnection of the subnetworks. Topology is at least as important as bandwidth and access protocols in determining the effective performance of integrated networks. Processes which need to communicate frequently should be directly connected, or connected to a common bus/ring, if at all possible. On the other hand, only two factors should really motivate dividing a network into subnetworks:

(*a*) the feasibility or cost of connecting all of the potential stations to the same physical network; and

(*b*) the ability of the single network to carry the total traffic load.

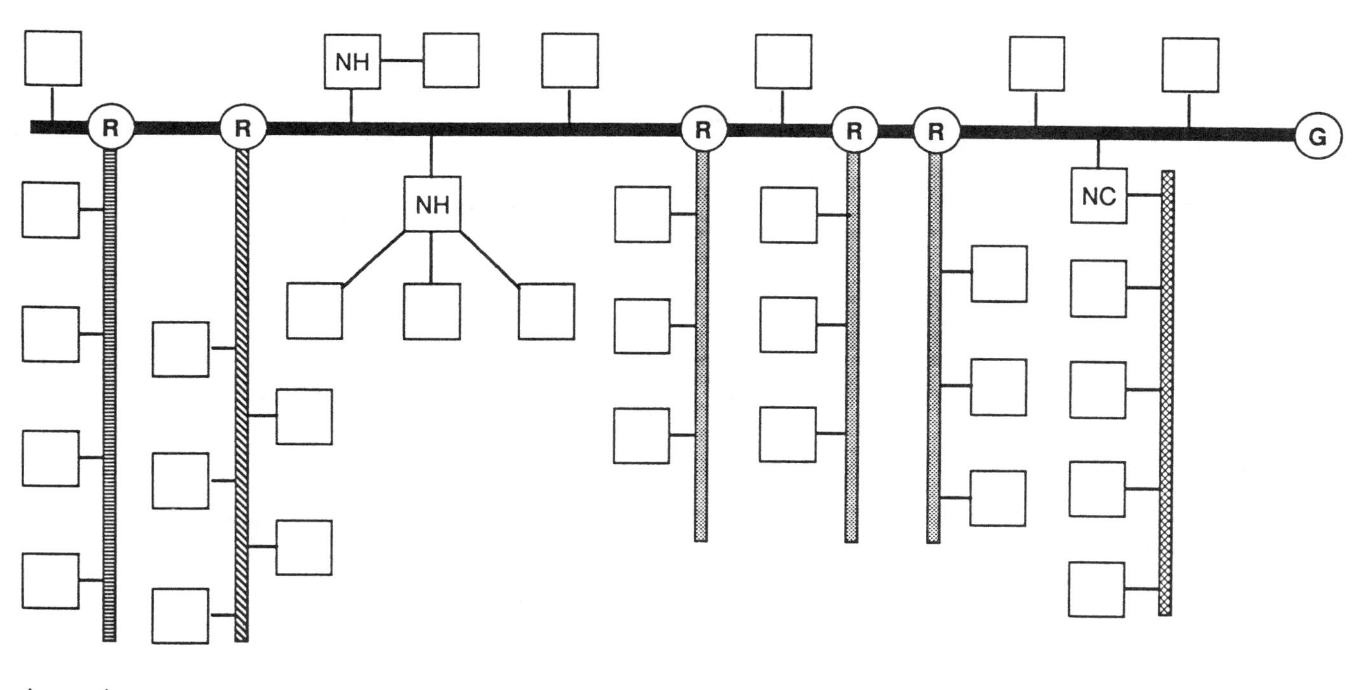

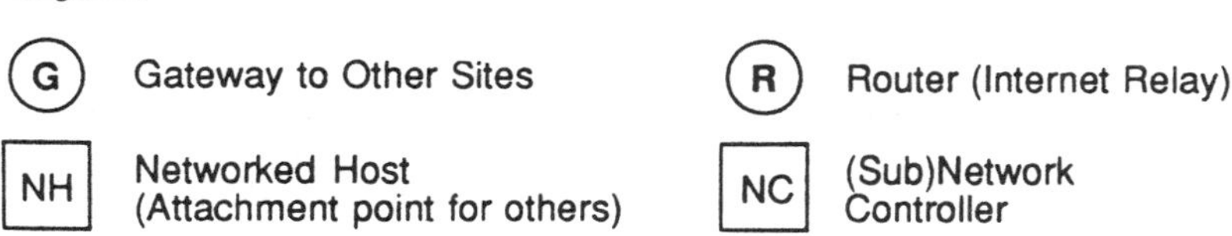

Figure 5. Ideal single-site CIM network.

Several varieties of bus/ring networks have limitations on the total number of stations which can be connected, or on the total cable length. When this limit is reached, partitioning is unavoidable. In addition, the performance of most bus and ring networks is inversely proportional to the number of stations or volume of messages placed on the network. When the performance of a subnetwork degrades the performance of the primary production management or data management functions using it, it is time to partition that subnetwork or replace the networking technology. The former is usually adequate, easier and cheaper, and adherence to the OSI model should make it invisible to the communicating processes.

Ideally, the generalized CIM network architecture in Fig. 4 will be implemented in the much more restricted form depicted in Fig. 5. There is a common 'spine' or 'backbone network' which connects to *all* subnetworks, although some of the subnetworks may be directly interconnected. This architecture guarantees that the maximum number of relays on any process-to-process connection is two. While this is not always practicable, it is, in our view, always the desirable goal for the network architecture of a single site.

5. Summary

In this paper, we have asserted that developing a system architecture is one of the key ingredients for a successful CIM implementation. We have argued that such an architecture has three interrelated management components: production, information and communications. We have discussed many of the problems impeding the integration of the functions within and across these components.

In production management, we examined the problems involved in automating and integrating human decision-making into a hierarchical, distributed, computer architecture. In addition, we pointed out the need for analysis tools to compare different designs before attempting to implement one. In data management, we discussed the impact of the CIM environment on data modelling, database design, and data administration. And, we argued for a hybrid architecture for administration of that data. In the area of communications, we stressed three principles that should guide the design of CIM communication systems. We used those principles in proposing a generalized architecture for CIM communications which is independent of the production and data management architectures.

Our conclusion is that a great deal of work must be done in the areas of system design, automated decision-making, and other manufacturing-related, technologies before the dream of computer integrated manufacturing becomes a reality.

References

ALBUS, J., BARBERA, A. and NAGEL, N. 1981, Theory and practice of hierarchical control. *Proceedings of 23rd IEEE Computer Society International Conference.*

BARKMEYER, E. 1989, Some interactions of information and control in integrated automation systems. In *Advanced Information Technologies for Industrial Materials Flow* (Springer-Verlag, New York) (in the press).

CHU, W. 1986, editor, *Distributed Systems*, Vol. II: *Distributed Data Base Systems* (Artech House Inc., Deham, Massachusetts).

DAY, J. and ZIMMERMAN, H. 1988, The OSI reference model, *Proceedings of the IEEE*, **71** 102–107.

DAVIS, W. and JONES, A. 1988, A real-time production scheduler for a stochastic manufacturing environment. *International Journal of Computer Integrated Manufacturing*, **1**, 101–112.

DAVIS, W. and JONES, A. 1989, A functional approach to designing an architecture for CIM. *IEEE Transactions on System 5. Man and Cybernetics*, **19**(3).

GEOFFRION, A. M. 1970, Elements of large-scale mathematical programming, Parts I and II. *Management Science*, **16**, 652–691.

JONES, A. and MCLEAN, C. 1986, A proposed hierarchical control model for automated manufacturing systems. *Journal of Manufacturing Systems*, **5**, 15–25.

JONES, A. and WHITT, N. 1985, editor, *Proceedings on Factory Standards Model Conference* (National Bureau of Standards, Gaithersburg, Maryland).

LIBOSVAR, C. M. 1988, Hierachies in production management and control, MIT Technical Report, LIDS-P-1734, Boston, Massachusetts.

LIU, J. 1985, The CAM-I advanced factory automation system. *Proceedings on Factory Standards Model Conference* (National Bureau of Standards, Gaithersburg, Maryland).

MAP and TOP Version 3.0 Specifications, Society of Manufacturing Engineers, Detroit, MI, USA, 1988.

MARK, L. 1987, The binary relationship model—10th Anniversary, University of Maryland Institute for Advanced Computer Studies, Technical Report UNIMACS-TR-87-50.

NOF, S. Y. and MOODIE, C. L. 1989, editors, Recommendations for future research directions in the automation of materials flow. In *Advanced Information Technologies for Industrial Materials Flow* (Springer-Verlag, New York).

OTTAWA, 1986, REFERENCE MODELS FOR MANUFACTURING STANDARDS, Internal Report N51, ISO 184/SC 5/WG 1.

SALEH, A. 1988, Real-time control of a flexible manufacturing cell. Ph.D dissertation, Lehigh University, Bethelehem, Pennsylvannia.

SU, S. 1986, Modelling integrated manufacturing data with SAM*. *Computer*, Jan., pp. 34–49.

THOMAS, G., BARKMEYER, E., BRADLEY, B., CHUNG, CHIN-W., EPSTEIN, R., MCCORD, R., REEDY, C., TEMPLETON, M., and THOMPSON, G., 1988, Heterogeneous distributed data systems for production use. ACM Computing Surveys Special Issue on Distributed Data Systems, Association for Computing Machinery, New York.

THE INTEGRATION OF WORK REDESIGN AND INFORMATION TECHNOLOGY

John J. Powers, Director, Management Services Division
and
Anthony T. Liotti, Director, IE Technology Unit

EASTMAN KODAK COMPANY
Rochester, New York

When Work Redesign and Information Technology are integrated, they have tremendous customer impact, and the nature of the IE discipline/associated skill sets are enhanced.

THE TIMES -THEY ARE A CHANGING

We are seeing a significant change in Industrial Engineering. There is a difference between what industrial engineers used to do and what they will be doing. This difference requires integration, particularly the integration of information technologies into our bag of skills, plus the systemic transformation of business and manufacturing processes - a whole new horizon for Industrial Engineering. We can be even more effective for our customers.

Our profession is in real limbo right now! Our traditional functions are being performed on the shop floor by manufacturing engineers and applications engineers, and in many cases, by the operating personnel themselves. We have done our job of transferring skills to others very well. However, the unique discipline we used to know as Industrial Engineering is fading away. And, in these days of the hard-nosed bottom line, Industrial Engineering is more and more being observed as a "soft" discipline, one whose value-added component is not crystal clear.

What if we were able to go in and look at all the elements of the business and manufacturing processes, take a broader view of the customer's problem, break down the walls and lead the sacred cows to slaughter,

Reprinted from *1991 IIE International Industrial Engineering Conference Proceedings*

integrate more technologies (behavioral, analytical and traditional), and finally add Information Technology to the Industrial Engineer's arsenal. This change in concept, this integration, is not only what the customer needs, but is the future of Industrial Engineering.

Most Industrial Engineers are already involved in the redesign of work systems, particularly in the areas of ergonomics and work flow. But what if it were possible to broaden our skills so we could participate in areas like identifying the needs of the customer, like influencing product and process design, like defining acceptable quality? What if we influence information systems so they could provide valuable strategic information in a timely manner to the people whose job it is to make key decisions? What if we could contribute to a changing corporate culture and the empowerment of the work force? Even effect management processes such as global sourcing, management of Research and Development, and sales and marketing decisions? We can do all these things and more, but we need a broader understanding of what integration is.

Integration applies to the complete business and manufacturing cycle and it involves a multiple set of disciplines from research and development, strategic planning, development, manufacturing, distribution, business processes, sales, and ultimately delivering product to the customers.

As we know it now, Industrial Engineering is heavily concentrated in manufacturing and represents typical IE activities like materials flow, methods analysis, facilities planning, material handling, analytic methods, and so on.

We see three directions in which we can grow this Industrial Engineering function and make it more valuable through integrating new skills and disciplines. One way is to grow in THE MANUFACTURING AREA, expanding our services by bringing in new skills and disciplines, (computer and information technology, organization development, interpersonal consulting skills) to create technology integration. This surely will give new dimensions to IE products and services.

We can also grow by expanding our customer impact potential through "customer site integration", being sensitive to the state of the customer's technical system and its components, as well as the social system and its components, understanding the situation from the customer's view and implementing change with a global system perspective.

Finally, we can expand by applying our traditional skills and our expanded set of integrated products and services just developed to every other area of the business system and looking for new opportunities from beginning to end.

In this paper we are going to be concentrating on Information Technology. But remember, the integration of Information Technologies is more than CAD/CAM, bits and bytes, hardware and software. We believe this particular type of integration gives a new perspective and a new future for the Industrial Engineer.

Why do we need this integration? Because business as usual is killing us. Our traditional weapons of process rationalization and automation are no longer producing the desired results. And, we are overlooking an even more powerful weapon - TIMELY INFORMATION.

It is the life's blood of business. But even though we live in this age of the computer, there is so much more opportunity and potential than we have taken advantage of. We are, in some cases, simply using our contemporary information systems to shuffle paper rather than getting the right information to the right place at the right time to make critical decisions in a timely manner.

We can begin to understand the potential value of our Information Systems if we define it as anything a computer can do.

For example, "flow" is a hot concept these days. The more we think about flow, the more we realize that every single piece of information about flow is contained somewhere in our Information Systems. And this information is neither remote nor inaccessible. These days it seems we are never more than a few feet from an Information System. But in order to use this information, we have to INTEGRATE it. Just as critical corporate activities cut across departmental lines, information should, too! We should discard old notions about how the computer can and should be used. An increasing number of American companies are discarding old ideas about how the computer should work and integrating Information Technology into their work design.

The Integration of Information Technology into Work Design currently goes under a number of names. The bottom line is that each of them uses Information Systems as a basic tool in work redesign. Three of these are:

- REENGINEERING
- CONCURRENT ENGINEERING
- Focus on TIME as a source of COMPETITIVE ADVANTAGE

Right now we will take a quick look at the basic characteristics of each of these three approaches.

REENGINEERING

The byword for REENGINEERING is "DON'T JUST COMPUTERIZE, BUT CHALLENGE...REDESIGN! The Reengineering concept is based on the premise that most work designs and organizational structures date back to the days before modern computer technology. The automation of existing processes is simply pouring concrete over the past, leaving existing processes in place. Simply using the computer to speed them up is not enough any more because the fundamental problems still exist. Reengineering would have you obliterate the past and start at square one. Challenge everything! Lead the sacred cows to slaughter! Ignore departmental boundaries! Rethink your businesses, processes and the systems that support them.

Combine the power of modern Information Technology and Industrial Engineering to achieve dramatic change and dramatic results.

The essence of REENGINEERING is looking at things from a brand new perspective, and breaking away from old rules and assumptions. After all, isn't challenge, creativity, and coming up with new ideas what IE is all about?

Sounds like a blueprint for an improved kind of Industrial Engineering, don't you agree? Today, more and more companies are adopting the principles of Reengineering because Reengineering has the ability to make large amounts of time, cost and capital simply disappear.

CONCURRENT ENGINEERING

The byword of CONCURRENT ENGINEERING is "DON'T TOSS IT OVER, BREAK DOWN THE WALLS". Too often, the Industrial Engineer lives in what is called a FUNCTIONAL SILO. We are educated and spend our careers in relatively narrow disciplines. As a result, we often tend to apply that ONE discipline to EVERY problem that confronts us.

CONCURRENT ENGINEERING applies to the product introduction cycle and is an outgrowth of efforts to meet Japanese competitors who are introducing three to four products to our one. Concurrent engineering means designing the process and the product at the same time. It is imperative that the IE gets involved at these earliest stages and become part of the broader process to be most successful because the costs of manufacturing and process flow are often locked in by early design decisions. Concurrent engineering is not really a technology issue. It is more a communications and a people issue. It calls for integrating across functional disciplines simultaneously rather than throwing things over the wall sequentially to people who may not share the same perspective.

IEs should be involved in facilitating the process in many ways, by providing the communications link between the involved organizations, by coordinating project timelines and milestones, and by contributing to the design of the manufacturing process. And that is CONCURRENT ENGINEERING -- DON'T TOSS IT OVER, BUT BREAK DOWN THE WALLS. Get out of FUNCTIONAL SILOS and move from a sequentially oriented, fragmented approach to concurrent product and process design.

TIME AS A COMPETITIVE ADVANTAGE

The third concept is the concept of TIME AS A COMPETITIVE ADVANTAGE. This concept holds that TIME IS A MORE CRITICAL COMPETITIVE YARDSTICK THAN TRADITIONAL FINANCIAL OR PRODUCTION MEASUREMENTS. Today, there is a new generation of companies using rapid-response strategies that they find more powerful than traditional strategies like economies of scale and cost reduction.

To these companies, time is everything. By reducing time consumption in every area of their business, they also reduce costs, improve quality, and most important, stay closer to their customer. The time-based manufacturer shortens his production run as much as possible because reduced run lengths mean more frequent cycles of the complete product mix and faster response to customer demands.

Sure, we have all read and seen example after example of textbook JIT operations. Not always obvious is the requirement that precise information is necessary at the right moment and at the right place. This cannot be accomplished without trust and reliance on information technology applications. It just does not work without it. A company that can bring new products to market three times faster than its competition certainly enjoys a huge competitive advantage.

To Kodak, and to the rest of the time-based competitive advantage community, time is everything. Save time, and the costs, quality, and responsiveness will take care of itself.

INDUSTRIAL ENGINEERING AND INFORMATION SYSTEMS

An essential part of this discussion about the integration of Industrial Engineering and Information Technology is the role the IE should play in the

Information Systems environment. The traditional Information Systems function is computer systems design. The traditional Industrial Engineering function emphasizes needs analysis.

How do you integrate the two? You will find that Information Systems and Industrial Engineering have a symbiotic relationship. One feeds off the other. Take an Information System project life cycle. Early in the development phase, IE support is valuable in helping clients develop their information management and system strategies. This is followed naturally with assisting in defining customer business requirements. At the end of the cycle, the focus is on helping to integrate the new system into the customer's operations.

If the Industrial Engineer is to really influence change with Information Technology, (s)he must do several things. The first is to get some IS know-how -- enough to employ commercially available tools to their work. The second is to identify areas of opportunity for their customers and encourage these customers to use some of these tools to simplify and improve their day-to-day business. Another area is to be aware of how customers are using existing tools. Many older systems are cumbersome and hard to learn. These systems may be candidates for a graphical user front-end -- an interim technique that extends the use of older systems while improving the ease of use for the end-user.

Finally, the IE needs to familiarize himself or herself with the key concepts of Information Technology, the corporate direction, and the "experts" within both local and corporate groups. They should attend overview presentations on areas of Information Technology, and read some of the professional journals that deal with business and Information Technology. In all areas, the IE should seek out and employ an IS partner for professional consulting and implementation. So break down the barriers that exist between the IE and IS communities! Get out of your functional silos!

If you think we have been describing an expanded role for the Industrial Engineer, a role as a partner with the IS community in support of the customer, you are absolutely right. Manufacturing and business process improvement requires the power of data analysis and information technology. And the new role of the Industrial Engineer is that of an INTEGRATOR. As a matter of fact, we have written a theoretical job description of the Industrial Engineer as an INTEGRATION ENGINEER.

THE INTEGRATION ENGINEER

AN INDUSTRIAL ENGINEER WITH THE ABILITY TO INTEGRATE, POSSESSING THE BACKGROUND AND THE CAPABILITY TO THOROUGHLY ANALYZE AND DEFINE A PROBLEM WITH THE CUSTOMER; TO ACT AS AN INTERFACE BETWEEN THE CUSTOMER AND THE SYSTEMS COMMUNITY; TO SELECT AND COMBINE VARIOUS TECHNOLOGIES INCLUDING INFORMATION TECHNOLOGY (THE BEST FIT SOLUTION FOR THE CUSTOMER). This Integration Engineer is change oriented; an agent of change, sensitive to the customer's social and business system with a specialty BUT willing to get out of the functional silo. (S)he is willing to test new theories and willing to seek out and partner to fully utilize the skills of others and look at the entire process.

Would you qualify under our definition? Perhaps -- but, remember, it will take a bigger than average perspective shift. It will take the ability to see and work across functional boundaries.

Now that we have defined what an INTEGRATION ENGINEER really is (an agent of change), where is the potential, the real opportunities? And how do you identify them? Where can this expanded, integrative view be put to advantage?

Look at the process cycle. Can you identify where the bottlenecks are for your company? The process variability? The decision points - where the right information should be for the most effective and timely decisions? One way to challenge the process is to utilize questions provided by the Index Group like: Where are the long sequential processes? Where is there excessive validity checking? Where are the high inventory levels? Where do a large amount of ad-hoc committees exist? These are the areas where a challenge can lead to significant improvements. All of these opportunities for change reflect INFORMATION BOTTLENECKS: Too much information -- too little information -- information with no value added. All of them could be redesigned to leverage the capabilities of information technology.

Let's review some of the Information Systems tools available to today's Industrial Engineer. The range of Information Technology covers everything that you can do with a computer or modern telecommunications. Right now, we are using many IT tools in our traditional functions -- for process modeling, production scheduling and control, materials management, and as analysis and modeling tools.

But what we really want to do is to move these tools beyond mere data gathering. We want to use them to design processes and turn these processes around faster. We should incorporate these tools right into the final solution.

IEs can also play a key role in helping business understand their needs, the constant changes in business and manufacturing processes, and the information people will need to do their job. They must align themselves with IT and IS units to help the business continuously integrate rapidly changing information technologies.

Remember, too, that the IT infrastructure supports end-user needs to access, process, and distribute information WITHOUT formal IS systems or formal, professionally built systems. IEs can help in developing these procedures, as well as advising on information security.

And end-users will be able to do more with commercial packages and the underlying infrastructure WITHOUT the necessity of traditional IS support. Most companies consider Information Technology an important key to their competitiveness. These are only a few of the areas where the IE and the IS functions can work together to integrate Information Technology. INFORMATION SYSTEMS AND INDUSTRIAL ENGINEERING CAN GET TOGETHER BUT IF THE INDUSTRIAL ENGINEER IS TO INFLUENCE CHANGE, (S)HE MUST HAVE SOME SYSTEMS TECHNOLOGY UNDERSTANDING.

INFLUENCING CHANGE

Let's take the concept of integration one step further. All improvements require change. The IE has a role in influencing change as well. It is true, as the philosopher Heraclitus said, that NOTHING ENDURES BUT CHANGE. But, at the same time, change is not universally welcomed. As a matter of fact, most companies have a built-in resistance to change. That is just human nature. Change is unsettling. And it is always easier and more comfortable to keep right on doing things the same old way. The following comment is typical: EVERY FEW MONTHS, OUR SENIOR MANAGERS FIND A NEW RELIGION -- QUALITY, CUSTOMER SERVICE, FLATTENING THE ORGANIZATION -- YOU NAME IT. WE JUST HOLD OUR BREATHS UNTIL THEY GET OVER IT, AND THINGS GET BACK TO NORMAL.

Yes, you will meet some obstacles in your efforts to influence change. Is your organization ready for change? Who is backing the proposed changes? Is there enough muscle behind your efforts? Are your people willing to change? Have they been properly motivated? Can they see what is in it for them? Who stands to be affected by the change? (Because they are likely to provide the greatest resistance.)

You may have to play a behind-the-scenes advocacy role helping to convince senior management of the change potential such as the power Information Technology can have when integrated with work design. And never forget that even though you are able to affect a process change, the behavior of the people will have to change before the process can change. This calls for some interpersonal, organizational development skills. You must persuade rather than instruct, convince rather than dictate. You need to help align the system. This will require a partnership. So you may need help from your Human Resources counterparts to revise incentives and recognition and reward systems.

TRAINING

The more people know about something new, the less their fears. This is especially true in the IT arena.

Training will help build a confidence level in Information Technology -- what it is, and what it can do for them. This should be vertical training stretching from management to staff to the shop floor. Remember, the integration of Information Technology is just as much a human change, or a behavioral change, as it is a technology change. Now, let's repeat something that is critical to our discussion.

INTEGRATION IS NOT A TECHNOLOGY ISSUE

The technologies are here waiting to be used. There is plenty of Information Systems know-how and plenty of Information Systems capabilities at our disposal. But that does not mean you should storm into the IS Department and try to take over. Cooperation and teamwork are essential to the effort. Our challenge is to integrate aspects of Information Technology, Information Systems, decision making and behavior change into Work Redesign.

We have presented some of the key concepts of this effort:

- STOP DOING BUSINESS AS USUAL!
- DON'T PAVE OVER THE PAST BY AUTOMATING IT!
- GET OUT OF OUR FUNCTIONAL SILOS!
- DON'T JUST COMPUTERIZE! CHALLENGE! REDESIGN!
- DON'T TOSS IT OVER -- BREAK DOWN THE WALLS.

- TIME AS A COMPETITIVE ADVANTAGE.
- PROCESS CHANGE WON'T WORK UNLESS BEHAVIOR CHANGE, TOO!

We have also learned more about the opportunities for change. Throughout this message is embedded the notion of teamwork -- teamwork in the broadest sense.

<u>SUMMARY</u>

We have only begun to explore the implications of integration. There are tremendous opportunities everywhere. And new life and new power for the Industrial Engineer. The Industrial Engineer of the future, regardless of the title that he/she carries, will be more and more concerned with the redesign of business processes using Information Technology. The companies that are successful in using Information Technology to redesign their processes will be the companies that will succeed in the future.

Sure, business redesign is difficult and risky but doing business as usual is even riskier. There are great opportunities before us. Think big! Be a leader! The Industrial Engineer is still the primary catalyst for change in the American Industry. AND THE TIMES THEY ARE A CHANGIN'.

John Powers is currently Director of Management Services Division, a division comprising the Industrial Engineering and Information Technology functions. He has B.S. and M.S. degrees in Industrial and Systems Engineering from the University of Florida. John is a member of the Research Advisory Committee for IIE, a member of the Council of Industrial Engineers, and past president of his local IIE Chapter.

Tony Liotti is a Unit Director of IE support to Kodak business units. Formerly, he was Director of the IE Technologies Unit. Tony has a BS Degree from Notre Dame and an MBA from Syracuse University. He has been active in the Rochester Chapter of IIE. In recent years, he has been on the program committee for the national IIE Managers Conference and was chairman of that conference in 1990.

The Benchmarking Bandwagon

As AT&T and Alcoa have found, this bandwagon can lead you on a trail to continuous improvement.

by Karen Bemowski, associate editor

SUPPOSE THAT MR. PUTTPUTT, AN AVID golfer, has the opportunity to attend a golf workshop that features some of the best golfers in the world. He has the basics down but wants to improve what is currently giving him the biggest headache—putting. Mr. Puttputt knows that one of the world's best putters, Mr. Eagle, is going to be there, so he reads Mr. Eagle's book and watches Mr. Eagle's videotape in anticipation—he doesn't want to waste valuable time asking already answered questions. While reading the book and watching the videotape, Mr. Puttputt analyzes his own putting stroke in his mind (or in the living room, if no one is around).

Finally, the day arrives. At the seminar, Mr. Puttputt meets Mr. Eagle, asks questions, and gets all the information he can (along with an autograph). On the way home, Mr. Puttputt mulls over all he has learned, deciding what will be most helpful in improving his putting. The very next morning he is on the green, applying what he learned in hope of becoming as good—or even better than—Mr. Eagle.

Mr. Puttputt's experience with Mr. Eagle is an example, albeit a simple one, of benchmarking. In more conventional terms, benchmarking can be defined as "measuring your performance against that of best-in-class companies, determining how the best in class achieve those performance levels, and using the information as the basis for your own company's targets, strategies, and implementation."[1]

Growing Interest

Benchmarking is by no means a new concept in the business world, but it's becoming a popular one. "There has been a substantial increase in interest in all phases of benchmarking, ranging from companies wanting to find out what benchmarking is to companies wanting to know how to actually implement it," said Robert C. Camp. Camp is well-versed in the subject; he started the benchmarking program at Xerox Corporation and wrote *Benchmarking: The Search for Industry Best Practices That Lead to Superior Performance.*

There are two main reasons for this increase, according to Camp. First, benchmarking has become a hot topic in the quality field; second, it is a criterion of the Malcolm Baldrige National Quality Award application. "In fact, benchmarking is fundamental to the entire award process because the process is for companies to, in effect,

AT&T's 12-Step Benchmarking Process

The benchmarking process used by AT&T's Material Management Services Division has 12 steps, which are divided into two distinct categories. Steps 1 through 6 are referred to as the first-things-first steps because they help prevent barriers that could hinder or even destroy the benchmarking process. Steps 7 through 12 are called process steps because they outline the process by which benchmarking is carried out.[1]

1. Determine who the clients are

The clients—the people who will use the benchmarking information to improve their processes—vary depending on their companies' organizational structures. AT&T has two sets of clients: process owners and planners, who are responsible for the continued improvement in their divisions, and business units, which are the divisions' end customers.

2. Advance the clients from the literacy stage to the champion stage

Clients are taken beyond just knowing what benchmarking is to visualizing how it will help them develop best practices. This helps develop client support and patience.

3. Test the environment

Time is spent with clients to determine the extent of their buy-in and commitment of resources, resulting in realistic expectations and exposing barriers.

4. Determine urgency

The sense of urgency within the client's environment is determined to see what degree of optimism it has for the project. Panic and a so-what attitude are formidable barriers to successful benchmarking. In these cases, benchmarking degenerates to the level of tours, and little objective fact-finding or self-appraisal is done. The most conducive situations for benchmarking are when the client is reengineering processes or striving to achieve its vision of becoming the best in class. The middle ground is made up of clients who want to evaluate themselves or continue improving in their areas.

5. Determine scope and type of benchmarking needed

The scope and type of benchmarking needed depends on the clients' sense of urgency, their environment, and their understanding of, and willingness to commit to, the benchmarking process.

The scope—the required time, people, resources, etc.—is proportional to the potential payback. For example, benchmarking a simple task takes a lot less time and manpower than benchmarking an operations process, but the payback will also be less. AT&T divides scope into four levels, as shown in Figure 1. The amount of space in each compartment represents the complexity and amount of time needed to implement the benchmarking project, as well as the potential payback.

▼ Figure 1. Scope

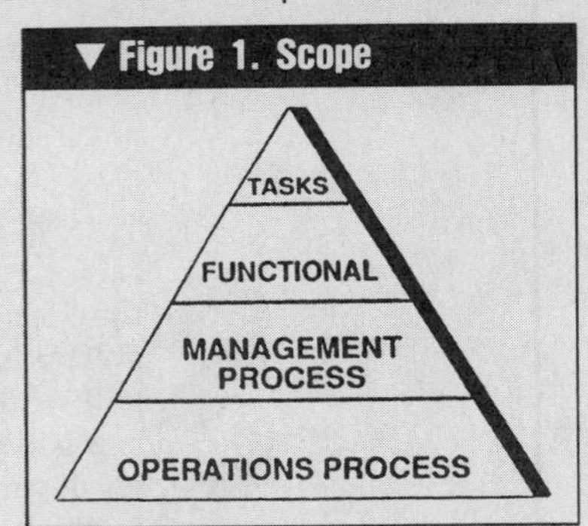

The type of benchmarking determines what type of organization will be benchmarked. AT&T benchmarks three types of organizations: best-in-class performers regardless of industry and location, internal-best performers, and competitive-best performers.

6. Select and prepare the team

Working with the client, a benchmarking team of six to eight members is selected. The team is responsible for putting the benchmarking proposal together and integrating the resultant recommended actions into the business plan so they can be implemented. However, only two to four team members actually visit the organization being benchmarked. All team members are fully trained in the benchmarking process.

7. Overlay the benchmarking process onto the business planning process

The team reaffirms that benchmarking is accepted by upper management as part of the business planning process.

8. Develop the benchmarking plan

The degree of organization and teamwork developed before the visit affects the team's effectiveness. When developing the benchmarking plan, the team:

- prepares a mission statement that formalizes expectations.
- prepares for data collection. Roles are assigned to team members, and the subject being benchmarked is thoroughly analyzed. The team identifies work processes to be studied and develops critical, open-ended questions.

measure themselves against the criteria and substantiate that they really are the best in their area," said Camp. "Benchmarking is the standard by which they will be judged as the best."

Thus, it is no coincidence that many companies with effective benchmarking programs have won the Baldrige Award, including Xerox, Milliken & Company, and IBM. However, the purpose of benchmarking is not to help win the Baldrige Award, but to improve performance. That is the reason why AT&T's Material Management Services (MMS) Division and Alcoa started their benchmarking processes. In fact, both started benchmarking before the Baldrige Award included the benchmarking criterion.

"AT&T went into benchmarking for the right reason: to improve our business and culture," said Darel Hull, manager of transportation and planning for AT&T's MMS Division. Hull was instrumental in developing the benchmarking process used in his division, which provides distribution services

- develops a profile for selecting the benchmark partners (i.e., the organization being benchmarked). The team determines which benchmark performance characteristics are most important to the client's interest, to the specific problems and work processes identified for benchmarking, and to any performance aspects of special importance (e.g., key measurements, similar volumes of activity, customer bases, product types, sales channels).
- does research. The team spends a considerable amount of time at a library or research center learning about the organizations being considered for benchmarking. This research is used not only to select an appropriate benchmark partner, but also to prepare properly for the site visit once the partner is chosen. It does not, however, replace the actual visit.
- develops scripts. The team develops written scripts to help organize and manage the site visit. These scripts include both the in-depth analyses of the functions, processes, tasks, etc. identified for benchmarking and the open-ended interview questions.
- describes present operations. The team describes how the client performs the function being benchmarked. Answering the script questions usually produces a solid description of current operations.
- indicates metrics. A well-documented set of metrics is included in the present operations description. Comparisons of metrics are used to select the benchmark partner and understand that organization's performance.
- sets up visits and protocol. With the planning done and the prospective benchmark partner chosen, the team contacts the prospective benchmark partner. A letter requesting a visit is followed by a phone call to discuss the intent in greater detail and to assess the prospects of benchmarking this organization. Once the organization agrees to be a benchmarking partner, it is sent the script to help it prepare for the visit. The sales, service, and other departments within the client organization are told who the benchmarking partner is since much can be learned from these departments and their protocol.

The visit is then made. The team understands that it is participating in an intelligence patrol, not a combat patrol. The team does not defend how its client performs the function being benchmarked. The team makes sure that each question is answered and documented.

9. Analyze the data

The information gathered is compared to the client's present operation to determine where improvements can be made. Where possible, findings and opportunities are quantified. To ensure a successful analysis, the team:

- organizes its visit documentation into flowcharts, narratives, matrix formats, comparison charts, etc. to clearly summarize the findings. This is done for each visit. The results are then integrated into one analysis summary, which is distilled into opportunities for quantification.
- makes sure the present operations description is accurate and effectively compared to the benchmark findings.
- makes sure that sound quality principles are followed.
- avoids follow-up visits to the benchmark partner (although revisits might be necessary to collect detailed data).
- identifies opportunities for improvement.

10. Integrate the recommended actions

The client takes the team's recommended actions and integrates them into the planning, budgeting, financial, service, and other applicable processes. How this is done depends on the client's vehicles for implementing and tracking change and service costs. The recommended actions are accounted for in the client's budget.

11. Take action

The client implements the action outlined in the various planning processes. The normal procedures for implementing change are followed. Ownership is assigned and progress is tracked for the improvement process.

12. Continue improvement

Once the opportunities are seized, the client makes sure that continuous improvement activities are in place by institutionalizing benchmarking in its planning and continuous improvement processes. Benchmarks are periodically recalibrated because they change as new leaders emerge.

[1]Steps 5 through 8 are described in greater detail in Robert Camp's book, *Benchmarking: The Search for Industry Best Practices That Lead to Superior Performance* (Milwaukee, WI: ASQC Quality Press, 1989).

for AT&T's business units.

Hull gets a lot of calls from companies that want to apply for the Baldrige Award. "They say, 'Geez, benchmarking is an important part of the award, so tell me about it. What is it?'" said Hull. "I think some companies are going to be in for a rude awakening when the award examiners come around. You just don't make a few phone calls, tour a warehouse or some other place, and come back and say, 'We were out at company X—everyone knows they are good—and we learned this and this.' It's not going to cut it. The Baldrige examining teams are well-schooled in the overall concept."

Alcoa, which primarily produces aluminum and aluminum products, also became involved in benchmarking to improve processes. "We benchmark for a purpose: to add value," said Tom Carter, vice president of quality at Alcoa. "We use benchmarking to understand what level of performance is really possible and to understand why the gap exists between our current performance and that optimum performance."

Both AT&T and Alcoa wanted to improve their performance to meet competition that, until recently, hadn't really been there. AT&T's unforeseen competition came when the telephone industry was deregulated by the government. "Since divestiture, we have been trying to quickly find our way in a very competitive market," said Edward Tracy, the operations vice president of distribution for AT&T's MMS Division. "Even our internal customers—19 autonomous business units—began asking questions on how our services and costs compared with outside suppliers'."

For Alcoa, the competition was out there, the company just didn't see it coming. Alcoa is as old as the aluminum industry itself, 102 years, and developed much of the industry's technology and processes—it even built many of its competitors' plants. Because of this involvement, said Carter, Alcoa developed a mind-set of not needing to look outside since there was nothing it didn't already know. "Back in the 1930s and 1940s that was probably true," Carter said. "In the early '80s, however, it wasn't. Customers came to our doorstep saying, 'This is what the competition is doing.' They were also saying, 'I need you, as a supplier, to be responsive in these areas.'"

Neither company buckled under the pressure of their new competitive environments; instead, they fought back. In 1988, after about a year of implementing other quality improvement processes, AT&T's MMS Division became aware of benchmarking. Wanting to learn more, it conducted research and hired an outside consultant. After further research and consultation with several other benchmarking experts and academicians, Hull designed, developed, and implemented a benchmarking process over the course of a year.

Alcoa's story is similar. In 1987, it recognized that the corporation wasn't working together as a whole. Employees weren't using the same language, and there wasn't a uniform approach to quality. Alcoa set out to remedy this. By the middle of 1988, it had launched a more structured approach to quality, which included the practice of benchmarking. Alcoa then benchmarked U.S. and foreign companies to design a formalized benchmarking process.

Their processes

Although AT&T's and Alcoa's benchmarking processes are structured differently, they share some fundamental concepts. (See the sidebars "AT&T's 12-Step Benchmarking Process" and "Alcoa's Six Steps to Benchmarking" for detailed descriptions of these processes.) Both processes stress the importance of:

- having an environment conducive to benchmarking.
- doing homework before benchmarking a company.
- using the information gained from the benchmarking project.

Create the right environment

Having an environment in which managers and employees are receptive to benchmarking is important, according to Tracy, Hull, and Carter. Both AT&T and Alcoa had quality processes in place that helped create this environment before they began benchmarking.

"If AT&T had not been into quality, I'm not sure we could have pulled off benchmarking because of the culture that is

Alcoa's Six Steps to Benchmarking

The benchmarking process used by Alcoa has six steps:

1. Deciding what to benchmark

The project sponsor (the owner of the product, process, or service) identifies potential topics to benchmark. To check the relevancy and validity of a topic, the following questions are answered:

- Is the topic important to the customers?
- Is the topic consistent with the Alcoa's mission, values, and milestones?
- Does the topic reflect an important business need?
- Is the topic significant in terms of costs or key nonfinancial indicators?
- Is the topic an area where additional information could influence plans and actions?

The output from this step is a purpose statement that describes the topic to be benchmarked and guides the activities of the benchmarking team.

Benchmark topics are also selected when teams working through Alcoa's eight-step quality improvement process find themselves asking, "Has anyone ever faced a similar problem? What was done about it?"

2. Planning the benchmarking project

A team leader is chosen. The project sponsor is the ideal team leader, but if that person cannot assume this role, he or she can designate the team leader. The leader, who will be responsible for seeing that the project is successfully completed, should have the authority to make changes in processes, products, and services based on benchmarking information.

Next, the team members are selected, based on the range of skills needed for the benchmarking project. The team's first task is to refine the benchmarking purpose statement by answering these questions:

- Who are the customers for the study?
- What is the scope of the study?
- What characteristics will be measured?
- What information about the topic is readily available?

Finally, the team submits to the sponsor a project proposal that includes all the information obtained up to this point. Once the project sponsor approves the proposal, the team moves to step 3.

Alcoa's benchmarking symbol evolved from two important aspects of the benchmarking process. The six ascending steps, corresponding to the six steps of the benchmarking process, represent the accumulation of both internal and external information that can be used by an organization to improve its performance. The surrounding hexagon, with the six step numbers, illustrates that, through repetition of the benchmarking process, there exists the potential to uncover many of the root causes of superior performance.

3. Understanding your own performance

This step is devoted to self-study. The team examines the factors that influence performance to learn which characteristics are most important and which are least important. The team also learns what data relate to the important characteristics and how to collect and measure those data. This process itself might reveal new ways to overcome specific barriers. The collected performance data create the baseline and structure for benchmarking comparisons.

4. Studying others

In this step, the team:

- identifies benchmarking candidates.
- narrows the list to a few candidates.
- prepares general and specific questions.
- decides the best way to get those questions answered.
- performs the study.

Also included in this step are some guidelines regarding ethical and legal issues that might arise during the study.

5. Learning from the data

The team analyzes the data collected, quantifies performance gaps, and identifies which pieces of information might be particularly useful for improving performance.

6. Using the findings

The team works with the project sponsor to determine how the benchmarking findings can best be used and what other organizations in the company can benefit from its work.

A Working Relationship

The relationship between the company doing the benchmarking and the company being benchmarked is usually not publicized. "Benchmark partners are not looking for recognition," said Darel Hull, manager of transportation and planning for AT&T's Material Management Services Division. "Both companies mutually value each other. It's a very private working relationship. When one professional sits across the table from another, there is a certain camaraderie, a certain bonding that's there. Benchmark partners like to talk about what they are doing, just as we do."

Both AT&T and Alcoa share information on their processes with the companies they are benchmarking, helping the companies improve their processes. Hull said that AT&T is planning to go one step further: "In 1991, we plan to have quarterly meetings and roundtable discussions with benchmarking partners about what we are doing and what we are learning."

needed," said Tracy. "You need to understand that benchmarking is a vital piece of the quality process. You need to understand quality principles, and you must have the necessary quality skills, structure, and environment in place."

"If don't have a quality process," added Hull, "you better take a realistic assessment of what the biases are in your company—how threatening it might be for whoever is responsible for a function to put himself under the test. If he finds a process that is drastically better than his own and is threatened by it, he might take the incentive and cover it up. One of the biggest problems that you must look out for when you get into benchmarking is managers who are defensive and want to use benchmarking to prove how good they are."

Carter said that, to have a successful benchmarking process, people must feel comfortable with learning about others who are better than themselves. He indicated that those involved with benchmarking usually progress through four phases:

1. Don't want, don't ask. People are happy with the status quo.
2. Don't want, but ask. People become involved with benchmarking just because it's popular, but they don't listen.
3. Want and ask. People ask and listen, but are uncomfortable and a bit defensive. Initially, people might be embarrassed to find someone doing something better, so they defend their actions. They might even try to find people who aren't as good as they are because it is gratifying.
4. Seek, desire, listen, and use. People have matured and are not defensive; they take a progressive stance. They have enough self-confidence to seek companies with better processes. They view finding better processes as opportunities for improvement.

Do the homework

Both AT&T's and Alcoa's benchmarking processes include steps for effective planning. "Many people think that benchmarking is simply going out and studying other companies," said Carter, "but you must do a lot of homework before you go out."

The homework includes selecting a benchmarking topic and benchmarking team, defining what functions or processes are to be benchmarked, and analyzing those functions in one's own company. The latter is a vital, but sometimes difficult, step, said Hull. "Finding out your own current practices can be as educational as finding out others'," he said. "But this step is the hardest to facilitate because people generally don't like spending the time documenting or flowcharting what they do.

"Seventy percent of the benchmarking project's success depends on how well you plan it. This includes selecting the appropriate benchmark partners. Obviously, if you go out and benchmark a dud, you're not going to learn a whole lot."

"A lot of people try to find companies that look very much like their own," said Carter. "There is some risk in that. There is a desire to copy what you see rather than understand and translate it. You can copy something without understanding it—but that's the short road to disaster. Robert Camp's book and other literature indicate there are probably a lot of companies *outside* your industry that, in fact, might be the best at something. We found that to be true."

Use what you have learned

Both Hull and Carter indicated that having the benchmarking information sitting neatly in a pile on someone's desk is not going to improve a company's performance. The benchmarking process must allow the company to transform what was learned into recommended actions, integrate those actions into its planning process, and then implement them.

"Our benchmarking process supplements a number of other processes we put in place," said Tracy. "It doesn't stand alone as something unto itself."

The rewards

Tracy feels that AT&T's benchmarking process has been "not just mildly successful, but enormously successful. What this process enables us to do is to identify voids. It's a structured discipline for analyzing a process to find improvement opportunities. Just 12 months ago I was skeptical. But, when I saw it in action, I realized the benefits of the process."

"I feel more confident about our ability to meet the challenges facing us this year than I did in previous years because benchmarking is generating, and will continue to generate, a lot of ideas," added Hull.

Alcoa is also pleased with the success of its benchmarking process. "Has Alcoa's benchmarking process produced value? Yes. Could it be improved? Yes," said Carter. "We have benchmarked administrative, manufacturing, and service functions in all different areas critical to our business. We have a lot of energy about going out and benchmarking, but we don't represent ourselves as being real mature in this whole process—it's just two years old. We have a lot to learn yet, but we feel good about where we are today."

Companies such as AT&T, Alcoa, and Xerox have found benchmarking to be an important quality improvement tool that, according to Camp, benefits everyone. "Benchmarking is a process that works," said Camp. "People are finding it quite successful, so the practice is expanding. I think that is excellent for the quality movement, excellent for the companies, and excellent for the country."

Reference

1. Lawrence S. Pryor, "Benchmarking: A Self-Improvement Strategy," *The Journal of Business Strategy,* November/December 1989, pp. 28-32.

BENCHMARKING THE SOFT AND HARD SIDES OF MANUFACTURING EXCELLENCE

Michael F. Stautner
General Manager
Engineered Products Div.
Goulds Pumps Inc.
Seneca Falls N.Y.

Joe E. Mize
Regents Professor
Department of IE&M
Oklahoma State Univ.
Stillwater, OK.

ABSTRACT

Successful example of dealing with the soft and hard sides of business excellence through the blending of benchmarked evolutionary management practices (JIT, MRP, CIM, etc.) and a high impact change process.

INTRODUCTION

The situation presented here is typical of many companies in the United States today. The practices that have been developed and used are not intended to be viewed as "the best or right way" but certainly worthy of understanding. The approach focuses on people, their empowerment, and a belief that "good practices will bring good results." The paper focuses on "how things are done" and uses 20/20 hind sight to better explain/understand the practices today than we could while we were in the beginning stages.

COMPANY HISTORY

It is important that you know a little about the company. All of the information is of public knowledge through annual reports and press releases. Goulds Pumps, Inc. designs, manufactures, and services pumps, motors, and accessories for industrial, agricultural, commercial and consumer markets. Industrial markets account for 64% of the company's sales. These include: chemical, petrochemical, refining, pulp and paper, utilities, mining and minerals. Goulds has manufactured pumps since 1848 and today the company is an acknowledged leader in its industry.

In early 1989 it was recognized that even though Goulds was financially stable and growing changes had to be made to meet:

* increasing competitive pressures,
* declining margins, and most importantly
* declining customer service levels in the largest division, Engineered Products Division (EPD)

SITUATION

This paper discusses the change process within that division starting in late 1989 THROUGH the end of 1990. EPD is located in rural up-state New York and constitutes one-third of the company as well as being the birthplace for the other two-thirds. 1990 division sales were approximately $200 million with 1200 employees. Starting in 1987 EPD's markets grew quickly. EPD focused on growth and experienced all of the "typical" problems: reductions in on time deliveries, lead times extended, excessive overtime, lowering margins, and unhappy customers. **The division was still the "major contributor" but needed to change. The challenge was "how to make it happen."**

CHANGES HAPPEN: 1990 SUCCESSES ANNOUNCED

Goulds announced records in sales, net earnings, earnings per share, and orders for 1990. These were the best results in the company's 142 year history. Said President Steve Ardia: "There is no question that we take tremendous satisfaction in reporting these record results. But we take equal satisfaction in reporting our progress toward long term goals...our commitment to Total Quality...support and participation of our people ...improved Customer service...dramatically cutting our business systems response times.". How was this done?

CHRONOLOGY OF THE PROCESS: NOV. 1989

Mike Stautner joined EPD in November, during the middle of the Annual Business Planning process. Clearly there were good plans and objectives for each department, all successful companies have them. What was revealed was a **"lack of consensus" by the division management team** relative to the four steps below:

```
1.WHERE ARE---->3.HOW ARE WE GOING---->2.OUR VISION
  WE TODAY?        TO GET THERE?          FOR 1995+
  *CUSTOMERS
  *MANAGEMENT TEAM        4.TIMING OF CHANGES??
  *ALL OUR PEOPLE           WHAT TO DO FIRST?
```

The key here is the "lack of consensus". Today it is easy to look back and recognize that almost EVERY MANAGEMENT ISSUE, TOPIC, OR DECISION that has been addressed deals with getting consensus on one or more of the four areas shown above.

This common sense and simple four step approach has been around for many years, and while questioned in some current literature, is still a very solid "starting point".

WHERE ARE WE TODAY? (Step 1) The eight members of the EPD management team were relatively new to their assignments, the longest 3 years, the next 17 months, the rest well under a year. The seniority within the team however averaged 9 years, with half under a year. This diverse group needed to become a team, to act as a cohesive unit. This began with a very comprehensive assessment, a rough "benchmark" on where they were TODAY.

They didn't do this using a tremendous amount of research, but LISTENED TO the expanded staff, a total of 65 managers, and gathered limited data. Small group meetings were conducted using flip charts to record the information. As suggested in "Break Through Thinking" by Nadler and Hibino [1] 80% of the situation was already well known and it was easy to define specifics.

The issues dealt with both the HARD and SOFT sides of a successful business:

HARD	SOFT
Credibility with:	Credibility with:
Customers	Division managers
The corporation	Our People
Stockholders	Ourselves
Financial performance	Make people feel GOOD
Controlling your appetite for new orders AND new improvement ideas	Communications
	Sales/Mfg. on same Team
	Sense of common purpose or vision

Looking at the HARD side, it is very easy to say that these issues could all be on the SOFT side, however the difference is that there are long standing ways to measure the HARD sides using BENCHMARKS both internally and externally. **IT WAS CONCLUDED THAT THE SOFT SIDE ISSUES MUST BE DEALT WITH IMMEDIATELY. THE HARD SIDE WOULD FOLLOW.**

The management group needed to become a team, more cohesive, more effective, and able to focus on the issues. They went OFF-SITE for three days of intense team cohesion development. A facilitator helped the EPD team develop "THEIR PROCESS" to reach consensus, increase the effectiveness of meetings, define how to relate to each other, and to define THEIR MANAGEMENT TEAM'S MISSION. That mission made THEM responsible to create/sustain the environment that will allow THEIR people to be successful.

The EPD team needed to "quickly" reach consensus against some traditional external benchmarks relating to all four of the steps: TODAY, VISION, HOW TO, and TIMING. The education of the Team had to be simple and thought provoking, offering a means to "stretch" and to self define terms instead of force feeding.

OPENING THE MIND for Steps 2, 3 and 4
Joe Mize has developed a series of time based relationships of business drivers and commonly used terms. The concept is called "Beyond CIE" and is shown on the next two pages and with further explanation at the end of the paper. Before you start let's define the purpose for reading it. Where is your specific department, where are those departments that you interface with, where is the company as a whole, would anyone agree with YOUR ASSESSMENTS, do you understand the terms and would others agree, where is your Customer base, where is the competition, what are the SOFT SIDES - the HARD SIDES?

DRIVING FORCES	1960's & 70's MRP/JIT	1980's CIM	1990's CIE	21st CENTURY BEYOND CIE
1 PRIMARY EMPHASIS	-REDUCE INVENTORY -IMPROVE DELIVERY	-SPEED DESIGN TO PRODUCTION	-DECREASE TIME TO MARKET	-DECREASE TOTAL BUSINESS RESPONSE CYCLE
2 PRIMARY MEASURES OF PERFORMANCE	-DIRECT LABOR -PRODUCTIVITY -SCRAP	-MACHN UTILIZATION -INVENTORY TURNS	-MFG VELOCITY -RESPONSIVENESS -DATA INTEGRITY -SYSTEM DISCIPLINE	-RESPONSIVENESS -ADAPTABILITY -VALUE ADDED PER FUNCTION
3 COST/ EFFECTIVENESS	-REDUCE PRODUCTION COSTS -FOCUS ON EFFICIENCY	-REDUCE DESIGN PRODUCTION CYCLE COSTS	-REDUCE WASTE IN ALL FUNCTION -FOCUS ON EFFECTIVENESS	-MEASURE OF TOTAL ENTERPRISE EFFECTIVENESS
4 QUALITY	-REDUCE REJECTS AND SCRAPS	-DESIGN QUALITY IN -BUILD QUALITY IN	-QUALITY FUNCTION DEPLOYMENT, SPC -COVERS ALL FUNCTIONS	-EXTEND TO SUPPLIER CHAIN -EXTEND FOCUS TO PRODUCT FUNCTIONALITY
5 PRODUCIBILITY	-DRIVEN FROM MFG	-LOWER THE WALLS BETWEEN MFG AND DESIGN	-CONCURRENT ENGR	-EXTEND TO CUSTOMER BASE
6 SCOPE	-MANUFACTURING -DISTRIBUTION -PURCHASING	-MANUFACTURING, DISTRIBUTION,PURCHASE -DESIGN	-ALL INTERNAL FUNCTIONS OF ENTERPRISE	-LINKAGES BEYOND THE ENTERPRISE
7 DIMENSIONS OF FOCUS	-VERTICAL, WITHIN SINGLE FUNCTION	-HORIZONTAL	-VERTICAL AND HORIZONTAL	- 3-DIMENSIONAL: VERTICAL, HORIZONTAL, DEPTH
8 TIME BEHAVIOR PERSPECTIVE	-STATIC	-STATIC	-INITIAL EFFORTS AT DYNAMIC BEHAVIOR	-COMPREHENSIVE DYNAMIC BEHAVIOR
9 TIME HORIZON	-DAILY,WEEKLY,MONTHLY PRODUCTION SCHEDULE -PRODUCTION CYCLE	-QUARTERLY,YEARLY ACTIVITY LEVEL -DESIGN CYCLE	-PRODUCT LIFE CYCLE	-FIVE YEAR AGGREGATE GLOBAL PERSPECTIVE OF PRODUCT FAMILIES
10 DATA/ INFORMATION	-ISOLATED, DIVERSE	-TOP/DOWN GRAND SCHEMES -PAIRWISE INTERFACING	-INTERCHANGE STANDARDS -INTERGRATION OF FUNCTIONAL CLUSTERS	-ENTERPRISE STANDARDS -COMMON DATA STRUCTURES -INTEGRATED, BUT DISTRIBUTED DATA/ KNOWLEDGE BASES
11 MANAGEMENT CONTROL	-FUNCTIONAL, DIVERSE -DISTRIBUTED, UNCOORDINATED	-FOCUS ON INTERFACES -DISTRIBUTED, COORDINATED	-ENTERPRISE LEVEL ARCHITECTURE	-DISTRIBUTED CONTROL WITHIN HEIRARCHICAL FRAMEWORK
12 USE OF FEEDBACK	-OPEN LOOP	-80% OPEN LOOP	-50% OPEN LOOP -ELEMENTARY "LEARNING"	-80% CLOSED LOOP -ADAPTIVE LEARNING
13 IMPROVEMENT EMPHASIS	-OPTIMIZE ONE VARIABLE AT A TIME, WITHIN A SINGLE FUNCTION	-ATTEMPT TO OPTIMIZE ONE VARIABLE AT A TIME, ACROSS PAIRS OF FUNCTIONS	-MULTI-VARIABLE SATISFICING ACROSS FUNCTIONAL CLUSTERS	-ENTERPRISE-WIDE, KEYED TO STRATEGIC BUSINESS OBJECTIVES

"BEYOND CIE" TEMPLATE
MAPPING CHANGE IN MANAGEMENT PRACTICES
JOE E. MIZE

DRIVING FORCES	1960's & 70's MRP/JIT	1980's CIM	1990's CIE	21st CENTURY BEYOND CIE
14 ORGANIZATIONAL STRUCTURE	-HEIRARCHICAL BY FUNCTION	-HEIRARCHICAL WITH LOWER WALLS -MATRIX	-HEIRARCHICAL WITH WINDOWS -STAFF FUNCTIONS MUCH SMALLER -DYNAMIC PROJECT ORIENTATION	-HEIRARCHICAL, ALL FUNCTIONS ACCOUNTABLE FOR DIRECT CONTRI. TO ENTERPRISE PERFORMANCE
15 BASIS FOR ORGANIZATIONAL STRUCTURE	-EXTREME SEGMENTATION OF FUNCTIONS -EXTREME DIVISION OF LABOR -INCREASING USE OF STAFF FUNCTIONS -ECONOMY OF INDIVIDUAL FUNCTION EXECUTION	-MATRIX STRUCTURES WITHIN MFG -ENTERPRISE-WIDE (HORIZONTAL) FUNCTIONS; e.g.PRODUCT LINE MGRS -COST CENTERS -SOME RE-COMBINING OF FUNCTIONS	-LIFE CYCLE OF PRODUCT LINE -SYSTEMIZATION OF INFO PROCESSING WILL ELIMINATE NEED FOR MANY STAFF FUNCTIONS -ORGANIZE ON BASIS OF VALUE ADDED ACROSS FUNCTIONAL CLUSTERS	-LIFE CYCLE OF PRODUCT FAMILIES -ORGANIZE ON BASIS OF VALUE ADDED ACROSS ENTERPRISE -ORGANIZE ENTERPRISE AS A FEEDBACK CONTROL SYSTEM
16 REWARD SYSTEM	-INDIVIDUAL INCENTIVES -FAILURE IS HEAVILY PENALIZED	-GROUP INCENTIVES -FAILURE IS PENALIZED, RISK IS TOLERATED	-PROFIT SHARING -PRUDENT RISK ENCOURAGED	-EMPLOYEE EQUITY -COMBINED GROUP/ INDIVIDUAL INCENTIVES -BREAKTHROUGH RISK ENCOURAGED
17 FACILTY CONFIGURATION	-ESSENTIALLY FIXED, INFLEXIBLE	-10% FLEXIBLE, THROUGH CELLS	-30% FLEXIBLE; MORE CELLS, VARIABLE IN-LINE FLOW	-PHYSICALLY RECONFIGURABLE THROUGH SOFTWARE
18 MAINTENANCE	-ADJUST AND REPAIR	-PREVENTIVE MAINTENANCE	-AUTOMATED MACHINE DIAGNOSTICS	-AMD INTEGRATED WITH SPC
19 INVESTMENT CRITERIA	-ROI, BY INDIVIDUAL EQUIPMENT ITEM -TANGIBLE FACTORS ONLY	-ROI FOR CELLS AND SYSTEM-WIDE ITEMS -SMALL CONSIDERATION OF INTANGIBLES	-ROI FOR FUNCTIONAL CLUSTERS -SYNERGISM CONSIDERED IMPLICITLY	-ENTERPRISE WIDE ROI -SYNERGISM EXPLICITLY CONSIDERED -STRATEGIC DECISION FACTORS

FOR FURTHER EXPLANATION THE END OF THE PAPER

The template "Beyond CIE" fit all of the needs for the team to "open our minds" and begin to think longer term. One of the key responsibilities of the Division Management Team (called Eagle 1) is to think strategically, get out 18 months to three years in front of the business. They had been running the company on a day-to-day reactive basis and needed to become more PRO-ACTIVE. In viewing "Beyond CIE" they needed to think about how they could "leap frog" the steps outlined. They certainly did not want to take 20 to 30 years to reach the 21st century.

The results of Eagle 1's review was as expected; all of the questions above needed to be answered. The responses were all over the map and the need to leap frog the process was a land-slide consensus. It would take too long to define this in detail, but the discussion/dialogue is the heart of the process.

Goulds had started a Total Quality process in 1988, and the steps in "Beyond CIE" seemed to fit very nicely. They started to relate the two and included the Malcolm Baldrige criteria's seven steps. This process again used the basic 4 step process and looked out 18 months. It also involved another group of people in the quality support structure to start "looking ahead".

At this point three months had been spent getting Eagle 1 to have a common focus and be forward looking. The next step was to formally start bringing more people into the process. The direction had to be simple and something that Eagle 1 was leading by example. The expression; "WALK the talk " was mentioned many times.

At the same time the book "Break Through Thinking" [1] became required reading. The seven principles outlined there helped further open the minds to "different" thinking, and perhaps thinking that said "IT IS NOT ONLY OK, BUT REQUIRED, THAT WE SHOULD DO IT **OUR WAY**".

"BREAK THROUGH THINKING" PRINCIPLES[1]

1. Each problem is unique, if for no other reason the people involved.
2. Focusing on the Purpose of doing something.
3. Look for the Solution-After-Next.
4. Each problem is part of a larger system.
5. Limit information collection, get 80% then do it.
6. People Design - users plan and "do it".
7. Betterment timeline - make a plan and monitor progress against it, continually improve.

Another step in the process of getting the expanded staff focused was a series of working meetings exposing them to a single page entitled "NEEDED IN THE NINETIES". It remains the focal point for specific actions.

"NEEDED IN THE NINETIES"

FOCUS ON:	DEFINE SUCCESS BY:
TOTAL QUALITY	THE CUSTOMER
SAFETY	FINANCIAL MEASURES
ON-TIME	TEAM WORK
CAPACITY	PERSONAL
-DEFINE IT	
-INCREASE FLEXIBILITY	
LISTENING TO OUR PEOPLE	LIMIT FIRE FIGHTING
ACTIVE PARTICIPATION	
ACTIONS TAKEN TODAY BENEFIT:	
SHORT TERM - 0-30 DAYS	BE PRO-ACTIVE
MID TERM - 60-90 DAYS	
LONG TERM - +120 DAYS	PARTICIPATION

Success in "Needed in the Nineties" starts at the bottom right, moves to the left and then up, eventually achieving success as defined. In an on-going business enough information is already known by the employees that "if you listen and participate" you will do the right things. Later better definitions of success will be required.

BIRTH OF THE "WAR ROOM" It was recognized that with all the progress that was being made a much more aggressive means to communicate was required. The organization had been very compartmentalized and there were many things that were going on that other parts of the organization didn't know about. A conference room was used to start this process.

The flow of the sales order moving through the business was flow charted on one wall. This wall became the "nuts and bolts" wall and allowed everyone to understand what other departments were doing. The other walls eventually were filled and named too. The Vision wall, the Strategies wall which includes planning, and the Customer and Products walls were created. At first this room was filled with words, ideas, things that "should be done". It has evolved in to be the nerve center of the Division.

PURPOSES OF THE EPD WAR ROOM

Awareness	Of what other's are doing.
Visibility	Show progress/actions to others.
Integration	Success is defined as a division and ACROSS departments, not within.
Ownership	Specific people and functions sign up and commit to performance/improvement.
Communicate	Provide a focal point to allow an exchange of ideas.

As the War Room became organized all members of management were provided "formal tours" by the Division Manager. One of the first groups scheduled was Union leadership. Clearly the intent was that ALL EMPLOYEES would be listened to and were welcomed to PARTICIPATE. This was a MAJOR CULTURAL CHANGE that both management and the employees are having difficulty with; the messages about change had to be communicated.

THE CHANGE PROCESS

As an organization, people began to learn about change. They needed/wanted to go back to the basics and understand what was needed to do things differently, do more with less, do it better. They needed the incremental changes also, but simultaneously needed to set the stage for "order of magnitude" improvements. They had set out to change the collective vision of the critical mass of the organization. The process is very conscious and EVOLUNTARY. Many adjustments are required to any plan; cut and paste is the name of the game. Establish a process with the doers, follow it, share the credit, take risks.

Change must start with each of us. All of us as employees participated in two days of Total Quality Awareness Training. One of the keys to that training is "change starts with me". That process includes senior managers and the brand new individual contributors. It starts at the top and the empowerment process means that the General Manager must stop running the business day-to-day and the immediate staff must pick it up. That process has two very important steps; THE TOP PERSON LETTING GO - AND - THE NEXT LEVEL PICKING UP; BOTH MUST OCCUR. This must occur level by level throughout the organization, and is one of the big reasons for shorter/flatter organizations. Each change is the same.

Level by level managers must stop contributing to barriers, find ways to get out of the way and shift from controlling to empowering. It is important to honor the past, those strengths that brought you to this point, and at the same time to challenge everything.

FORMING TEAMS

The process continues with managers accepting feedback, people who identify problems help solve them, and adequate training being provided. As part of the Total Quality Process a problem solving training program was purchased. It was tailored to fit the company culture and training started in July of 1990. This seven step process is growing in use and helps control the cultural urge to fire-ready-aim. Various types of team work are being practiced.

The most unique is the "cell". This is the same idea as a machining cell except this is being applied in the administrative areas. The people responsible for some well defined group of tasks are co-located, but no organizational changes are made. The team members still report to their departments. The idea is that "TOGETHER THEY GET THE JOB DONE" AND the rest of the organization supports them. The size of the cells vary from 3 to 25 people and in each case the results have far exceeded everyones expectations. This concept is expanding.

Formal Corrective Action Teams use the seven step process and work on problems that have been suggested and formally evaluated for impact.

Functional Teams are formed as part of a normal work group; the seven step process is optional. They pick their own success measurements and problems to work on. The team members come from various departments.

Ad hoc Task Teams are used as needed and normally don't formally use the seven step process, although the experience with the process has created the ready-aim-fire mentality. Get the facts first.

INTERNAL CUSTOMERS AND SUPPLIERS

The change process goes beyond teams and must impact each of the departments. Using the concept of internal customers and suppliers, each department identifies their three key suppliers and what represents the single most important parameter with each supplier. The supplier then begins their own measurements and reports on their performance. In many cases this was only available on a monthly basis, the trend is to make it weekly, and then daily. The specific parameters are changed as control is established.

The parameters chosen by each department relate specifically to the overall objectives of the division and focus on the HARD sides relative to customer service. On-time performance was first and reductions in lead times were second. A war on waste follows in 1991.

OVERALL RESULTS

The financial results (summarized earlier) occurred BECAUSE of the overall improved feedback throughout the division. THE BELOW IMPROVEMENTS WERE ACHIEVED DURING A 15% INCREASE IN SALES AT RECORD HIGH LEVELS. Only a sample is showed, note the trend lines, historical perspecitive, good arrows, and comments. Specific improvements are:

ASSEMBLY FEEDBACK A key to success is closing the loop from the assembly area back through the entire organization and eventually to the customer. Many of the design requirements are customer/order specific and as a good supplier EPD must ask the correct questions at the time of quote and order entry to ensure that the divisional capabilities match up to the needs of the customer. Closing the loop allows the division to say "no" when it cannot meet the customer's requirements and also tells the rest of the division what assembly needs to do error free work. "Doing it with Data" is key. The visibility and comprehension provided allows ownership.

Figure 1. Simple trend of Corrective Actions needed.
Figure 2. Praeto distribution of who took action.
Figure 3. Sample - department improvement trend
Figure 4. Praeto distribution of sample department.

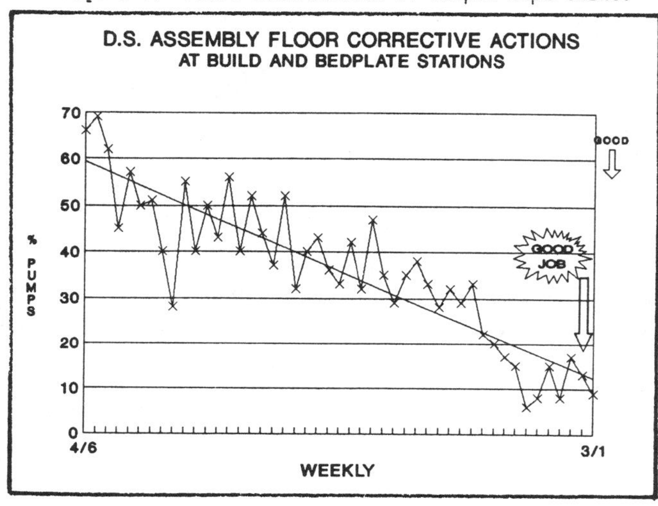

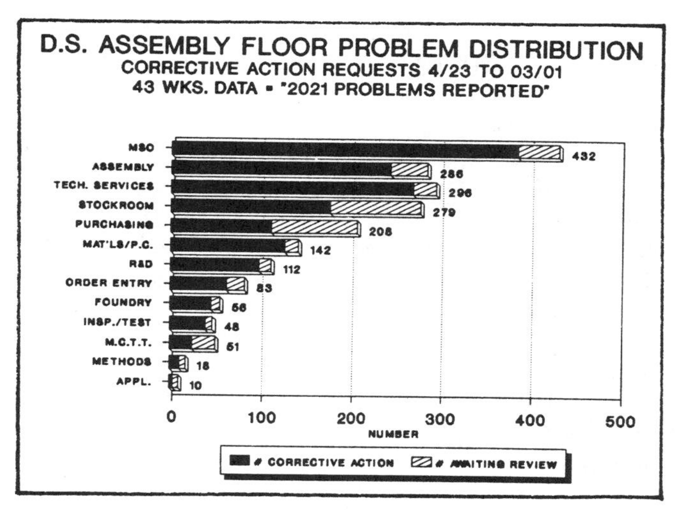

Figure 2. Distribution of Assembly Dept. Action Request

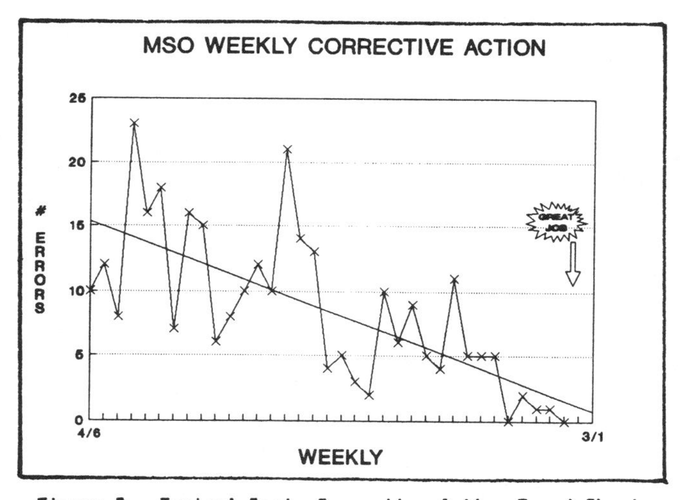

Figure 3. Typical Dept. Corrective Action Trend Chart

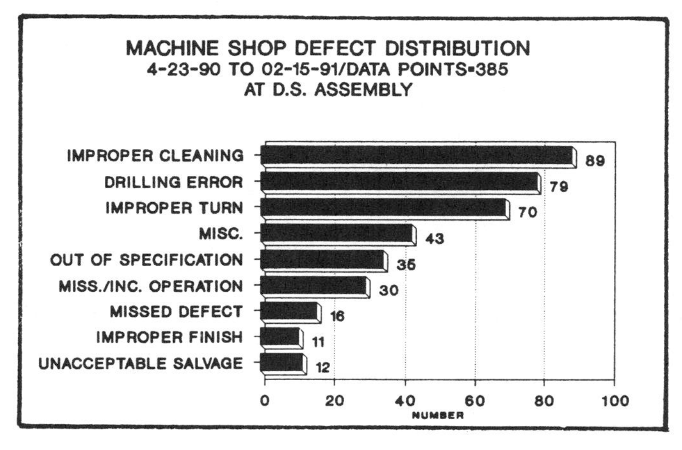

Figure 4. Distribution of Machine Shop Root Causes

← Figure 1. Assembly Dept. Corrective Action Trend

ORDER ENTRY LEAD TIMES: In a make-to-order business it is essential that the administrative portion of the cycle does not eat up more than it's fair share of lead time. On high spec products the reduction has been from a high of 38 days to 18 days. On commercial products the reduction was from 15 days to 8 days. This was accomplished by forming 3 teams internal to the department and the customer/supplier measurements.

INSTALLATION DRAFTING BACKLOG: To many customers the installation drawings are more important than the actual pump. The drawings are required by the customer to do additional engineering. The backlog in the drafting department rose to a high of 5000 hours and was reduced by forming 3 teams focused on: installation only, regular drafting, and a team that moves back and forth between the two. Presently, the backlog remains under 500 hours even though the complexity and number of pumps requiring drafting has increased.

FACTORY ORDER CHANGE NOTICES: Figure 5. The division's efforts to better define it's informational needs has resulted in fewer orders requiring changes. The peak was 3000 per month and is now running under 1000 and declining.

DELINQUENT PUMP ORDERS: Figure 6. The number of delinquent pumps peaked at over 20 days of sales and are presently at 6 days and declining. Additionally, the average length of time they are late has been reduced by 60%. On-time performance is the highest it has been in over 3 years and continues to improve.

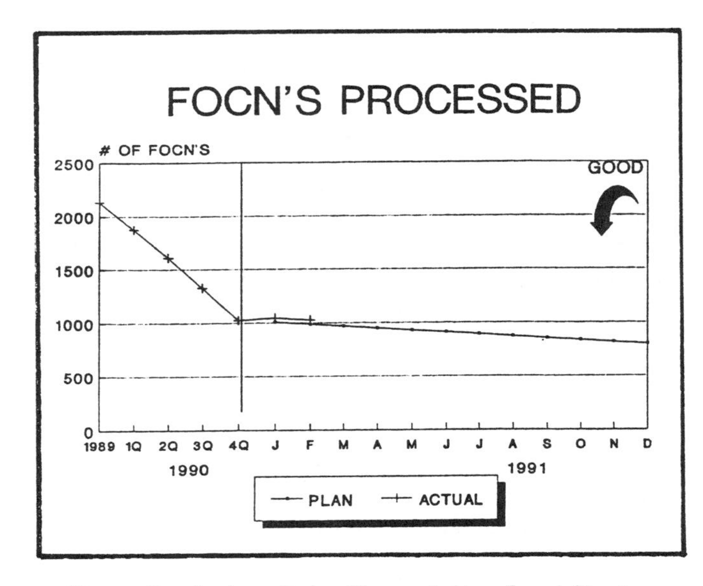

Figure 5. Factory Order Change Notice Trend Chart

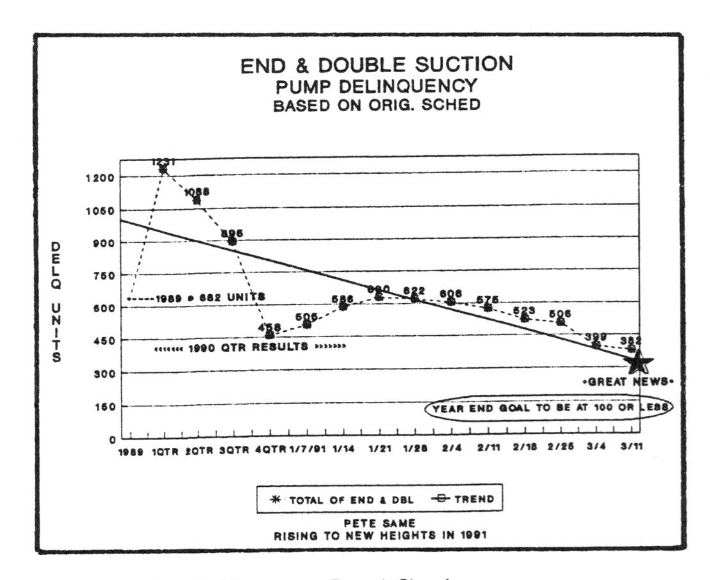

Figure 6. Delinquency Trend Chart.

LONGER TERM CHANGES - REVITALIZATION

Most of the actions discussed to this point are dealing with two areas:

1. Creating the environment to deal with the SOFT issues THROUGH Total Quality, team work and listening to everyone.
2. Increasing effectiveness relative to the HARD issues over the next rolling 12 months.

Obviously longer term success requires the breakthroughs discussed earlier. A formal small planning team was established in early 1990, it's charter being to define success and the related processes for the late 1990's. The process is called Transformation and Revitalization. This 4 man team has completed their assessments, defined the plan outline and is presently in the design phase for major changes. Input from the line organizations has increased exponentially as the environment has been created. Ownership of the plan is in the LINE ORGANIZATION. The planning team is the nucleus to get it on paper and make it visible to the organization. A concept similar to the War Room is being use to make this happen.

SUMMARY

The division management team learns daily. The process of continuous improvement is relentless and the empowerment of all employees grows daily. Listening takes more time than talking, but it also produces better results. The investment in time is well worth it. The results at Goulds Pumps EPD are because of the people in every part of the organization, from the newest to the oldest, everyone is contributing. The improvements will continue because of a sound process, one that is based on trust and facts. The Total Quality process has made "a world of difference" in the last year, and it will continue to do so.

EXPLANATION OF "BEYOND CIE" TEMPLATE MAPPING CHANGE IN MANAGEMENT PRACTICES

Manufacturing managers have been under a severe frontal assault in recent years relative to the magnitude and scope of fundamental changes being thrust upon them. Traditional performance measures (e.g., machine utilization) are being challenged. The fundamental concepts underlying traditional management practices (push vs. pull) are being debated. The scope of the manufacturing manager's job has been enlarged to require consideration of the strategic business issues. Acceptable quality and reliability standards have been shattered. Technological advancements have buried the manager in ever greater depths of data. Personnel management practices have been revolutionized, with "empowerment" replacing "control" as the preferred mode. Organizational strucutres are being flattened. Data/information accessibility by everyone throughout an organization renders the traditional hierarchy impotent at best.

Coming to grips with all these pervasive forces of change has proven to be exceptionally challenging for many managers. A contributing factor is the difficulty of relating each of them to all of the other, and of seeing a common framework against which the myriad changes may be portrayed.

Humans are better able to make intelligent decisions about the future if they can understand the historical forces and context of the factors with which they must deal. What is needed is a means of portraying the set of factors within a relevant time-placed progression.

During the 1960's, manufacturing management practices were dominated by a "push" mentality. Forecasts were made and inventories were relied upon throughout all phases of the production/distribution process to accommodate lack of synchronization within the system. This time period saw the birth of MRP systems.

The 1970's saw the MRP movement flourish, but also challenged by the on-slought of the JIT movement. Much controversy still exists regarding the appropriateness of "pure" JIT concepts.

The buzzword of the 1980's was "CIM". The promises of CIM were great, but the promised results have been difficult to achieve.

Now that the 1990's are upon us, we are being told that for "integration" to really work, it must encompass the entire enterprise. This, of course, resulted in the "mother of all acronyms", CIE, for Computer Integrated Enterprise.

Using these general time-frames as convenient fences, we can portray the evolution of sereral aspects of manufacturing management practice. The accompanying table attempts to characterize each of several "Driving Forces" across four arbitrary time segments.

Two advantages occur from considering these parallel movements through time. First, the matrix provides a convenient means of getting one's mind around a majority of the forces impacting an organization. We can consider how one of the forces influences, or is influenced by, other forces. Secondly, the matrix provides a benchmark for assessing one's organization in terms of these factors. Goulds Pumps took advantage of this capability, the results being reported in this paper.

FOOTNOTES

[1] "BREAKTHROUGH THINKING"
Gerald Nadler and Shozo Hibino 1990,
Prima Publishing & Communications
P.O. Box 1260GN
Rocklin, CA 95677
916-624-5718

BIOGRAPHICAL SKETCH

Mike F. Stautner is General Manager of the Engineered Products Division of Goulds Pumps Inc. located in Seneca Falls, N.Y.. He has spent the last 25 years in operations management in the semiconductor, aerospace and automotive industries before joining Goulds in November of 1989. He was with Garrett, one of the Allied Signal companies, for ten years; TRW Electronics and Motorola. He received his BSIE from Penn State and his MBA from Loyola Marymount of Los Angeles.

Joe E. Mize is Regents Professor of Industrial Engineering and Management at Oklahoma State University. Dr. Mize is a leading consultant to manufacturing firms in strategic planning for moving toward a CIM environment. He is a Fellow and Past President of IIE, and is a member of the National Academy of Engineering.

ENGINEERING AND SYSTEM DESIGN: OPPORTUNITIES FOR ISE PROFESSIONALS

Wolter J. Fabrycky
Lawrence Professor of Industrial Engineering and Operations Research
Virginia Polytechnic Institute and State University
Blacksburg, Virginia

The emergence of life-cycle and concurrent engineering design methodologies is creating numerous opportunities for industrial and systems engineering professionals (ISE's). System integration approaches based on these new methodologies are presented first, followed by discussions of several areas for the effective application of industrial and systems engineering expertise.

I. SYSTEM LIFE-CYCLE ENGINEERING

Engineering activities of analysis and design are not an end in themselves, but are a means for satisfying human wants. Thus, engineering has two aspects. One aspect concerns itself with the materials and forces of nature; the other is concerned with the needs of people.

All products, systems, and structures which have utility (the capacity to satisfy human wants) are physically manifested. It follows that utilities are created by altering physical factors. The purpose of engineering is to determine how the physical environment may be altered to create the most utility for the least cost, in terms of product design cost, production cost, and product service cost [9].

The System Life-Cycle Concept

Products, systems, and structures are designed and developed in accordance with a process which is not as well understood as it might be. System life-cycle engineering is suggested as a integration approach for bringing competitive products into being in such a way as to minimize their deficiencies and life-cycle cost [1]. This integration involves design and development efforts to

1) Transform an operational need into a description of system performance parameters and a preferred system configuration through the use of an iterative process of functional analysis, synthesis, optimization, definition, design, test, and evaluation;

2) Incorporate related technical parameters and assure compatibility of all physical, functional, and program interfaces in a manner that optimizes the total system definition and design; and

3) Integrate performance, producibility, reliability, maintainability, manability, supportability, and other "specialties" into the overall engineering effort.

Fundamental to the practice of life-cycle engineering is an understanding of the system life cycle. The life cycle of a product, system, or structure begins with the identification of a need and extends through conceptual and preliminary design, detail design and development, production and/or construction, distribution, customer use, support, and then phaseout and disposal.

The life-cycle concept is universal in its applicability. It originates with the perception of a need and terminates with product phaseout and disposal. Between these end points there are two major life-cycle phases. The first is the acquisition phase including the several iterative steps necessary

Reprinted from *1989 IIE Integrated Systems Conference Proceedings*

to define the need, perform the design, test and evaluate, and finally produce and distribute the product. The utilization phase follows and involves activities required to deal with the product in being. These include operating, maintaining, modifying, retiring, and product disposal.

A life-cycle approach for bringing competitive products and systems into being must go beyond consideration of the life cycle of the product itself. It must simultaneously embrace the life cycle of the manufacturing system as well as the life cycle of the product service system. Accordingly, there are three concurrent life cycles progressing in parallel, as is illustrated in Figure 1 [1].

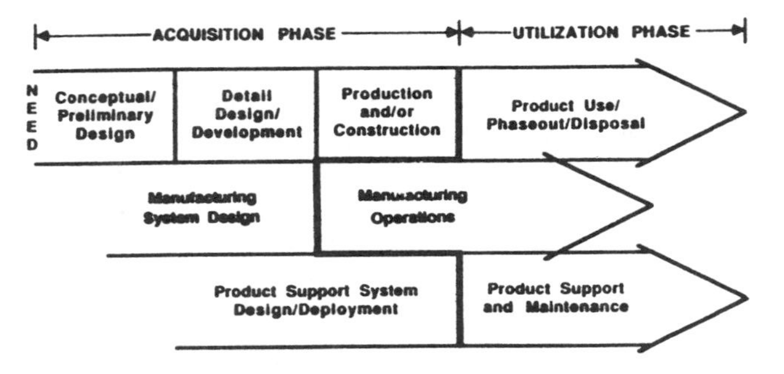

Figure 1. Product, Process, and Support Life Cycles

The need for a new product comes into focus first, initiating the product life cycle. Conceptual and preliminary product design follows need determination, with simultaneous consideration for manufacturing system design; activities best pursued before detail product design as a parallel life cycle. Product support system design, too often omitted as a design imperative, is the third life cycle. It should be synchronized with product design and manufacturing system design [4].

The Life-Cycle Design Process

A detailed presentation of the elaborate technological interactions which must be integrated over the coordinated life cycles is shown in Figure 2 [1]. The progression is iterative from left to right and not serial in nature, as might have been implied by Figure 1. The paragraphs which follow briefly describe the life-cycle phases of conceptual design and others shown in Figure 2.

Conceptual Design (Figure 2, Block a). Regardless of the product, system, or structure, design begins with an identified need based on a "want" or "desire" arising out of a perceived deficiency. An individual and/or organization identifies a need or a function to be performed, and a new (or modified) product is conceived and produced to perform that function.

Preliminary Design (Figure 2, Block b). Preliminary design begins with the technical baseline for the product as defined in the feasibility analysis, and proceeds through the translation of established system-level requirements into detailed qualitative and quantitative design requirements. This includes the process of functional analysis and requirements allocation, the accomplishment of trade-off studies and optimization, system synthesis, and configuration definition in the form of detailed specifications.

Detail Design and Development (Figure 2, Block c). The detail design phase begins with the concept and configuration derived through the primary design activities identified in Block b. When an overall design configuration has been established, it is necessary to progress through further definition leading to the realization of hardware, software, personnel, data, and support capability.

The Life-Cycle Approach

The life-cycle approach is applicable to both small and large-scale products and systems. Further, the process is applicable to many different categories of systems, (e.g., an airplane or missile system, a ship system, an electronic system, a manufacturing or production system, a structure or facility, etc.). Although the nature of the requirement may vary from one system to the next, the process is essentially the same.

Traditional engineering design has focused mainly on the acquisition phase of the life cycle. But, recent experience indicates that a properly coordinated and functioning product or system, which is competitive in the marketplace, cannot be achieved through efforts applied largely after it comes into being. Accordingly, it is essential that engineers be sensitive to operational feasibility during the early stages of product development and that they assume the responsibility for life-cycle design which has been largely neglected in the past [4].

Good design for a product's primary function often produces side effects in the form of operational problems. This is due to exclusive consideration of the primary function, rather than to the more challenging problem of designing in the face of the

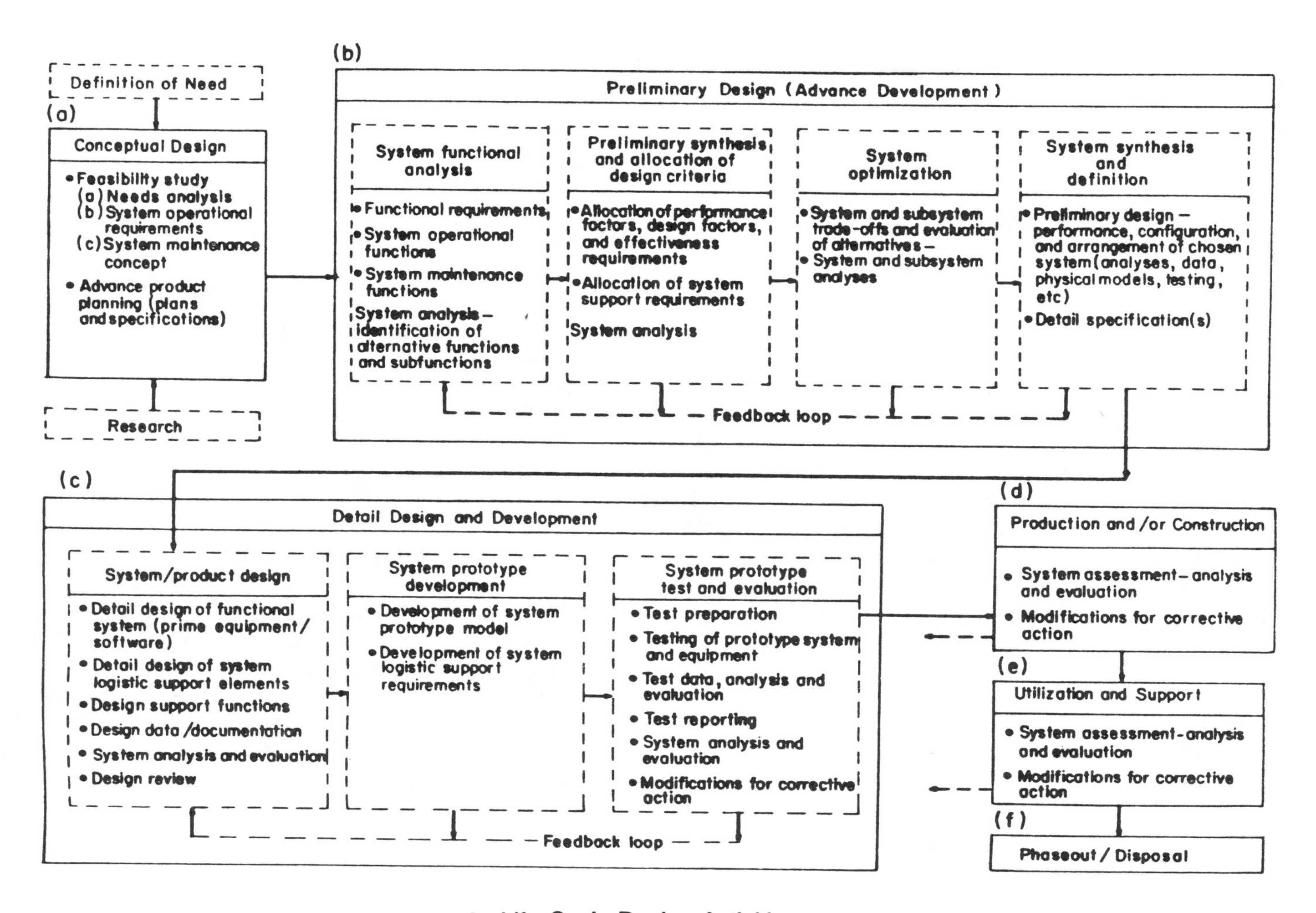

Figure 2. Life-Cycle Design Activities and Interactions

several "ilities". Enough specialized knowledge exists to solve this problem. The impediment to its solution is the integrated use of what is known in a systematic manner. This integration capability exists within the industrial and systems engineering profession.

II. OPPORTUNITIES FOR ISE PROFESSIONALS

Industrial and systems engineers collectively constitute a modern and potent profession, capable of greater contributions to engineering and system design. As defined by the Institute of Industrial Engineers [3],

> "Industrial engineering is concerned with the design, improvement, and installation of integrated systems of people, material, information, equipment, and energy. It draws upon specialized knowledge and skills in the mathematical, physical and social sciences together with the principles and methods of engineering analysis and design, to specify, predict and evaluate the results to be obtained from such systems".

Greater contributions to engineering and system design can be made by ISE's by getting involved earlier in the life cycle, promulgating concurrent engineering, developing computer-aided capability, evaluating engineering designs, extending engineering economics, and seeking quality outcomes. These opportunities are discussed in the sections which follow.

Getting Involved Earlier

Industrial engineering found fruitful involvement during its formative years in production/construction (Figure 2, Block d). Such pioneers as Taylor (shoveling) and Gilbreth (brick laying) were followed by others applying more modern approaches [3]. With the advent of operations research, industrial engineers took on more of a systems approach to production, but also became heavily involved with the optimization of systems during utilization (Figure 2, Block e), and to a lesser extent with phaseout and disposal (Figure 2, Block f) [2] [3] [6].

Although it is unlikely that ISE's can contribute significantly to detail design and development (Figure 2, Block c) due to shallow preparation in the engineering sciences, the opportunities in conceptual and preliminary design (Figure 2, Blocks a and b) are excellent. Furthermore, great benefit can be derived from the application of knowledge earlier in the life cycle, as is illustrated in Figure 3. Fully two-thirds of the life-cycle cost of a product or systems is committed by the time conceptual and preliminary design is completed. This commitment reaches about 80% upon completion of detail design and development.

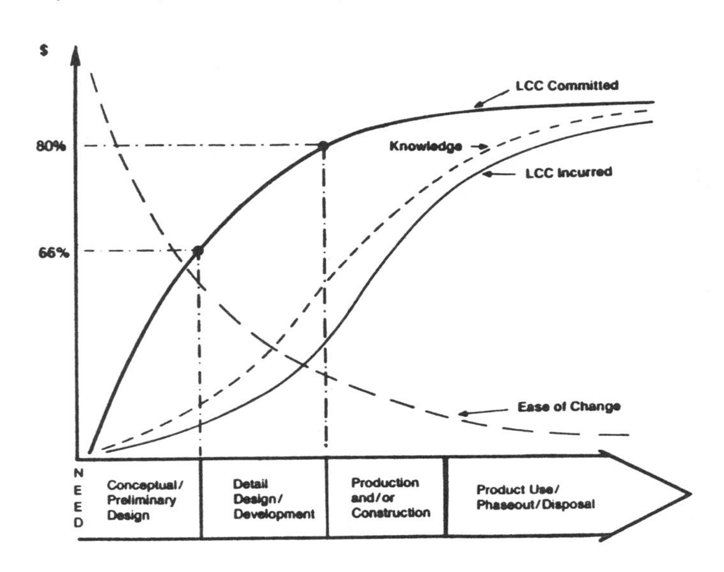

Figure 3. LCC Committed and Incurred

ISE's, because of their broad training, can contribute successfully to decision making in conceptual and preliminary design. The objective is to narrow the "gap" (shown in Figure 3) between available knowledge and life-cycle cost committed. Additionally, the cost-effectiveness of extra investments in knowledge acquisition is a natural for ISE's to evaluate.

Promulgating Concurrent Engineering

Concurrent engineering is an emerging methodology based on the system life-cycle approach. It has great promise for assuring the competitiveness of new products, structures, and systems. As defined in a recent report by the Institute for Defense Analysis [10],

> "Concurrent engineering is a systematic approach to the integrated, concurrent design of products and their related processes, including manufacture and support. This approach is intended to cause the developers, from the outset to consider all elements of the product life cycle from conception through disposal, including quality, cost, schedule, and user requirements."

The concurrent engineering design process emphasizes functional analysis, the allocation of requirements, synthesis, trade-offs and optimization, and so on. These activities (by themselves) are not new to ISE's, nor is the approach. However, in many instances, designers will start identifying "hardware" for a given system without first fully identifying essential manufacturing requirements [7]. Further, the designer will often ignore reliability, maintainability, and life-cycle cost factors. The emphasis must be on a concurrent process discipline necessary for the orderly development of the product, the manufacturing process, and the support system in a coordinated manner [1].

It is the commercial understanding of concurrent engineering that is the most incomplete. Unlike the Department of Defense, commercial firms do not assume life-cycle responsibility for the products they produce (except for warranties). Although desirable outcomes during use are the ultimate proof of product quality, short term thinking often overshadows the longer term benefits in prospect from an acquired product reputation. ISE professionals can be of great help in fostering the longer term thinking.

Developing Computer-Aided Capability

The communication and coordination needed to bring the product, the process, and the service system along in a coordinated manner is not easy to achieve. Progress in this will likely be facilitated by new technologies making possible the more timely acquisition and utilization of design information. CAD/CAM is only one of these technologies. Others need developing which can integrate relevant activities of the enterprise over the spectrum of life cycles. The most promising of these is Computer-Aided Concurrent Engineering Design (CACED), illustrated schematically in Figure 4.

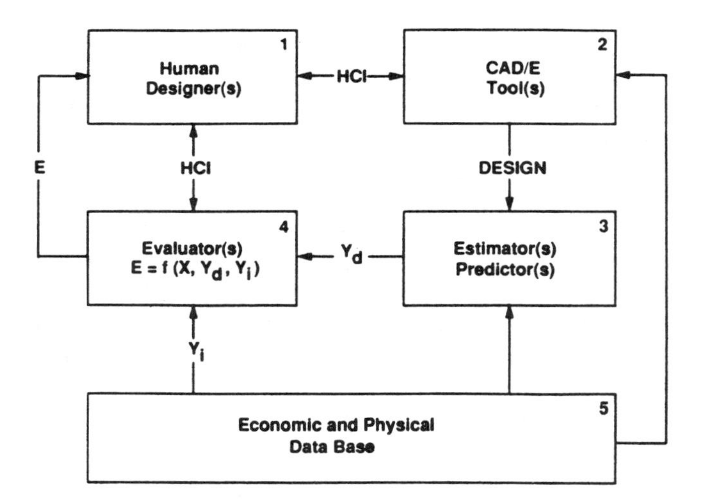

Figure 4. A Schematic for CACED

Opportunities for ISE's exist in Block 3 (estimation and prediction), Block 4 (design evaluation), Block 5 (economic and physical data base development), and in developing the HCI (human computer interface). Networking considerations for CACED workstations within the organization can also be the concern of ISE's. With the advent of information system networking, and the increased availability of computer-aided X (where X = design, cost estimating, process planning, etc.), "facts" become more accessible and hence of diminishing relative value. The ability to apply facts (knowledge) in a creative, and more rapid manner, has become more important.

Evaluating Engineering Designs

Design evaluation in terms of life-cycle cost and system effectiveness is an activity open to ISE's with the appropriate analytical background. This evaluation includes numerous factors, some of which are illustrated in Figure 5 [1].

Block 4 in Figure 4 shows the placement of evaluation within CACED, but evaluation is often needed outside the domain of a workstation. In either case, the design dependent parameter approach is proving to be of considerable value. The evolution of this approach is shown in Figure 6 [1].

The design dependent parameter approach is a mathematical way to link design actions with operational outcomes. It utilizes the design evaluation function in Figure 6. From the following definitions of terms in the function, the opportunities for ISE's (particularly those with good operations research backgrounds) should be evident.

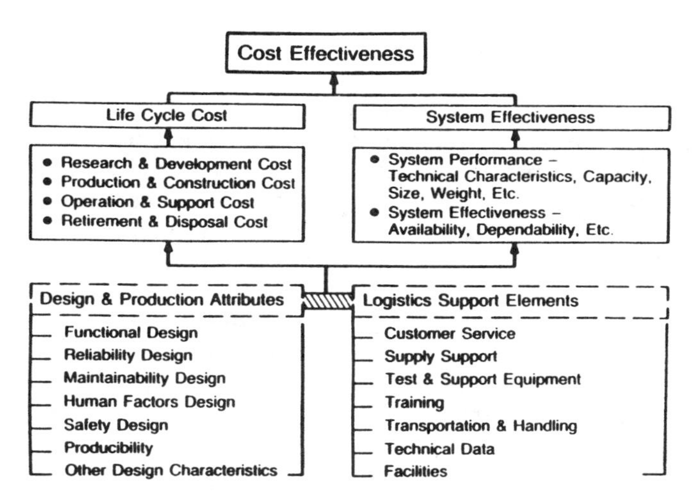

Figure 5. Factors to be Evaluated in Systems Design

References	Functional Form	Application
Churchman, Ackoff, & Arnoff (1957) [2]	$E = f(x_i, y_j)$ E = system effectiveness x_i = variables under control y_j = variables not subject to direct control	Operation
Fabrycky, Ghare, & Torgersen (1984) [6]	$E = f(X,Y);\ g(X,Y) \lesseqgtr B$ E = system effectiveness X = controllable variables Y = uncontrollable variables	Operations
Blanchard & Fabrycky (1981, 1990) [1]	$E = f(X,Y_d,Y_i);\ g(X,Y_d,Y_i) \lesseqgtr C$ E = evaluation measure X = design variables Y_d = design dependent parameters Y_i = design independent parameters	Design Optimization

Figure 6. Design Evaluation Function Evolution

X = design variables (e.g., number of deployed units, armor thickness, retirement age, repair channels, rated thrust, pier spacing, etc.)

Y_d = design dependent parameters (e.g., weight, reliability, design life, capacity, producibility, maintainability, etc.)

Y_i = design independent parameters (e.g., cost of money, labor rates, material cost per unit, shortage cost penalty, etc.)

This design evaluation function, with its design dependent parameters and design independent parameters, facilitates system optimization.

Extending Engineering Economics

Opportunities for the more effective application of engineering economics derive from the life-cycle process. The life-cycle approach holds great promise for helping engineering economics reach its full potential to evaluate design, manufacturing, and operations concurrently [5].

Engineering economics has always been associated with time; the time value of money, receipts and disbursements over time, etc. The central "model" in engineering economics is the money flow diagram depicting estimates of income and outlay over time. Accordingly, engineering economics and the product life cycle are on the same "dimension". Missing from many applications of engineering economics is a complete view of the life cycle.

Life-cycle engineering economics is a good opportunity area for ISE's. It can be maximally effective during conceptual and preliminary design. Economic outcomes cannot be influenced much after the design is completed. Furthermore, economic considerations with life-cycle connections must be made during manufacturing system design and support system design in a concurrent manner.

Seeking Quality Outcomes

Quality outcomes are desirable "effects" derived from design "causes". Progress in gaining the needed understanding of cause and effect is not progressing rapidly enough. More must be done to discover the essential links between design and usability. ISE's can contribute insight and provide the needed leadership.

It is well known that quality cannot be inspected into a product. Engineers have turned to statistical process control as a better approach. However, quality must first be designed in. This opens a new area for ISE's; that of understanding and applying parameter design ideas due to Taguchi [8]. An opportunity exists to bring experimental designs and indirect experimentation into product design and development.

Quality outcomes have been achieved by production workers concurrently producing and "inspecting"; this in lieu of leaving quality concerns to others. In an analogous way, quality can be even more greatly enhanced when ISE's find organizational and informational ways of assisting designers to concurrently consider factors upon which quality depends (reliability, maintainability, operating efficiency, etc.). The goal is to facilitate the embodiment of quality responsibility in the design person.

Production workers cannot impart quality beyond that intended and specified by design engineers. Design engineers can influence quality from the very beginning of the product life cycle. When design is done in a truly concurrent manner, benefits far exceeding those obtained by production workers can be expected [1] [4] [10].

III. SUMMARY AND CONCLUSIONS

In these times of intensifying international competition, commercial firms are searching for ways to gain a sustainable competitive advantage in the marketplace. Advertising alone is not sufficient. One promising strategy is to adapt life-cycle development methodology to proposed new products, their required manufacturing activities, and product support systems.

Product development embracing concurrent life-cycle methodology has a excellent chance of enhancing consumer satisfaction, corporate identity, and company profitability through the integration of important design considerations (appearance, cost, performance, and quality). But these desiderata may not be attainable unless the completeness inherent in the methodology is recognized, nurtured, and promulgated. ISE's have the appropriate expertise and are in an excellent position to provide the needed leadership.

IV. REFERENCES AND BIOGRAPHICAL

Selected References

[1] Blanchard, B. S., and W. J. Fabrycky, Systems Engineering and Analysis, Second Edition, Prentice-Hall, Inc., 1990.

[2] Churchman, C. W., R. L. Ackoff, and E. L. Arnoff, Introduction to Operations Research, John Wiley, Inc., 1957.

[3] Emerson, H. P., and D. C. E. Naehring, Origins of Industrial Engineering, Institute of Industrial Engineers, 1988.

[4] Fabrycky, W. J., "Designing for the Life Cycle". Mechanical Engineering, Vol. 109, No. 1, Januay, 1987.

[5] Fabrycky, W. J., "System Life-Cycle Engineering: A Framework for Life-Cycle Engineering Economics," Proceedings, Fall Conference, Institute of Industrial Engineers, Boston, December, 1986.

[6] Fabrycky, W. J., P. M. Ghare, and P. E. Torgersen, Applied Operations Research and Management Science, Prentice-Hall, Inc., 1984.

[7] Nevins, J. A. and D. E. Whitney, Editors, Concurrent Design of Products and Processes, McGraw-Hill, Inc., 1989.

[8] Taguchi, S., and D. Byrne, "The Taguchi Approach to Parameter Design", Proceedings, ASQC Quality Congress, 1986.

[9] Thuesen, G. J., and W. J. Fabrycky, Engineering Economy, Seventh Edition, Prentice-Hall, Inc., 1989.

[10] Winner, R. I., J. P. Pennell, H. E. Bertrand, and M. M. G. Slusarczuk, The Role of Concurrent Engineering in Weapons Systems Acquisition, Report R-338, Institute for Defense Analysis, December, 1988.

Biographical Sketch

Wolter J. Fabrycky received the Ph.D. in Engineering in 1962 from Oklahoma State University, the M.S. in Industrial Engineering in 1958 from the University of Arkansas, and the B.S. in Industrial Engineering in 1957 from Wichita State University. Dr. Fabrycky taught at Arkansas and Oklahoma State and then joined Virginia Tech in 1965 where he served as Chairman of Systems Engineering, Associate Dean of Engineering, and then as Dean of Research for the University. He is now the John L. Lawrence Professor of Industrial Engineering and Operations Research.

Fabrycky is co-author of five Prentice-Hall books. These are: Systems Engineering and Analysis (1990), Engineering Economy (1989), Procurement and Inventory Systems Analysis (1987), Applied Operations Research and Management Science (1984), and Economic Decision Analysis (1980). He also co-edits the Prentice-Hall International Series in Industrial and Systems Engineering.

Dr. Fabrycky was elected to the rank of Fellow in the Institute of Industrial Engineers in 1978 and the rank of Fellow in the American Association for the Advancement of Science in 1980. He is listed in Who's Who in Engineering and Who's Who in America. Fabrycky is a Registered Professional Engineer in Arkansas and Virginia.

Reprinted with permission from *Proceedings of Manufacturing International '90*.

MANUFACTURING SYSTEM REDESIGN: AN INTEGRATED APPROACH

R. Sairam
W. Silver Inc.
El Paso, Texas

J. P. Hsu
Department of Mechanical and Industrial Engineering
University of Texas at El Paso
El Paso, Texas

ABSTRACT

In most of the manufacturing operations, the existing system is one which was designed decades back. The dynamics of the market has changed so much that, what was applicable ten years back is not applicable anymore. This necessitates a thorough examination of the existing manufacturing system and redesign to tailor the manufacturing facility to the current market needs. There is a real need to be more competitive in the market for the American manufacturing companies than ever before. The current approach of increasing productivity by adding capacity of the same kind to the existing system should be changed. Productivity increase should be achieved through planned introduction of new methods and redesign of the manufacturing system. In redesigning the manufacturing system a structured integrated approach should be followed for successful implementation. The main objective of this paper is to outline an integrated approach to system redesign so that the company can be more competitive in the market place.

The structured procedure talks about setting of Corporate objectives, developing a clear marketing strategy, designing effective manufacturing modules, justifying and implementing the system. Integration of cellular manufacturing, single minute exchange of dies, just-in-time production with kanban control and its importance in system redesign is discussed. The impact of modelling and simulation and its utility during the design process is explained. This paper recommends a total systems approach in which many facets of the business will have to be changed simultaneously for successful implementation. Modular factory approach and customer supplier relationship are highlighted. Cost benefit analysis is discussed. Task force implementation approach and its advantages are listed. Results of a successful pilot implementation have been reviewed.

INTRODUCTION

A manufacturing system is designed to meet the market needs. Changes in the market situations therefore entail a continual evaluation of the manufacturing system. The critical question then arises as to whether this is being done. A study showed that only 5% of the manufacturing units in the United States carry-out this evaluation. The objective of this paper is to outline a plan for a close examination of the present system, structured redesign method and an approach to implementation. The main objective in redesigning the system is to make the company more competitive in the market and to respond to the dynamic changes in the market conditions with overall productivity gains.

EXAMINATION OF THE PRESENT SYSTEM

Systems approach involves critical review of the existing system. The main focus is on establishing the current performance ratios and the current procedures. This is the basis for defining the corporate objectives and system redesign needs. This examination process also indicates the level of response to the market changes and how well the market demand is being met. One of the important issues in the examination process includes the data collection methods. An accurate data collection procedure should be defined and followed. Data integrity and accuracy to certain extent determine the success or failure of the project.

SYSTEMS APPROACH

The main objective in redesigning the system after examination is to tailor the manufacturing facility for the market needs. Structured redesign process talks of a total systems approach. This approach requires simultaneous changes in multidisciplinary areas. One of the important design cardinal being not to consider individual methodology (JIT) or technology (Automation) as a separate entity in isolation from the total system objective. The systems approach follows a clear path as indicated in the master flow chart.

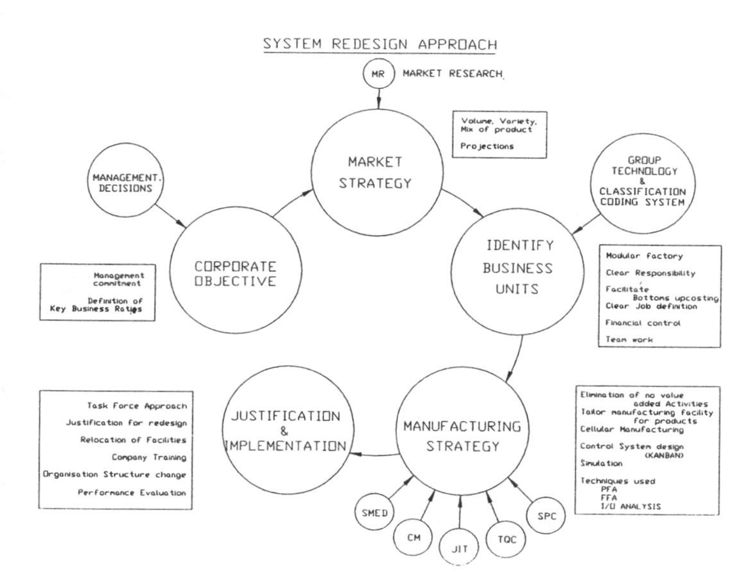

Based on the data collected during the examination process, the corporate objectives will have to be defined. System redesign and implementation requires changes in many functional areas of the organization depending on the existing conditions. These changes require full commitment and support from the top management.

CORPORATE OBJECTIVES

Before starting the design process, the corporate objectives will have to be defined. These objectives provide the basis for making decision during the design and implementation process. Corporate objectives can be broadly classified into two major categories viz., Financial objective and Manufacturing objective. Key business ratios form the financial objectives. Typical ratios include stock turn ratio, return on sales and capital, sales per employee and percent reduction in cost.

Similarly on the manufacturing side objectives such as the quality standard, cellular manufacturing, simple production control, utilization of resources and the manufacturing lead time for products will have to be defined. Once the corporate objectives are made clear the next step is to develop a clear marketing strategy. Using market research and other forecasting techniques the volume, variety and the mix will have to be defined. These projections are the volume basis for system design. The normal practice is to design the system to take care of atleast twenty percent variation.

BUSINESS UNITS

After determining the volume, variety and the mix of the products the next step is to identify, group and define them into manageable business units. Group technology, classification coding systems and production flow analysis are extensively used in forming product based or customer based business units. By doing so it is possible to have clear responsibility and ownership of problems. These key elements are not built into the existing structure with large specialist departments. Creation of business units automatically establishes a customer supplier relationship. Customer supplier relationship forms the basis for bottoms up costing in which all costs are incurred rather than being allocated. The question then arises as to how to integrate commonly shared resources into the business unit concept. For example a plating shop may be used by business units A,B and C. If unit A uses it for 70% of the time then plating shop will be part of unit A. Customer supplier relationship is formed with unit A as the supplier and units B and C as the customer. Units B and C have the option of using unit A or any outside vendor based on cost and delivery. This will force the manager of unit A to make sound business decisions. The overall impact of this concept is to have the highest performance levels on all activities thereby reducing lead times and overhead costs.

MANUFACTURING STRATEGY

The manufacturing strategy adopted should not look at issues like plant layout and production control system as separate entities. The main focus is on integration of popular concepts like cellular manufacturing, just-in-time production technique, single minute exchange of dies and total quality control. Such a methodology has the advantage of building in effective manufacturing systems rather than superimposing the above mentioned concepts on to an existing system designed with stand alone objectives.

Cell Design

The design process includes cell Design, production control system design and simulation of the system to evaluate the design process. The main objectives are elimination of no value added activities, smooth flow of material, reduced set-up time, minimum work in progress and maximum resource utilization. Input/Output analysis technique, production flow analysis, SMED and group technology are extensively in CM design. During the design process a cell manning procedure will have to be defined. Usually the system approach requires that the employees be multi-skilled and cross trained to handle multiple tasks.

Control System

Different manufacturing system requires different forms of production control systems. A single production control for a variety of products manufactured within a factory does not work because of the unique product characteristics and the volume, variety and the mix needs of the market. During the process of designing the business units we classify the products into High volume low variety, Low volume high variety and non-repetitive products. Kanban production control system can be used for the first two categories provided we take care of change-over time from one model to another model. Assembly manufacturing of non-repetitive nature with long lead time products are best controlled with the conventional MRP II system. Classifying the products and identifying a particular production control system for that category determines the effectiveness of the control. Production control with Kanban requires that we do not follow blindly a standard single or dual card system. Instead a hybrid system could be customized for that particular control loop. This type of approach takes the best out of MRP II and the kanban control system.

Simulation

Simulation modeling is a strong analytical technique for predicting and evaluating the performance of a designed manufacturing system. The statistical results which we get from these simulations help in redesigning and modifying the control system. Using the simulation packages like SLAM II modular production control systems should be developed. These modules can be assembled later to simulate different manufacturing systems. This assembled model can then be simulated for obtaining the statistical results. Typical control modules include single card module, dual card module and simple container control module. Variations in control from the conventional systems can be programmed during the assembly process. These programs help in predicting the effect of production bottlenecks, variability in processing times and delays in material movement.

JUSTIFICATION AND IMPLEMENTATION

Justifying a new system in place of an existing system requires that we identify the cost and benefit sources. After identification the next crucial step is to quantify the benefit levels. Accurate pre audit and post audit needs to be carried out. While making comparisons to justify, we need to ensure the appropriateness of the comparison. Procedures based on sound analysis will have to be followed in justifying new system. Costs associated with the system design, tooling change, control system change, machine relocation, organization structure change and training cost will have to considered. On the benefits side we need to quantify the benefits from reduced WIP, reduced cycle time, reduced scrap, reduced handling, elimination of no-value added activities, reduced manufacturing lead time and improved quality. Factors should be arrived at for evaluating the impact of improvements which cannot be quantified but which affect the growth of the business.

One approach in design of a system is to have different departments redesign their sub systems in their area. These sub systems are usually designed with strategies to meet the departmental goals. The amalgamation of these different departmental goals may conflict with one another and not necessarily be in-line with the corporate objectives. A better approach is the task force approach wherein a team consisting of members from different disciplines will work together to redesign the system. These members are trained in modern manufacturing systems engineering concepts. These members are usually relieved of their duties and work exclusively on the task force. This prevents vested interests in their departments and their sub systems. Senior management should provide the leadership in setting up and coordinating the activities of the mixed teams from different functions. The teams activities must be reviewed periodically against set targets over a period of time.

A comprehensive analysis and a pilot implementation approach was followed. Results of implementation of this approach in the pilot module are very encouraging. For one particular module The profit before tax was turned around from a negative 15% to a positive 10 to 15%. The stock turn ratio was increased form 10 to 44, the change-over time was reduced from 108 hrs to 9 hrs for the entire module for a product model change. Because of the successful pilot implementation the same approach is being followed for redesigning the other business units.

BIBLIOGRAPHY

1. SCHONBERGER,R.J,1983,Applications of single-card and dual-card kanban. Interfaces,13,56-67.

2. MONDEN.Y., Jan 1981, What Makes the Toyota Production System Really Tick, Industrial Engineer,36-46.

3. VILLEDA, R.,RICHARD DUDEK, and MILTON L.SMITH, 1988, Increasing the production rate of a just-in-time production system with variable operation times. International Journal of Production Research,26,1749-1768.

4. HUANG,P.Y., REES,L.P, and TAYLOR,B.W.,1983, A simulation analysis of the Japanese just-in-time technique (with kanbans) for a multiline, multistage production system. American Institute for Decision Sciences,14,334-338.

5. SHINGO,S., 1981,Study of 'Toyota' production system from industrial engineering viewpoint (Toyota: Japan Management Association).

6. MONDEN.Y.,1983, Toyota Production System: Practical Approach to Production Management (Norcross, Georgia: Industrial Engineering and Management Press).

Additional Readings on Design Issues in Systems Integration

Beeckman, Dirk. 1989. CIM-OSA: Computer-integrated manufacturing - open system architecture. *International Journal of Computer Integrated Manufacturing*, Vol. 2, No. 2, pp. 94-105.

Brown, John S. 1991. Research that reinvents the corporation. *Harvard Business Review.* January-February, pp. 102-111.

Chang, Lee C. 1989. An engineering approach to computer-integrated manufacturing. *Proceedings 1989 IIE Integrated Systems Conference*, Norcross, GA: Institute of Industrial Engineers. pp. 61-65.

Dutton, Barbara. 1990. An interview with Jerre L. Stead (CEO, Square D. Company), *Manufacturing Systems.* April, pp. 36-43.

Eyrich, H. G. 1991. Benchmarking to become the best of breed. *Manufacturing Systems.* April, pp. 40-47.

Graefe, Udo and Vince Thomson. 1989. A reference model for production control. *International Journal of Computer Integrated Manufacturing.* Vol. 2, No. 2., pp. 86-93.

Hammer, Michael. 1990. Reengineering work: Don't automate, obliterate. *Harvard Business Review.* July-August, pp. 104-112.

Moss, S. P. 1989. A management and control architecture for factory-floor systems: From concept to reality. *International Journal of Computer Integrated Manufaturing.* Vol. 2, no. 2. pp. 106-113.

Zuboff, Shoshana. 1991. Can research reinvent the corporation? *Harvard Business Review.* March-April, pp. 164-175.

Competitive Benchmarking: What It Is and What It Can Do for You. Stamford, CT: Xerox Corporation. 1987.

V. IMPLEMENTATION ISSUES IN SYSTEMS INTEGRATION

Many industrial engineers engaged in systems integration view implementation as the most common cause of project failure. After the designs have been finalized and approved, there remain many obstacles before their full implementation can be achieved. Many factors contribute to this situation.

Often, the management of a firm will give their approval to a comprehensive system integration plan, and then balk at approving specific expenditure requests. Market fluctuations, changing sentiments among members of the Board of Directors, and a lengthy, clumsy approval process for capital expenditures are some of the contributing factors.

Another major obstacle to successful implementation are the behavioral issues involved. An aggressive system integration plan will usually call for significant widespread change to occur within a short period of time. Particular managers are threatened, touch-labor people are always skeptical, and many staff functions are combined, eliminated, etc.

Still another factor inhibiting successful implementation is the sheer magnitude of managing and coordinating the tens of thousands of individual activities that make up the dozens of projects.

The articles included in this section, and some of those included in Section VIII, were selected to give insight into how one might approach implementation issues.

Implementing Integration

Some pointers on the right approach to implementing integrated systems and a few pitfalls to avoid.

Brian Maskell
European Installation Manager
Xerox Computer Services
London, England

Integrated computer systems can put in the hands of the executive those tools really needed to control the business. The task of finding new systems to help run the business can be extremely difficult, however. Companies devote many months of hard work before making a decision and then, once the decision is made, the task of implementing the system begins. This can be equally difficult and often much more costly.

Most companies using computers have a variety of independent modules or applications, each doing a particular job. There may be general ledger, accounts payable or material requirements planning (MRP). Some companies provide links among these modules so that accounts payable updates the general ledger at month's end or MRP interfaces with the procurement system.

With a fully integrated system, all business functions are maintained on one set of data files and are updated instantaneously throughout all the modules. In fact, the idea of modules or individual applications becomes redundant because the user has available the most up-to-date information from all the functions all the time.

Implement a standard system in a standard way.

Integration of data means that all departments are using the same information. The information used by the Sales Department to make delivery schedules is the same information used by shop floor control. Budgets are developed from the same costing data that manufacturing uses when shipping material to the shop. This data is available online to all authorized users

With such a system, there is no reason to have time-consuming monthly disagreements between manufacturing and financial people over the true value of work-in-process. The figures will be derived from one set of data instead of the previous two or more.

This integration has a three-fold payoff. Executives have available, for the first time, information enabling them to control the business effectively. Reports aren't four weeks out-of-date produced by systems three steps removed from the real activity of the company. Information is timely, accurate and in a form that is usable. And reports aren't contradictory because they are based on the same information.

Instead of creating conflict, the interdependence of departments can be used for the benefit of the company. Major departments of all companies are dependent on each other, but that frequently leads to fruitless conflict because each department is gathering and processing its information differently.

Finally, there is a significant improvement in productivity because information is only entered once, cutting out wasteful duplication of effort. The information is more accurate and up-to-date because it's entered by the department using it.

Examples, please

There has been an emphasis during the past 10 years on the implementation of MRP systems. These packages take the Production Plan for each product to be manufactured and "explode" it against the Bill of Materials for that product. From this explosion, the purchasing department can place orders with their suppliers, knowing how much material is needed and when. The production department can plan its personnel requirements, machine utilization, and requisitions for material so that all factors are available at the right time.

MRP systems have shown significant benefits for a large number of companies. Even so, in many companies there is a fundamental problem: How valid is the Production Plan?

The Production Plan is determined by the cooperative efforts of Marketing, Production, Finance and Engineering.

Integrated systems impact every sector of a company.

Traditionally, Marketing and Production have been very critical of each other. They look at the Plan from very different perspectives.

Finance and Manufacturing are often in open conflict. Each presents different figures derived from different sources and processed by different systems using different assumptions. The result is a Production Plan that's at best a compromise and at worse a victory for one side. To make matters worst, the company doesn't have an effective and timely method of analyzing the results and its impact on the various areas of business.

Enter integration

With an integrated system, the picture changes. The Sales forecast for current products is produced directly from the Sales History available to all concerned. Forecasts can be evaluated immediately by Finance, using Marketing's information, to produce revenue outlooks, inventory value forecasts, cash flow forecasts, scrap provisions, etc. The same information can be used by Production to do a "rough cut" capacity plan and master production schedule.

To assist with this planning, an integrated system can be used to perform analyses such as goal-seeking, sensitivity analysis, and simulations. In the past, most companies couldn't do this at all; the more sophisticated companies did it once a year. With an integrated system, these analyses can be done once a month or several times during the monthly Production Plan review.

The Production Plan that is produced will be the very best available. It was developed in agreement with the contributing departments using the same information. The full effect of any change is readily analyzed in each area in a timely and thorough manner.

Then the new plan can be fed straight into the Master Production Schedule that drives the MRP system. Actual customer orders can be used to update the system as they are entered in Order Entry. The company will be making the right products at the right time, purchasing the right materials in the right quantity at the right time, and controlling working capital in the best way.

Elements for success

It's because changes affect every department that you cannot afford to have an integrated system fail. A successful, on-time implementation is assured, however, if the following points are observed.

• **Top management commitment.** These are not just computer systems. They are a complete set of tools for controlling the business. They handle the day-to-day requirements and provide long term information so vital to executives. The company's top managers

The 'install and hope' approach is irresponsible.

must have an active involvement in the implementation because the new system will change the way the company does business. If top managers aren't the prime movers, the new system will fail.

It's the CEO's job to provide the leadership required to establish and maintain the new way of thinking about the business. It's not necessary to understand the technical details, but it is necessary to set the direction. The CEO will chair the regular management review meetings where senior managers from major functional areas resolve issues and guarantee that the project is on schedule. Senior management must be accountable for the success of the implementation. The most effective way of assuring this is to have the system justification spell out specific benefits to be achieved and then have the appropriate functional managers commit to achieving those benefits.

If the full responsibility for the implementation is left to the MIS department and middle level management, it will fail. Don't get me wrong, that isn't because these people are incompetent but other priorities will prevail. When a crisis occurs, the project team members will be called away to fight the fire, the project plan will slip, and other personnel will become disillusioned and cynical.

If the CEO and other top management are the driving force, the project will move in the right direction, the resources will be made available, interdepartmental problems will be resolved quickly and the benefits of the new system will be achieved.

• **Make the project your own.** There is a tendency to regard an integrated system as a very technical problem requiring a specialist—person or company—to do the implementation. Business consultants—such as me—are somewhat to blame for this view. It's wrong.

The project must be taken on by the company that is going to use the system. If your company cannot handle the implementation, then it will not be able to use the system after it is installed. It's valuable to get assistance from people who specialize in this area, but the responsibility for implementation must lie with the internal project manager. From the outset, the project manager must see that the implementation isn't given to outside consultants or hardware and software vendors.

The project manager must be from one of the major functional areas and have a thorough understanding of the needs and workings of the business—the type of person who takes charge, who takes responsibility. A common pitfall in making this choice is picking a systems person—responsibility for the payback of the system lies with the line managers, not the MIS department.

Another don't in picking the project manager is, don't use an outside consultant. Outside consultants can make a valuable contribution but the role of project leader must go to a permanent employee. If you need to hire a consultant, use that person's broader experience to solve the more strategic issues, to provide education, and to identify potential pitfalls before they become disasters.

The project manager must devote full time to the implementation, not part to it and part to putting out other departmental fires. If the only people capable of doing the job are indispensable in their existing jobs, that is all the more reason to take them out for a six month period so that someone else can be trained. A company that has an indispensable employee runs a grave risk if the person quits, retires, has an accident or illness.

How valid is your Production Plan?

And finally, pick someone with a proven track record of success within the company. It's not wise nor fair to give a job like this to an inexperienced employee, no matter how well qualified. There is the tendency to think that managing a project of this type requires a bright-eyed young "whizz-kid." A recently hired person, fresh out of college, won't have much credibility with line managers, supervisors, buyers and systems people. Such a person won't know who to see—who has the real power—to get things done.

• **Implement a standard product.** Historically, companies wrote their own software specifically tailored to their needs. For most companies, this is a thing of the past. Today, it makes more sense to go to a software house and buy a standard system that meets your needs. This approach is equally viable whether the package is for microcomputer or mainframe.

The mistake many companies make is that, having bought some software, they set about modifying it. As a rule, it's important to implement the standard system as purchased. A standard system implemented in the standard way has a much higher chance of working successfully the first time and providing a fast payback. Of course, there are some qualifications to this. Obviously, it's necessary to choose software that provides the most important features for the company. If you do a great deal of modification, there is a high risk of the implementation taking longer and costing more than planned and failing to meet the objectives.

The company must be prepared to change its procedures and methods to gain the advantages of the new system. If a company wants to implement a new system without making any significant changes to its work methods, then the project is unlikely to provide the hoped for benefits. And software must be purchased from a proven, reputable vendor. Make sure that the software has been successfully implemented at a number of sites. Beware of buying future developments.

• **Include Information Center products.** This may seem to contradict the previous point, but in fact it's complementary. Information Center (IC) products are tools that can be used for nonstandard reports, inquiries and subsystems to the standard product after implementation. These products include high-level languages to enable additional features to be added to the standard system quickly. The efficiency of these tools comes from their use of "English-style" terms and many on-line checks and helps. Ideally, they should be simple enough to use that a manager, accountant or other business professional can easily create reports and queries. They should also be powerful enough that the MIS staff can write complex subsystems in a very efficient way with them.

With these tools, the company has the best of two worlds. They can implement the standard system and still have the flexibility to customize the system with additional features. These features can be added without touching the original programming and therefore the company doesn't run the risk of ruining the standard system when adding features. This could prevent the warranty from being voided, too.

Don't pick an outside consultant for project manager.

Even with IC tools, it's still advisable to implement the system initially in the standard form as much as possible. An element of compromise is required—there will always be a requirement for customized features but if they become substantial during the initial implementation, then the project runs the risk of delay and difficulties.

• **Education and training.** The importance of education is usually underestimated; the budget for it is usually too low. A company will happily invest thousands of dollars in hardware, software and consultants but be unwilling to invest in training and education of the people who are expected to get results from the new system. Every company pays lip-service to the idea that "people are the most important resource." This maxim is rarely put into practice when implementing a new system.

An integrated system will impact every part of the company. The benefits will only be realized if the company's personnel have a clear understanding of the reasons for the required changes and their role in the new environment. Fears must be allayed and people must be convinced to be fully cooperative. These are the goals of education.

Education creates an understanding of the system and why it's being brought into the company. It clarifies what attitude changes are necessary and why. It's purpose is to teach the users of the system how to do their jobs better under the new environment.

Three groups of people need to be educated. *Top management and key personnel* need to be introduced to the concepts and objectives of the system. Key personnel, in particular, need to be taught the features and implementation procedures of the system. They will consist of the implementation team, supervisors, middle managers, planning staff and systems people.

The third group—*the users*—requires a thorough training program at an early stage of the project. Training should be done by key personnel with managers, supervisors and others teaching their own staffs. While the primary emphasis will be on the new working procedures and methods, there is also an important need for the users to be educated in the concepts and reasons for the new system and its approach.

Several companies run regular education seminars for top management and key personnel. A detailed explanation of the new system's features is usually available from the appropriate software supplier or hardware vendor. Some excellent classes are available on video tape to augment both key personnel and user training.

• **Pilot implementation.** There are three possible ways of testing the new system: doing a parallel run, installing it and hoping for the best, or trying a pilot implementation. In practice, a parallel run is not very practical. It's too time consuming for the personnel involved, there is no direct comparison drawn between the old and new systems, and it is often not done properly because of the pressures involved.

The "install and hope" approach is irresponsible when the company is implementing a system as important and all embracing as this.

The purpose of a pilot implementation is to establish that the system is working properly and to guarantee that the people using the system have had sufficient training. The system will, of course, have already been exhaustively tested prior to the pilot. Still, it's important to see it working in real life before pushing ahead to full implementation.

The pilot area that is chosen can be a particular product range, a small location, or one part of a larger location. It should be representative of the company's business and yet small enough to allow the detailed attention that is necessary.

There are enormous benefits to be gained by implementing an integrated, closed-loop business system. These benefits can be felt in every area of the company; they provide executives, analysts, accountants and managers with the real information they need to effectively control the business. The challenge is to achieve these benefits by successfully implementing the system and learning how to really use these new tools. The success of a system as powerful as this is dependent on the implementation and therefore that task must be approached with care and professionalism. The six key features outlined here have proven time and again to be the elements of success when approaching this complex task. (MS)

A catalytic implementation methodology for CIM

STUART SMITH and DAVID TRANFIELD

Abstract. The paper describes a research project focusing on the managerial and organizational problems of implementing computer applications associated with CIM and the steps toward CIM. The research's major finding is that successful implementation requires a revolution in management thinking in many companies. This revolution requires management to re-examine its manufacturing strategy in the light of the opportunities for quantum leaps in competitiveness offered by new technology. The paper goes on to outline a methodology for implementing smooth revolutionary change. It identifies the main phases of the change process and the pace at which these should be carried out. The paper concludes with a methodology for propagating change throughout the organization.

1. Introduction

Widespread experience reported in the research literature suggests that the success rate in introducing computer applications to manufacturing is limited. Voss (1985) reports that although most firms (86%) using CAD/CAM or FMS made some productivity gains, just over half (57%) had been able to reap the full benefits associated with flexibility. Similarly, a survey conducted by Cranfield Institute of Technology for the Engineering Council found disappointing results. A quarter of companies surveyed were dissatisfied with the returns on their CAD/CAM investments and the figures were even worse for solid modelling, where one third said their installations were not living up to expectation. These findings were further confirmed in Ingersoll Engineers' (1986) survey of companies involved with CIM installations, where only approximately a half were seen as successful and profit improvement was patchy.

Further evidence is emerging indicating that the greater the level of integration of systems, the greater the problems in successful implementation. Waterlow and Monniot (1986) in their study of the state of the art in computer-aided production management, found that the greater the level of integration demanded by CAPM technologies, the lower the likelihood of successful exploitation.

Many reasons have been given for these difficulties: problems of getting 'machines to talk to each other', inappropriate systems architecture, lack of a longer term view of how the integrated system should evolve, etc. It is not the purpose of this paper to add to this technical debate, but to look at some of the more human, managerial and organizational problems that contribute to the disappointing results experience by companies attempting to implement CIM.

Many consultants (e.g. Hartland-Swann 1986) argue that the application of computers to manufacturing situations should be the final stage in the process of planning and implementing CIM. The 'Simplify, Automate and Integrate' recipe for CIM is based on the observation that most companies' manufacturing arrangements have evolved over long periods of time, and are usually the result of series of 'bolt on' adaptations, rather than being the product of a well thought through design. The exhortation to simplify and automate before integrating is aimed at making companies rethink their manufacturing arrangements prior to applying computers. It is essentially a systematic argument for getting the manufacturing system 'right as a whole' rather than trying to increase the efficiency of sub-systems.

Our research into the problems that manufacturing companies have experienced in introducing new technology supports this view, and suggests that not only is it necessary for companies radically to rethink their manufacturing strategies and arrangements in their steps towards CIM, but also they need to adopt implementation methodologies which incorporate approaches to managing change which are appropriate to the radical changes required.

2. Background and preliminary research

There is widespread agreement that the low success rate is the result of difficulties in implementation rather than inherent problems in the technology itself. What is

Authors: Professor Stuart Smith and Dr David Tranfield, Change Management Research Unit, Sheffield Business School, Sheffield City Polytechnic, Sheffield, UK.

Reprinted courtesy of Taylor & Francis from *International Journal of Computer-Integrated Manufacturing*, vol. 2, no. 3.

more, the most significant implementation problems are seen as being associated with the management and organizational issues involved rather than technical commissioning or resistance at a shop floor level.

New technology by itself does little to increase competitiveness; it is how it is exploited that counts. Substantial changes in competitiveness are only achieved by management fundamentally rethinking how new technology can be exploited in developing business strategy.

However, our early discussions also suggested that it was not just a matter of policy that was critical. Senior management needed not only to understand and be committed to capital investment and its business effects, but also to think through and support the organizational and managerial changes necessary for its implementation. This meant bringing together functions which had for years existed separately, for the new technology was not respectful of traditional boundaries. As we talked to more and more companies about their experience, it became clearer that the second industrial revolution is more than simply a technological revolution. It necessitates a revolution in how managers think and organize manufacturing.

The argument that a shift in strategic thinking regarding the role of manufacturing is needed, is not novel. Many writers in recent years (e.g. Schonberger 1986, Skinner 1985, Hayes & Wheelwright 1984, Hill 1985) have argued that the demise of Western manufacturing companies reflects an inability to rethink the role of manufacturing as a strategic competitive tool. This is corroborated in our experience, by the fact that despite these exhortations and the wide usage of the term 'manufacturing strategy' in the literature, very few companies explicate one in practice. Clearly, further integration within the manufacturing system, with other business functions, and with suppliers is both made possible by new technology and is a requirement for its successful competitive exploitation. All these arguments point to the need for the development of a strategic methodology for implementing CIM.

Our research revealed that although some companies had encountered technical problems in installation and commissioning, most, like those cited in other research, did not regard these as crucial in determining overall success. Frequently, companies reported difficulties with suppliers and the problems of operating at the forefront of technological development, but these seemed to be problems which most of them were familiar with and had experience of handling. By far the most significant issue that pervaded our discussions was the notion that the exploitation required a radical change in how management think and organize manufacturing. Some seemed to have been successful in bringing about this revolution, others less so. It seemed less difficult in greenfield implementations, in smaller companies and in situations where there had been some powerful external threat to make radical change more palatable.

3. The key determinant of success

The key determinant of success was the way in which the companies had tackled the issue of managing change, and particularly whether they had managed to break the inertia of traditional-taken-for granted ways of thinking. If they had seen the implementation as a step-function change and had adopted a change methodology suitable for achieving this type of organization-wide change, then they were likely to succeed. Conversely, if they had simply seen new technology as a way of making the existing organization more efficient, then they were unlikely to reap the potential rewards.

In addition there were those who recognized that opportunities for quantum leaps in competitiveness were being offered, but did not recognize that this could only by achieved by quantum leaps in management thinking and organization. These companies tended to adopt change methodologies in line with their previous experience of implementing technology. That is, they regarded new technology as creating small-scale knock-on effects which were best coped with in piecemeal fashion, as they emerged. However, success was seen as 'patchy' with this approach, confirming the research findings reported earlier in this paper.

The main point seemed to be that the impact of applying computers and revised production methodologies to manufacturing companies was so extensive as to require changes of a different order. Manufacturing was not simply being asked to be more efficient, but to transform what was possible, and in doing so to challenge not only existing custom and practice, but also to place new demands on its capacity to manage step-function change. The catalyst required to bring this about was the change required in management outlook in implementing CIM.

4. Theoretical perspective: the theory of change

Because implementing CIM involves organizational transition, the main theoretical perspective that we applied to the research was taken from the theory of change. This is a relatively new and diffuse field, but Smith (1982) outlines two distinct ways in which change may be conceptualized. Firstly, there is morphostatic change which preserves order by treating disturbance as external noise requiring minor adjustments or blocking out. Change in a morphostatic sense is therefore incremental. Secondly, there is morphogenic change which treats disturbance as information about internal conditions and suggests that the system should respond by altering orders. In this way, change in a morphogenic sense produces a logically different order than that which came before.

Our thesis is that this simple framework possesses much credence in explaining the difficulties being encountered in implementation. In the past, technological change in organizations has been introduced by managers usually on an implicit or explicit morphostatic model. This has been effective where the technology requires only an incremental readjustment, such as a limited redesign of jobs or some skills training, within an overall system which remained the same. However, where the technology does not just make the existing system more efficient, but transforms what is possible and has implications for the organization on a wide variety of fronts including corporate strategy, organization structure, job design, management attitudes, etc., then clearly a morphostatic model becomes inappropriate. What is required is that management does not just go through the morphostatic loop: what are the disturbances?—what do we need to adjust? but instigates a change programme inside the system designed to question the basic order and bring about a revolutionary or morphogenic change: what is our current situation?—where are we going?—what are the steps to get there?

This suggests a 'vision building' approach to change requiring a clear view of the horizon toward which the organization is moving. Morphogenic change can only be planned and achieved when such a 'vision' of the future has been developed. Morphogenic change is revolutionary change on a wide variety of fronts. The actual change itself becomes a series of incremental steps toward a 'vision of the future', thus 'smoothing' the step-function.

Morphogenic change is almost always seen as uncomfortable by those inside the system. This kind of change generates high levels of insecurity. In this sense, one can understand managerial reluctance to inaugurate morphogenic change. Revolutions, by their nature, signal the overthrow of existing organizational arrangements, and involve the transition to a new order. Individuals may end up with new 'empires', reduced 'empires' or no 'empires' at all, which can lead to caution among managers, and the need for new skills and work arrangements can lead to resistance elsewhere in the organization. Not only this, but the embeddedness of most managers within the existing system often makes it difficult for them to develop sufficient distance to diagnose the extent and scope of the changes required and how they might be achieved. Given all of this, it is not surprising that fundamental change is so difficult to achieve, that organizations contain so much inertia, and often tend to continue as they are.

Unfortunately, as many companies appear to have found out to their cost, computers by themselves do not create morphogenic change—the second industrial revolution. Computers applied to existing facilities and systems merely make them more efficient, achieving morphostatic change. Quantum leaps are achieved only when management fundamentally reapraise how computers can be exploited in enhancing the competitive position of the company, defining the technological requirement to facilitate this, whilst at the same time redesigning the organization where required. From our research, companies achieving successful implementation were seen as implicitly or explicitly using a morphogenic change model rather than a morphostatic model. Furthermore, these companies were prepared to treat implementation as multi-disciplinary distinguishing between implementation (incorporating both user satisfaction and business benefit) and technical installation. Lastly, successful implementers rethought business operations first and considered technology last in the implementation process.

We found this approach to thinking about different models of change useful in making sense of what companies were telling us about implementation. In addition, it provided the theoretical underpinning of an outline methodology, the aim of which was to bring about planned morphogenic change—'a smooth revolution'.

5. Applications studied

Nine examples of CIM were studied. These included various aspects of CIM, ranging from one company making its first step toward CIM by investing in MRPII, to another with a multi-million pound investment in a flexible assembly system to manufacture diesel engines. We also studied two examples of flexible mechanical handling systems, one in cutting and handling glass from a float plant, and one chemically treating sub-assemblies in a composite shop; three FMS in a machine shop of an aerospace company, in a detail shop in an aerospace company, and in a milling shop in an aerospace company. Indeed, this latter application was to be linked to an existing Local Area Network which connected a factory management system and a CAE system.

Six applications were entirely historical in the sense that they had been implemented and were operational. Two were in the process of being implemented, and although technically commissioned, were not fully operational. A further application was in the early stages of implementation, where the decisions to invest were still being taken. The sample chosen to examine was not 'designed' but was dependent on identifying companies who were prepared to collaborate and provide sufficent access to investigate the application in depth. However, we do consider that the sample was sufficiently random to serve the purpose of the study.

The study took place over the period September 1986 to July 1987. Data was collected by interviewing, participant observation, and by examining company documentation which was made freely available by the companies. Altogether, nine applications were selected, located in seven different companies throughout the UK. Approximately 50 interviewees contributed to the study, including 12 managing, manufacturing, or other directors of the collaborating companies. Other interviewees comprised line managers down to supervisory levels, project leaders, members of steering groups, and specialist advisers. Approximately 350–400 hours of interviews were conducted throughtout the project.

6. Approach to developing a methodology

Several questions exercized our minds during the research:

what is a realistic, useful outcome from the research?
is it possible to develop an 'action recipe' for implementation?
if so, what are useful components of the recipe?
what are the main issues it should address?

It is clear from everyday experience such as following recipes to make cakes or assemble do-it-yourself furniture that it is impossible to cover every aspect in a guide to practical action, such that success is always guaranteed. Implementing CIM is a vastly more complex act than baking a cake or assembling a wardrobe. Clearly, any recipe for action is bound to be limited in scope and will inevitably rely on the inherent intelligence, experience and skills of the implementors in practice.

In examining everyday experience it is apparent that the most useful feature of methodologies or recipes is that they provide a conceptual map to guide practice. A useful map contains sufficient detail to inform the practitioner what to do and how to do it, but inevitably some details are omitted. For example, most recipes do not inform you how to crack an egg, or drill a hole. In fact, the most useful recipes for action focus on the tricky, more difficult bits and skirt over the parts where people have sufficient experience to be successful without detailed guidance.

The other basic features of recipes worth noting is that they focus on what and how. They inform the user about what needs to be attended to and how this should be achieved. The do-it-yourself wardrobe instructions for action usually contains a comprehensive list of parts and a method for transforming them into a wardrobe. The cake recipe contains a list of ingredients and a method for mixing them and transforming them into a cake.

We decided to take these observations from everyday experiences and use them to shape the design specification for a useful implementation methodology.

(1) The methodology should focus on those aspects of implementations which are the most 'tricky' i.e. out of the ordinary compared with other experiences of implementing technology.
(2) The methodology should provide a conceptual framework for thinking about and making sense of the implementation process, incorporating the 'best' features from the implicit and explicit methods used by the companies studies.
(3) The methodology should highlight the issues that need to be attended to in implementation. It should have the capacity to address a wide range of technological, managerial and organizational dimensions.
(4) The methodology should suggest how the transformation, i.e. implementation, can be achieved. It should be a methodology for achieving morphogenic change.

7. Elements of a methodology

From our sample we were able to distill nine elements which were keys to a successful implementation methodology. Not all of our examples exhibited all of these elements, but each of these elements recurred sufficiently regularly to justify their inclusion in this overall list. Sometimes respondents referred to these elements as serious omissions in the implementation process. Specific case examples have been left out in the interests of brevity.

(1) Business driven—the one and only reason for investing in AMT was to improve the competitiveness of the company. Any other reason was peripheral to this.
(2) Back to basics rethink—implementation of new technology usually involves step-function (morphogenic) change rather than incremental (morphostatic) change. Step-function change required primarily a fundamental analysis of the business situation and reappraisal of business objectives along various dimensions of competitiveness. A technological horizon and an organization design and redesign are important elements of the future vision which has to be built as part of the rethinking process.
(3) Top management driven—our own research

agrees with many others that the best way to achieve morphogenic change is top down. Usually this involves at least one senior management 'champion' in the first instance, but needs to receive much wider senior management support if it is to be successful.

(4) Front end—back end; business and managerial issues are best considered at the 'front end' of the change process, rather than being the tail on the technological dog. Implicitly or explicitly, successful implementations of change worked to a model which emphasized first, the specification of a business and management strategy, secondly, the ascertainment of a manufacturing strategy, and then only at the back end of the process, the development of a technological horizon capable of delivering.

(5) Everybody on board—any effective implementation must be cascaded down throughout the organization, particularly the management hierarchy. Without this, morphogenesis cannot be achieved and new technology is implemented into a situation of traditional structures, attitudes and work practices. This is a situation from our evidence where it is most likely to fail.

(6) Time scales—both short and long term are important, particularly as change is brought about in the step-function revolutions followed by periods of relative stability. More is said about this below.

Whilst all of these elements were important, we were surprised to hear engineers as well as managers talk to us about this importance of cultural change as being the key issue for successful implementation. Increasingly we have come to view culture as the shared, taken-for-granted (TFG) assumptions of the members of the company. For example, companies who had introduced just-in-time (JIT) production methodologies reported the importance of changing TFGs embedded in the previous batch scheduling philosophies.

Culture is a difficult concept to grasp when defined as shared TFG assumptions, and a number of points are worth noting. Whilst on the one hand it is the shared TFGs that make social and organizational life possible, on the other it is these same TFGs that provide structure and security and hence inhibit change and adaptability. Often they are implicit and can only be inferred from behaviour, being rarely up for scrutiny and making overt appearances only when some infringement occurs. However, they do tend to permeate the whole of organizational life, giving a sense of overall purpose, and enabling members of the organization to identify with the whole, as well as being a source of control over individual behaviour. Successful implementation, from our experience of the manufacturing sample, requires that any implementation methodology addresses these company-wide issues prior to installation, for changes in these TGS, such as changes in attitudes to quality, costs, productivity, lead times, customer service, etc. are key to institutionalizing morphogenic change.

Whilst the above gives a list of 'ingredients', it does not specify the process whereby they can be organized to produce the desired outcome. This is the job of the next section.

8. Outline of the methodology

Our research aimed to produce a form and scope of a practical methodology for implementation. Our initial thoughts had been to produce a detailed methodology probably related to project management. However, experiences in researching the companies involved led us to the conclusion that the development of a more strategic methodology than mere project management was needed, for two reasons.

Firstly, strategic variables were of prime concern to management and it was this strategic impact which made AMT implementation different from other forms of technological innovation. Secondly, successful implementation seemed to be governed by whether it was seen, implicitly or explicitly, as part of the strategic response of the company, central to business performance. Strategy was certainly the 'tricky' bit that our methodology needed to address in the first instance.

In the event, only one of the companies we studied had adopted a formal methodology for implementing AMT, and essentially this was a project management methodology for ensuring agreement by the wide-ranging group of people involved in the implementation. However, all of the other companies had evolved implicit methodologies on the basis of their previous experience of implementation, and their view of particular needs of the new technology they were introducing. It quickly became clear from the managers involved that whilst they were enthused by the approach taken by themselves, there was a great need for a systematic methodolgy.

Our methodology begins with an audit of the current company business situation and manufacturing strategy, which incorporates a clear assessment of what is going well, what opportunities there are, and where problem areas exist. Sometimes this audit is the response to a crisis and sometimes the response to new knowledge indentified with a 'champion' of a particular cause. Inevitably, this audit involves the exploration of various alternatives along a variety of competitive dimensions (cost, productivity, quality, lead time, etc.) and forms the basis for the development of manufacturing strategy.

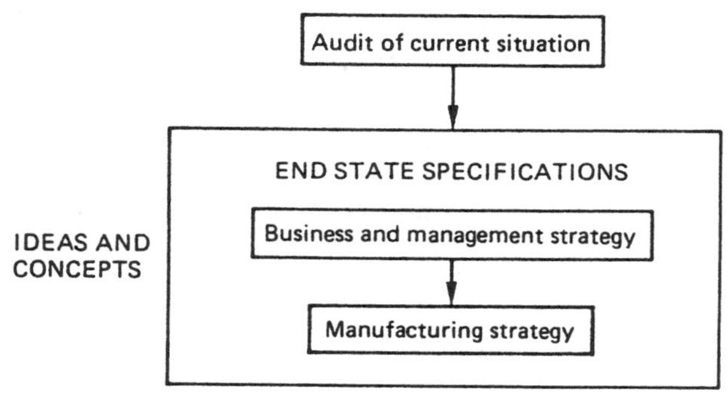

Figure 1.

Developing a manufacturing strategy frequently demands that senior manufacturing managers fundamentally rethink their approach to manufacture, considering the kinds of questions outlined earlier. In this way, the methodology begins with ideas and concepts about business and manufacturing strategy in relation to environmental opportunities, pressures and constraints (Fig. 1).

Once top management has agreed the current situation and the way forward, the methodology moves into a design for delivery phase in both organizational and technological terms. On the organization side, the methodology involves restructuring and culture change, both designed to bring about an organizational delivery system to exploit the new technology to the full and contribute to the agreed end-state. At the same time, production managers, engineers and computer scientists need to be investigating and designing a detailed technological horizon to deliver the manufacturing strategy (Fig. 2).

Finally, the methodology is delivered into the company by the development of three interrelated support strategies, the objective of which is to deliver the designs outlined earlier. These three are outlined in Fig. 3 and

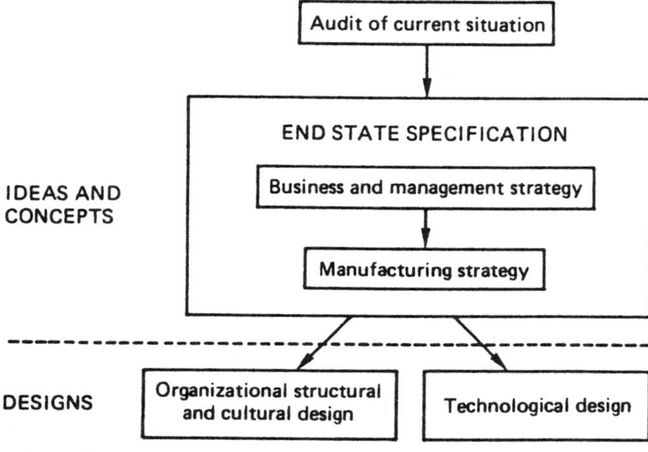

Figure 2.

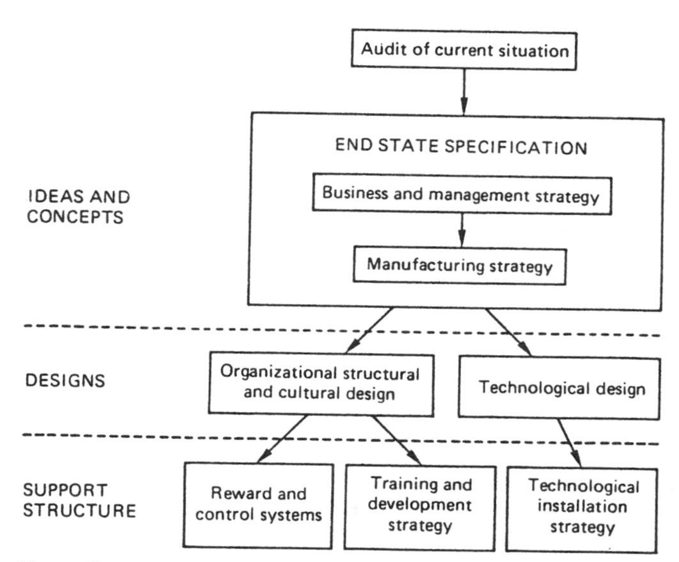

Figure 3.

involve on the one side a strategy for technological installation, and on the other, structural and cultural change. These are brought about by changing regard and control systems, and by devising a training and development strategy to support these changes. Nowhere is the importance of the latter better illustrated than in the implementation of MRPII, where attitude change as well as understanding is vital it the new system is to avoid problems of data capture.

9. Making the methodology practical for managerial use

Whilst the methodology briefly outlined above provides a guide for implementation, it does not identify the key managerial activities involved, nor does it mention time scales and the pace of change. From our research we have been able to identify the following four key managerial activities in implementation.

(1) Audit and Vision Build—this has been outlined above and is a top management activity which involves looking outward from the company and identifying market standing on key competitive dimensions. This analysis and vision building activity may use Boston Consulting Group (SWOT) analysis or stakeholder analysis. In any event, the main aim is to identify a business plan complete with a set of central values (no more than 4–6) with which company members can identify.

(2) The Cascade—once the initial work has been done and agreed by the top management, then the 'message' (both business plan and central values)

can be cascaded through the management team. This may be done after structural reorganization, and will usually involve the use of training and development as well as change in reward and control systems. A cascade is usually best handled by a series of workshops in which senior managers present a view of the future, and workshop delegates are invited to comment, discuss, and look at the implications for their own areas of work, as well as feedback upon the ideas given to them. This cascade structure and may consume large quantities of resource.

(3) The Sprint—time scales are of vital importance in implementation. Change usually comes about in short bursts of revolution followed by periods of relative stability. Pacing the change can be as important in success as the content of the change itself. Sprints may be said to have the following characteristics:

    (*a*) they are managed from the very top of the company;
    (*b*) they are focussed on one or several key dimensions of competitiveness central to the survival and development of the company, e.g. cost reduction, quality improvement, lead time reduction, etc.;
    (*c*) they involve all members of the company and are seen to do so;
    (*d*) they have measurable targets of success, e.g. costs remain the same for the next x months whilst new technology is installed; quality improved by $x\%$ whilst turnover doubles in the next y months; stock turnover increases x-fold in the next period; and
    (*e*) they have a fixed time frame in which targets must be achieved.

The sprint is the first company-wide activity in managing strategic change inside the organization. It requires clear direction from the senior management of the company using the methodology outlined earlier. Sprinting is an activity which can engender much support and create much enthusiasm within the company. Sprints consume much resource and may be associated with a 'cascade'; however, sprints are not sufficient if change is to be sustained over the medium and long term and the organization is not to slip back into crisis once the initial burst of energy created by the sprint is dissipated.

(4) The Performance Ratchet—the aim of the sprint is to bring an ailing company back 'up to speed' and make it competitive once more. It is the prime mechanism for bringing about morphogenic change. However, if sprinting encapsulates the idea of morphogenic change, then the performance ratchet encapsulates the idea of morphostatic change.

The energy, commitment and enthusiasm from management and organization members is difficult to sustain over a protracted period. Indeed, the rate of change during the sprint would itself prove a major stumbling block to organisational effectiveness were it to continue unabated for a great length of time. Once the company is launched on its new path, with the accompanying technological advance, a further mechanism is required to ensure that the technology is exploited to maximum business effect. The performance ratchet does this by ensuring that minor incremental (morphostatic) changes are made by those most appropriate to initiate and support them.

The performance ratchet usually takes the form of senior management requesting from operating divisions or sections of the company the kind of performance improvements that they (the operating divisions or sections) believe that they could initiate and develop over the next time period, coupled with estimates of the degree of certainty that these performance improvements might be achieved. These may or may not be approved depending on how well they fit into an overall business plan, but the establishment of a performance ratchet such as this has the effect of continuing the impetus for change, and involving those concerned with the delivery of change in defining the targets. Performance ratchets are not the operating plan of manufacturing departments. They are concerned with change rather than delivery, and provide the link between strategic direction and operations during times of morphostatic change.

These four activities form the practical basis of an implementation methodology for managers.

10. Conclusion

The methodology which has been described outlines the strategic orientation required to establish a direction and pace for technological change toward CIM. It specifies the form and scope of an implementation methodology and sets out the strategic dimensions important for success. Particularly, it focuses on the need for a revolutionary orientation on a wide yet integrated set of fronts, which emphasize the development of a management and manufacturing strategy, and an organization redesign, as variables of importance equal to technological installation.

A feature of strategic methodologies, however, is that by definition they tend to be non-specific. In many ways, they resemble the derivation of an equation specifying relations between variables without assigning values to them. In this way, work is required to further refine and develop the methodology derived, specifying the details in various stages and applying the methodology to various types of company and technology.

Lastly, a series of studies is required to test and validate the methodology in live company environments. It is our view that this will produce a much needed methodology for quick and effective implementation, of use to consultants, practising managers, and project teams.

References

HARTLAND-SWANN, J. 1986, CIM: hit or myth, *Industrial Computing*.

HAYES, R. H. and WHEELWRIGHT, S. C. 1984, *Restoring our Competitive Edge: Competing Through Manufacturing* (Wiley, Chicester).

HILL, T. 1985, *Manufacturing Strategy: The Strategic Management of the Manfacturing Function* (Macmillan, London).

INGERSOLL ENGINEERS 1986, *Management Attitudes to Change: Results of a MORI Survey* (Ingersoll Engineers, Bourton, Rugby).

SCHONBERGER, R. J. 1986, *World Class Manufacturing: The Lessons of Simplicity Applied* (The Free Press, New York and London).

SKINNER, W. 1985, *Manufacturing: the Formidable Competitive Weapon* (Wiley, Chichester).

SMITH, K. K. 1982, *Philosophical Problems in Thinking about Organsational Change in Goodman & Associates Change in Organisations* (Jossey Bass, San Francisco and London).

VOSS, C. 1985, Success and failure in advanced manufacturing technology, Working Paper, Warwick University.

WATERLOW, G. and MONNIOT, J. P. 1986, A study of the state of the art in computer aided production management in UK industry, ACME Directorate, Science & Engineering Research Council, Swindon.

HOW TO GUARANTEE CIM PROJECT MANAGEMENT SUCCESS

By

Dale A. Smith
Dale Smith & Associates, Inc.
Dallas, Texas

The objective of the presentation is to provide a recipe for CIM Project Management success which includes proven project management tools and techniques. The intent is to offer tools to avoid the risks of project failure at the outset before the program is launched. As a result of attending this presentation, individuals will come away with a complete set of project management aids which will equip them to build the framework of a successful, professional project.

The presentation will also deal with the basic pitfalls to be avoided when progressing down the CIM path. The conclusion of the presentation will revolve around a risk evaluation exercise identifying areas of exposure before a project is started.

This presentation will provide a recipe for CIM Project Management success which includes a proven project management approach, tools, and techniques so that companies facing this challenge can learn from our experience. As a result of attending this presentation, individuals should come away with a complete set of project management aids which will equip them to build the framework of a successful CIM implementation.

The presentation offers a chronological path for successful project management and control. A detailed discussion of each tool used will take place, giving real examples of how the tool is to be applied in a practical environment. The basic tools presented include such things as: establishing project organization and structure, constructing a CIM strategic plan, project control and monitoring, education planning and execution, status reporting, problem reporting, operations analysis, conversion methodology, procedure manual creation, software evaluation and selection, integration and project evaluation. The presentation will also highlight some of the factors affecting the success of these tools.

PHASE I.

Our first step is the establishment of a project organization and the development and publication of a detailed CIM strategy. The early assignment of responsibilities and identification of critical items will eliminate conflict, encourage teamwork, facilitate communications, and establish accountability at the outset.

The project implementation strategy will direct attention to critical activities and eliminate wasting resources on problems and issues that do not affect schedule, cost, or technical performance.

The above process should be accomplished through the use of a proven implementation methodology which includes tools previously developed. These plans should be evolved through a project team training program to ensure ownership and accountability at the project team and task team levels.

Conducting an extensive manufacturing environment upgrade such as a Computer Integrated Manufacturing (CIM) environment will be a major undertaking which will require many levels of organizational involvement. The project organization should include a steering committee, implementation project team, special task teams from the functional areas to be involved, and experienced implementation and technical support consultants.

The steering committee should be made up of senior management staff. The implementation project manager advises this committee of the progress of the project on a regular

Reprinted from *1988 IIE International Industrial Engineering Conference Proceedings.*

basis. The steering committee is responsible for project guidance, financial support, and project participation.

The project team should be made up of key representatives from each business discipline, including the Data Processing Department.

Special task teams should be formed as needed to address detailed implementation tasks, such as developing desk procedures, conducting "hands-on" training and providing on-going support. The teams will be led by the supervisor most affected, and staffed by personnel designated as their area's primary interface.

Experienced implementation consultants should be used to provide project start-up and management direction; assisting personnel in the formulation of detailed action plans with a goal of making the project team self-sufficient at the earliest possible time. Consultants also provide an objective review of project team direction and accomplishment on a regular basis.

The education strategies required to effectively implement the CIM concept and operate a totally integrated environment are based on the concept that CIM is a total management system dealing not only with hardware and software, but mainly with culture. You should recognize that it is not just the installation of a data processing software system. You should further recognize that CIM encompasses Marketing, Production, Engineering, Finance, Materials, etc., and as such has to be a company wide "Team" effort. Education is recognized as a critical factor in determining implementation success or failure. You should understand that any short-cuts in the education process in the beginning would cause ineffective implementation and system utilization requiring future re-education resulting in additional costs and/or delayed benefits.

The education strategy establishes the need for education of each facility in total, beginning with senior management and working through the entire organization utilizing targeted materials and educational media on a "need to know" basis. Users at all levels need to understand how this new "way of life" will assist them in their day-to-day functions.

The initial efforts should be dedicated to educating senior management and the implementation team to perpetuate the transfer of knowledge throughout the facility.

Education is distinguished from training in that it presents industry standard concepts, principles, and technologies on an overview basis, whereas training teaches specific features of the system being used and the desk procedures to perform business functions utilizing the new system available.

The first step in the education planning process is to conduct an education needs assessment. The objective here is to assess the level of knowledge, expertise and system application skills which the operational groups within the company must acquire to enhance their short term and long term capabilities.

The information gathered forms the basis of a well defined statement of educational requirements for the organization. The content of this statement of requirements is as follows:

\ Identification of key educational deficiencies and opportunities to be addressed. The objective here is to identify conceptual materials that would maximize understanding of the tools provided by the new system.

\ A summarized analysis of educational requirements with a structured education program to meet those requirements.

\ Identification of who should be included in the formal education process from top management down through the organization. A class matrix format allows the generation of education days by class and by job function. This provides a budgeting tool to project both costs and man-hours required to complete the program.

The education plan should be a complete array of conceptual educational offerings. Classes should be customized to reflect the environment and meet specific educational goals of each function utilizing examples of products, part numbers, bills of material, routings and function goals. It is our experience that the materials become much more meaningful and effective when presented utilizing company situations and products. Although the basic intent of the classes should be of a conceptual nature as compared to software specific training, it is helpful to show the students examples from their actual environment.

Since the CIM project will be viewed as a full business integration rather than just a computer system installation, the involvement and support of all personnel is essential to its overall success. A well organized and targeted publicity program linked closely with education must be structured and continued throughout the life of the project to help meet its objective.

Necessary funds to carry out this publicity program must be part of your project budget.

PHASE II.

In this phase your company should develop and execute an analysis of each business function to be impacted. We feel that effective resolution of manufacturing management problems requires systematic analysis of the functional areas within the organization. Records could be incomplete or inconsistent. Personnel often have limited knowledge or understanding of operations outside their own areas of responsibility. The analysis methodology provides an effective method of gathering information from diverse sources and assembling information into a coherent picture of the organization.

The objectives of the operations analysis are:

\ Rapidly study and document current information handling as it relates to all aspects of the business.

\ Capture live samples of all relative documents, forms, reports, etc., and record related volume statistics.

\ Identify and document all concerns and/or weaknesses which would hamper the successful implementation of a computer integrated manufacturing concept.

\ Identify specific benefits which could be achieved through the implementation of a computer integrated management system.

The operations analysis yielded the following benefits:

\ Establishes the necessary benchmarks upon which to base improvement plans and performance measurements.

\ Provides the foundation for developing the functional specifications required for hardware/software selection.

\ Provides a clear understanding of the symptoms and problems of each functional area requiring prerequisite or post implementation attention.

\ Identifies specific benefits which could be achieved through the implementation of the action plans developed.

\ Provides a realistic view of the magnitude of the task at hand.

PHASE III.

As a result of the operational analysis, specific systems requirements will be incorporated into the Request for Quotation (RFQ) document.

An extensive rough draft of an RFQ document should be prepared from industry standard requirements and the basic requirements developed in the strategic planning document. The rough draft will be reviewed by the project team members and other operational personnel to ensure that it includes all specific requirements derived from the operational analysis. The RFQ document should be developed using a computer assisted decision support model with each item being weighed by level of importance to the business, and each module being balanced for equal weight regardless of the number of items within it. Each module should further be weighed by level of importance in relation to the critical needs to the business. These factors will be loaded into the decision model and subsequently used to rate each vendor's response, resulting in a percent of fit by module and overall fit for each vendor. Both users and data processing personnel will review the final RFQ document to set the stage for their further assistance in final selection and contract negotiations.

Pre-screening and First-cut

The RFQ document will be issued to vendors as a result of our telephone pre-screening survey of the most prominent suppliers of CIM support hardware and software. The selected vendors should be chosen based upon their ability to operate within the pre-defined environment. The second requirement to be met by the vendors is a professed integrated offering of all elements under consideration.

Vendor Responses to RFQ

Each of the selected vendors should be provided a copy of the RFQ document and requested to respond within three weeks.

The vendors should be asked to respond to each functional spec with a point value depending on their ability to fulfill the requirements of each element.

Vendors should also be asked to enter modification and development estimates where necessary and bidable.

RFQ Response Evaluation

Upon receiving the RFQ input from all vendors, the Project Manager, Data Processing Representative, and Industry Consultant should complete a preliminary review of each response. The data should be loaded into the decision support model in order to provide stack ranked visibility for presentation to the overall project team.

The vendor responses should be entered into two principle worksheets:

1. Software/Hardware Evaluation/ Comparison

2. Software/Hardware Investment Responses

The details in these two worksheets should be rolled up providing a:

\ Vendor/Module Percent Fit Comparison by Application Module

\ Overall Vendor Totals Comparison

\ Overall Vendor Ranking Table

Second-Cut

The Overall Vendor Totals Comparison Report provides the percent of fit in total for each vendor, a summary of the estimated modification man-weeks and cost, and the investment summary totals for each cost element. A ranking table should be subsequently produced showing the relative ranking for each of the critical factors of the evaluation. Totaling the points of the ranking provides an overall weighed ranking. At this time you should be able to identify all critical concerns about each vendor's offering. With this information in hand, a second-cut can be made to narrow the field. Up to this point vendors should be anonymous to the evaluators to assure unbiased appraisals.

Semifinalist Vendor Presentations

The remaining vendors should be requested to present a one-day full review of their proposed solution at your facility. An agenda organized in the same order as the RFQ is recommended and should be followed by the presentors. Members of the project team and other concerned functional personnel should attend their particular sections of the vendor presentations. Evaluation critique forms should be developed for each module and provided to the attendees. Reviewers should forward their evaluations to the principal functional representative on the project team for consolidation into the final evaluation for each module. All evaluators should be encouraged to ask appropriate questions to qualify the vendor as to how they would accomplish the specific tasks identified in their RFQ response.

Finalist Selection

As a result of the vendor presentations, the team elects to concentrate its final evaluations upon one or more vendors.

Selected Software/Hardware Vendors

The remaining finalists are requested to conduct on-line demonstrations at their facility. These presentations will serve to further qualify the vendor's capabilities relative to your needs, as well as to allow you to compare the user friendliness of each offering. Additionally, vendor supplied references for similar manufacturing companies should be queried by telephone as to their satisfaction. A software/hardware evaluation and selection questionnaire for telephone/user visit screening should be prepared for this purpose. A final comparative summary and recommendation document should be prepared highlighting your findings of the major strengths and weaknesses of each of the finalist's offerings, justifying your project team's unanimous recommendation.

PHASE IV.

The Operations Analysis phase provides identification and prioritization of implementation prerequisites. In order to provide corrective action and follow-up of these existing conditions to assure a successful implementation, an implementation prerequisite phase should be established. As a result of our research we have learned that many companies falsely expected improved disciplines and significant return-on-investment to be an automatic result of installing the CIM concept.

Installation on the computer is a "far cry" from implementation into the daily functional user environment. Those that have truly succeeded in this endeavor have found that a significant amount of operational improvement can and must be made prior to bringing the new system on-line. Otherwise, the new system will have no better credibility or opportunity for success than the old one.

In order to effectively prepare the environment for implementation, a benchmark for the business must be established through the Operational Analysis. This process captures the current operating environment, volume statistics, business problems, and their relative impact on the business. Prioritized corrective action and performance measurement programs can then be put in-place to bring these problem areas in line in preparation for system turn-on.

Primary elements and issues to be considered in prerequisite activities include, but are not limited to:

\ organizational issues

\ data accuracy clean up, e.g., inventory integrity, bills of material integrity, process and routings

\ establishing policies and procedures

\ physical plant requirements and modifications

\ preparation for hardware/software installation

\ critical operational problems

\ defining and implementing performance measurements

PHASE V.

To effectively carry out the many activities required to assure a timely and highly user accepted turn-on of each system feature, a closely integrated six element process is recommended. This consists of detailed implementation planning, business/system analysis, demo simulation, policies and procedures, cut-over pilot, system turn-on and pre & post implementation operational effectiveness evaluations.

Detailed Implementation Planning

The detailed project implementation plan directs attention to critical activities and eliminates wasting resources on problems that do not affect schedule, cost, or technical performance. Your plan should be an ever expanding form of the master plan in your strategy document.

Business/System Analysis

Once the software selection is made, the software supplier's system documentation should be compared with the Operations Analysis documentation to identify data elements to be defined, forms to be designed and ordered, transactions and reports to be tested and used, conversions, modifications, development and interfaces to other systems, etc. Each of these and many other decisions need to be reflected in the implementation plan.

Demo Simulation

Your implementation methodology should guide you through a systematic definition, test, documentation and training process within a mini-company demo prior to live turn-on.

The demo simulation begins immediately following new hardware/ software installation shakedown.

A mini environment should be configured to reflect the functional world defined by the operations and systems analyses. This data base should be copied at different stages of simulation to allow for restart should subsequent tests fail and contaminate the data. Where possible, the simulation should be conducted as a walk through of the functional flow defined by the Operations Analysis. Users can be trained without fear of contaminating live data. Conversions, reports and transactions will first be tested in the demo. Since the simulation reviews the entire startup process, mechanical conversions and fail safe/recovery plans are tested to assure proper operation.

The demo simulation approach is used for the initial introduction of all new elements in the system prior to its release into the live production environment.

Policies and Procedures

The implementation teams must continue the accumulation, publication, and incorporation of needed policies and desk procedures in concert with the demo simulation process.

The intent of this activity is to document the final main guidelines for running the business using the new system.

The system should be documented sufficiently to allow for the effective on-going operation, trouble shooting, and recovery of the hardware, software, and functional operational procedures.

Needed policies and procedures must be identified and prioritized during the operations analysis, business/system analysis and demo simulation phases. Appropriate information is accumulated and then documented in a published company standard format. Approval and implementation of the resulting guidelines are based upon company policy.

Separate central libraries should be established for system and general business documentation. Wherever possible, the software supplier's documentation is the principal source. Where supplier documents do not adequately describe a business/system environment, an internally generated desk procedure, users/operators manual, or reference guide should be created using system input/output where possible.

The system should be documented as an on-going activity throughout the life of the project to minimize the traditional implementation panic and incomplete results.

Cut-over Pilot

Once all the preparatory activities are completed, a live volume test of the new system can be conducted for a short, predetermined period of time. This pilot run begins with a full data conversion and validation, allowing the entire user community to exercise the system in actual daily activities, complete a period closing cycle, and generally provide the project team and senior management with the required visibility of the company's readiness to fully implement the new system. This step assures that all preparatory actions necessary to yield an optimum system initialization and on-going utilization has been accomplished and that the turn-on plan is effectively executed.

As a result of this cut-over pilot, a GO/NO GO review should be conducted by the project team, independently reviewed by your support consultant with the results presented to the steering committee. A formal management approval or decision to delay or turn-on should be obtained.

System Turn-on/Cut-over

With all of the aforementioned activities complete, you can confidently cut-over to the new (full) system. After cut-over you will only encountered a few minor problems which are easily resolved. The system should now be functioning well with no problems.

Pre & Post Operational Effectiveness Evaluation

A periodic review of the on-going effectiveness of the project implementation must be conducted to ensure that goals and objectives are achieved as planned.

The continued use of this structured project evaluation service is recommended to ensure that the project stays on track during the implementation. Your implementation support consultant should stay with your project until the system is fine-tuned and you become totally self-sufficient.

FACTORS AFFECTING YOUR SUCCESS

The level of success of any project is directly related to positive as well as negative influences. An important positive aspect includes management support exhibited by the President and his staff. These individuals must give the project the level of importance needed to sustain the project through the duration. They must also remain active throughout the life cycle of the project. This, tied to user enthusiasm, support and dedication to the cause, will keep up project momentum and participation while the users performed their regular jobs in addition to project tasks.

Biographical Sketch

Dale A. Smith, President, Dale Smith & Associates, Inc. Mr. Smith has over twenty seven years professional experience in Materials Management, Data Processing, and Manufacturing Industry Consulting which offers in-depth experience in the design and implementation of Manufacturing Planning and Control systems. He has held both line and staff positions in Materials Management and Data Processing with total project responsibility for the implementation of several "Class A" MRP II systems.

Currently he is President of Dale Smith & Associates, a management consulting and education firm, which specializes in and is dedicated to quality management consulting in the area of successful MRP II, CIM, DRP, and JIT systems implementation and related education/training.

Prior to becoming an independent consultant, Mr. Smith was District Manager for a major software firm where he was responsible for assisting clients through the successful implementation of their manufacturing software products. He has served as a consultant and educator to the manufacturing industry, was Manager of Material Systems, Supervisor of Manufacturing Information Systems, and a Senior Analyst. His well-rounded experience offers a healthy blend of user and data processing talent.

Mr. Smith is certified (CPIM) by the American Production and Inventory Control Society, as well as an active member and guest speaker of the organization. He graduated from Washington University, St. Louis, Missouri, with a Bachelor of Science degree in Industrial Management.

Managing the Introduction of Advanced Manufacturing Technologies

JACK R. MEREDITH
College of Business Administration, University of Cincinnati, Cincinnati, OH 45221

STEPHEN G. GREEN
The Krannert Graduate School, Purdue University, West Lafayette, IN 47907

This article offers guidance based on a number of case studies to help manage the introduction and implementation of new technologies in manufacturing organizations. First, a behavioral and organizational framework is offered to aid in understanding the dynamics of technological change and to provide a basis for the recommendations. Then the recommendations are detailed, related back to the framework, and substantiated with insights from the case studies.

INTRODUCTION

Demands placed on industrial manufacturers throughout the world have escalated significantly in the last decade. New requirements include higher quality, faster response, and greater variety, typically all at a reduced cost. It is clear that the level of competition has increased substantially in this short period and that only the most responsive, best-managed firms will survive.

One major competitive strategy for many firms is the adoption of new, sophisticated, computer integrated manufacturing systems. These advanced technologies frequently offer consistent quality, quick response, high variety, and low unit cost all at the same time. Although initially expensive, the major impediment to their implementation has not been the capital investment required but rather the difficulty of managing their successful introduction and operation in the firm.

There are many obstacles to overcome in the process of implementing these advanced technologies [1, 2]. Because of their extensive level of integration, the hardware and software interfaces are particularly difficult. The solution to one hardware problem often creates new hardware problems in half-a-dozen other places. With software, the new problems created are invisible, frequently not being discovered until the final testing of the system. Furthermore, hardware solutions often create software problems, and vice versa. Frequently the best solution to a hardware problem is a software change, and the best solution to a software problem may be a hardware change.

But the successful integration of these systems transcends the solution of difficult hardware and software problems. Our studies indicate that the most difficult problems have been managerial and lie in the organizational and behavioral area. Whereas management was typically prepared for difficult hardware problems, and was expecting frustrat-

Reprinted with permission from *Manufacturing Review.*

ing and expensive software problems, they were invariably unprepared for the most difficult of all "people" problems. Based on these experiences, we have developed a number of recommendations for managers when introducing advanced technology into their firms. A simple framework is described next to give context to the organizational and behavioral recommendations that follow.

A FRAMEWORK FOR VIEWING NEW TECHNOLOGY INTRODUCTION

Current literature [3, 4, 5] on the introduction of new technology and its effect on the organization suggests that a somewhat different process is at work than the normal introduction of straightforward change into an organization's systems. First, the introduction of these new technologies is extremely complex, and yet, can be synergistic with other elements of the organization. In addition, significant changes to the working milieu are inevitable. Moreover, there are important "ripple effects" where change in one area of the organization leads to changes in other areas. In spite of these challenges, our experience indicates that the impact of new technology can often be moderated by various managerial interventions during the technology implementation process.

A simple framework to help understand the dynamics of new technology introduction is illustrated in Fig. 1. This framework will be referred to during the discussion of the recommendations. The elements of the framework are briefly described below.

The firm operates in a competitive environment for its products or services, the nature of which affects its strategy and its competitive advantage relative to other firms in its market. The introduction of new technology is often spurred by the competitive position and strategy of the firm relative to this competitive environment [6]. The strategy may be one of growth and market domination or one of defense and protection of markets.

At this point, a new technology is adopted to help implement the strategy of the firm. It may be either reactive or proactive and there may be a lengthy, formal justification process required or just an informal strategic approval of the technology [7, 8].

When the firm introduces new technology into its organization, that technology brings with it certain objective, physical characteristics, such as performance abilities, input requirements, and maintenance standards. If the benefits of the technology are to be attained, these characteristics must be recognized and dealt with; they cannot be ignored. Yet, they have a number of direct consequences for how the organization will look and operate in the future.

Figure 1 also illustrates that the impact of these organizational consequences of new technology typically results in tradeoffs at a number of levels. For example, the individual or group will lose certain desirable aspects of work because of the introduction of new technology but, at the same time, because of that technology, may gain in other ways. Clearly, the individual is significantly affected by the impact of the new technology.

Finally, the framework suggests that management can engage in a number of strategic interventions during the implementation process that can ameliorate the negative effects of the new technology and enhance its positive effects. Strategically-planned actions taken by management to train workers to use the new technology effectively and to

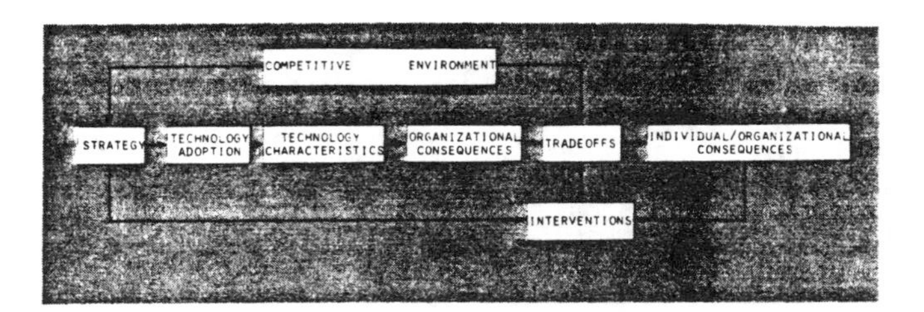

FIG. 1. Framework for new technology adoption

redesign work so it capitalizes on the opportunities presented by the new technology can turn minuses into pluses.

RECOMMENDATIONS FOR MANAGERS

The recommendations that follow are based on a reading of the literature compared with the experience we gained from a series of six case studies conducted in a variety of manufacturing firms. The case studies were conducted relatively simultaneously over a period of ten months and the data were analyzed at the conclusion of the overall study. Not all the results are presented here [9]; instead only the most significant managerial findings related to technology introduction are discussed.

The case study firms were selected intentionally to exhibit a wide range of characteristics. Sales among the firms varied from five million dollars to five billion dollars and the technologies varied from hard automation, consisting of a transfer line for producing a high volume of standardized parts, to extremely flexible automation, including computer-aided design connected to a laser-cutter as well as a flexible manufacturing system (FMS). Some of the technologies were just being installed while others had been in operation for a number of years. All the firms competed in different markets, ranging from petroleum cable to auto and truck parts to machine tools. For futher details, see Table 1.

Recommendation 1: Do Not Try to Sneak Up on Your Employees

The whole process shown in Fig. 1 begins with the strategy you have selected. Be clear in your own mind *before* implementation of what you are trying to accomplish strategically with the new technology, state these goals clearly to your employees, and then strive to achieve them. This may help you anticipate how you want to manage the implemen-

Table 1. Comparison of case situations

Firm	*Product line*	*Plant size (1000 ft²)*	*No. of employees*	*Unionized?*	*New technology*
saw co.	sawblades	20	40	no	CAD/CAM
gear div.	drivetrain components	750	1400	yes	FMS
OEM plant	automotive suspension components	48	35	no	transfer line
tool sub.	machinery components	88	80	no	FMS
die plant	machinery components	50	100	yes	FMS
wire sub.	steel cables	80	140	no	automated process equipment

tation process as well as employee reactions. Do not try to hide the technology impacts, either positive or negative. One of our case firms attempted to minimize the negative impacts of a new technology and lost credibility with their employees.

If the intent of management is to reduce the work force or to centralize authority, the new technology can lead to impoverished work experiences. On the other hand, that same new technology can also be a source of significant job enrichment. What should be recognized, however, is that the technology strategy management adopts is often linked to the competitive environment the firm is facing. A firm that is beleaguered and defending itself in a very competitive market may adopt new technology in an attempt to retrench, cut costs, and maintain market share. In such a defensive posture, the introduction of new technology may well lead to negative job consequences, such as loss of job security and reduced autonomy. In contrast, a firm focused on growth and capturing new market share may adopt new technologies as part of that growth strategy. In this case, the technology may be accompanied by an enriched job experience with greater job security and less stress.

Even in the face of a defensive strategy, however, the firm can adopt some helpful policies. For example, a number of firms have publicized their policy that no one will be laid off because of the success of a new technology. Clearly, this type of wise intervention by upper management significantly improves the likelihood of successful technology implementation.

Recommendation 2: While the Technology is Changing, Change the Organization

It has been suggested that during the introduction of new technology, changes should be kept to a minimum and making multiple changes at the same time (such as both product and process) should be avoided. Yet, our experience has shown that attempting to keep the same organizational infrastructure — systems, policies, departments, rules — in the face of significant process change will almost guarantee failure of the technology.

Certain types of organizational consequences of new technology have been so widely observed in practice, independent of the type of firm, that they appear to be "unavoidable," regardless of the decisions management makes during the implementation process. For example, advanced manufacturing technologies are invariably accompanied by:

- accelerated time frames — lead times, response times, and production flow times are all frequently reduced with the new technologies, sometimes by as much as a factor of ten;
- a tighter coupling of production with external parties — in addition to customers, suppliers and distributors are also more tightly linked with the firm to improve quality and cut lead times;
- new ways of producing — the manner of making the product is invariably altered in a number of ways;
- control focused on output rather than activity — because of the changed manner of work, output is more closely monitored; and
- increased levels of organic structure — communication is more frequent and work flows are smoother and more coordinated.

As an example of the organizational consequences, common wisdom [10, 11, 12, 13] holds that the new technologies will bring increased stress, reduced autonomy, and greater boredom. The complexity, speed, and importance of the equipment increase the stress, while the automatic reporting and operation reduce the worker's autonomy and increase the amount of boring monitoring that must be done.

However, our case results contradict this wisdom. When the firms were proactive in the implementation process by redesigning jobs and instituting team responsibility, working conditions actually improved. Individual autonomy was replaced with greater *team* decision making and autonomy, and job combining and enrichment eliminated boredom from the monitoring of machines' functions.

Management needs to recognize the inevitable technology consequences and prepare themselves, their organization, and the employees not just to accommodate them, but to be proactive by anticipating and capitalizing on them. Even effective up-front planning will not allow the firm to avoid these organizational consequences; however, it *can* better prepare the managers and employees to face them through needed improvements in the organization's infrastructure.

In every one of the study cases, the organizational structure of the firm changed significantly with the introduction of the new technology [14, 15]. Department functions were integrated, reporting responsibilities were altered, measurement and incentive systems were updated, and so on. In sum, the organizational structures became increasingly organic, offering a major potential benefit to alert firms.

Recommendation 3: Embrace the Negative Consequences

Recognize that the "individual and organizational consequences" of new technology shown in Fig. 1 are symptoms of fundamental changes to the organization's form and operation as demanded by the new technology. Embrace these consequences and, as abovementioned, take action to capitalize on them as opposed to trying to overcome or merely endure them. If certain negative consequences appear inevitable, given the strategy, move to accomplish them as soon as possible and move on; uncertainty breeds rumors and dissipates the possible benefits of new technology.

There is nothing inherent in the functional characteristics of a technology that is necessarily hostile to workers. The impact of a new technology on the individual worker depends a great deal on what management is trying to accomplish with the technology. It is management's job to enhance the impacts of technology on the employees and the organization.

It was found, for example, that poor implementation tended to increase employee stress significantly. But if the firm was well-prepared for the technology and trained the employees properly, a smooth implementation of the technology actually decreased worker stress because of the workers' awareness of the stronger competitiveness of the firm.

Similarly, an important misperception of the effect of new technology has been a reduction in job security. Again, this need not be the case since a number of the firms in the study found that the increased competitiveness of the company increased the demand for other positions within the firm. In two of the cases that were facing the prospect of bankruptcy, the successful new technologies made the workers feel *more* secure about their jobs rather than less secure.

Recommendation 4: With More Computerized Communication, Increase the Amount of Face-to-Face Communication

A common expectation contradicted by the case results was that face-to-face communication would be replaced with electronic communication. Because of the greater need to integrate activities throughout all functions, face-to-face communication significantly increased rather than decreased. Computer communication also increased significantly as the workers had to interface with the system.

One benefit of this increased amount of communication was the elimination of a number of common buffers in the production system, such as inventories and lead time. In terms of the framework in Fig. 1, it is important for management to provide additional means for increased communication of all types between workers when new technologies are being implemented and operated.

Such infrastructural changes were frequently the most valuable benefit the firm derived from the technology, particularly during the implementation period. And the tighter integration with external functions, especially customers and suppliers, was an added benefit of major importance, contributing to better quality, faster customer response, and increased market share.

Recommendation 5: Attack the Strongest Force for Stability, the Existing Infrastructure

The most rigid, inflexible force poised against the intrusion of the new technology is not the equipment about to be replaced, nor even the employees who might be laid off, but the organization's existing infrastructure. For many years, this set of systems and procedures has protected the firm from the chance instances of foolish decisions and natural disasters, and it automatically continues this protection, even when change is necessary.

Unfortunately, the new technologies demand extensive, strategic levels of change. Our study results confirmed the aggregate wisdom of the literature [16, 17, 18, 19, 20] relating to the need for tighter integration of the firm's functions and systems under the new technologies. Specifically, it was found that firms' existing infrastructures were particularly inadequate to support the new technologies. Furthermore, the "organizational time lag" to modernize the infrastructure of systems, policies, and rules was a significant problem because the expensive equipment remained idle until the infrastructure could support it. As noted, this lag was found to be not only a problem in terms of time delay but also in terms of organizational "inertia," as the existing systems and procedures tended to protect the current infrastructure from change.

Recommendation 6: Prepare to Fight the Same Old Problems

In spite of the newness of the technology, our case firms found that the most difficult implementation issues revolved around problems they had been having even with their old technology. But with the additional requirements and time pressures of the new technology, these old problems became fatal. For example, one firm had always had vendor quality problems with castings but managed to "fix" the castings so they were acceptable. With the new technology, however, no one was there to fix the castings before they were processed and the automated machinery could not handle the bad quality. In another case, maintenance personnel had always resisted giving an estimate of repair time but now, with the new technology, repair time estimates were mandatory.

One of the perplexing aspects of implementing new technology is that it *is* new, that is, unknown. Thus, management does not have a well-developed experience base to draw upon in managing the impact of the newly adopted technology. We now know, however, that its introduction often resurfaces some *old* management problems that have received considerable attention in other contexts and literatures. For example, the implications of shifting the control focus from activities to outputs has been discussed in the control literature [21, 22]. Similarly, the implication of moving to organic structures is a classic topic in organization theory [23, 24]. Thus, recognizing that new technology often results in organizational consequences that have already been widely studied provides management with access to literature and ideas that may be helpful in managing the impact of these technologies.

Recommendation 7: Keep the Production Utilization of the New Equipment Low

Many of our cases pointed to the importance of letting the employees experiment and "play" with the new technology to see what it could do. In those cases where this was allowed, the equipment eventually was able to approach the theoretical maximum capacity of the machine, and even to perform beyond its theoretical performance limits. However, where management pressured the workers to maximize the utilization of the technology as quickly as possible, it never did reach either its stated capacity or performance limits.

One researcher of new technologies [25] has indicated that management still frequently employs the old, inappropriate measure of equipment utilization as an indication of the contribution of the new technology to the firm. Our case results indicate that better managers now avoid this trap and recognize the value of early experimentation and flexibility with the technology. Nevertheless, once the new technologies were fully implemented they began to be used intensively. Thus, high utilization was indeed a characteristic of these new technologies, but not during the intentionally protracted implementation and learning stage.

Recommendation 8: Do Not Expect to Eliminate Jobs

As noted in Fig. 1, new technology offers many tradeoffs. Recognize that the old "job" may appear to be eliminated by the technology but that what appears to be elimination may be transformation into another form. And even though one task may be taken over by the technology, others often arise in its place.

Management should be aware that these tradeoffs can often lead to a net gain in the overall quality of work life for employees. For example, the commonly accepted wisdom [26] has been that skill levels will be reduced with a corresponding reduction in job fulfillment. It has been recognized that maintenance will be better off and enjoy higher status, but all other line tasks will be worse off as the machines take over the majority of the work.

Our case results dispute this belief. Although manual skills were reduced, the opportunities to exercise cognitive skills increased dramatically. The better-managed firms took advantage of this opportunity to broaden and enrich the job responsibilities of the workers.

Also, a job rich in individual and manual skills may be transformed into a richer group and cognitive skills job. For example, the individual may lose autonomy in his or her work because of the introduction of technology, but because of the complexity of manufacturing technology and the need for quick decisions during production, the work group may find its autonomy and cohesiveness actually increased. Other examples are:

- cognitive skill use versus manual skill use;
- cognitive, social variety in work versus operational, manual;
- broader, richer tasks versus more narrow, routine tasks;
- decreased danger, aversiveness in some work tasks versus increased danger, aversiveness;
- greater job security versus reduced job security;
- reduced stress versus increased stress; and
- broader, faster communication versus narrow, infrequent communication.

Recommendation 9: Use the New Technology to Monitor Your Workers

With the new computerized technologies, equipment and worker activity monitoring and report generation can be made virtually automatic. A managerial issue of concern in the literature [15] is the possibility of managers measuring worker results through the automatic reporting systems of the new technologies, thus changing the existing evaluation and control process. Our case results showed that this indeed happened but, rather than being a negative attribute, was actually viewed very positively by the workers. That is, instead of being measured in terms of *what* they were doing they had the freedom to do as they wished, as long as the results were what was specified.

Recommendation 10: Change Your Strategy As You Go

Expect the unexpected and be prepared to change your strategy. In a number of cases, new technology brought unexpected benefits that allowed the firm to change from a defensive posture to a growth posture. Be prepared to recognize these opportunities by constantly assessing the state of your competitive environment and the strategic implications of the technology.

Also, stay flexible and responsive to both problems and opportunities with the new technology. Monitor the organizational and individual consequences and intervene as shown in Fig. 1 to address the issues that will inevitably arise.

CONCLUSION

In conclusion, it appears that some guidance and warnings can be sounded for firms looking to introduce new technology.

- Be clear in your own mind *before* implementation what you are trying to accomplish strategically with the new technology. This may help you anticipate how you want to manage the implementation process. If certain negative consequences appear inevitable, given your strategy, move to accomplish those as soon as possible and move on; uncertainty breeds rumors and dissipates the possible benefits of new technology.
- Assess the functional characteristics of a new technology and attempt to anticipate the "unavoidable" organizational consequences of it. Embrace these "consequences" and take action to capitalize on them as opposed to trying to overcome or just endure them.
- Treat the disease, not the symptoms. The firm needs to recognize that the individual and group consequences of new technology are symptoms of fundamental changes to the organization's form and operation. To attack the negative consequences of new technology, management must address the organizational consequences of the technology and intervene as appropriate.
- Expand your view of "the job." New technology offers many tradeoffs. Recognize that the old "job" may appear to be eliminated by the technology but what appears to be elimination may be transformation. A rich individual or manual skills job may be transformed into a richer group or cognitive skills job.
- Be prepared to change your strategy. In a number of cases, new technology brought unexpected benefits that allowed the firm to change from a defensive posture to a growth posture. Be prepared to recognize these opportunities by constantly assessing the state of your competitive environment and the strategic implications of the technology.
- Stay flexible and responsive to both problems and opportunities with the new technology. Monitor the organizational and group consequences and take action to address the issues that will inevitably arise.

It is clear that new technologies will be transforming the factories of the future. The nature of that transformation, however, will depend more than ever on the skills and vision of the managers and employees who will be implementing those technologies. Furthermore, the organizational changes required by the technologies will extend far beyond the production area and will demand total organizational involvement. Only with this breadth and depth of involvement can these factory-of-the-future technologies effectively contribute to the overall competitive position of the firm. MR

REFERENCES

1. Rosenthal, S R (Winter, 1984). A survey of factory automation in the U.S., *Operations Management Rvw* **2:** 3-15.
2. Meredith, J R (1987). Implementing the automated factory, *Manufacturing Systems* **6:** 1-13.
3. Adler, P (Fall, 1986). New technologies, new skills, *California Management Rvw* **XXIX:** 9-28.
4. Walton, R E and Susman, G I (March-April, 1987). People policies for the new machines, *Harvard Bus Rvw* **65:** 98-106.
5. Young, A, Levi, D and Slem, C (Nov. 1987). Dispelling some myths about people and technological change, *Industrial Engineering* **19** (11): 52-55, 69.
6. Porter, M (1985). *Competitive advantage,* The Free Press, (New York)
7. Meredith, J R, and Hill, M M (Summer, 1987). Justifying new manufacturing systems: A managerial approach, *Sloan Management Rvw* **28:** 49-61.
8. Meredith, J R (1986). *Justifying new manufacturing technology,* Industrial Engineering and Management Press, Norcross, GA.
9. Meredith, J R, and Green, S G (1987). A reexamination of organizational and behavioral changes in the factory of the future, *Working Paper MATI-87-1,* Graduate Center for the Management of Advanced Technology & Innovation, University of Cincinnati, Cincinnati, OH.
10. Blumberg, M and Gerwin, D (1984). Coping with advanced manufacturing technology, *J of Occupational Behavior* **5:** 113-130.
11. Salvendy, G (1986). Human aspects in integrated manufacturing systems, *International J of Production Research* **24:** 749-762.

12. Sheridan, T B (1986). Supervisory control. In: *Handbook of human factors* (G Salvendy editor), Wiley, New York.
13. Susman, G I and Chase, R B (1986). A sociotechnical analysis of the integrated factory, *J of Applied Behavioral Science* **22** (3): 257-270.
14. Asendorf, I and Schultz-Wild, R (1983). Work organization and training in a flexible manufacturing system: An alternative approach, *proceedings of the conference on design of work in automated manufacturing systems,* Pergamon Press, Oxford, UK.
15. Trist, E (1981). *The evolution of socio-technical systems,* Occasional paper No. 2, Ontario Quality of Working Life Center, Toronto, Ont.
16. Damanpour, F, and Evan, W M (1984). Organizational innovation and performance: The problem of 'organizational lag', *Administrative Science Quarterly* **29** (3): 392-409.
17. Gerwin, D (1985). Organizational implications of CAM, *Omega* **13** (5): 443-451.
18. Gerwin, D (1981). Control and evaluation in the innovation process: The case of flexible manufacturing systems, *IEEE Transactions in Engineering Management* **EM-28** (3): 62-70.
19. Meredith, J R (May, 1986). Automation strategy must give careful attention to the firm's 'infrastructure', *Industrial Engineering* **18** (5): 68-73.
20. Weick, K E (1976). Educational systems in loosely-coupled systems, *Administrative Science Quarterly* **21** (1): 1-19.
21. Merchant, K (1985). *Control in business organizations,* Pitman, Boston.
22. Ouchi, W (1977). The relationship between organizational structure and organizational control, *Administrative Science Quarterly* **22**: 95-113.
23. Burns, T, and Stulker, G (1961). *The management of innovation,* Tavistock, London.
24. Daft, R (1983). *Organizational theory and design,* West Publishing Co., St. Paul.
25. Jaikumar, R (1986). Postindustrial manufacturing, ***Harvard Bus Rvw* 64** (6): 69-76.
26. Blumenthal, M (1984). *Computerized manufacturing automation: employment, education, and the workplace,* OTA-CIT-235 U.S. Congress, Office of Technology Assessment, Washington, DC.

Mr. Meredith is Professor of Operations Management and Director of the Industrial and Operations Management programs in the College of Business Administration at the University of Cincinnati. He is also Director of Administration of the college's new Graduate Center for the Management of Advanced Technology and Innovation. His research interests are in strategic planning, justifying, and implementing advanced manufacturing technologies.

Mr. Green joined the faculty at the Krannert School of Management at Purdue University in 1987; prior to that he was a professor at the University of Cincinnati. His research interests focus on leader-member interactions, control in organizations, and the management of industrial innovation. His work has been supported by the NSF and the Center for Innovation Management Studies. Mr. Green has published in many journals and has won several teaching awards.

The challenge of CIM is 80% organizational

by
Charles M. Savage

When manufacturing companies venture into CIM, they almost always become uncertain and frustrated when the promises and the reality of the process do not match. As more companies look to CIM for technological solutions, they experience a growing number of human, managerial, and organizational problems; the technological miracle can become a human debacle.

A 1985 survey conducted by the Yankee Group indicates that 76% of the manufacturing industry respondents identified human and organizational issues as major hurdles to CIM. Moreover, a recent study conducted by Booz, Allen & Hamilton asked the manufacturing companies what their major problems were in addressing CIM. All of the respondents mentioned the people challenge. Surveys conducted by the National Research Council and the Automation Forum, among others, have also supported these findings. Although the challenges of CIM were once viewed as 80% technical and 20% organizational, practitioners now agree that clearly the reverse is true (also see Tom Gunn's article in this issue of CIM REVIEW).

To further address and explore the human issues, the Human Systems Development Group at Digital Equipment Corp developed a presentation entitled "The Human Side of CIM." It was shown at DECWORLD in October 1987 to more than 1,200 senior executives. Many of them filled out short questionnaires identifying their major human, managerial, and organizational challenges.

Challenging Implications

When reviewing the results of the questionnaires, it was noted that 75% of those involved in CIM efforts also have just in time (JIT) and total quality assurance (TQA) projects under way. The interplay between these three is critical, because JIT and TQA help expose the real issues CIM must address. Without these three, CIM often amounts to little more than a computerization of the company's contradictions, confusions, and inconsistencies.

The 80 respondents to this questionnaire offer a peek into the minds of some managers. Exhibits 1 through 4 summarize the highlights of their responses. (Note: Response percentages listed do not add up to 100%. Respondents were asked the general questions listed; answers were not influenced by interview format. Therefore, percentages listed represent the number of times respondents mentioned these items. Weightings are relative; they approximate the importance of the responses.)

Look at Question One and its responses shown in Exhibit 1. The responses—resistance to change, turf protection, job security—imply that CIM has some rather far-reaching organizational ramifications. Some respondents cited fear of losing control, upper management's failure to understand what CIM is or how much of it is

Charles M. Savage *is a CIM consultant with Digital Equipment Corp. He has worked for many years with corporate presidents, division managers, and other senior manufacturing executives who are concerned about improving their business strategy, information integration, and organizational productivity. He has been pioneering the discussion of Fifth Generation Management as a means to tap the creative capabilities of management and leverage advances in computer-based technology.*

Exhibit 1. ***What Are the Major Human, Managerial, and Organizational Issues Faced by Companies Trying to Implement CIM?***

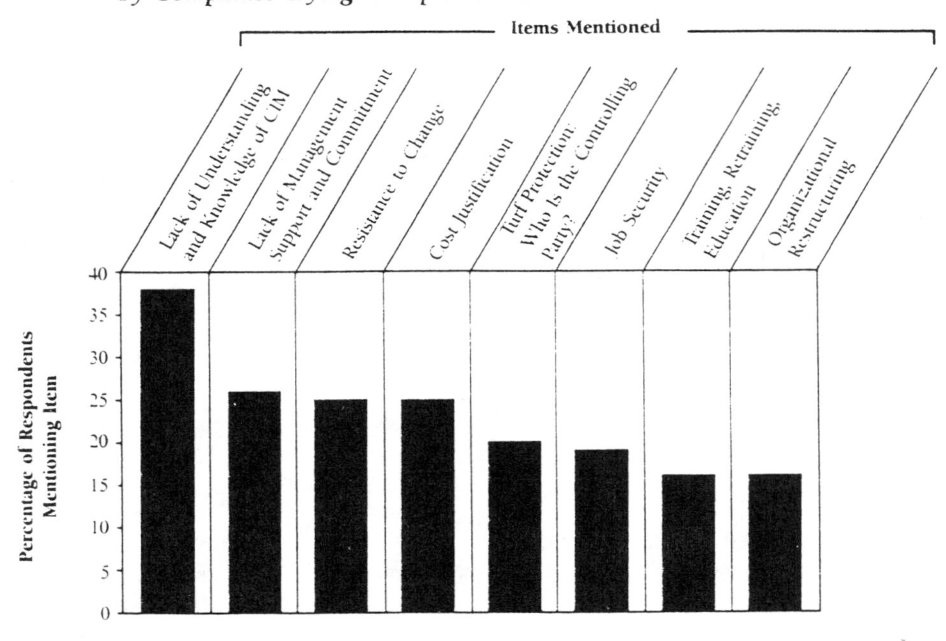

managerial rather than technical, cross-functional conflicts, and power drift.

Clearly, the implications of CIM extend far beyond the technical realm. But what are those implications? How do they challenge the present structure of most manufacturing organizations? Look at Question Two, which is presented in Exhibit 2. Do words like *participative, honesty* and *openness, equality,* and *team attitude* apply to your organization? If not, you're not alone. Most US companies with formal hierarchies based on top-down commands operate with less benefit from these concepts than many would like. But the plot thickens. As you scan some of the other responses to Question Two, you'll see phrases like *reorganization* and *culture change.* As George III once observed, "I think there's more than a tea party going on."

One respondent's answer as to why CIM raises such far-reaching organizational issues was, "You must realize that shared information leaves your department's performance open to appraisal by outsiders." Shared information is probably the concept central to the conflict between traditional organizations and the new, networked organizations; the new technology brings with it a new set of dynamics and values. The transformation of data into useful knowledge is now as important, if not more so, as the transformation of raw material into finished goods. Although the product touches only a few people, information about it touches almost everyone. Sharing knowledge and information between departments, therefore, becomes vital to doing the job; in fact, it is the job.

Lifeblood. Distrust inhibits the flow of significant information—the very lifeblood of the organization. If people cannot trust the accuracy and timeliness of information, everyone suffers. Faith in the information and in one another is essential.

Sequential flow

Management has traditionally followed industrial production-line logic. Work and power are divided and subdivided to manage the sequential flow of raw material from work in process to finished product. An organization that operates this way may be depicted as layers of boxes connected by lines, with the boxes stacked in the shape of a pyramid.

This is management by a command-and-control mentality from the top, where the great-

Exhibit 2. ***How Will an Integrated and Networked Environment Necessitate Changes in Human, Management, and Organizational Style and Approach?***

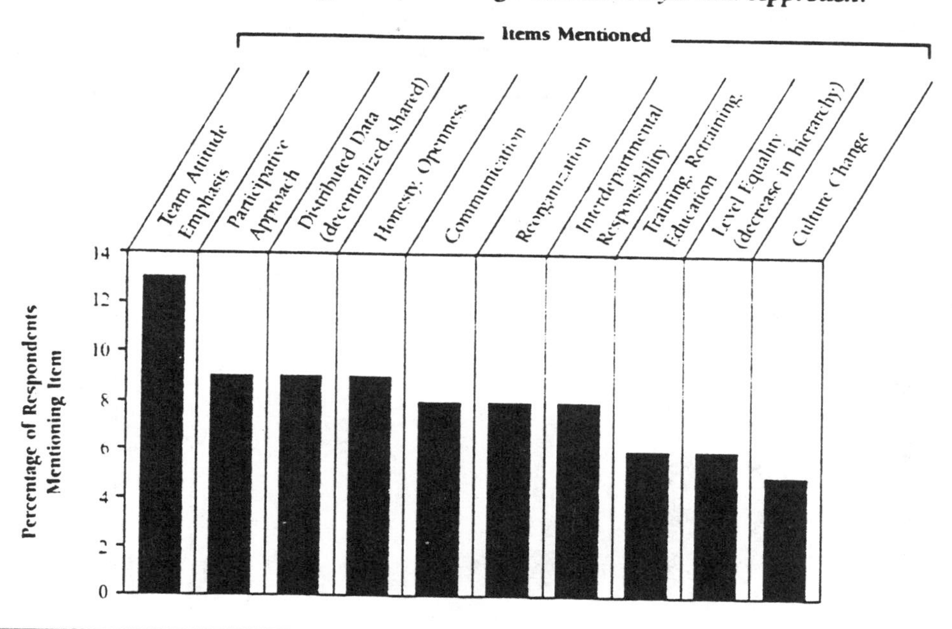

est amount of power—supposedly in the form of knowledge—is hoarded. Sharing information with other departments or functions has traditionally made people vulnerable to attack or to subversion of power. Is it any wonder that in such an environment distrust is the rule?

Cross-functional networking (the essence of CIM) changes that. With such innovations as JIT and TQA, networking is also changing the shape of the typical manufacturing organization.

Companies are finding that they need fewer levels of management. The hierarchical structure has been transformed into nodes in the cross-functional network. Professionals are becoming more interactive. There are more decision point nodes in companies, and the professionals in the organization who interact on them play an increasingly important role in the life of the enterprise.

On the other hand, an organization that wants to overlay networking technology on a traditional management structure is in for many surprises. Forcing new technologies into existing structures may create a Pandora's box.

The information age. The US manufacturing industry is undergoing a fundamental change; there has been a shift from the industrial age to the information age. Consequently, the task of management is changing. Instead of managing a routine environment, people must now manage variety and ambiguity. In a routine environment, automation has replaced human beings. In a truly integrated (not just interfaced) CIM environment, the task is to enhance management capabilities and to encourage a dynamic flow of creativity and knowledge between people. There will be more demands in the new environment than had been conceivable in the traditional industrial organizations.

How are companies addressing the issues and making the necessary changes? The responses to Questions Three and Four suggest some direction. (These two questions and their responses are presented in Exhibits 3 and 4.) Education and training headed the list of responses, along with a team approach to the problem. Communications has also emerged as an important tool in managing the complex issues involved in implementing CIM. Some companies have even fallen back on trial and error. As one individual wrote in answer to Question Four, "Make serious mistakes and get it better the second time."

Exhibit 3. ***What Tools, Techniques, or Approaches Are Companies Finding Useful in Their Human, Management, and Organizational Integration Efforts?***

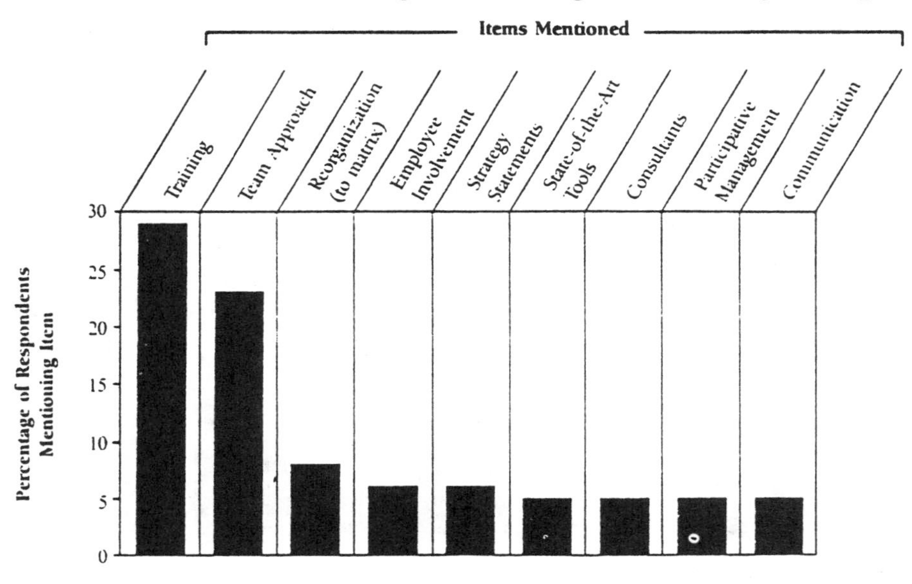

A willingness to take risks is important if the new networking culture is to flourish. Companies successfully making the shift to networking, however, have identified specific business visions and values that guide the organization through the transition period and beyond. "Promulgating a vision of the future" is the way one respondent put it.

The Big Picture

Digital has established a manufacturing strategy called Manufacturing's Big Picture. This strategy establishes a priority for corporate goals, which are customer satisfaction, financial integrity, and product quality. Digital has also cited elements that hold the organization together.

Exhibit 4. ***What Are Some of the Effective Ways Companies Are Addressing These Issues?***

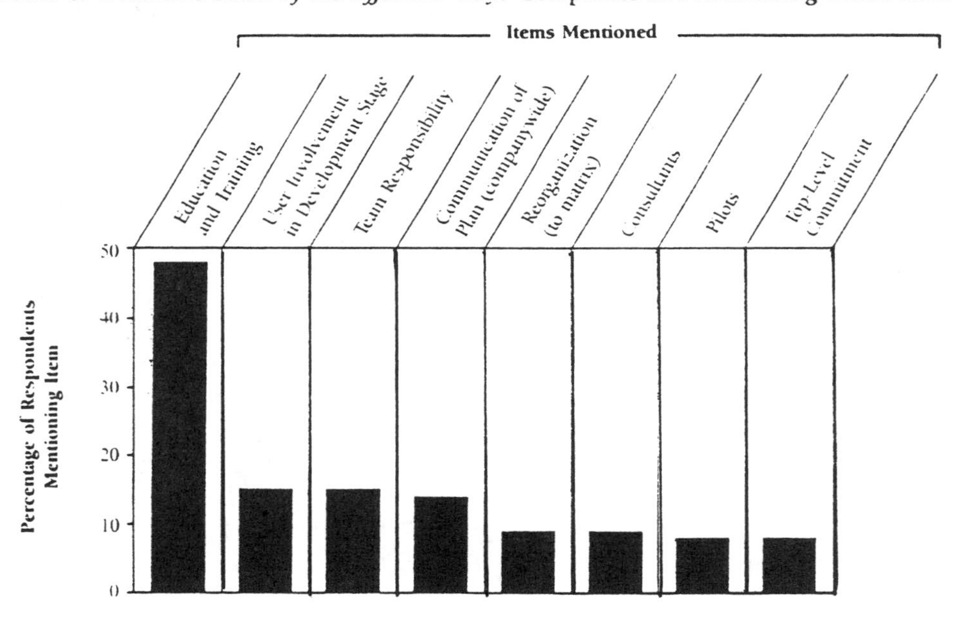

Step by step

Ten key steps are important in proceeding from vision to reality.

Step 1. Communicate the goals of the strategy. This may sound obvious, but it is imperative to begin any automation process with openness for all involved.

Step 2. Identify values that support the expectations of the organization. From my observations over the years, it is critical that a networked organization establish a sense of generosity and openness between people, so that when they share information, it will not be used against them.

The corollary value here is trust. Trust is an absolutely essential element in a networked organization. In the past, distrust has often been built into the very fabric of the organization, as people and functions check and recheck each other. Techniques like JIT are now eliminating this kind of inefficiency.

Integrity is important, along with assuming responsibility for one's actions and decision making, instead of waiting to be told what to do. For example, I've been challenged to understand a problem, help others understand, and even fight for a budget to have it addressed. In the process, I've involved and depended on many other people. Ironically, I've also used and demonstrated my talents more effectively than I would have if I'd been the Lone Ranger.

Teamwork, in turn, enables people to value each other's differences, such as gender, age, and cultural or educational background. It recognizes explicitly that differences are important.

In the new networked organization, leadership comes from an understanding of responsibility, individual differences, management, and how to work with a team to get the job done. In order to accept differences and work as a team, it is necessary to be willing to adopt another person's point of view and not be too sure of yourself. Discipline and enthusiasm are two other important values in a networked organization.

Step 3. Set quantifiable goals. Leading practitioners of CIM are picking four to six goals as a means to keep people focused. Digital also uses concrete measurements, such as the reduction in the cycle time and on-time delivery, to focus staff efforts and move the organization toward collective success.

Step 4. Create cross-functional CIM teams to simplify and integrate processes, automating only when necessary.

One of the most costly lessons of the past few years is that automation for its own sake results only in the automation of confusion. Simplifying the organization is far more important. Cross-functional CIM teams help the participants understand one another's constraints as they work on concrete business or technical problems, such as order processing or engineering design management.

Step 5. Develop products and processes concurrently—similar to the way computers perform parallel processing. For example, Digital recently needed to produce a new controller for its industrial VAX line. Instead of assembling a design team at one location and working on the process in a linear fashion, team members at locations all over the world worked on different facets of the process simultaneously, using the worldwide network to coordinate efforts.

"Simplifying the organization is important"

Design layout took place in Ireland, business management was based in Massachusetts, and the chip manufacturer was in California. As a result, the product cycle time was reduced from three years to 18 months.

Step 6. Implement a networked infrastructure that entails a variety of schemes based on local-and wide-area-networking technology. Although the connectivity provided does not supply full integration, CIM does need a networked environment. As companies plan their networking implementation, they should pay particular attention to the ease with which it can be dynamically reconfigured. They will soon find they are adding, deleting, or moving nodes to match their business cycles.

Step 7. Sort out and leverage basic information by sharing a common language throughout the organization so that every term has a consistent meaning. Often, the same word has different meanings or shades of meaning, which results in duplication and confusion.

For instance, when is a project really a project? When it has a number? When it has a budget? Until basic terminology has been established, computers will only magnify the confusion. We are slowly coming to appreciate the role and significance of data administration—the real challenge of the 1990s.

Step 8. Build supporting knowledge bases. Yesterday's experience stays with us today—something to remember when storing the information. Engineers in product design often do not use previous designs, which could have saved them time and effort, because it is too laborious to sort through them.

However, when the designers create a new part, they also create a new part number, which means more inventory and more storage space. This only increases complexity, a trend we need to reverse. Again, the past appears in the present. A poor job of managing key information will haunt an organization for years to come.

Step 9. Redefine human systems to support and even lead the overall integration of the organization. This entails revising reward systems to encourage open communication, restructuring accounting systems to capture true costs rather than burdening direct labor, and hiring multi-talented people who can grow in their careers rather than fit into narrowly defined jobs.

In the redefined organization, employee involvement facilitates goals and values from the CEO to the maintenance staff. Programs that support this involvement may include job enrichment, continuing education, and training. People are the primary resource of the organization, and they deserve ongoing commitment and investment, not just lip service in the annual report.

Step 10. Provide organizational development support. This brings us full circle because it is really the first step toward networking as well as a vital part of maintaining the vision and values of the organization. It means a diagnostic approach to problems. It means nurturing the development of teams, supporting cross-functional networking, and improving labor-management relations to increase the quality of human interaction.

Growing Pains

Organizations will experience growing pains, but it will be in the most positive sense of the term. It is easier for some companies than for others to change from a traditional industrial culture to a networked one. There is risk, pain, and challenge, but eventually, a defined measure of success will emerge.

Question authority

Remember that one of the most powerful tools of the information age is the question. The thoughtful question is likely to become the key management resource of the 1990s. Is there a better place from which to begin than where you sit?

Are you ready to face the organizational, managerial, and human challenges of CIM? On a scale of 1 to 10, how much do you trust the people with whom you work? Would you be able to share with others what you know and do not know? How would it help your organization if there were more humility in yourself and your co-workers? Do you appreciate the differences among people in your organization and build on their individual strengths? How would you articulate the vision of your organization? Honest answers to these questions will effectively address 80% of CIM issues—the organizational issues.

Focused efforts

Our visions can become focused efforts. The 10 steps previously discussed and summa-

Exhibit 5. *The 10 Steps Toward Implementing CIM*

1. Communicate the strategy.
2. Identify the values.
3. Set quantifiable goals.
4. Create cross-functional teams.
5. Develop products and processes concurrently.
6. Implement a networked infrastructure.
7. Sort out basic information.
8. Build supporting knowledge bases.
9. Redefine human systems.
10. Provide development support.

rized in Exhibit 5 focus on and support the process of cross-functional networking and integration.

What major concerns are your CIM efforts up against? What people breakthroughs have you achieved? I would like to learn from your experiences and share them with others with the appropriate credit. Please write to me at CIM Review, c/o Auerbach Publishers, One Penn Plaza, New York NY 10119. ▲

The Testbed and Phased Project Approaches to CIM

Often a company fails in its CIM efforts because it understood the technology but failed to understand how to plan for it.

Barry W. Sheldon
Senior Manager
Deloitte Haskins & Sells
San Jose, CA

Can computer integrated manufacturing work in your company? The failures in CIM haven't been due to failed technology but more often to a lack of understanding of what CIM is, how to define its usage, and what approaches can be used to successfully implement it.

There are two CIM approaches you should be familiar with—testbed and phased project. Successful CIM implementations have stressed information systems, often with automated equipment a secondary issue. In manufacturing, there is a significant disparity in how sophisticated present systems are, and how these two development approaches can transcend this disparity.

The testbed approach is designed to exercise various design elements of CIM and take advantage of a preimplementation "hands-on" prototyping of desired CIM applications and system interfaces prior to any full-scale commitment to a particular solution.

The phased project approach is a more long-term method to achieve systems development and CIM programs with increased emphasis on meeting the company's business needs.

Four general design parameters need to be considered to make a successful transition into CIM:

• Specifications and designs for each of the CIM information modules—inventory, shop floor control, etc.—must be coupled in so they can react flexibly with one another in a timely and cost effective manner.

• CIM architecture often requires a modular, building block type of design with each workcell being linked to its peers, but not necessarily dependent on them. Consequently, the ability to handle progressively changing data and event relationships is needed in the design.

• CIM architecture is an evolving technology that requires a data processing environment that can easily accommodate connectivity to a wide variety of systems and products, be readily expandable to handle increasing transaction volumes and additional functionality, and be capable of supporting interfaces with emerging manufacturing processes.

• Cost exposure needs to be partitioned into discrete, nondependent investments so one center can abandon a portion of the project, workcell plan, design approach, etc. without jeopardizing a large capital outlay.

The testbed approach is a key element in user-designed, bottom-up manufacturing system development. It provides system users with the ability to determine their own

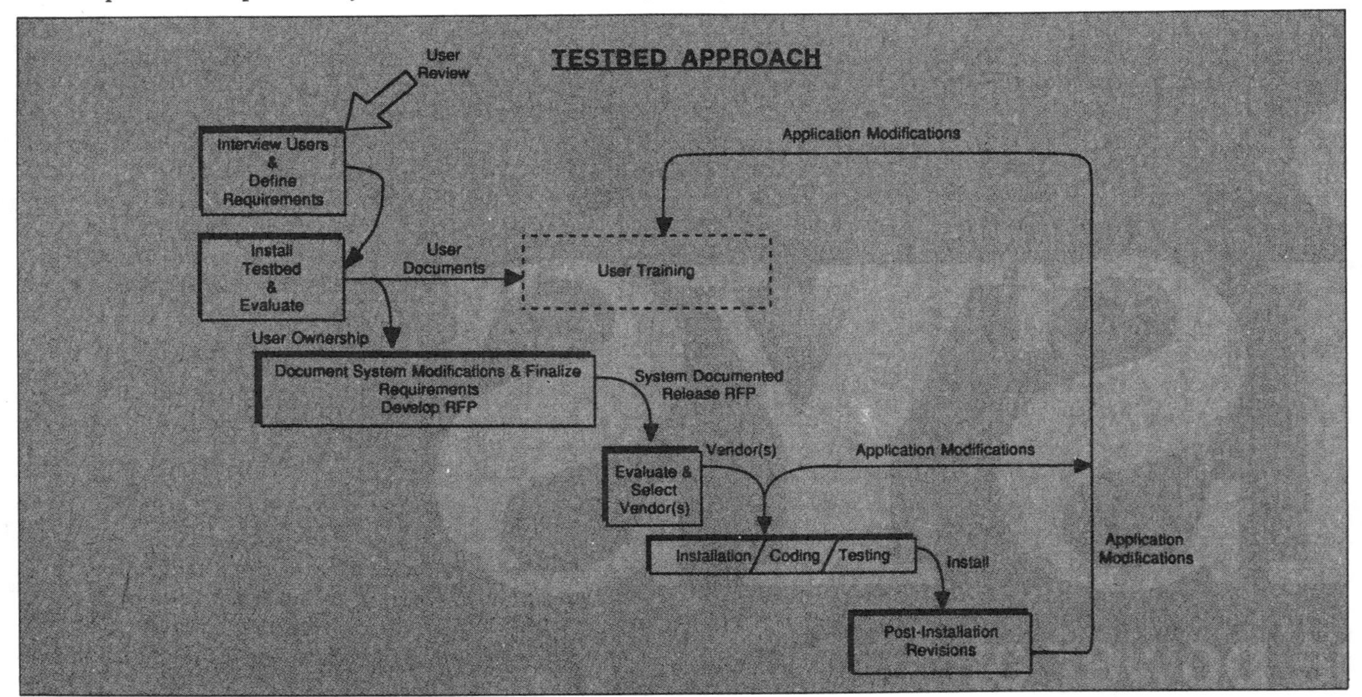

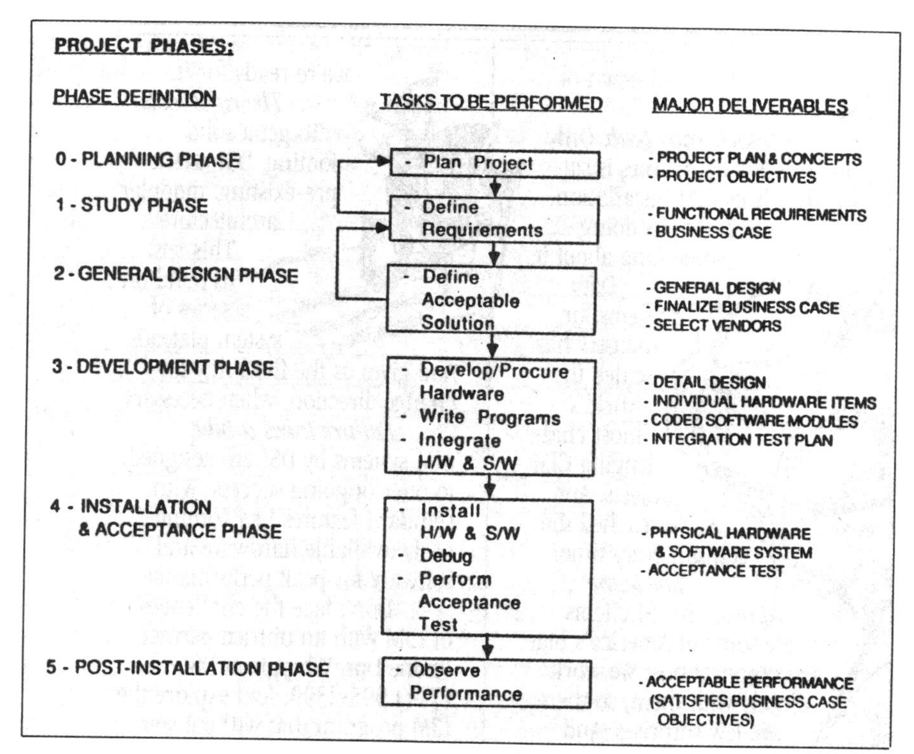

system interfaces—through data element input and output, screens, reports, etc.—by simulating interfaces with planned or existing software and automated applications. The testbed approach provides immediate user feedback on the feasibility and effectiveness of a specific automation approach to an operation, process or workcell plan.

Based on the CIM criteria, a testbed can be a logical methodology for satisfying the parameters and, at the same time, fully accommodating the user requirements for a specific CIM application environment. This approach allows a company to evaluate suggested CIM concepts, to select the one that seems best suited to its particular environment, to provide a venue in which the hardware and software can be functionally demonstrated and the opportunity to use a fully operable production resource.

The approach consists of:

• **Interfaces to selected, existing systems**—This component allows any user with the correct security profile to access data from any of the data bases residing in external systems linked to the testbed system. The data base interface will handle any data conversion necessary to conform data extracted from embedded company systems not in the format(s) required in the testbed system.

• **Transparent application pass-through**—This allows the testbed system to reside, transparently, between the user community and the mainframe applications. The testbed system will monitor each transaction for applicability insofar as its own applications are concerned. If the transaction is one the testbed system needs for its internal use, the transaction is copied into the testbed system data base and then forwarded to its original destination. If the transaction is not required, it is forwarded without copying. This allows the user the comfort of continuing to use familiar screens while feeding the testbed system the information it requires.

• **In-line application integration**—This component enables applications to use information on the testbed system and the company's other systems simultaneously and transparently—appearing as a single session—while maintaining context in each session. This facility also allows the user to switch between applications on the testbed system and the company's other systems as two or more separate sessions, causing the testbed to be an "application gateway."

• **Performance monitoring**—The testbed system contains a number of performance monitoring capabilities including data base performance; line, terminal, and overall communications performance; and overall system performance monitoring with processor, memory and storage device input/output performance.

• **Test platform design and implementation**—The testbed platform facility enables users to prototype new applications, run them in a simulated environment and, when satisfied with the design, generate the code to turn them into testbed production entities. This facility is used to design and implement the filters to convert the data stream element necessary to the testbed environment, to build applications to retrieve information from the company's systems, and to generate new production systems for use by the manufacturing and material operations. Additionally, this development tool can be employed to develop user-customized front-ends to off-the-shelf software acquired in support of the testbed.

In general terms, the testbed approach is a design tool for understanding the impact of CIM on an operation's present systems.

The phased approach to CIM differs from the testbed approach by taking a "corporate-directed strategy" viewpoint instead of a sometimes more subjective "user-designed" perspective.

The phased approach identifies specific and agreed-upon deliverables during each phase of the project. It requires significant management support but provides a single focal point for the company when dealing with large and/or diverse CIM teams. It considers both the company's product and information/programming needs and consists of:

• **Planning**—In this initial phase, the management team develops an Initial Business Proposal (IBP) to determine if the products or value-added operations are feasible and prepares a plan for program development.

A financial analysis based on the IBP's cost estimates, product forecasts and a review of the program's budget then is completed. During this phase, an interfunctional review is conducted to reach an agreement with marketing, service and manufacturing on the operation's feasibility. It results in funding for the first phase and selection of an implementation site.

• **Study**—The study phase results in a commitment of development funds and the decision to include the operations in the company's strategy and operations plans.

• **General Design**—In this phase, the company needs to reach a point of commitment to market the product or program and agrees to proceed with development.

• **Development**—The principle goals of this phase are to complete design verification testing and release an announcement about the product and/or program.

• **Installation and Acceptance**—The goals of this phase are to release the product and/or program for distribution and verify the volume production capability for a product or to complete the program's development and testing.

• **Post Installation**—In this final phase, the product is reviewed to determine if its performance, quality, reliability, availability, service criteria and cost projections are being met under volume production and if the program can be transferred to the field. Central support would be terminated once the program is transferred.

The phased approach should be considered by companies with limited or antiquated information systems. These companies will be embarking on a longer "journey" to successful CIM implementation than those with established systems and will need the organized methodology that the phased approach can deliver.

These two CIM approaches and the criteria used in selecting either can be summarized in a matrix. Either approach will work but a realistic assessment of each company's information systems is essential.

CIM requires companies to justify long-term investments in versatile automation. Many find committing to this investment difficult in light of well-publicized CIM failures. The risk of failure can be reduced, however, by applying either the testbed or phased project approach to CIM. The pressure for higher quality, increased flexibility and improved response will continue to increase, and CIM will be a viable solution to alleviate these pressures. Companies need to *move forward now* to meet the future—the CIM future. MS

Additional Readings on Implementation Issues in Systems Integration

Eshleman, Richard. 1990. The maturing expectations of CIM. *Managing Automation*. December, pp. 6-7.

Hanson, Russell G. 1987. Managing the human aspects of CIM implementation. *Proceedings, Autofact '87 Conference*. Dearborn, MI: Society of Manufacturing Engineers.

Meredith, Jack R. 1987. Managing factory automation projects. *Journal of Manufacturing Systems*. Vol. 6, No. 2, pp. 75-91.

Viola, Michael R. 1989. Manageable steps to factory automation. *CIM Review*. Spring, pp. 57-59.

VI. CONTROL ISSUES IN SYSTEMS INTEGRATION

One of the shortcomings in the formal education of most industrial engineers is the lack of exposure to the concepts of feedback control theory. Implementation of a system is too often regarded as the finish of an effort, rather than the beginning. All systems must be maintained, monitored, and continuously upgraded. The systems involved in a system integration initiative should be designed so as to monitor the total actual performance of the company, detect when significant deviations have occurred, and prescribe appropriate corrective action.

In the articles selected for this section, there is considerable emphasis on performance measures. It is critical that a system measure the *appropriate* performance measures so that management attention may be focused on the factors that really matter.

Once an industrial engineer "sees the light" regarding performance measures, there is often a tendency to design a control system that is too detailed and too complex. Many such systems generate minutely detailed reports on every performance measure the designer can think of. It is important to isolate those system variables that really matter and to provide precisely the information needed for managerial decision making. Simple, elegant reporting systems are far superior to complex, cumbersome ones.

The articles in this section cover several aspects of control systems. Included are systems dealing with financial performance, labor performance, and many other types of system measures.

COST MANAGEMENT AND PERFORMANCE MEASUREMENT IN INTEGRATED SYSTEMS

Robert Bonsack and Joseph Donnelly
EDS Corporation
Consulting Division
300 East Big Beaver Road
Troy, MI 48083
(313) 524-8326

INTRODUCTION

Performance measurement and cost management have always been closely linked. Though the outputs of the cost management system have been considered a major tool for the management of manufacturing, this is becoming less true today than in the past. We realize that the true management of manufacturing operations is carried out through the management of activities and resources which ultimately result in the incurrance of cost. Because of this ultimate tracking to the cost impact of operating decisions, the subjects of performance measurement and cost management are conveniently discussed together.

One of the principal challenges to systems designers in manufacturing, particularly when a great deal of integration of factory operations with engineering and accounting systems must occur, is how best to achieve the integration of cost management and performance measurements with the integrated systems of manufacturing.

In order to explore this subject further, we will examine the revolution that is going on in cost management and then the evolution that is typical in the development of a computer-integrated manufacturing environment. We will then discuss how the cost revolution and the CIM evolution can proceed in an orderly manner concurrently and with a high degree of success.

COST MANAGEMENT REVOLUTION

Understanding the revolution going on in the area of cost management requires a definition of the three primary roles of cost management systems. The first and most traditional role is external reporting. This includes the valuation of inventories, providing monthly, quarterly, and year-end financial statements for shareholders, bankers, and the board of directors. These reports typically show high-level, aggregate numbers which reveal very little detail about operations, but provide a measure of overall financial well-being.

The second role is product costing. This information provides the basis for competitive analysis, pricing decisions, and refinement of product design. The third role is process or operations monitoring and control. This provides insight for operating management for the adjustment of production processes, scheduling decisions and distribution.

The overall purpose of the cost management systems in each of these roles is score keeping, decision support, and cost reduction. Unfortunately, with the exception of the external reporting role, traditional cost systems do not fulfill these requirements adequately. Information is aggregated at too high a level, is untimely, and in many cases, is inaccurate and thus insufficient to provide decision support or the identification of cost reduction initiatives.

It is in recognition of these short falls that a revolution is underway in the cost management area. One of the leaders of this shift in the accounting paradigm is Professor Robert Kaplan of Harvard University, who in 1987 said, "It is worse to be precisely inaccurate than to be approximately correct." His point was that heretofore, improvements in cost accounting had been aimed at precision rather than correctness. Precision can be characterized by the number of digits to the right of the decimal point, whereas correctness can be described in terms of the importance of the numbers to the left of the decimal point.

The cost revolution centers around a new approach to cost management called activity-based costing. This concept portrays the flow of costs to products based on the activities which are required to produce those products, rather than flowing directly from the source of cost to the products, based on various allocation algorithms. It is the activities which cost money, not the products. The control of cost is therefore achieved through the control of activities and the trigger mechanisms or drivers which cause those activities. By analyzing the activities associated with the various business processes in a manufacturing enterprise, costs can be more accurately associated with not only products but also customers, and activities can be analyzed in terms of their contribution or lack of contribution to the value of the products being sold.

The root cause of this revolution in cost management is the changes that are occurring in the manufacturing plant. Lot sizes are approaching one; direct labor is approaching zero; many factories are operating without traditional work orders. More operations in the factory are "process" oriented; artificial intelligence and expert systems are providing added control and overhead costs are a much larger percentage of total cost than in the past.

Performance measurements are changing. The purpose of performance measures is to provide actionable information and analysis, not score keeping. Managing labor is no longer done as a result of efficiency reports, but rather based on head count. There is increasing need to know the cost of tooling changes, engineering revisions, re-work and other non-value-added costs. The emphasis is on knowing what drives costs and triggers costly activity. The goal is continuous improvement in operating performance. Everything about traditional cost management is changing. Concurrently, everything about factory operations is changing.

CIM EVOLUTION

As a factory moves toward higher levels of automation and systems integration, it will approach a computer-integrated manufacturing operation. Consider the following physical and operational changes and how they might affect performance measurement and cost management. Factory layouts will move from traditional, functional departmental organizations to cells and flexible manufacturing systems. From there, islands of automation will be developed and ultimately highly integrated "assembly highways" could characterize the factory layout.

Material handling and storage will move from pallets and forklifts to transfer devices and conveyors. Ultimately, robotics and automated guided vehicle systems, automated storage and retrieval systems will dominate both the factory floor and the warehouse.

Manufacturing control systems will deal with shorter planning cycles, less inventory, fewer transactions, synchronous pull scheduling systems, and almost no paperwork. Local area networks will link machines and computers to people in a "lights out" factory environment.

Given these dramatic changes in the physical and operating characteristics of the factory, how will cost management and performance management be effective?

THE PROCESS OF COST MANAGEMENT AND PERFORMANCE MEASUREMENT DURING THE EVOLUTION OF CIM

For purposes of this discussion, though somewhat over-simplified, we can describe the evolutionary process of CIM in four broad stages; product and process rationalization, systems adjustments, islands of automation, and focused factory. Let us examine each of these stages and the impact that each has on cost and performance reporting.

During the product and process rationalization stage, management must examine what products to keep, what processes to out-source, and what operations to automate. This rationalization must consider internal capabilities, both near-term and projected, changes in the marketplace, the voice of their customers, and cost. Considerable special analysis may be required in several of these areas due to the short-falls of current performance and cost measurement systems. Often one of the first steps in answering these questions is to develop an activity-based cost model to simulate the new factory and to help rationalize the design of future operations.

The second stage, systems adjustments, may include the implementation of just-in-time, kanban systems, and total quality management practices. Implementation or revision of MRPII systems may be undertaken and changes may be made in performance measures. It is during this phase that the old cost system might be brought up to a "best possible" state short of major overhaul.

During this stage, the activity-based costing system may be enhanced and expanded with a refined database. Old performance measurement and cost systems may be constrained from further development or enhancement.

The third stage is the development of islands of automation in the factory. This will cause changes in shop floor control practices and the establishment of cell controls. This stage is often characterized by a great deal of education and training to facilitate these rather dramatic changes.

With these changes in the factory, typically the activity-based costing system must be revised to reflect new processes and revised cost drivers. This is the point where responsibility and accountability adjustments are made or refined to reflect changes in organization and related responsibilities.

The final stage in the CIM evolution is the focused factory. This is characterized by the completion of factory integration, the fine-tuning of systems, and the refinement of performance measures. It is at this point where the ultimate cost and performance measurement systems decisions have to be made. Whether the off-line, activity-based costing system will remain as the cornerstone of factory control, or whether it will be integrated with other accounting systems, will be decided. The degree of success to be achieved in evolving both an automated factory and an integrated cost and performance measurement system is a function of the thoroughness of the planning and the approach taken to the system's issues.

APPROACH

At the outset of the CIM evolution, certain key decisions need to be made. To what degree is the integration of systems desirable and cost-effective? As it relates to the cost and performance measurement system, what formal design approval is needed, and who should "own" these systems? If an interim activity-based costing system is to be implemented off-line, how precise need the system be? Should the initial design be rudimentary and simple, or should one strive for full complexity to reflect factory evolution?

In addition to these decisions, a number of basic truths must be recognized. To assure success, these new systems and performance measures need a champion with enough authority and prestige to minimize the risk of failure. A major paradigm shift will be required. There must be a change in the traditional set of management's mind, major changes to performance measures, and major changes to the role of cost management in the overall scheme of things.

Another major step in the approach is to pay particular attention to and analyze overhead costs. This should be done with the focus on both today's costs and projected costs for the future. The analysis should include what major activities need to be performed, and the cost of those activities. The cost drivers and the links of those to activities and to products need to be defined both for now and in the future. Activity-based costs need to be computed in order to focus on current, non-value-added costs and their elimination, and to support longer-term product and process rationalization decisions. As with so many analyses, it is important that correctness is not overshadowed by the drive for precision. Typically, detailed analysis of 80 to 85% of the overhead costs will be sufficient to set the stage for future operational controls.

SUMMARY

The integration of cost management and performance measurement systems, while enduring the process of factory evolution, can be achieved through four basic steps. Step 1: Use activity-based costing to help describe and rationalize the new plant profile. Step 2: Bring the old systems up to "best possible" levels through refinement short of major overhaul. Step 3: Implement full activity-based costing systems and performance measurement systems off-line, and disconnect old systems. Step 4: Integrate activity-based costing with other business information systems.

Important to the success of the development of such systems is the recognition that this is not just a project for the accountants. It requires participation and ownership by all functional areas throughout the company, particularly sales, engineering, purchasing, production control, manufacturing, and distribution, ... and of course, senior management.

INTEGRATED CHANGE MANAGEMENT: 2000+

Donald E. White
Associate Professor, Industrial Engineering
California Polytechnic State University
San Luis Obispo, CA 93401

ABSTRACT

Accelerating technological, competitive, and business change has dramatically impacted the ability of firms to manage rapid, turbulent, and often chaotic, change within firms. By the year 2000 and beyond, effective change management will be a critical survival issue for firms. Paper presents 16 future needs for managing chaotic change and an Integrative Change Management Methodology which meets those needs. Case study highlights given vividly demonstrate the value of the Methodology for chaotic future environments.

I. INTRODUCTION

In the 1990s, chaotic change will become more common as firms encounter turbulent competitive pressures, rapid technological advances, and other accelerating environmental change. Multiple simultaneous change-needs quickly surface, but urgency, internal conflicts, and limited resources render traditional change-management approaches useless. By the year 2000 and beyond, effective management of chaotic change will be a critical survival issue for firms.

The purpose of this paper is to present future needs for managing such chaotic change, and to demonstrate that effective management of chaos is possible using the Integrative Change Management Methodology presented. This Methodology is tightly linked to the President of the firm, creates an interdisciplinary team approach to change, empowers individuals with a clear direction, and achieves synergistic results. Case study highlights demonstrate the use of the Methodology in rapidly changing, complex, chaotic change environments.

Methodology assumes that required changes have a specific focus so as to be definable as "change-projects". Example change-projects might include: implementing CIM, JIT, TQM, new/modified facilities, new/redesigned products, new technologies/R&D projects, and new management systems.

Sections II, III and IV respectively review the key environmental change forces, summarize the traditional vs. future implications of those forces, and present 16 future needs for managing change. Sections V and VI summarize the Integrative Change Management Methodology, and describe how the Methodology meets the 16 future needs. Section VII gives case study excerpts from experience in applying the Methodology. Final conclusions are given in Section VIII.

II. ACCELERATING ENVIRONMENTAL CHANGE

Success for firms via traditional methods is mostly history! The basic problem is summarized in Figure 2.1. Accelerating technological, competitive, and business change has dramatically affected the viability of using traditional approaches for managing change. The increasingly rapid pace of change has dramatically reduced the time available for implementing change-projects. Recent best-selling books -- such as "Thriving on Chaos", "Re-inventing the Corporation", "Megatrends", "The Change Masters", "The Renewal Factor" - - all communicate the

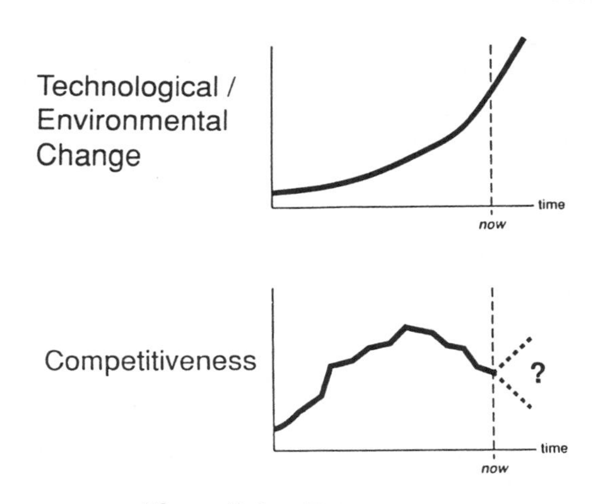

Figure 2.1 "THE PROBLEM"

Reprinted from *1991 IIE International Industrial Engineering Conference Proceedings*

dealing with change management.

Key environmental change forces include:
1. Accelerating Technological Advances
2. Mounting Competitive Pressures
3. Increasing Quality Expectations
4. Turbulent Environmental Change
5. Increasing Speed of Change

1. Accelerating Technological Advances
Technological advances in the form of new technologies, new capabilities to existing technologies, and new scientific knowledge, are accelerating with a resulting dramatic impact on firms. The exploding pace of technological change, as depicted in Figure 2.1, is mind-boggling. Examples of technologies which did not exist, or were embryonic, 10 years ago include:

- computer technology
  (microcomputers, desk-top publishing, supercomputers)
- high-technology industrial products
  (CAD, CAM, CAE, electronic ignitions)
- high-technology manufacturing
  (robotics, CNC, CIM)
- intelligent information systems
  (expert systems, AI, neural networks)
- materials
  (superconductivity, ceramics, composites)
- telecommunications
  (fiber optics, LANs, computers, voice recognition)
- laser applications
  (computers, optics, manufacturing, medical)
- biotechnology
  (genetic engineering, molecular electronics)
- medical technology
  (magnetic imaging, genetics, recombinant therapeutics)
- consummer electronics/products
  (CD players, HDTV, camcorders, automated tellers)

All of these technologies (except the last two) will be used in the manufacturing of other products by the year 2000 according to a recent SME report [1].

In all cases, these technologies have resulted in new capabilities, new opportunities for competitive advantage, and new potential threats from other firms using as competitive weapons. Such technological advances can quickly create product substitutes for existing products, or produce cost efficiencies over traditional manufacturing processes.

2. Mounting Competitive Pressures
Mounting concern over the declining competitiveness (also illustrated in Figure 2.1) of U.S. firms in domestic and international markets grows daily. As a recent Scientific American article [2] indicates:

- "The alarm expressed by U.S. policymakers over the nation's loss of competitiveness in global high-technology markets is not misplaced. The U.S. share of the world semiconductor market, for example, dropped from 50% in 1984 to 37% in 1988."
- "America's share of the world market for consumer electronics products (e.g. C.D. players)...has dropped ... Japan's share has risen."
- "Japan now dominates the numerically controlled machine-tool industry."
- "Japanese companies are far ahead of American ones in the commercialization of high-definition television."

Global competitors have also made major inroads in industries other than the high-technology industries (e.g., automotive, textile and steel).

Further, many U.S. firms are moving into other firm's markets within the U.S. Past suppliers, vendors, or even customers have become competitors. Also, firms are increasingly utilizing strategic information systems/ decision support systems as competitive weapons [3]. Often competitors are agile and sophisticated. Building on their strategic strengths, they plan and move swiftly into competitor's markets.

3. Increasing Quality Expectations
Quality excellence is a key source of competitive advantage. Total Quality Management (TQM) and continuous improvement have dramatically improved the competitiveness of many firms, and the quality expectations of the customers in their industries. Quotes from a recent seminar [4] on TQM reflect its importance:

"...the TQM driver is survival"
"...TQM will be the central focus of companies in the 21st century"

TQM is an essential tool for improving on-going operations. However, major change-needs induced by technological and environmental change must also be implemented without disruption to on-going operations, and with continued focus on quality excellence.

4. Turbulent Environmental Change
Firms are hit by numerous sources of change. These sources and their crushing survival impact are illustrated in Figure 2.2.

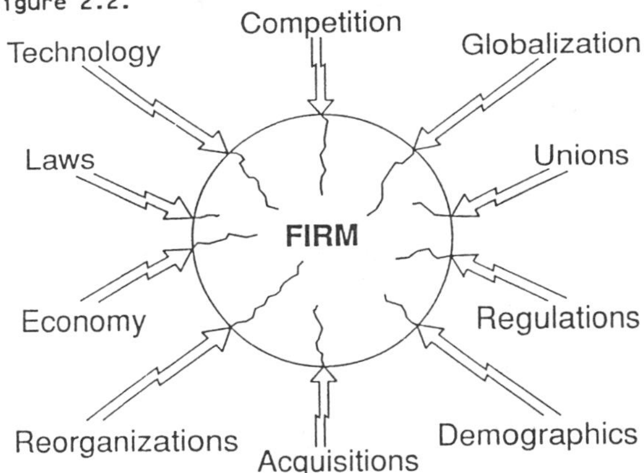

Figure 2.2 NUMEROUS CHANGE SOURCES

Businesses moving into international markets must quickly adapt to new environments. Potentially lucrative new opportunities are the attraction, but unfamiliar business environments often present many unknown threats. Quick effective responses are critical.

Shifting demographics are in constant transition, including the changing mix of incomes, ages, genders, ethnic backgrounds, and habits/preferences. Fortunately these can be anticipated and planned for. Firms must maintain constant surveillance and adapt to the demographic transitions.

Regulatory/legal changes vary by industry and can have a sudden, and dramatic impact on firms. Recent examples include changing pollution standards, food and drug industry regulations, transportation industry deregulations, and telecommunications industry divestiture decree. Traditionally regulated firms, emphasizing full services and high reliability, encounter much difficulty trying to respond post-deregulation to new agile competitors, emphasizing narrow services and low costs. Regulatory transitions in the future will continue to create environmental discontinuities for firms.

5. Increasing Speed of Change

Traditionally, change came at a slow, stable, constant rate. For recent transitions over the past 10 to 15 years, the only thing "constant" has been the constant increase in the speed of change as depicted in Figure 2.1. The SME recent report [5] states :

> "A frightening aspect of this future is the speed of change...the sum of all knowledge accumulated throughout history will more than double by the turn of the century."

Another measure is the tremendous reduction in technology and product life cycles, as shown in Figure 2.3. Traditionally, the time from primary research to product commercialization was about 20 years. Now 5 years is more typical. Product life cycles traditionally have been 10 to 15 years. Now most are 1 to 5 years, and forecasted to become shorter.

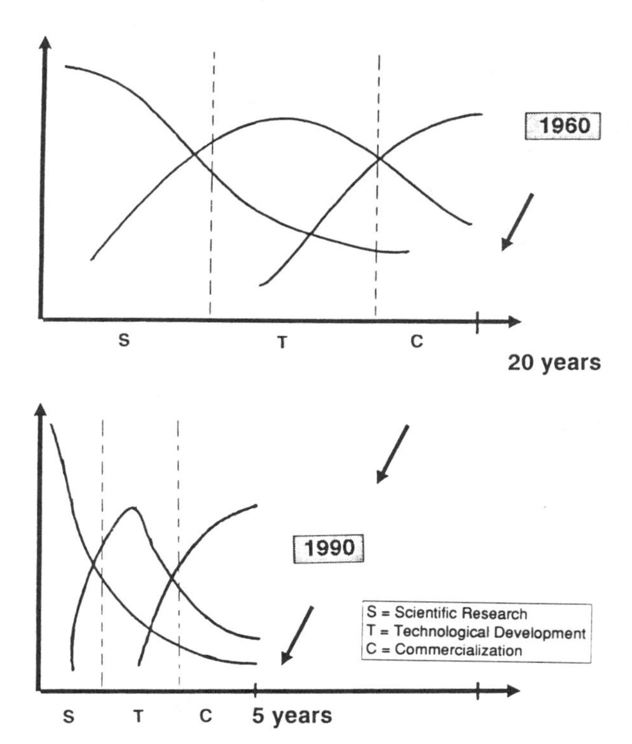

Figure 2.3 ACCELERATING TECHNOLOGICAL CHANGE

The accelerating pace of change has attracted the attention and heightened the sense of urgency for most (if not all) major firms.

III. TRADITIONAL vs. FUTURE IMPLICATIONS

1. Traditional vs. Future Organizations

Companies, both traditionally and today, have organized themselves into formal organizational structures to effectively conduct their businesses. Behavioralists label traditional organizational structures as "mechanistic" or "bureaucratic" structures. First introduced to improve order and efficiency, ideal bureaucracies [6] featured: (1) clear-cut division of labor, (2) positions arranged in hierarchy of authority, and (3) formal systems of impersonal rules, standards and procedures. Ideal bureaucracies are based on clearly aligned formal authority. Advantages included orderly, fair, and highly efficient operations. More recent views of appropriate organizations propose a more dynamic view which is a function of the firms' external and internal environment [7]. The extremes are "mechanistic" structures and "organic" structures. An organic structure is characterized by: (1) ambiguous and changing lines of responsibility, (2) decentralized informal authority, and (3) informal open systems with few rules and procedures.

Schermerhorn [8] states that "mechanistic structures tend to work best under stable conditions", and that "organic structures do better when dynamic environments require flexibility in responding to changing conditions". From 1945 to 1970 stable conditions and predictable changes were prevalent. Firms had time to react, plan-for, and implement new technologies and changes in operations. Since 1975, the pace of change has been accelerating to the point where firms are grasping to just maintain, let alone increase, their competitiveness.

2. Traditional vs Future Product Development

Traditionally, most U.S. companies have used a sequential process (as shown in Figure 3.1) for product development. Such a process is used for new products, modifications to existing products in current markets, or adaptation of existing products for new markets. Product ideas might come from Sales, Customer Service, or Marketing for current technologies, and R&D for new technologies. The sequential, orderly, function-to-function process has been described as "throwing it over the wall". It is analogous to a relay race, where the function whose responsibility is next in line would then take over the next stage of

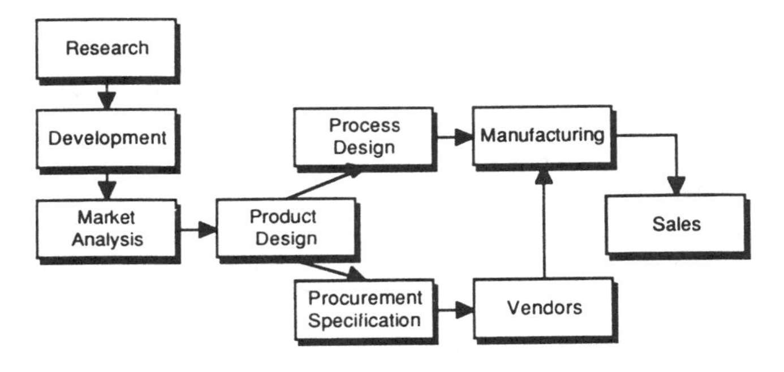

Figure 3.1 TRADITIONAL PRODUCT DEVELOPMENT

development (or leg of relay race), and then throw it over the wall (pass relay baton) to subsequent function. This sequential process works fine, except when both time-to-market and quality excellence are essential. Now more interdisciplinary team approaches are surfacing, such as project management [9,10] or concurrent/simultaneous engineering [11,12].

3. Traditional vs. Future Strategic Planning

Strategic planning is the management process of analyzing the firm, its environment, and creating the firm's strategic direction. Traditionally and today, the planning process determines the major objectives of the firm, selects strategies needed to achieve them, identifies resources required, and guides implementation.

Traditionally, under stable, slowly changing conditions, the firm's environmental assessment and strategy implementation were much easier. History provided an excellent basis for establishing future strategies. Implementing strategies normally only required assigning them to the separate functional organizations for implementation, as illustrated in Figure 3.2.

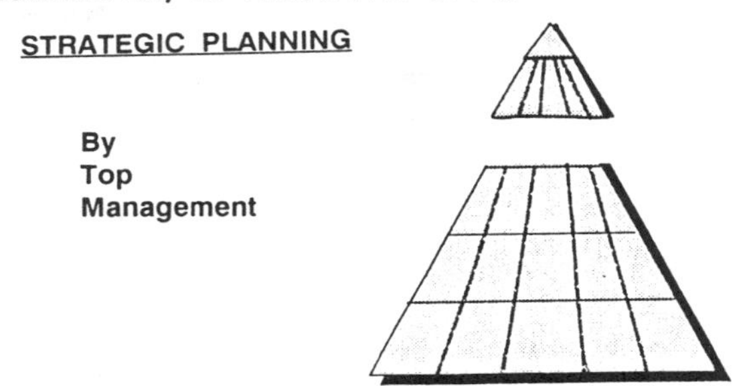

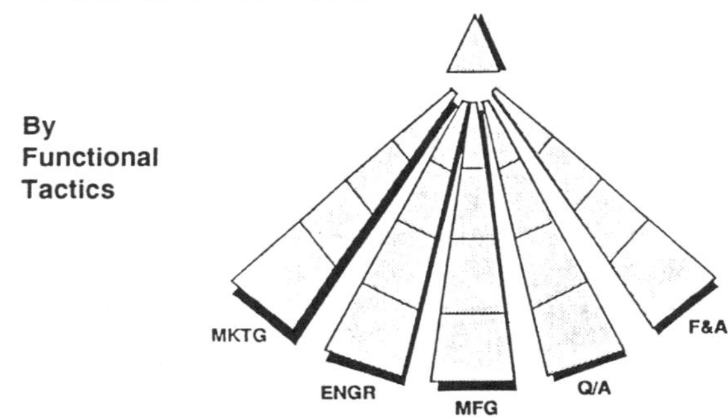

Figure 3.2 TRADITIONAL STRATEGIC PLANNING

4. Traditional vs. Future Interdisciplinary Emphasis

Traditionally, change implementation centered on each functional organization independently doing their part. However, technologically induced change usually affects multiple functional organizations. The accelerating pace of change, coupled with the quality imperative, dramatically exacerbates the problem. Thus, several sources have pointed out the critical importance of cross-functional integration. For example [13,14]:

- "The creation of the manufacturing engineering discipline has only served to build a wall between engineering disciplines, i.e. Manufacturing Engineering and Product Engineering. Our charge for the rest of the century is to figure out how to tear down the wall."
- "Integration of various organizational elements and even vendors is a requirement of winning manufacturing product development...(which) should be implemented on a product-by-product basis; the product team should be broad-based; the customers' needs and wants should be known to the entire team; marketing, sales, product designers, process designers, purchasing, manufacturing, and vendors should all have representation on the product team; and the product team should live with the product from preliminary design through successful production."
- "Integration is the key to future success... integration of people, equipment and information, all tied into one smooth operation."
- "No function will stand alone in the organization of 21st century companies. The need for integration... integration across all business functions, is driving out isolated unilateral approaches. Interdependence is key and bound to grow."

The criticality of cross-functional integration is clear. However, "tradition" presents a formidable barrier.

5. Traditional vs. Future Dimensions of Change

At the next level of detail, one finds multiple dimensions of change contributing to the problem and becoming more critical for the future, as shown in Figure 3.3. The number of simultaneous changes and their urgency grow exponentially. Typically top management authorizes and budgets numerous major change-projects; frequently all are "top priority" and all are "urgent". However, limited resources inhibit a firm's ability to simultaneously pursue all important change-projects. Flatter organizations mean than resources are strained, with many individuals frequently working on several change-projects. Thus, conflicts increase between functional organizations over differences in priority assumptions, over not respecting the urgency (and even the need) for a change-project, and sometimes over not even recognizing that the project had been authorized. Increased need for integration is the general cry, but increased chaos is frequently the reality.

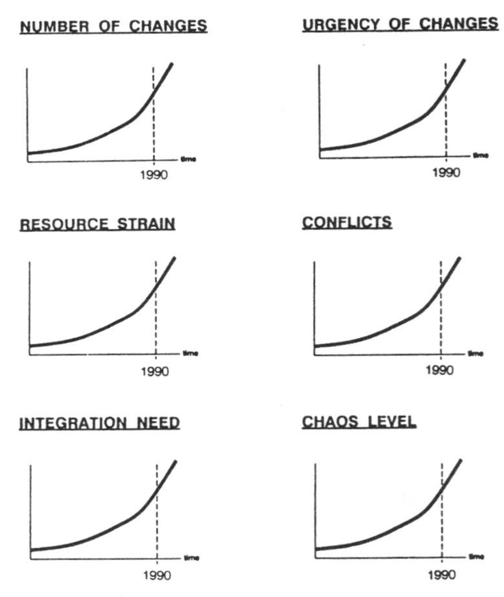

Figure 3.3 IMPACT OF TRANSITIONS: MULTIPLE DIMENSIONS

IV. MANAGING CHANGE: Future Needs

Technological, competitive, and business change has made integrated strategic and operations change-management of firms a survival imperative. Recent total quality and productivity management practices have significantly improved the on-going operations of firms. Yet the accelerating pace and urgency of change has resulted in chaos, rather than smooth, orderly change-management.

Thriving on chaos is not realistic! Managing chaos is. Future needs for managing chaotic change in the 1990's and beyond are as follows.

OVERALL NEEDS:

1. Top management commitment.
2. Simple, flexible, robust change management process capable of operating in chaotic environments.

INTEGRATION NEEDS:

3. Smooth integration of strategic change with on-going operations.
4. Organizational integration vertically (top-down), horizontally (across functions), and externally (customers, suppliers partners).
5. Simple, flexible, justification and prioritization process which considers multiple criteria and relative importance of projects across functions.

COMMUNICATIONS NEEDS:

6. Clear, open, free-flowing communications at all levels and across all functions.
7. Clear communication of project authorization and priority for each change-project to focus and align resources on most critical needs.
8. Clear change-project objectives, clear individual roles, and early identification/resolution of problems for each change-project.
9. Interactive, flexible, Integrated Change Management (ICM) Database/System.

HUMAN RESOURCE NEEDS:

10. Clear authority and responsibility at the lowest appropriate level.
11. Energized, empowered, self-directed teams.
12. Team-based reward/compensation/motivation.

CONTINUOUS IMPROVEMENT NEEDS:

13. Continuous improvement of the time, quality, and productivity for managing change.
14. Continuous technological/product innovation, and the incubation of high potential change-projects.
15. Continuous benchmarking of firm to its top competitors, and the identification of strategic change priorities.
16. Managing resources/organizations as intelligent and continuously learning.

V. INTEGRATIVE CHANGE MANAGEMENT METHODOLOGY

The Integrative Change Management Methodology presented next considers the impact of turbulent external transitions and is founded on meeting the 16 future needs for managing change just specified. Methodology is tightly linked to CEO or President of firm, creates an interdisciplinary team approach to change, empowers individuals with a clear direction, and achieves synergistic results. Key components of Methodology are:

1. Executive/Management Steering Teams
2. Integrated Project Life Cycle Phases
3. Multi-Criteria Project Prioritization
4. Interdisciplinary Project Teams

Figure 5.1 illustrates the interrelationship between these components. As shown, a disorderly array of multiple simultaneous projects appear in all phases of completion and from all functional organizations. Methodology transforms the chaos into a systematic, yet highly flexible, implementation process for change-projects.

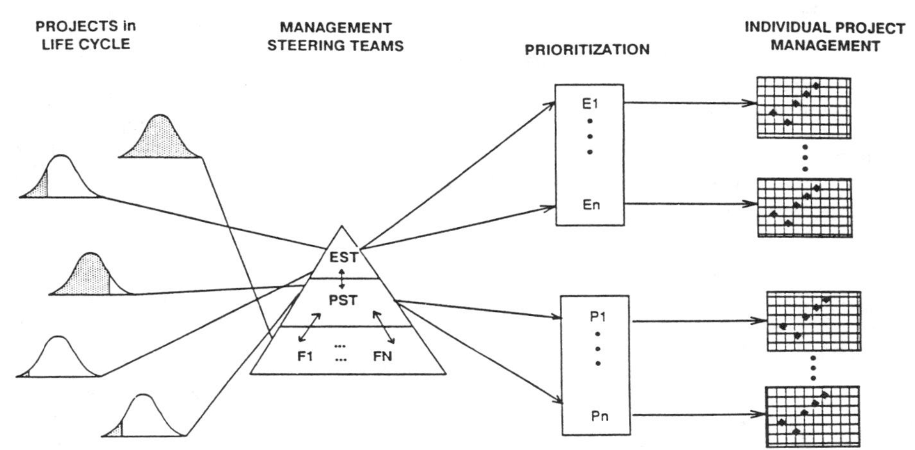

Figure 5.1 **METHODOLOGY OVERVIEW**

1. Executive/Management Steering Teams

Cross-organizational steering teams, called Executive Steering Team (EST) and Project Steering Team (PST), are responsible for maintaining the integrity of the firm's project management process, and integrating it with traditional sources of change. EST, PST and (in some applications) lower level teams are responsible for projects at their corresponding priority level. Team roles include: reviewing/approving/prioritizing projects, facilitating communications, and insuring resources for all phases of each project's life cycle. Specific purpose, membership and responsibilities for each team are given in Figures 5.2 and 5.3.

2. Integrated Project Life Cycle Phases

Project life cycle framework (shown in Figure 5.4) is an important component of Methodology which separates change-projects into natural phases of completion: (1) Think, (2) Study, (3) Research, (4) Plan, and (5) Implement. As projects are transformed through the phases, the degree of uncertainty about the change-project (e.g. market expectations, technological feasibility, financial projections) starts high, but decreases over time as the specifics of the project develop. Development

PURPOSE

1. To approve, prioritize, monitor "Top 30" level projects from Study through Implement Phases.
2. To ensure integrity of Methodology.

MEMBERS

- President/CEO (Chair) - CFO
- Marketing & Sales VP - Manufacturing VP
- Technology, R&D VP - Administration VP
- PST Chair (non-voting) - Regulatory/Legal Counsel

RESPONSIBILITIES

1. Prioritize and approve "Top 30" projects.
2. Review selected existing and new projects monthly.
3. Reprioritize monthly or as needed.
4. Select/ensure project staffing.
5. Reject projects not following guidelines.

Figure 5.2 EXECUTIVE STEERING TEAM (EST)

PURPOSE

1. To approve, prioritize, monitor PST-level projects from Study through Implement Phases.
2. To resolve resource conflicts at PST-level.
3. To make recommendations to EST.

MEMBERS

1. Director level members appointed by V.P.
2. Typical functions:
   - Marketing - Production
   - Sales North or South - R&D
   - Information Services - Regulatory
   - Financial Operations - Planning

RESPONSIBILITIES

1. Prioritize and approve PST-level projects.
2. Review selected existing and new projects monthly.
3. Reprioritize monthly or as needed.
4. Recommend/select project staffing.
5. Reject projects not following guidelines.
6. Recommend projects be elevated to "Top 30" level.

Figure 5.3 PROJECT STEERING TEAM (PST)

expenditures vary inversely to the uncertainty and the phase of completion. Expenditures start low and increase during later phases. Thus, the largest expenditure flow occurs when the uncertainty is lowest. Each phase contains a review point to release the change-project to next phase, revise its scope, place it on hold, or terminate it. This ensures the firm's resources are being expended most effectively.

Think Phase purpose is to translate the idea, improvement thought, or business opportunity into short memo for evaluation by functional manager. Change projects meriting more study are assigned for further evaluation.

Study Phase purpose is to generate sufficient information to prioritize the change-project among all the other projects, and to release resources from all functions needed to complete the Research Phase. Study Phase results are forwarded to, and reviewed by, PST for their action, or elevation to EST. The PST or EST sets priorities and allows resources to be allocated according to their priority.

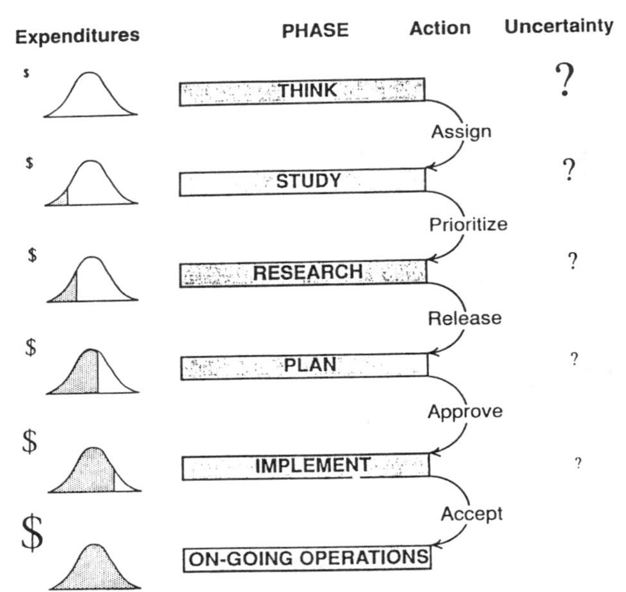

Figure 5.4 PROJECT LIFE CYCLE PHASES

Research Phase purpose is to develop the Business Case for the change-project which contains sufficient information to justify and authorize the project and prepare detailed project plan. Business Case includes: definition of Cost, Schedule, and Performance (CSP) for project, cost/benefit analysis, and functional resource requirements/responsibilities. Clear definition of Performance includes scope and quality requirements. After approving the Business Case, the PST or EST formally authorizes project, assigns project team members, ensures resources, and releases project to Plan Phase.

Plan Phase purpose is to complete a project implementation plan showing how the activities outlined in the Business Case will be accomplished. Project team members prepare detailed Responsibility Matrix and Schedule/Constraints Chart showing all project tasks, their schedule and interdependencies, and individual responsibilities. Project Manager accepts responsibility for project CSP. PST or EST approves the implementation plan and helps with any unresolved issues.

Implement Phase purpose is to complete the project within the specified CSP of change-project. After completing the last task, the appropriate functional manager customer signs off, thereby accepting the completed change-project into on-going operations.

3. Multi-Criteria Project Prioritization

Multi-criteria project justification and prioritization process is used to justify and prioritize all projects. Process is comprehensive, considers multi-criteria cost/benefit measures (both quantitative and qualitative), is common for projects from all sources, and establishes a serial ranking of all projects.

Communicating the prioritized list of change-projects helps avoid problems, and quickly resolve conflicts. Competition for resources is intense. Unless projects are prioritized, persuasive project champions, parochial interests, or other factors may cause critical resources to be misallocated. Project performance suffers, as well as increased employee frustration and reduced job satisfaction, when work priorities are constantly changing. Communicating priorities shows alignment of management, ensures high-potential change-projects receive essential resources in both their early and late stages, and simultaneously provides an actionable tool for resolving conflicts between projects being implemented.

Prioritization process has three key elements: (1) multiple levels of responsibility, (2) guidelines, and (3) ranking method. Figure 5.5 illustrates the 3 levels of responsibility: functional manager, PST and EST. All EST level projects have higher priority than PST level projects, which in turn are higher priority than functional department level projects.

Management Level	Functional Manager	Project Steering Team (PST)	Executive Review Committee (EST)
Top 20 Projects	Recommends to PST	Recommends to EST	Approves
Other Prioritized Projects	Recommends to PST	Approves	
Departmental Projects	Approves		

Figure 5.5 PRIORITIZATION RESPONSIBILITIES

Prioritization guidelines include:

- On-going operations take precedence over change-projects.
- Emergency work (e.g. safety, facility crises) takes precedence over projects.
- Change-projects are prioritized at earliest phase in life cycle where they may compete for resources.
- EST and PST project priorities are reviewed and published monthly.
- EST and PST prioritized projects take precedence over non-prioritized and functional department projects if resource conflicts arise.

Priority ranking method used, which was called "Nested Class & Rank Method", is illustrated in Figure 5.6. Alternative approaches are possible which vary over a wide range of complexity and sophistication. Method used is powerful, practical, simple to use, and actively involves EST and PST members in (often lively) discussions on relative merits of projects.

4. Interdisciplinary Project Teams

Interdisciplinary project teams are led by a Project Manager whose critical role is outlined in Figure 5.7. These teams use simple tools (more complex as needed) to create clear direction and commitment of team members for the management of individual change-projects. The Responsibility Matrix shown in Figure 5.8 is the central tool used by project teams on every project. It serves as the key communications tool with the project team members, as well as multi-level management teams (EST, PST). Other tools include the Schedule and Constraints Chart and Monthly Status Reports. Each are in standard formats to facilitate effective communication with project

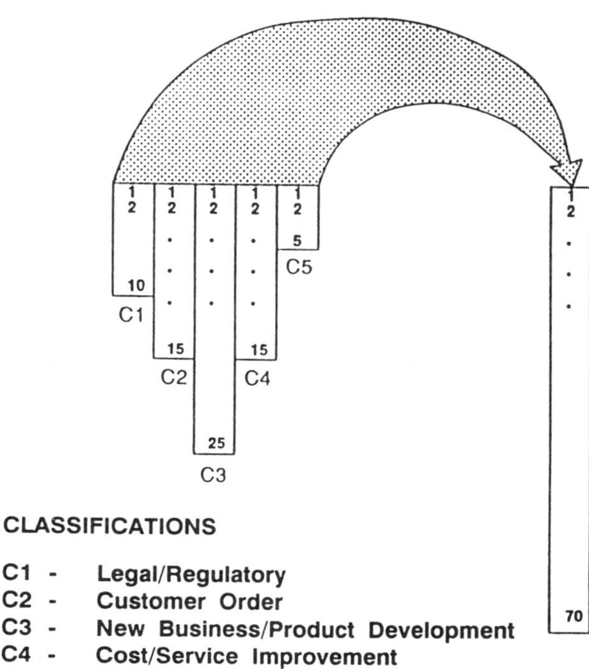

Figure 5.6 PRIORITIZATION: Nested Class & Rank Method

team, organization, and steering teams. Also, depending on the complexity of the individual project, a hierarchy (i.e. 2nd or 3rd levels) of responsibility matrices may be appropriate and useful. Here, for example, a task from the Responsibility Matrix of the overall project would be exploded into a 2nd level Responsibility Matrix of its own.

ROLE

1. Project "CEO" with CSP as Objective.
2. Authority crosses functional lines and levels of management.
3. Final responsibility and accountability for success and effective management of project.

RESPONSIBILITIES

1. Recruit/assemble/motivate project team.
2. Lead/manage/inspire project team.
3. Prepare/communicate project documentation.
4. Ensure clarity/commitment of project/task objectives with team members.
5. Encourage early problem identification.
6. Resolve problems/conflicts at lowest level, or escalate to required level.
7. Present end-of-phase reviews and status reports.

Figure 5.7 PROJECT MANAGER

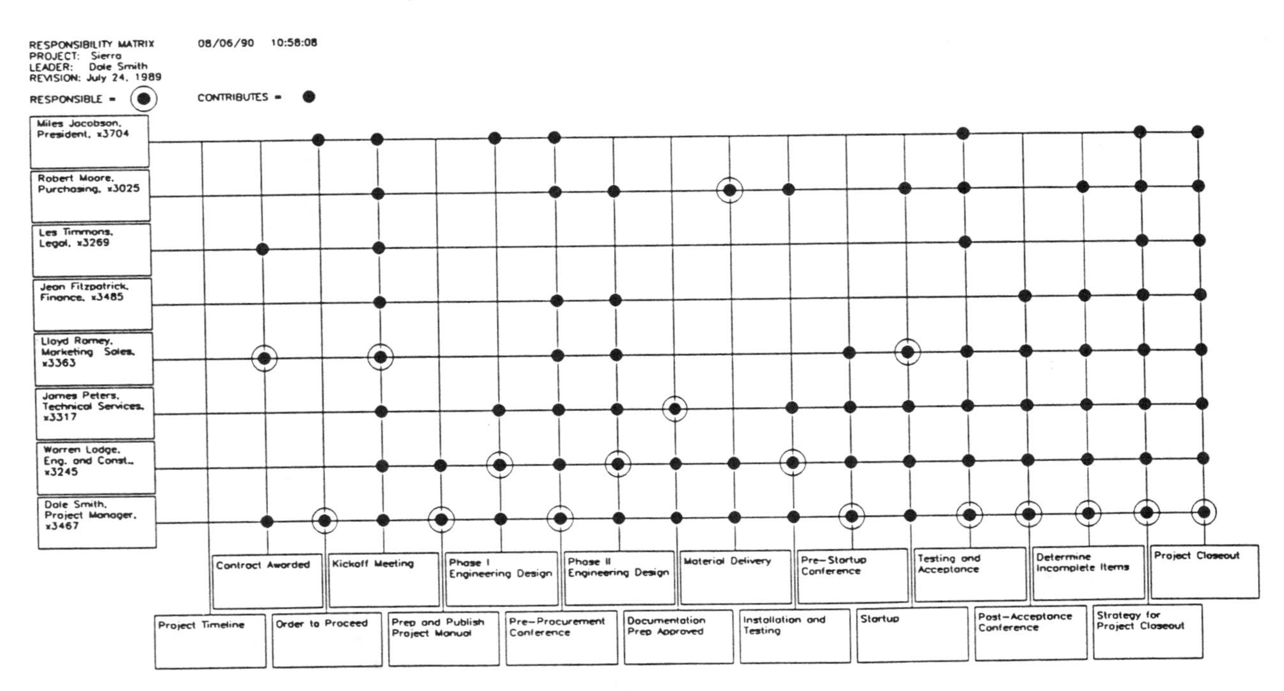

Figure 5.8 RESPONSIBILITY MATRIX

VI. METHODOLOGY MEETS FUTURE NEEDS

The Integrated Change Management Methodology presented in Section V. strongly supports all change management future needs identified in Section IV. How components of the Methodology are related to each identified need is summarized next.

OVERALL NEEDS:

1. Top Management Commitment. Methodology is driven by top management. President/CEO chairs the EST on which all VPs are members. EST approves/prioritizes/monitors top priority change-projects, delegates other projects, ensures project resources, resolves (as needed) resource conflicts, and ensures integrity of Methodology. Methodology is tightly linked to top management.

2. Flexible, Robust Process. Methodology flexibility and robustness are key benefits. Project teams working on change-projects need clear communication of priorities so that the focus continues to be on the most critical needs. During periods of turbulence or chaotic events, a temporary change in priorities can quickly shift needed resources to focus on unforeseen needs. Project teams/members can quickly adjust their efforts.

INTEGRATION NEEDS:

3. Integrating Strategic Change and Operations. Smoothly integrating strategic change projects into on-going operations with minimal disruptions to the functional operations is essential, and critical for quality excellence. For example, switching to a higher technology manufacturing process of a product must be transparent to the product's customers. Having functional representatives from Engineering, Manufacturing and Marketing involved in the project from the start ensures a smooth transition with a quality and customer focus.

4. Vertical/Horizontal/External Integration. Vertical integration (i.e. up and down the organization) is ensured by top executive/project team interfaces and by the clear authorization, prioritization and communications on projects. Horizontal integration (i.e. across functional organizations) is ensured by the cross-functional representations of the EST, PST, and project teams and the clear communications on projects throughout the firm. Integration with external stakeholders (e.g. customers, suppliers, partners) can be ensured by their representation on, or with, the project teams. Also, the clear authorization and prioritization of projects enables project managers and team members to attain required project support and needed resources from functional managers. The various phases of the project provide a clear indication regarding what level of support is needed. The ICM Database/System can further facilitate horizontal and vertical integration.

5. Justification and Prioritization Process. All functional organizations are represented on the EST and PST teams. All pertinent strategic, financial, and operational criteria are considered in approving change-projects, and then prioritizing the change-projects relative to each other. This critical process then provides the clear direction and essential basis for all functional personnel and project team members to make day-to-day decisions for quickly resolving resource conflicts and expeditiously implementing change. Priorities can always be revised by the EST or PST, given the appropriate need, but the priorities stimulate a bias for action.

COMMUNICATION NEEDS:

6. Open Communications. Clear, open, free-flowing communications is essential for a firm to be capable of fast responses to required changes. The EST, PST, Project Teams, and the published availability of all change-project information to each other, and to all functional managers, should satisfy even the most unquenchable desire to be well-informed.

7. Communicate Priorities/Align Resources. Establishing and clearly communicating project authorization and priority decisions by the EST and PST gives an overall organizational focus to the firm's critical change needs. It clearly aligns people across all functional organizations, and it allows resources to be released to the most critical needs. The use of different phases of projects and the ICM Database/System facilitates and clarifies the resource needs.

8. Project Objectives/Roles/Problem Resolution. Written change-project objectives for cost, schedule, and performance (CSP) and a published project plan provides clear change-project direction so a common understanding is available for EST, PST, Project Teams and all functional managers. This facilitates a clear understanding by project team members of their roles, responsibilities, commitments, and priorities. Starting with this common understanding at the outset, it is much easier for project team members and project managers to identify potential problems early, so time is allowed for problem resolution.

9. Interactive Integrated Databases/Systems. Several technology/computer support systems are desirable. The ICM Database/System provides an interactive system accessible to the entire organization (i.e. EST, PST, project managers team members, functional managers/personnel). This clearly communicates approved projects and their priorities, individual project objectives, project plans, project status, and project manager/team member/functional manager roles, authorities and responsibilities.

RESOURCE NEEDS:

10. Authority/Responsibility. Implementing change through the Integrated Change Management Methodology is based on the philosophy of delegating complete authority and complete responsibility for each change project down to the lowest appropriate level, the project manager. That project manager then: (1) should be viewed by the entire organization as the "Project CEO", (2) should not be incumbered by further approval requirements, and (3) should have complete freedom to successfully implement the change project, subject only to the priority of that project and potential conflicts with higher priority projects.

11. Energized/Empowered/Self-Directed Teams. Project teams, under the leadership of each project manager, must be formed, become energized, and become self-directed. Being empowered by top management through the EST or PST can be a powerful motivating force for the project manager and project team. Tight control, close supervision, or bureaucratic-style supervision of the project manager and/or team members would be an absolute disaster. This style must be replaced by a loose-reins supervision style which allows achievement-oriented teams capable of high performance, innovative approaches, and breaking-down barriers to truely achieve their potential (e.g. like Peter Drucker's vision for "Orchestrated Business Teams" [15]).

12. Reward/Motivation Structure. The motivational environment is key too. A powerful motivational environment will exist when the project team has complete freedom and knows that rewards and successes of project team will be shared with each member. Responsibility will be readily accepted without reservation. The Responsibility Matrix used by each project team also provides a powerful motivational tool. During the initial project meetings and team formation, tasks required to complete the project are created by the team members. When team members volunteer their responsibility for a given task they created, they internalize that responsibility and become committed to that task and the project. Team members develop high expectations of each others' commitments, and mutually hold one another accountable.

CONTINUOUS IMPROVEMENT:

13. Continuous Quality Improvement. Constantly monitoring, removing bottlenecks, and continuously improving the time, quality, and productivity for managing change must become an integral part of the new Methodology. The Methodology itself should incorporate the TQM concepts which have become part of the on-going operations continuous improvement. For example, as each project is closed out into on-going operations, a brief audit is done by the customer, project team, and others involved with the project. Successes and problems based on their team's experience are recorded and recommendations/conclusions are made. These may be implemented into the Methodology immediately, or saved for a Pareto-type analysis of several projects in the future. These successes/problems/recommendations/conclusions are available for future project teams via the ICM Database/System.

14. Continuous Innovation/Incubation. Technological and competitive innovation is becoming an ever increasing imperative of firms. The use of the project phases, together with EST/PST project reviews, encourages innovation by allowing high potential change-projects to be prioritized at an early stage, and thereby obtaining resources for their incubation and for the assessment of their ultimate potential.

15. Continuous Benchmarking. Firms should continuously benchmark themselves versus their top competitors and even firms outside their industry: (1) to identify additional change-projects needed, (2) to help assess strategic change priorities, and (3) to help assess the Methodology used versus the best change-management approaches elsewhere. This will further refine and improve the practices of the EST, PST and project teams, and improve the quality of change-management.

16. Managing Intelligence. Methodology will work best when strong communication linkages exist between all functional organizations. Knowing what changes should be made to products requires the collective intelligence of all (e.g. Engineers and Market Researchers with customers). All functions (e.g. Sales, Marketing, Manufacturing, Engineering, R&D) should be connected by an interactive intelligence information network. All personnel, including project managers, team members, functional managers/employees and EST/PST members, must be considered as learning resources and organizations. Starr has suggested [16] that firms must really be viewed as "managing intelligence and continuous learning", rather than data or information. This will be an important concept in the future. Open communications and continual learning are essential for the collective intelligence to grow and flourish.

VII. CASE STUDY EXCERPTS

Methodology has been applied within many companies in diverse industries. These applications, while currently at various stages of full implementation, are moving towards the comprehensive and tightly integrated Methodology described above. Five excerpts from these case applications follow.

Excerpt 1: Firm's Survival Threatened
The first case is a Fortune 500 sized firm which manufactures large, high-cost industrial products. Traditionally the firm held a dominate position in a stable and growing market. Suddently, over a 2 year period, several external forces bombarded the firm including: dramatic regulatory change, technological advances, major new competitors entering the market, and shifts in international markets. Initially, the impact of these external forces were denied or rationalized. Later, after continued precipitous decline in market share (see Figure 7.1), reality was accepted. Ensuing panic and uncontrolled chaos followed.

After several false starts, the Methodology was sold and implemented. The chaos became manageable. Real progress was achieved. By 1989, they began their first profitable year since 1982. The firm had turned around this competitive threat, regained much of its market share losses (amounting to 1/3 of its peak market share), and lowered its breakeven production volume by three-fourths (3/4). Case details are available [16].

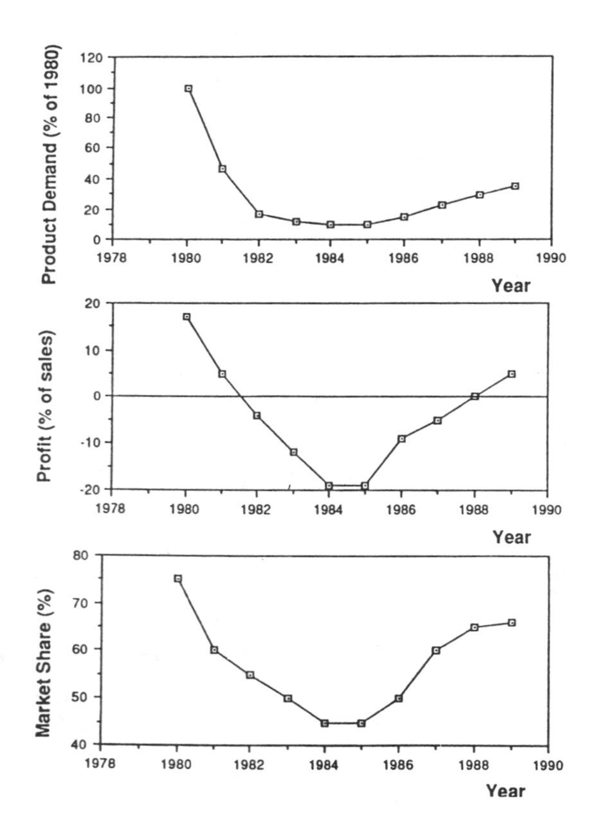

Figure 7.1 DEMAND/PROFIT/MARKET SHARE

Excerpt 2: Strategic Plan Implementation
Post-divestiture telecommunications industry firm's strategic plan identified major strategies and functional tactics. Functional VPs assumed responsibility and sent numerous assignments down the chain of command - - all urgent, all top priority, all assuming and needing the resources of other functions. Quickly, conflicts ensued and were elevated back up the chain of command for resolution. Forward progress stopped due to conflicting priorities, limited resources, bureaucratic processes, and unclear authorities and responsibilities. Action taken was: implementing Methodology, top executive commitment, training of all key personnel (including executives), and establishing clear open communication of strategic direction, change-projects, priorities, and responsibilities. The situation was reversed within one year.

Excerpt 3: Reducing Time-to-Market
High technology firm, which implemented the Methodology, dramatically reduced its time-to-market for developing new products. For example, one project involved the development of a new technology-based product requiring new concepts, new equipment, and demanding customer quality expectations. The team included: 14 people from all key functions and 2 outside vendors (future suppliers). The EST-assigned project priority (within the top 5) was communicated to all functions. The project team was given clear direction, complete freedom, and a reward structure based on team success. The team quickly became cohesive and self-directed. Driving through project barriers, the team completed the product with quality results, within budget, and with a 25% reduction in time-to-market (18 months vs. 24 months previous best case). Project has become the internal benchmark for the firm.

Excerpt 4: Urgent Field Failures Smoothly Handled
Firm utilizing the Methodology was hit with 18 major field failure items resulting in major customer problems. Before utilizing the Methodology, such an incident would have taken 4 months to restore, would have stopped progress on other change-projects, and taken over 6 months to get back to normal operation. Operating with Methodology, the 18 field failure projects were put to the top of the EST priority list, other priorities and resources were adjusted, all EST projects and many PST projects continued, while lower priority projects were put on hold. Field failure projects were completed in 2 months and the firm was back to normal operation within 3 months -- about half the time it would have taken before Methodology was installed.

Excerpt 5: Shrinkage Mystery
An unusual phenomena occurs in most applications of Methodology. When projects are prioritized for the first time, about 1/3 of the change-projects originally envisioned to be "urgent" or "top priority" simply disappear within the first 4 to 5 months. Various reasons are cited (e.g. "couldn't really define the project", "bad idea when evaluated", "project not feasible"). Expect 1/3 shrinkage of projects!

VIII. CONCLUSIONS

Several conclusions should be noted.

1. Accelerating technological, competitive, and business change has dramatically reduced the time available for, and complexity of, implementing change-projects.

2. Traditional approaches for managing change are no longer feasible in chaotic environments.

3. Sixteen (16) future needs have been identified for managing change in the 1990's and beyond.

4. Integrative Change Management Methodology presented meets all future needs identified, and has proven successful in practice.

5. Applications have shown that concentrating on keeping process simple and clear will lead to best results.

6. Near panic conditions within some firms have demonstrated value of Methodology in chaotic environments.

7. Open communications, clear priorities and direction, self-directed project teams, and loose-tight controls work best in chaotic environments.

8. All 16 needs are "Critical Success Factors" for the most effective use of the Methodloogy in the expected chaotic environments of the 21st Century.

ACKNOWLEDGEMENTS

Author thanks his wife for her valuable support, patience and understanding, and thanks his friend and colleague John Patton (President of Cadence Management Corporation in Portland, Oregon) for his valuable input.

BIOGRAPHICAL SKETCH

Dr. Donald E. White (Associate Professor, Industrial Engineering, California Polytechnic State University, San Luis Obispo, CA) joined Cal Poly in 1987 after 19 years increasingly responsible industrial experience in diverse functions at Pacific Telesis, ARCO, Exxon, and Bell Labs. Education: B.S., Mechanical Engineering, U.C. Berkeley; Ph.D., Systems Engineering, Case Western Reserve University; and MBA, Pepperdine University. Architect and coordinator for Cal Poly's new unique, joint MBA/MS in Engineering program in Engineering Management.

REFERENCES

1. Detlef K. Koska and Joseph D. Romano, "Countdown to the Future: The Manufacturing Engineer in the 21st Centure", Society of Manufacturing Engineers (SME) Report, SME, Dearborn, Michigan, 1988.

2. Robert B. Reich, "The Quiet Path to Technological Preeminence", Scientific American, October, 1989.

3. Charles Wiseman, Strategy and Computers: Information Systems as Competitive Weapons, Dow Jones-Irwin, Homewood, Illinois, 1985.

4. Steulpnagel, Thomas, "TQM: 21st Century Survival Requirement", Cal Poly University Seminar, May, 1990.

5. Detlef K. Koska and Joseph D. Romano, Op. Cit.

6. Sample treatments on bureaucracy include Alvin Gouldner, Patterns of Industrial Bureaucracy, 1954; Robert Merton, Social Theory and Social Structures, 1957; both Free Press, New York.

7. Tom Burns and George Stalker, The Management of Innovation, Tavistock, London, 1961.

8. John R. Schermerhorn, Management for Productivity, 3rd edition, Wiley, 1989.

9. David I. Cleland, "The Age of Project Management", Project Management Journal, March, 1991.

10. Donald E. White, John R. Patton, "Integrative Multiple Project Management", Project Management Institute Conference Proceedings, Calgary, October, 1990.

11. Richard H. Walklet, "Continuous Improvement and Simultaneous Engineering", Automotive Engineering, October, 1989.

12. M. Carl Ziemke, Mary S. Span, "Warning: Don't Be Half-Hearted in Your Efforts to Employ Concurrent Engineering", Industrial Engineering, February, 1991.

13. Detlef K. Koska and Joseph D. Romano, Op.Cit.

14. James A. Tompkins, "The Importance of Product Development to Winning Manufacturing", Industrial Management, July/August, 1989.

15. Keely Brunner, "Teamware: Supporting the Project-Driver Enterprise of the Future", Project Management Institute Conference Proceedings, October, 1990.

16. Martin Starr, "How 'Fast Response Organizations' Achieve Global Competitiveness", IEEE Engineering Management Review, December, 1990.

17. Donald E. White, John R. Patton, "Concurrent Project Management Methodology Reverses Firm's Competitive Decline", IEEE International Engineering Management Conference Proceedings, Santa Clara, October, 1990.

Within the next five years, every company will have to redesign how it measures its business performance.

The Performance Measurement Manifesto

by Robert G. Eccles

Revolutions begin long before they are officially declared. For several years, senior executives in a broad range of industries have been rethinking how to measure the performance of their businesses. They have recognized that new strategies and competitive realities demand new measurement systems. Now they are deeply engaged in defining and developing those systems for their companies.

At the heart of this revolution lies a radical decision: to shift from treating financial figures as the foundation for performance measurement to treating them as one among a broader set of measures. Put like this, it hardly sounds revolutionary. Many managers can honestly claim that they – and their companies – have tracked quality, market share, and other nonfinancial measures for years. Tracking these measures is one thing. But giving them equal (or even greater) status in determining strategy, promotions, bonuses, and other rewards is another. Until that happens, to quote Ray Stata, the CEO of Analog Devices, "When conflicts arise, financial considerations win out."[1]

The ranks of companies enlisting in this revolution are rising daily. Senior managers at one large, high-tech manufacturer recently took direct responsibility for adding customer satisfaction, quality, market share, and human resources to their formal measurement system. The impetus was their realization that the company's existing system, which was largely financial, undercut its strategy, which focused on customer service. At a smaller manufacturer, the catalyst was a leveraged recapitalization that gave the CEO the opportunity formally to reorder the company's priorities. On the new list, earnings per share dropped to last place, preceded by customer satisfaction, cash flow, manufacturing effectiveness, and innovation (in that order). On the old list, earnings per share stood first and almost alone.

In both companies, the CEOs believe they have initiated a sea change in how their managers think about business performance and in the decisions they make. Executives at other companies engaged in comparable efforts feel the same – rightly. What gets measured gets attention, particularly when rewards are tied to the measures. Grafting new measures onto

Robert Eccles is a professor of business administration at the Harvard Business School. This article grows out of research and course development for a new first-year course, "Information, Organization, and Control."

1. Ray Stata, "Organizational Learning – The Key to Management Innovation," *Sloan Management Review*, Spring 1989, pp. 63-74.

an old accounting-driven performance system or making slight adjustments in existing incentives accomplishes little. Enhanced competitiveness depends on starting from scratch and asking: "Given our strategy, what are the most important measures of performance?" "How do these measures relate to one another?" "What measures truly predict long-term financial success in our businesses?"

Dissatisfaction with using financial measures to evaluate business performance is nothing new. As far back as 1951, Ralph Cordiner, the CEO of General Electric, commissioned a high-level task force to identify key corporate performance measures. (The categories the task force singled out were timeless and comprehensive: in addition to profitability, the list included market share, productivity, employee attitudes, public responsibility, and the balance between short- and long-term goals.) But the current wave of discontent is not just more of the same.

One important difference is the intensity and nature of the criticism directed at traditional accounting systems. During the past few years, academics and practitioners have begun to demonstrate that accrual-based performance measures are at best obsolete – and more often harmful.[2] Diversity in products, markets, and business units puts a big strain on rules and theories developed for smaller, less complex organizations. More dangerously, the numbers these systems generate often fail to support the investments in new technologies and markets that are essential for successful performance in global markets.

Such criticisms reinforce concern about the pernicious effects of short-term thinking on the competitiveness of U.S. companies. Opinions on the causes of this mind-set differ. Some blame the investment community, which presses relentlessly for rising quarterly earnings. Others cite senior managers themselves, charging that their typically short tenure fosters shortsightedness. The important point is that the mind-set exists. Ask almost any senior manager and you will hear about some company's failure to make capital investments or pursue long-term strategic objectives that would imperil quarterly earnings targets.

> Managers' willingness to play the earning game calls into question the very measures the market focuses on.

Moreover, to the extent that managers do focus on reported quarterly earnings – and thereby reinforce the investment community's short-term perspective and expectations – they have a strong incentive to manipulate the figures they report. The extent and severity of such gaming is hard to document. But few in management deny that it goes on or that managers' willingness to play the earnings game calls into question the very measures the market focuses on to determine stock prices. For this reason, many managers, analysts, and financial economists have begun to focus on cash flow in the belief that it reflects a company's economic condition more accurately than its reported earnings do.[3]

Finally, many managers worry that income-based financial figures are better at measuring the consequences of yesterday's decisions than they are at indicating tomorrow's performance. Events of the past decade substantiate this concern. During the 1980s, many executives saw their companies' strong financial records deteriorate because of unnoticed declines in quality or customer satisfaction or because global competitors ate into their market share. Even managers who have not been hurt feel the need for preventive action. A senior executive at one of the large money-center banks, for example, grew increasingly uneasy about the European part of his business, its strong financials notwithstanding. To address that concern, he has nominated several new measures (including customer satisfaction, customers' perceptions of the bank's stature and professionalism, and market share) to serve as leading indicators of the business's performance.

Discontent turns into rebellion when people see an alternative worth fighting for. During the 1980s, many managers found such an alternative in the quality movement. Leading manufacturers and service providers alike have come to see quality as a strategic weapon in their competitive battles. As a result, they have committed substantial resources to developing measures such as defect rates, response time, delivery commitments, and the like to evaluate the performance of their products, services, and operations.

In addition to pressure from global competitors, a major impetus for these efforts has been the growth of the Total Quality Movement and related programs such as the Malcolm Baldrige National Quality Award. (Before a company can even apply for a Baldrige Award, it must devise criteria to measure the performance of its entire operation – not just its products – in minute detail.) Another impetus, getting stronger by the day, comes from large manufacturers who are more and more likely to impose rigid quality requirements on their suppliers. Whatever the stimulus, the result is the same: quality measures represent the most positive step taken to date in broadening the basis of business performance measurement.

Another step in the same direction comes from embryonic efforts to generate measures of customer satisfaction. What quality was for the 1980s, customer satisfaction will be for the 1990s. Work on this class of measures is the highest priority at the two manufacturing companies discussed earlier. It is equally critical at another high-tech company that recently created a customer satisfaction department reporting directly to the CEO. In each case, management's interest in developing new performance measures was triggered by strategies emphasizing customer service.

As competition continues to stiffen, strategies that focus on quality will evolve naturally into strategies based on customer service. Indeed, this is already happening at many leading companies. Attention to customer satisfaction, which measures the quality of customer service, is a logical next step in the development of quality measures. Companies will continue to measure quality on the basis of internally generated indexes (such as defect rates) that are presumed to relate to customer satisfaction. But they will also begin to evaluate their performance by collecting data directly from customers for more direct measures like customer retention rates, market share, and perceived value of goods and services.

Just as quality-related metrics have made the performance measurement revolution more real, so has the development of competitive benchmarking.[4] First, benchmarking gives managers a methodology that can be applied to any measure, financial or nonfinancial, but that emphasizes nonfinancial metrics. Second (and less obvious), it has a transforming effect on managerial mind-sets and perspectives.

Benchmarking involves identifying competitors and/or companies in other industries that exemplify best practice in some activity, function, or process and then comparing one's own performance to theirs. This externally oriented approach makes people aware of improvements that are orders of magnitude beyond what they would have thought possible. In contrast, internal yardsticks that measure current performance in relation to prior period results, current budget, or the results of other units within the company rarely have such an eye-opening effect. Moreover, these internally focused comparisons have the disadvantage of breeding complacency through a false sense of security and of stirring up more energy for intramural rivalry than for competition in the marketplace.

Finally, information technology has played a critical role in making a performance measurement revolution possible. Thanks to dramatically improved price-performance ratios in hardware and to breakthroughs in software and database technology, organizations can generate, disseminate, analyze, and store more information from more sources, for more people, more quickly and cheaply than was conceivable even a few years back. The potential of new technologies, such as hand-held computers for employees in the field and executive information systems for senior managers, is only beginning to be explored. Overall, the range of measurement options that are economically feasible has radically increased.

Veterans know it is easier to preach revolution than to practice it. Even the most favorable climate can create only the potential for revolutionary change. Making it happen requires conviction, careful preparation, perseverance, and a decided taste for ambiguity. As yet, there are no clear-cut answers or predetermined processes for managers who wish to change their measurement systems. Based on the experience of companies engaged in this revolution, I can identify five areas of activity that sooner or later need to be addressed: developing an information architecture; putting the technology in place to support this architecture; aligning incentives with the new system; drawing on outside resources; and designing a process to ensure that the other four activities occur.

Developing a new information architecture must be the first activity on any revolutionary agenda. Information architecture is an umbrella term for the categories of information needed to manage a company's businesses, the methods the company uses to generate this information, and the rules regulating its flow. In most companies, the accounting system implicitly defines the information architecture. Other performance measures are likely to be informal – records that operating managers keep for themselves, for instance – and they are rarely integrated into the corporate-driven financial system.

The design for a new corporate information architecture begins with the data that management needs to pursue the company's strategy. This may sound like a truism, but a surprising number of companies describe their strategies in terms of customer service, innovation, or the quality and capabilities of

2. Donald A. Curtis, "The Modern Accounting System," *Financial Executive*, January-February 1985, pp. 81-93; and H. Thomas Johnson and Robert S. Kaplan, *Relevance Lost* (Boston: Harvard Business School Press, 1987).

3. Yuji Ijiri, "Cash Flow Accounting and Its Structure," *Journal of Accounting, Auditing, and Finance*, Summer 1978, pp. 331-348.

4. Robert C. Camp, *Benchmarking* (Milwaukee, Wisconsin: ASQS Quality Press, 1989).

their people, yet do little to measure these variables. Even time – the newest strategic variable – remains largely underdeveloped in terms of which time-based metrics are most important and how best to measure them.

As part of this identification process, management needs to articulate a new corporate grammar and define its own special vocabulary – the basic terms that will need to be common and relatively invariant across all the company's businesses. Some of these terms (like sales and costs) will be familiar. Others, however, will reflect new strategic priorities and ways to think about measuring performance. For example, both a large money-center bank and a multidivisional, high-technology manufacturer introduced the use of cross-company customer identification numbers so they could readily track such simple and useful information as the total amount of business the company did with any one customer. It sounds elementary and it is – as soon as you start to look at the entire measurement system from scratch.

Uniformity can be carried too far. Different businesses with different strategies require different information for decision making and performance measurement. But this should not obscure the equally obvious fact that every company needs to have at least a few critical terms in common. Today few large companies do. Years of acquisitions and divestitures, technological limitations, and at times, a lack of management discipline have all left most big organizations with a complicated hodgepodge of definitions and variables – and with the bottom line their only common denominator.

Developing a coherent, companywide grammar is particularly important in light of an ever-more stringent competitive environment. For many companies, ongoing structural reorganizations are a fact of life. The high-technology company described above has reorganized itself 24 times in the past 4 years (in addition to a number of divisional and functional restructurings) to keep pace with changes in its markets and technologies. Rather than bewail the situation, managers relish it and see their capacity for fast adaptation as an important competitive advantage.

One high-tech company has reorganized 24 times in the past 4 years to keep pace with changes in its markets.

A common grammar also enhances management's ability to break apart and recombine product lines and market segments to form new business units. At a major merchant bank, for example, the organization is so fluid that one senior executive likens it to a collection of hunting packs that form to pursue business opportunities and then disband as the market windows on those opportunities close. The faster the company can assemble information for newly formed groups, the greater the odds of success. So this executive (who calls himself the czar of information) has been made responsible for developing standard definitions for key information categories.

How a company generates the performance data it needs is the second piece of its information architecture. Not surprisingly, methods for measuring financial performance are the most sophisticated and the most deeply entrenched. Accountants have been refining these methods ever since double-entry bookkeeping was invented in the fifteenth century. Today their codifications are enforced by a vast institutional infrastructure made up of professional educators, public accounting firms, and regulatory bodies.

In contrast, efforts to measure market share, quality, innovation, human resources, and customer satisfaction have been much more modest. Data for tracking these measures are generated less often: quarterly, annual, or even biannual bases are common. Responsibility for them typically rests with a specific function. (Strategic planning measures market share, for example, while engineering measures innovation and so on.) They rarely become part of the periodic reports general managers receive.

Placing these new measures on an equal footing with financial data takes significant resources. One approach is to assign a senior executive to each of the measures and hold him or her responsible for developing its methodologies. Typically, these executives come from the function that is most experienced in dealing with the particular measure. But they work with a multifunctional task force to ensure that managers throughout the company will understand the resulting measures and find them useful. Another, less common, approach is to create a new function focused on one measure and then to expand its mandate over time. A unit responsible for customer satisfaction might subsequently take on market share, for example, or the company's performance in human resources.

Unlike a company's grammar, which should be fairly stable, methods for taking new performance measures should evolve as the company's expertise increases. Historical comparability may suffer in the process, but this is a minor loss. What matters is how a company is doing compared with its current competitors, not with its own past.

The last component of a corporate information architecture is the set of rules that governs the flow of information. Who is responsible for how measures are taken? Who actually generates the data? Who receives and analyzes them? Who is responsible for changing the rules? Because information is an important source of power, the way a company answers these questions matters deeply. How open or closed a company is affects how individuals and groups work together, as well as the relative influence people and parts of the company have on its strategic direction and management. Some companies make information available on a very limited basis. At others, any individual can request information from another unit as long as he or she can show why it is needed. Similarly, in some companies the CEO still determines who gets what information – not a very practical alternative in today's world. More often what happens is that those who possess information decide with whom they will share it.

Advances in information technology such as powerful workstations, open architectures, and relational databases vastly increase the options for how information can flow. It may be centralized at the top, so that senior executives can make even more decisions than they have in the past. Or it may be distributed to increase the decision-making responsibilities of people at every level. The advantages of making information widely available are obvious, though this also raises important questions that need to be addressed about the data's integrity and security. In principle, however, this portion of the information architecture ought to be the most flexible of the three, so that the company's information flows continue to change as the conditions it faces do.

Determining the hardware, software, and telecommunications technology a company needs to generate its new measurement information is the second activity in the performance revolution. This task is hard enough in its own right, given the many choices available. But too often managers make it even harder by going directly to a technology architecture without stopping first to think through their information needs. This was the case at a high-tech manufacturing company that was growing more and more frustrated with its information systems planning committee. Then the CEO realized that he and the other senior managers had not determined the measures they wanted before setting up the committee. Equipped with that information, the committee found it relatively easy to choose the right technology.

Once the information architecture and supporting technology are in place, the next step is to align the new system with the company's incentives – to reward people in proportion to their performance on the measures that management has said truly matter. This is easier said than done. In many companies, the compensation system limits the amount and range of the salary increases, bonuses, and stock options that management can award.

In companies that practice pay-for-performance, compensation and other rewards are often tied fairly mechanically to a few key financial measures such as profitability and return on investment. Convincing managers that a newly implemented system is really going to be followed can be a hard sell. The president of one service company let each of his division general managers design the performance measures that were most appropriate for his or her particular business. Even so, the managers still felt the bottom line was all that would matter when it came to promotions and pay.

The difficulty of aligning incentives to performance is heightened by the fact that formulas for

Formulas that tie incentives to performance look objective – and rarely work.

tying the two together are rarely effective. Formulas have the advantage of looking objective, and they spare managers the unpleasantness of having to conduct truly frank performance appraisals. But if the formula is simple and focuses on a few key variables, it inevitably leaves some important measures out. Conversely, if the formula is complex and factors in all the variables that require attention, people are likely to find it confusing and may start to play games with the numbers. Moreover, the relative importance of the variables is certain to change more often – and faster – than the whole incentive system can change.

For these reasons, I favor linking incentives strongly to performance but leaving managers free to determine their subordinates' rewards on the basis of all the relevant information, qualitative as well as quantitative. Then it is up to the manager to explain candidly to subordinates why they received what they did. For most managers, this will also entail learning to conduct effective performance appraisals, an indirect – and invaluable – benefit of overhauling the measurement system.

Outside parties such as industry and trade associations, third-party data vendors, information technology companies, consulting firms, and public accounting firms must also become part of the performance measurement revolution. Their incentive: important business opportunities.

Industry and trade associations can play a very helpful role in identifying key performance measures, researching methodologies for taking these measures, and supplying comparative statistics to their members – so can third-party data vendors. Competitors are more likely to supply information to a neutral party (which can disguise it and make it available to all its members or customers) than to one another. And customers are more likely to provide information to a single data vendor than to each of their suppliers separately.

Public accounting firms have what may be the single most critical role in this revolution.

Consulting firms and information technology vendors also have important roles to play in forwarding the revolution. Firms that specialize in strategy formulation, for example, often have well-developed methods for assessing market share and other performance metrics that clients could be trained to use. Similarly, firms that focus on strategy implementation have a wealth of experience designing systems of various kinds for particular functions such as manufacturing and human resources. While many of these firms are likely to remain specialized, and thus require coordination by their clients, others will surely expand their capabilities to address all the pieces of the revolution within a client company.

Much the same thing is apt to happen among vendors of information technology. In addition to helping companies develop the technological architecture they need, some companies will see opportunities to move into a full range of services that use the hardware as a technology platform. IBM and DEC are already moving in this direction, impelled in part by the fact that dramatic gains in price-performance ratios make it harder and harder to make money selling "boxes."

Finally, public accounting firms have what may be the single most critical role in this revolution. On one hand, they could inhibit its progress in the belief that their vested interest in the existing system is too great to risk. On the other hand, all the large firms have substantial consulting practices, and the revolution represents a tremendous business opportunity for them. Companies will need a great deal of help developing new measures, validating them, and certifying them for external use.

Accounting firms also have an opportunity to develop measurement methods that will be common to an industry or across industries. While this should not be overdone, one reason financial measures carry such weight is that they are assumed to be a uniform metric, comparable across divisions and companies, and thus a valid basis for resource allocation decisions. In practice, of course, these measures are not comparable (despite the millions of hours invested in efforts to make them so) because companies use different accounting conventions. Given that fact, it is easy to see why developing additional measures that senior managers – and the investment community – can use will be a massive undertaking.

Indeed, the power of research analysts and investors generally is one of the reasons accounting firms have such a crucial role to play. Although evidence exists that investors are showing more interest in metrics such as market share and cash flow, many managers and analysts identify the investment community as the chief impediment to revolution.[5] Until investors treat other measures as seriously as financial data, they argue, limits will always exist on how seriously those measures are taken inside companies.

GE's experience with its measurement task force supports their argument. According to a knowledgeable senior executive, the 1951 effort had only a modest effect because the measures believed to determine the company's stock price, to which incentives were tied, were all financial: earnings per share, return on equity, return on investment, return on sales, and earnings growth rate. He believed that once the financial markets valued other measures, progress within companies would accelerate.

Would managers be willing to publish anything more than the financial information the SEC requires?

Investors, of course, see the problem from a different perspective. They question whether managers would be willing to publish anything more than the financial information required by the SEC lest they reveal too much to their competitors. Ultimately, a regulatory body like the SEC could untie this Gordian knot by recommending (and eventually requiring) public companies to provide nonfinancial measures in their reports. (This is, after all, how financial

5. "Investors: Look at Firms' Market Share," *Wall Street Journal*, February 26, 1990, pp. C1-2.

standards became so omnipotent and why so many millions of hours have been invested in their development.) But I suspect competitive pressure will prove a more immediate force for change. As soon as one leading company can demonstrate the long-term advantage of its superior performance on quality or innovation or any other nonfinancial measure, it will change the rules for all its rivals forever. And with so many serious competitors tracking – and enhancing – these measures, that is only a matter of time.

Designing a process to ensure that all these things happen is the last aspect of the revolution. To overcome conservative forces outside the company and from within (including line and staff managers at every level, in every function), someone has to take the lead. Ultimately, this means the CEO. If the CEO is not committed, the revolution will flounder, no matter how much enthusiasm exists throughout the organization.

But the CEO cannot make it happen. Developing an information architecture and its accompanying technology, aligning incentives, working with outside parties – all this requires many people and a lot of work, much of it far less interesting than plotting strategy. Moreover, the design of the process must take account of the integrative nature of the task: people in different businesses and functions including strategic planning, engineering, manufacturing, marketing and sales, human resources, and finance will all have something to contribute. The work of external players will have to be integrated with the company's own efforts.

Organizationally, two critical choices exist. One is who the point person will be. Assigning this role to the CEO or president ensures its proper symbolic visibility. Delegating it to a high-level line or staff executive and making it a big piece of his or her assignment may be a more effective way to guarantee that enough senior management time will be devoted to the project.

The other choice is which function or group will do most of the work and coordinate the company's efforts. The CEO of one high-tech company gave this responsibility to the finance function because he felt they should have the opportunity to broaden their perspective and measurement skills. He also thought it would be easier to use an existing group experienced in performance measurement. The president of an apparel company made a different choice. To avoid the financial bias embedded in the company's existing management information systems, he wanted someone to start from scratch and design a system with customer service at its core. As a result, he is planning to combine the information systems department with customer service to create a new function to be headed by a new person, recruited from the outside.

What is most effective for a given company will depend on its history, culture, and management style. But every company should make the effort to attack the problem with new principles. Some past practices may still be useful, but everything should be strenuously challenged. Otherwise, the effort will yield incremental changes at best.

Open-mindedness about the structures and processes that will be most effective, now and in the future, is equally important. I know of a few companies that are experimenting with combining the information systems and human resource departments. These experiments have entailed a certain amount of culture shock for professionals from both functions, but such radical rethinking is what revolution is all about.

> Combining information systems and human resources is a culture shock for both departments. But that's what revolution is all about.

Finally, recognize that once begun, this is a revolution that never ends. We are not simply talking about changing the basis of performance measurement from financial statistics to something else. We are talking about a new philosophy of performance measurement that regards it as an ongoing, evolving process. And just as igniting the revolution will take special effort, so will maintaining its momentum – and reaping the rewards in the years ahead.

Reprint 91103

Teaming Up for Performance

By establishing performance measurement teams, Martin Marietta Missile Systems achieved dramatic improvements in quality, cost and schedule performance.

Vladimir J. Mandl
Martin Marietta Electronics
& Missiles Group
Orlando, FL

Faced with a critical Army review of its operating practices in 1986, Martin Marietta Missile Systems nearly flunked. Three years later, the Orlando, FL company received the highest score ever awarded by government auditors.

The phenomenal transformation has earned Martin Marietta the highest recognition the US Army gives its contractors: entrance to the coveted Contractor Performance Certification Group. As of February 1990, the company became the first defense contractor to have earned the distinction of being certified for both Army Missile Command and the Armament Munitions & Chemical Command and the first to have the entire facility enrolled in the program.

The program recognizes contractors who consistently deliver quality products, apply process controls, use preventive and proactive audit procedures and demonstrate continuous efforts to improve quality.

The turnaround is the result of a dynamic and unique process, one that continues to reap rewards for the company, the government and the taxpayers. The process is called Performance Measurement Teams (PMT) and is largely responsible for dramatic improvements in the company's quality, cost and schedule performance.

In the first three years of the program, average first-time yield has risen to 93 percent, manufacturing rework dropped 81.4 percent, overtime decreased by 61 percent and performance to standards improved 86.2 percent, see *Figure 1*. These improvements were attained while output was increased by more than 45 percent, and manpower requirements were reduced by 31.1 percent *Figure 2*. Moreover, performance to contract delivery schedule reached an unheard of 100 percent.

PMT philosophy and process

This record is the result of a highly disciplined philosophy of personal involvement practices from the president's office on down to the line personnel. Management firmly believes that line supervision must be provided with every tool required to perform its job efficiently. Without it, the organization cannot succeed.

The fundamental tool is PMT, a formula born in the belief that people want to succeed and be recognized. When provided with the responsibility, data, authority and accountability to solve problems affecting their community, they will achieve outstanding results.

PMTs consist of all the employees in a given work area, plus various support personnel, such as industrial engineers. Each team conducts a weekly one-hour meeting held in a conference room away from the shop floor. Here, the discipline calls for compulsory attendance by all team members, including the area general foreman, and assigned product support and quality personnel.

Each meeting begins with the week's performance review, and continues with discussion of goals and objectives, along with problems facing production flow. As problems are identified, ideas for solutions are brainstormed and action items are assigned to team members for resolution.

Percent	Before PMT	1989	Improvement
Yield	87.0	98.3	13.0
Mfg Rework	4.3	0.8	81.4
Performance	58.2	86.2	48.1
Overtime	20.0	7.7	61.0

Figure 1 Missile Systems quality tracking resulted in highest quality rating awarded by US Army.

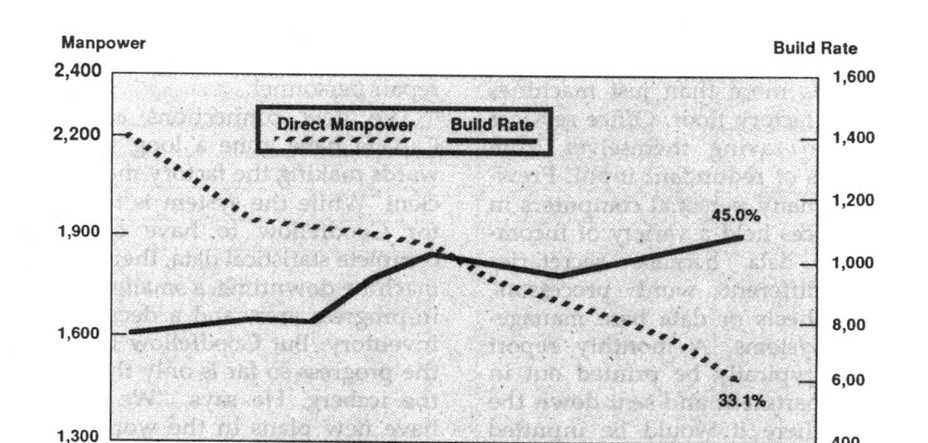

Figure 2 Personnel involvement in product quality output results in drastic build rate increase and reduction in manpower requirements.

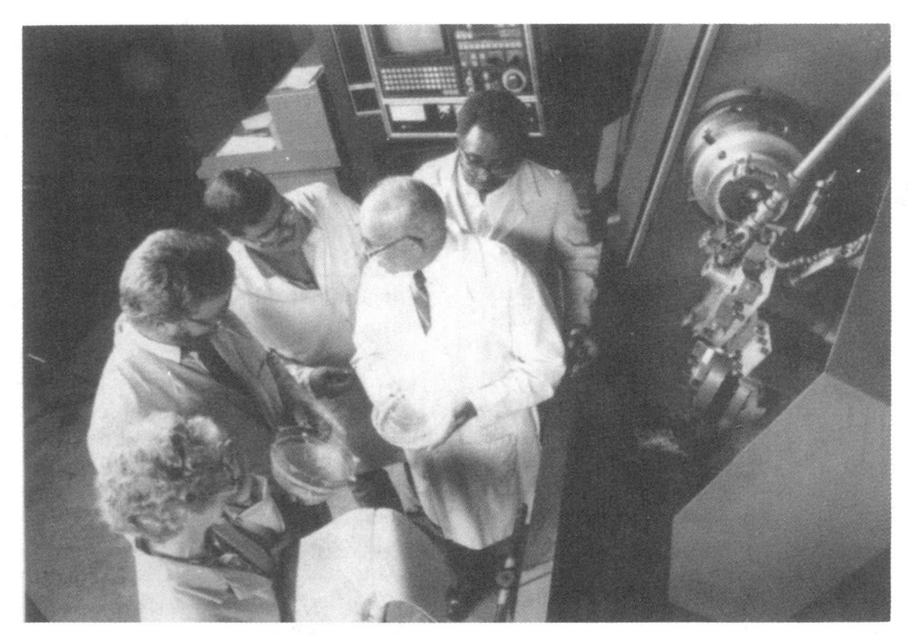

Over 200 PMTs are involved in Martin Marietta Missile and Electronic Systems Group's quality improvement effort.

Two key ingredients are accurate and timely work center measurement data, and the authority for each center to propose and make changes.

While the teams are responsible for reviewing the real-time performance data and the setting and reviewing of team goals, management has a responsibility to provide timely response to supervision's requests. This, in turn, calls for the development of a two-way communications and reporting system, where, while direction comes from the top, proposed improvements proceed from the production lines upward through the various levels of supervision and management until resolution is provided.

No finger-pointing

PMT is effective because it places responsibility and accountability where it belongs. There's no such thing as pointing fingers. Each PMT work center's performance is measured in standards vs. actuals and displayed against a goal the team believes it can achieve.

Each work center PMT board displays the same data. It tracks performance and goals for inspection tests and yield results, scrap and cost performance, lost time/overtime status, rework levels, material review board results, audit and quality reviews, delivery schedule records and action item status. This regimen is ironclad, and all data are updated weekly. In this manner, there is real-time control on what's happening to the product.

The work center's entire chain of command, in turn, is measured against the same criteria. This data shows up as one of eight summary charts on the operations director's display reflecting the total organization's performance curve. Charts for production operations also are maintained in the office of the vice president.

Involving everyone

Total employee involvement, or "ownership," is central to the Performance Measurement Teams. "PMT cannot exist without ownership," says Bob Keymont, Missile Systems vice president of production operations and the creator of PMTs. "The expertise is in the hands of the people on the line. Just look at how long they have been in the position and the experience they have. There is continuity, lessons learned, and realization of mistakes made. You must involve everyone. Everyone owns a part of the company; it is a culture."

As a new and continually developing culture, PMT provides its participants with real-time, measurable goals. It places the burden for the execution almost entirely in the hands of first line supervision, reflecting the conviction that properly trained, led and involved, line personnel can provide a real insight and expertise to resolve performance roadblocks.

One additional factor that provides team strength is the direct involvement of operations support—manufacturing engineering, industrial engineering, production control, facilities, test—in PMT activities, both on the manufacturing floor and in team meetings. Moreover, manufacturing support personnel assigned to a work center are physically located on the floor, further facilitating the exchange of views, ideas and suggestions as these affect product quality and performance.

In Keymont's view, ownership motivates line personnel to excel at a given job. The evidence rests in the performance charts discussed as part of a set agenda each week at team meetings and publicly displayed at work centers throughout the facility. The team members can track and see the result of their joint labor, knowing they can contribute and be listened to when their concerns are voiced.

Action item resolution

One of the keys to the system's success is that no item is allowed to hibernate. While practice has shown that about 85 percent of the problems are solved at the team level within a week to 10 days, no item is allowed to go unresolved for more than 30 days, unless there is a formal plan for its solution. Items that cannot be resolved at the team level—typically capital-related issues—are "bubbled up" to upper levels, again reflecting the same time-frame tracking and reporting requirements as those practiced at lower echelons.

The PMT process, because of its formal reporting system, requires resolution of action items regardless of where they end up within the management hierarchy. Once an item is placed on the agenda, it is recorded, charted, time-bracketed, publicly displayed and assigned to a given individual or group, who must report on its status within a set time period. The item remains on the agenda until the problem has been resolved.

For a team to function to its full potential, management backup and prompt response is a must. Practice has shown that unless there is a quick response to an inquiry, the

SPC stations, introduced into operations by PMT/PST efforts, place responsibility for quality output at the employee's workstation.

team becomes lax and unresponsive. As one director said, "I make it a practice to visit my teams at least once a month. This helps both of us. I get a better understanding of how the team functions and what they need, and they have an opportunity to get real answers to what is going on, and what I can, and cannot, do for them. Plus, the visits serve to convince the team that management is backing them up."

Management backup, coupled with informed and participating engineering support personnel and trained line supervision, has inspired the line personnel response needed to effectively pursue the PMT process. PMT, in the words of an hourly missile mechanic, "has given us a sense of pride in our work. When we succeed, our team succeeds, our company succeeds, our country succeeds." A fork-lift operator expressed like views when he stated, "PMT gives management and hourly people an opportunity to review and solve potential problems before they have a chance to impact productivity. The bottom line is teamwork."

Team training

Right at the beginning of the PMT process, it became evident that one of the keys to successful team performance was the proper training of the team leaders. The process evolved from relatively basic motivational class sessions for line supervisors into a PMT specific training matrix, covering leadership, motivation, communication skills, team building techniques, problem solving, team coaching, recognition and reinforcement. This training series may range from 25 to 54 hours.

Once the team leader is trained, it becomes his or her job to train the team, and encourage its members to take the opportunity to view and discuss their performance and work-related problems in an open forum, both on the manufacturing floor and in their weekly team meetings.

On the manufacturing floor this requires learning work techniques, adherence to procedures and discussion of potential or real problems without apprehension. The evidence of this happening is in the action items displayed on PMT performance boards located in the work center area.

Among the major contributors to successful team operation are team facilitator coaches. They function as observers, initially helping the team leaders, most of whom have had little or no experience in standing up before a group and making it work as a unit. As the groups begin to function as a team, facilitator attendance is reduced to occasional visits and consultations.

Once team leaders are trained in meeting mechanics, convinced of the value of sharing responsibility for product quality and confident of total management support of the process, generating team response becomes a matter of time. Team members, in the view of one supervisor, become, "all the experts I need to resolve a line problem. I have all the resources I need at hand, and if I need someone from

Quality Circles	PMTs
Voluntary workforce participation	Mandatory workforce participation
Voluntary circle leaders	Designated team leaders
Reactive management participation	Proactive management participation
Informal meeting structure	Formal meeting structure
Randomly selected issues	Job-related issues
Circle effectiveness not measured	Real-time team measurement system established
Circle goals nonexistent	Defined/measurable team goals
No formal recognition program	Team recognition based on performance

Figure 3 Unlike quality circles PMTs are geared to resolve and track job-related issues.

Courses	Length	V.P.	Dir	Mgr	GF	Supv	Support	Coord	Coaches
PMT Leader Training Overview									
Mod 1 History, Process Overview	(1 hr)	X	X	X	X	X	X	X	X
Mod 2 Roles/Resp/Mers	(1 hr)				X	X	X	X	X
Communication Skills	(16 hrs)								
Mod 1 Presentation Skills	(4 hrs)				X	X		X	X
Mod 2 DISC Review	(4 hrs)				X	X	X	X	X
Mod 3 Motivation	(4 hrs)				X	X	X	X	X
Mod 4 Active Listening	(4 hrs)				X	X	X	X	X
Teambuilding Techniques Situational Leadership	(4 hrs)				X	X	X	X	X
Problem Solving Kepner-Tregoe	(24 hrs)	X	X	X	X	X	X	X	X
Coaching & Reinforcing People Strategies	(8 hrs)			X	X	X		X	X
Total Hours		25	25	33	54	54	42	54	54

Figure 4 PMT training reaches every level of management.

the outside, I bring them in, and the people love it. They know the upper echelons care and support them. The process makes life so much easier. I find acceptance, solutions and relief from pressure—all in one place."

Besides training the team leaders, company officials faced the challenge of providing timely and accurate data so essential to the process. Prior to the PMT inauguration, performance was measured at area manager levels, with several work centers reflected in the data. To track individual teams required an additional breakdown of the data base. This type of effort, relatively new to any industry, produced its own benefits. The use of state-of-the-art data bases, graphic software and high-speed plotters led to a significant reduction in costs and improved efficiency of the total performance measurement data system.

Performance recognition

Employee recognition is a cornerstone in continuing PMT contributions to the company's record of quality excellence. It is the lifeblood that motivates the teams to ever-improving performance.

Since the PMT process was inaugurated, outstanding performance is being recognized through Team-of-the-Month and Team-of-the-Year competitions, and a monthly periodical designed to highlight outstanding events within the PMT arena. Team-of-the-Month awards are made in three categories: the manufacturing fabrication shops, final assembly and test, and manufacturing support operations. Winning teams are honored by top management at several formal breakfasts where team leaders are given plaques for display in the work areas, and group photographs are taken and displayed at strategic points throughout the company.

The selection of the Teams of the Year award is based on documented records of consistent quality improvement. The process starts with director-level recommendations of the five best performers in their areas to the vice president of production operations. Each candidate team then presents its accomplishments and rationale for top spot selection. From this competition, finalists are chosen from each area to present their case to the company president and his staff, who decide on the winners for the year. The winning teams are honored at a banquet where each member receives a gold watch. The banquet is attended by family members, corporate and company management, union leadership and customer representatives.

Internally, PMT accomplishments are recognized through *Production Operation's Shop Talk (POST)*, a monthly publication where news of both group and individual accomplishments is reported.

Migrating to management

Tracking performance results through statistical methods has produced a domino effect that now reaches across virtually every discipline within the company. Aside from production operations, PMT has been implemented in business operations, management systems, quality assurance and engineering.

Highly visible evidence of the impact PMTs have had on the company is the creation of white collar Process Simplification Teams (PST). Using the PMT approach, process simplification represents the next level of personal involvement: the development and implementation of a prevention-based system through simplification of processes.

While relatively new, the PSTs are multi-functional, because they respond to and resolve problems reaching across disciplines. Each team is assigned specific issues and sets and negotiates its goals and milestones. Tracking results weekly, they work towards a process simplification model.

Regardless of the task, each team first analyzes and depicts the work flow in question, applying common sense to each step. The primary question is whether the step can be performed more efficiently. Finally,

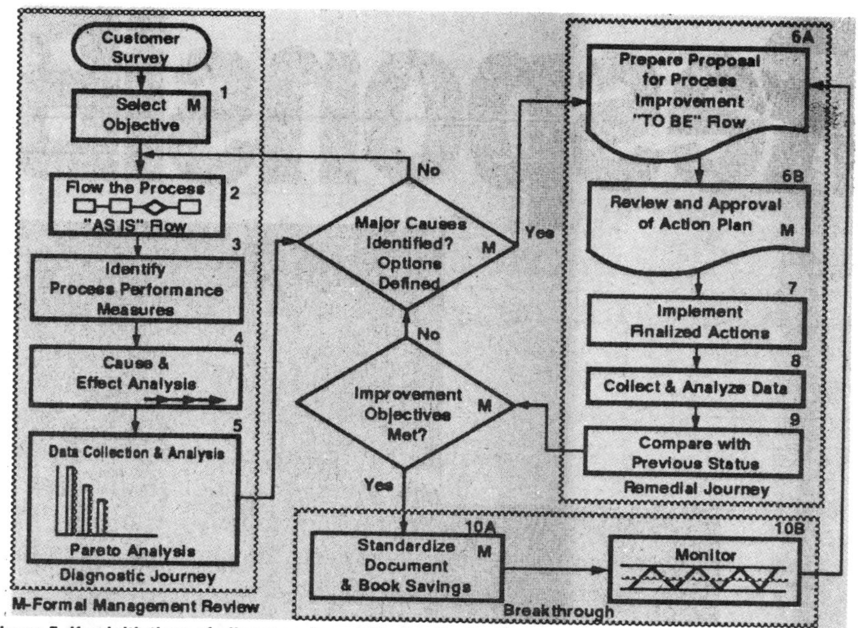

Figure 5 Key initiatives of all companies within the Martin Marietta Missile and Electronic Systems Group are methods improvement and process simplification.

team members implement changes that may range from replacement of a factory process to changes in policies involving the company.

An organizational approach

PMTs and PSTs complement each other, providing the basis for an organizational approach to productivity improvement through capital commitment.

While PMTs are tied to given processes to produce hardware, they also have the right to call for changes above and beyond their normal scope of work. This provides opportunity for Process Simplification Teams to step in and examine these additional opportunities, which can involve products and services supplied, and set up those higher goals and objectives as well as plans for their implementation.

This mutually beneficial effort has facilitated the introduction, or expansion of new paperless systems and technologies into the total operation. Among these is Statistical Process Control (SPC). This initiative was successfully introduced in Missile Systems Production Operations manufacturing, primarily because of positive and aggressive PMT reactions to the system's operation in machining centers. Manufacturing took further ownership of SPC by extending the system's implementation in electrical and product development areas.

SPC has become a vehicle to significantly improve process and hardware performance. For instance, the process is used to identify deteriorating machine tool performance before hardware defects occur. Nearly 100 SPC stations, representing considerable capital commitment, are in place within Missile Systems, placing more responsibility for quality output where it belongs—at the employee's workstation.

Today, more than three years and 70 teams later, Missile Systems Production Operations has the proof that PMT is an effective process to achieve world-class quality while slashing costs and meeting schedules. The process is being applied throughout the company and has been adopted, to one degree or another, by other Martin Marietta operations in Denver, Baltimore, Colorado Springs, Michoud (LA), Orlando and Ocala, FL, with approximately 200 teams in operation. In addition, the concept is being flowed down to the company's suppliers.

The PMT process provides every individual in the work force with a forum to voice, act on and take pride in task ownership. PMT disciplines provide each team with the criteria, the talent and the means by which to set goals and measure progress. The process, by virtue of its structured reporting and communication systems, reaches and produces positive results at every level with the company. Furthermore, the PMT process, because of its adherence to and understanding of requirements, error prevention methodology and measuring systems, has become a major element in the corporation's commitment to total quality management. MS

Reprinted by permission of the Society of Manufacturing Engineers from the *Journal of Manufacturing Systems*, Vol. 6, No. 3.

Economic Measure of Productivity, Quality and Flexibility in Advanced Manufacturing Systems

Young Kyu Son, Chan S. Park, Auburn University, Auburn, Alabama

Abstract

Productivity, quality and flexibility are critical measures of manufacturing performance for justifying the investment in integrated manufacturing and production systems. The objective of this research is to define, quantify and incorporate these three measures. The conventional productivity measure was improved so that it could be used in integrated manufacturing production systems. Quality and flexibility measures were newly defined and quantified. These three measures were integrated for the evaluation of a manufacturing system as a whole.

Keywords: Advanced Manufacturing Systems, Productivity, Performance Measure, Manufacturing Flexibility, Economic Analysis.

Introduction

U.S. industries recently have had serious problems with manufacturing performance. In an effort to solve this problem, various manufacturing principles and technologies have been introduced such as group technology (GT), just-in-time (JIT) systems, flexible manufacturing systems, and so forth. These integrated manufacturing and production systems (IMPS) have the potential for improving manufacturing performance. Yet, the adoption of IMPS in manufacturing industries is relatively slow. The IMPS requires a large initial investment under a long-term, uncertain environment. Without proof of economic justification, few firms will be willing to take the financial risk.

Productivity, quality and flexibility are critical measures of total manufacturing performance.[1-6] In this paper, productivity indicates the efficiency of converting inputs (resources) to outputs. Quality refers to the degree of excellence in making products. Flexibility measures the adaptability to various changes in manufacturing environments. The need for an improved measure of productivity has been addressed in the literature.[7] Furthermore, how to measure quality and flexibility have been the primarily issues in justifying the IMPS, but little work has been done in measuring these elements due to the difficulties in quantifying them in dollar terms.

The objective of this paper is therefore to offer a definition of and to quantify productivity, quality and flexibility, and to combine them to obtain an overall performance measure of a typical manufacturing activity. Such a measure will aid firms in making the investment decision on advanced manufacturing systems. A case example is then presented to illustrate the process of obtaining the necessary data and quantifying the proposed measures.

Review of Related Literature

Productivity has long been the only measure of manufacturing performance. Productivity is defined by the Bureau of Labor Statistics as the value of goods manufactured divided by the amount of input. There are at least two distinct types of productivity ratios at the firm level, total productivity and partial productivity.

Total Productivity = Total Output/Total Input

Partial Productivity = Total Output/Partial Input

For many years, most manufacturing companies have used partial productivity, such as labor productivity.[8,9] The use of partial productivity, such as labor, can also result in serious misunderstandings. For example, a company buys a higher quality

Editor's Note: This paper was awarded the 1987 Alfred B. Bodine/SME Award for studies in machine tool economics. This award is presented annually to a business or engineering graduate student and administered by the SME Engineering Education Foundation.

raw material that significantly reduces the man-hours necessary for processing. Suppose this increased cost of using higher quality material equals the savings in labor cost from reduced man-hours. Although the firm's total productivity does not change, labor productivity would naturally increase since a worker now can produce more of the same product in less time. Using the labor productivity index as a guide, the company would note an increase in productivity, i.e., gains indicated by increased labor productivity may not actually be gains at all. This type of fallacy is inherent in all partial productivity measures. Therefore, some measure of total productivity must be used for most company wide management decisions.

Total productivity which combines labor, capital, raw material and miscellaneous productivity was developed.[7] Energy cost was separated from miscellaneous costs to form energy productivity, and attempts were made to incorporate productivity measures into the firm's accounting systems.[8] These approaches have shortcomings, however.

First, the labor cost calculation included only direct, not indirect labor. With automation, machines replace labor. A worker can handle many machines and the worker usually spends more time in solving problems than in operating the machines. This trend results in a shift from direct to indirect labor; therefore, labor productivity should include indirect and direct labor costs. Second, the designation "miscellaneous productivity" lumps together too many categories. For accurate evaluation of various IMPS changes, "miscellaneous productivity" must be separated into categories such as energy efficiency floor space reduction, increased use of computer software, and so forth. Third, but more importantly, little work has been done to predict and estimate future productivity. Productivity should not only measure the past or current manufacturing performance, but also predict the future efficiency resulting from the introduction of IMPS.

Product quality has become one of the most important elements of manufacturing performance, as seen in the success of Japanese manufacturing. The literature on quality control is voluminous, but that dealing with quality in economic (monetary) terms is sparse.[10,11] Quality needs to be expressed in dollar terms rather than solely in statistical terms for better understanding. Quality should also be related to productivity.

Despite good productivity and quality, some firms report declining profitability.[3] With the frequent changes in internal and external factors which affect manufacturing systems, another important measure, flexibility, has been recognized. Flexibility was defined as an economic response to system changes[12] and classified into eight elements of machine, product, process, operation, routing, volume, expansion and production.[13] An attempt was also made to measure flexibility.[14] Because direct quantification is difficult, an indirect approach was attempted by comparing an FMS with a transfer line.[15] The value of flexibility was neither measured explicitly nor related to productivity and quality.

A widespread belief is that the higher the flexibility, the lower the productivity.[16,17] Another common belief is that improving quality increases costs and decreases productivity.[3] Recently, however, some authors suggested that the three measures may have positive relationships to one another.[18,19] Unfortunately, little research has been performed to support these observations. The relationships among productivity, quality and flexibility should be investigated for the proper measurement of performance and a sound decision on the introduction of IMPS. In summary, review of the literature indicates that:

- Productivity measures need to be improved to properly account for IMPS manufacturing characteristics.
- Little work has been done to predict and estimate future productivity.
- Quality and flexibility are not properly defined and quantified.
- Relationships among productivity, quality and flexibility need to be investigated.

Measurement of Productivity

Productivity is a measure of manufacturing performance which indicates a firm's efficiency in converting inputs to total outputs. The output (O_T) by any manufacturing system is usually expressed in units of physical volume, such as pieces, tons, and any other measurable units. These physical units must be weighted in some manner so they can be added together. Therefore, a definition of output is the summation of all units produced (not units sold) times their market price. Note that productivity is concerned with the efficiency of converting inputs to outputs, thus, units sold cannot be used. The use of market price (rather than profit) is the most common for expressing the total output in monetary terms. Four different types of partial productivity measures may be considered: labor, capital, material and overhead. These measures are integrated to form the total productivity.

Partial Productivity Measures. *Labor productivity* measures labor performance required to produce total output. This productivity is useful in manned cellular manufacturing systems or labor intensive industries. We have chosen to define labor productivity for a given period (P_L) as

$$P_L = O_T / C_L \quad (1)$$

where:

C_L = labor cost.

A detailed description of C_L equation is in the Appendix. The O_T and C_L values are in terms of constant or deflated dollars. (Hereafter refer to the Appendix for detailed equations of various costs.) With automation, direct labor decreases and shifts to indirect labor. Hence, labor cost includes the indirect as well as the direct labor cost.

Capital productivity measures the efficiency of capital that is invested on equipment and buildings that are used in producing the output. This productivity measure is especially useful in unmanned cellular manufacturing systems or capital intensive industries. The capital input factor is defined as the sum of the annuity values calculated for each asset on the basis of its base year cost, productive life, and the firm's cost of capital. In this paper, we have chosen to define capital productivity for a given period (P_C) as:

$$P_C = O_T / CC \quad (2)$$

where:

CC = the service cost of using invested capital.

Material productivity measures the efficiency of raw material use. This measure is useful when material cost is a large part of the total cost. We have chosen to define material productivity for a given period (P_M) as

$$P_M = O_T / C_R \quad (3)$$

where:

C_R = raw material cost.

Raw material cost usually includes costs of both direct and indirect material.

Overhead productivity is the efficiency of all resources except labor, capital and material. This group of inputs includes machines, tools, floor space, and computer software, if any. Machine cost may include expenses such as energy (power and fuel), maintenance, repair, insurance and property tax. Tool cost may occur from monitoring tool wear and potential breakage. Floor space cost may consist of energy, maintenance, repair, insurance and property tax. Therefore, in this paper we define overhead productivity for a given period (P_O) as:

$$P_O = O_T / OH \quad (4)$$

where:

OH = overhead cost.

As the level of automation increases, many significant changes are expected in the overhead. Insurance and property tax increases are usually due to the high initial investment required for the advanced machine. Utility, maintenance and repair costs also increase because 24 hour operation is possible and preventive maintenance is emphasized to enhance product quality. Tool cost may also increase if tools are replaced before average tool life is reached. Although regular replacement does not optimize tool life, it minimizes tool and machine breakdowns and improves process quality through preventive inspection during tool changes.

Floor space cost may decrease because of reduction in floor space required for machines, work-in-process inventory (WIP), tool-and-fixture storage, and support facilities. Space reduction results from grouping machines into cells, reducing WIP, attaching tool magazines to machines, reducing the number of conventional fixtures by using flexible fixtures, and reducing the number of workers. Computer software cost is bound to increase because more software would be needed. Thus, any increase or decrease in the overhead affects the overhead productivity.

Total Productivity Measure. We have chosen to define total productivity for a given period (TP) as:

$$TP = O_T / (C_L + CC + C_R + OH) \quad (5)$$

Total productivity integrates the partial productivity measures and may be used to measure the total manufacturing efficiency. Equation (5) is similar to the productivity measure by Craig and Harris.[7] Obvious trade-offs exist among partial productivity measures. Consider, for example, the investment of an automated machine that replaces labor. Labor input decreases while capital input increases. If the total output is constant, the use of labor productivity and capital productivity signals the contradictory manufacturing efficiency, as seen in Eqs. (1) and (2). If the reduction in labor cost is the same as the increase in the capital cost, total productivity shows no change despite the actual changes in labor and capital inputs. Therefore, partial productivity measures are still useful as supplementary measures.

Measurement of Quality

Quality is a measure of manufacturing performance which indicates the degree of perfection in making products. Two different types of quality measures are considered, i.e., process and product. Each of these quality measures is a partial quality measure and is integrated to form a total quality measure.

Partial Quality Measures. *Process quality* is the ability of processes to make good products with a small prevention cost. The prevention cost is the in-process inspection cost for checking and maintaining quality before final inspection. We define process quality for a given period (Q_S) as:

$$Q_S = O_T / C_P \tag{6}$$

where:

C_P = prevention cost.

A reduction in prevention cost increases process quality. For example, the use of machine vision systems, which are attached to machine tools, can be a good example.

Product quality is the degree of excellence of finished products. Product quality increases as product defects decrease. The defects can be expressed in terms of failure cost. This failure cost may indicate loss due to failure of finished products to meet quality standards set by both a company and customers. In this paper, we define product quality for a given period (Q_P) as:

$$Q_P = O_T / C_F \tag{7}$$

where:

C_F = failure cost.

As we expect, failure cost disappears if there are no defects. For example, the purpose of computer aided text and inspection (CATI) in IMPS is to minimize failure cost and improve product quality.

Total Quality Measure. We define total quality for a given period (TQ) as:

$$TQ = O_T / (C_P + C_F) \tag{8}$$

Total quality integrates the partial quality measures and may be used to measure the total quality performance. A tradeoff exists between the two partial quality measures. Assume, for example, that tight process control to reduce product defects raises prevention cost, but lowers failure cost. If the total output is constant, the use of partial measures indicate a conflicting signal in quality performance, since process quality decreases but product quality increases. On the other hand, if the increase in the prevention cost equals the decrease in failure cost, total quality indicates no change despite these cost input changes. Therefore, partial quality measures are still useful as surrogate measures of total quality.

Measurement of Flexibility

Flexibility is a measure of manufacturing performance which indicates a manufacturing system's adaptability to changes in manufacturing environments. Four different types of flexibility measures are considered: equipment, product, process and demand.

Partial Flexibility Measures. *Equipment flexibility* is the capacity of equipment to accommodate new products and some variants of existing products. If the equipment has excess capacity, then idle time for the equipment is inevitable. In our paper, equipment flexibility is measured in terms of idle cost, i.e., the opportunity cost for equipment underutilization. Equipment flexibility measures the opportunity of equipment to add value to raw materials, whereas capital productivity measures the efficiency of converting the capital invested in the equipment into total output. In this paper, we define equipment flexibility for a given period (F_E) as:

$$F_E = O_T / C_I \tag{9}$$

where:

C_I = idle cost of equipment.

Equipment idle time should be reduced whenever possible to increase equipment flexibility. Idle time, however, may not always be a critical problem, as seen at the Japanese automaker Toyota.[6,20] Toyota usually maintains overcapacity of equipment so that when demand increases, extra workers are needed only temporarily.

Product flexibility is the adaptability of a manufacturing system to changes in product mix. A frequently changing market demands new products and variations of existing products. These changes in product mix result in a shorter life cycle of an individual product and a smaller lot size as variety increases. Smaller lot size means higher setup cost as shown in the economic lot size models.[21] Reducing setup cost in spite of small lot size is often viewed as the key to increased product flexibility. Group technology, rapid exchange of tools and dies (RETAD), and flexible fixtures are good examples of setup reduction methods. For this reason, we define product flexibility for a given period (F_P) as:

$$F_P = O_T / A \tag{10}$$

where:

A = setup cost.

Setup cost is in fact an opportunity cost. Therefore, product flexibility would measure the opportunity of adding value to products with the changes in product mix.

Process flexibility is the adaptability to various changes in part processing, such as in equipment and tools breakdowns, random access of product mix, process schedule, and so forth. Process flexibility measures process utilization. Poor responsiveness to the changes results in a large WIP. The WIP is directly proportional to waiting time or costs of parts processed. If the responsiveness is good, waiting cost in processes decreases. In this paper, process flexibility for a given period (F_S) is defined as:

$$F_S = O_T / C_W \tag{11}$$

where:

C_W = waiting cost of parts processed.

Waiting cost is another opportunity cost like set up cost. Process flexibility indicates a foregone opportunity to add value to materials processed. Poor process flexibility (a large waiting cost or a large WIP) is a primary source of various manufacturing problems including poor quality. These problems do not become obvious until WIP is reduced. One example of a WIP reduction method is the Japanese Kanban technique which controls the production amount at each process with a card (Kanban) so that WIP is minimized.

Demand flexibility is the adaptability to changes in demand rate. In our discussion, we recognize the two types of demand: customer demand for finished products and manufacturing system demand for raw materials. Demand flexibility may be measured in terms of inventory costs of finished products and raw materials. Inventory costs consist of carrying cost and shortage cost. Decreases in inventory costs due to higher inventory turnover indicate reduced differences between supply and demand with respect to raw materials and finished goods, i.e., good responsiveness to internal and external demands. With this reasoning, we have chosen to define demand flexibility for a given period (F_D) as:

$$F_D = O_T / H \tag{12}$$

where:

H = inventory costs of finished products and raw materials.

Inventory cost is an opportunity cost because inventories tie up large amounts of capital. Also, inventory cost disappears when demand is forecast exactly. Demand flexibility measures the opportunity of adding value to products according to job scheduling and demand forecasting. The ultimate goal of JIT is known to maximize demand flexibility.[20]

Total Flexibility Measure. We define total flexibility during a given period (TF) as:

$$TF = O_T / (C_I + A + C_W + H) \tag{13}$$

Total flexibility may replace partial flexibility measures and can be used as a global measure of the opportunity of a manufacturing system to add value to products. As we have seen before, obvious trade-offs exist between partial flexibility measures. For example, assume that a conventional milling machine is replaced by a CNC milling machine with a larger capacity and automatic tool change. Setup cost—and possibly waiting cost—decrease, but idle cost increases. If total output is constant, product flexibility and process flexibility increase, but equipment flexibility decreases. The partial flexibility concept may not measure the total manufacturing adaptability to various changes. Partial flexibility measures are only useful as surrogate measures to help explain the changes in total flexibility, as seen earlier in productivity and quality.

Measurement of Performance

Integral Manufacturing Performance. Now that we have defined total productivity, total quality and total flexibility, the next step is to come up with a global measure that reflects these three attributes. Our goal is to maximize manufacturing performance by increasing productivity, quality and flexibility at the same time. Productivity, quality and flexibility have some correlation to each other, whether positive or negative. Mass production with poor quality of product increases productivity because of mere increase in total output. Quality and flexibility, however, usually decrease because of relatively large defects and product inventory. The use of only one total measure (productivity, quality or flexibility) over the others is not desirable in evaluating the manufacturing performance as a whole. With these conceptual relationships in mind, we propose a global manufacturing performance measure as follows:

Let I_P, I_Q and I_F be the input factors of total productivity, total quality and total flexibility, respectively. Then, we define the integral manufacturing performance measure for a given period (IMP), which incorporates these three as:

$$IMP = O_T / (I_P + I_Q + I_F) \tag{14}$$

The values of I_P, I_Q and I_F are easily obtained from Eqs. (5), (8) and (13).

Relationships Among Productivity, Quality, and Flexibility. *General Relationships* - Productivity and flexibility are considered to have a reciprocal relationship.[16,17] This observation may have originated from the characteristics of conventional job shops and flow shops. Job shops can easily adapt to the special requirements of filling different orders from customers, because general purpose equipment is grouped by function. Increasing flexibility, however, decreases productivity by lengthening throughput time, increasing production cost and decreasing output. Flow shops are mass production systems with specialized equipment organized along lines. Therefore, increasing productivity often decreases flexibility, because adapting dedicated equipment in lines to different orders from customers is difficult.

We should note that the reciprocal relationship between productivity and flexibility is in terms of partial rather than total measures. By using total measures, however, the relationship can be reversed. For example, the flexibility advantage of a job shop is restricted to equipment (i.e., partial) flexibility. Other partial measures, such as product flexibility and process flexibility are usually poor because of long setup and large WIP.

Mathematical Relationships - We can examine the mathematical relationships among total productivity, total quality, and total flexibility from our proposed *IMP* in Eq. (14). By taking the inverse of Eq. (14), we obtain:

$$\begin{aligned} 1/IMP &= (I_P + I_Q + I_F)/O_T \\ &= 1/TP + 1/TQ + 1/TF \\ &= (TP * TQ + TQ * TF + TF * TP)/ \\ &\quad (TP * TQ * TF) \end{aligned} \tag{15}$$

Therefore, we can rewrite our IMP as:

$$IMP = (TP * TQ * TF)/(TP * TQ + TQ * TF + TF * TP) \tag{16}$$

When we take a partial derivative of *IMP* with respect to variable *TP*, we obtain the following relationship:

$$\frac{\partial\, IMP}{\partial\, TP} = \left[\frac{1}{\frac{TQ + TF}{TQ \times TF}\; TP + 1}\right]^2 > 0 \tag{17}$$

The derivative is referred to as marginal *IMP*. In other words, the marginal *IMP* in Eq. (17) is viewed as the increment in integral manufacturing performance per unit increase in total productivity, where both *TQ* and *TF* remain unchanged. The derivative being positive implies that as *TP* increases, so does *IMP*. We can observe the relationship similar to this when we take partial derivatives of *IMP* with respect to *TQ* and *TF*. *Figure 1* shows the relationship between *TQ*, *TF*, and *IMP* when *TP* = 1 (or 100%).

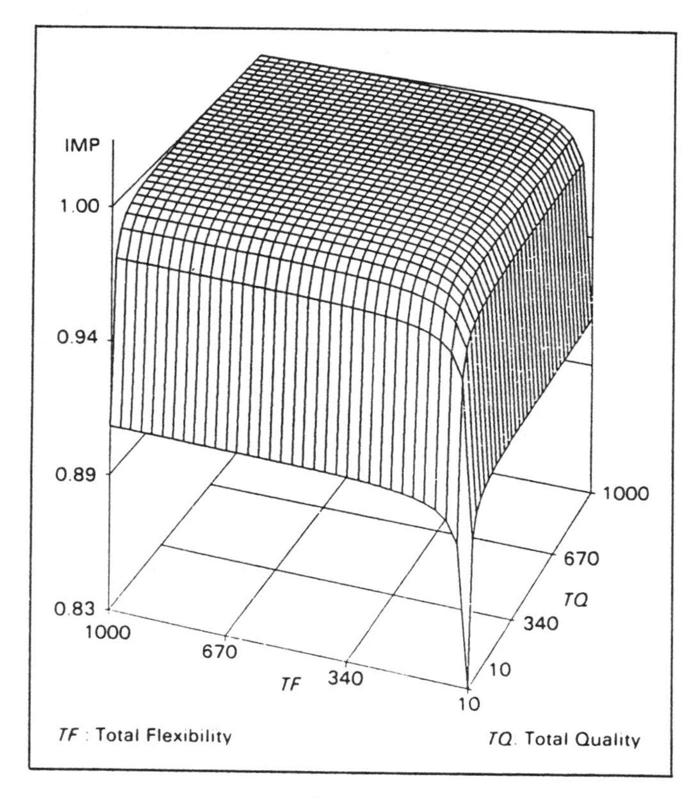

Figure 1
Relationship Among *IMP, TQ, TF* When *TP* is Fixed

The *IMP* value increases as either *TQ* or *TF*, or both increase. The rate of increase, however, decreases as these values increase. We can also derive the mathematical relationships between total and partial measures in a similar fashion. Finally, *Figure 2* summarizes the hierarchy of various performance measures and their cost elements discussed in this paper.

Case Example

A hypothetical flexible manufacturing system (FMS) and a job shop (JS) are compared to illustrate the performance measures developed earlier. The FMS is a GT based manufacturing cell and a simple form of IMPS. For simplicity, we will assume that each functional group in a JS has only one major machine.

Layout of Each System. *Figure 3* shows the layout of each system. Both systems include an inventory area, a load/unload area, a boring machine, a lathe, a milling machine, a drilling machine, a washing machine, an inspection machine, a conveyor belt, a tool-and-fixture storage area, and a facility-supporting area. The FMS, however, has more advanced manufacturing equipment, such as direct numerical control (DNC), and a computer is added to the system.

Each system's inventory area has two sections:

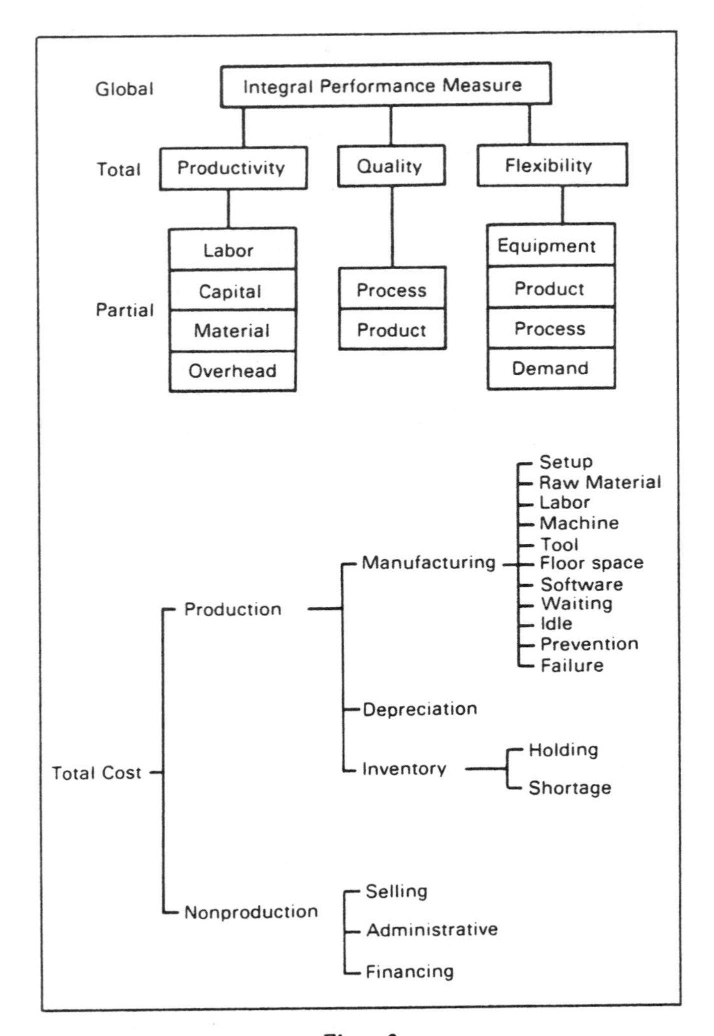

Figure 2
Hierarchy of Performance Measures and Their Cost Elements (in our case example, both selling and administrative costs were not included since these costs do not depend on level of automation)

one for raw materials and the other for finished goods. The FMS inventory system is an automatic storage and retrieval system (AS/RS). The tool-and-fixture storage area is in the center of the machine group to reduce traveling time. Because GT and an AS/RS are used, the FMS requires much less factory space than a JS requires. Although the manufacturing systems considered are hypothetical, companies with similar layouts can be found in manufacturing engine frames for locomotives, wear plates for centrifugal pumps, and so forth.

Assumptions in the Case Example. Specific assumptions were made for our case example as follows:

- Three different types of parts are processed in each system. The operation sequences of each part are already determined: for part 1, lathe, milling, drilling, washing, and inspection; for part 2, boring, milling, washing, and inspection; and for part 3, boring, lathe, drilling, washing, and inspection.

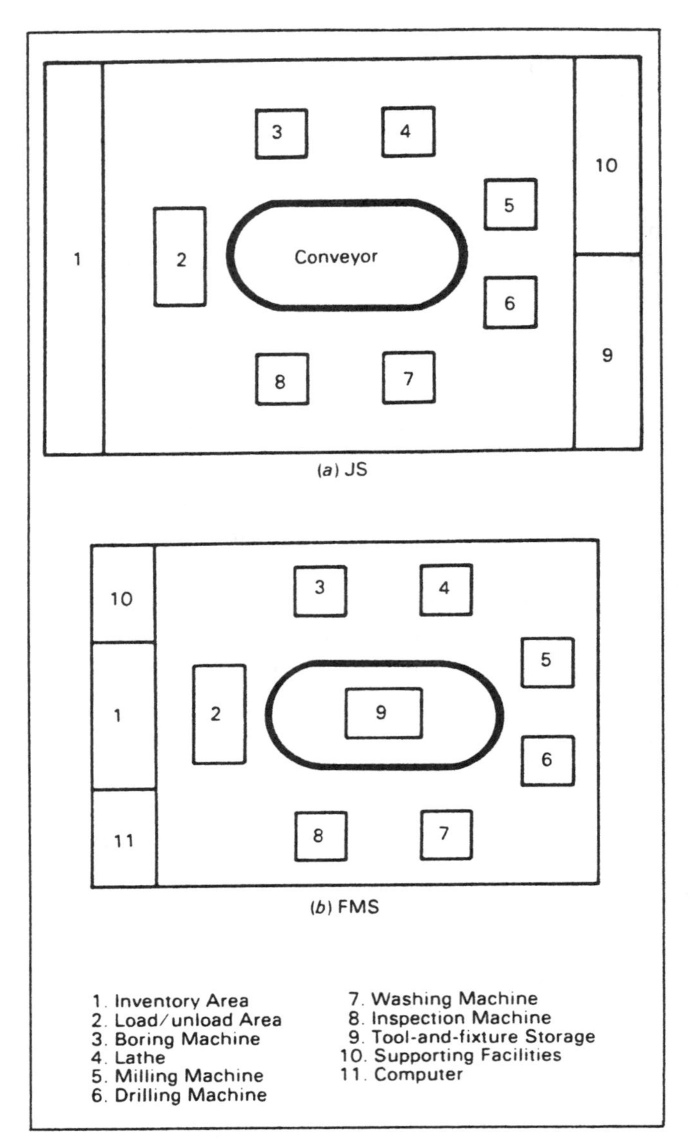

Figure 3
Layout of Hypothetical Manufacturing Systems

- Seven different tools are used, but only one tool is used for a part in a process. The same tool can be used for other parts in the process or for a part or parts in different processes.
- Demand for each part follows the Poisson distribution - each part's parametric value differs from another's.
- The initial inventories of raw materials and finished goods are zero, respectively. No raw materials shortage occurs. If a finished goods shortage occurs, they are all back ordered.
- All physical assets such as equipment and buildings were purchased in 1982, and the accelerated cost recovery system (ACRS) is used as the depreciation method.
- The money used to purchase the assets was borrowed from only one source with 12% interest rate compounded annually. The loan should be repaid in five years.

- The input data for the simulation and the additional data for the cost evaluation are available from various departments in the company.
- The planning horizon is a month, i.e., 20 days or 19200 min (20 days x 2 shifts/day x 8 hr/shift x 60 min/hr).

Simulation Model. The simulation model is written in SIMAN, an especially useful simulation language for modeling manufacturing systems and for communicating with a user-written FORTRAN program. The details of the simulation model (such as input/output procedures, computer programs and statistical testing procedures) are not provided in this paper, but can be found in Reference 22. Instead, we provide a brief description of the simulation procedures.

In the FMS and a JS simulation procedure, each part arrives randomly in the load/unload area with an interarrival time from the raw material section of the inventory area. The part is palletized there during load time. After the part is loaded, the conveyor belt transports it to the first work station according to a predetermined job schedule. If the work station is idle, the part is immediately processed; otherwise, the queue space is checked. If the queue space is enough, the part enters it; otherwise, balking occurs. The balked part is marked and returned to the work station, traveling full circle on the conveyor belt. Then, the above procedure is repeated until the balked part joins the queue space. The conveyor belt is assumed to have enough space for balked parts.

During part processing, the following are checked: tool-and-fixture setup, machine and tool breakdowns, machine maintenance, and tool wear.

- If the part type differs from the previous one, another whole new setup should be made in the JS; in the FMS, however, only quick adjustments, after the initial setup, are required for the different part types.
- If the machine suddenly fails with the specified probability, the part machining stops and the machine is repaired. Then, the part is reprocessed for the remaining machine time.
- If the tool breaks with the specified probability, the part processing also stops and a new tool is installed. Then, the part is reprocessed for the remaining machine time.
- If the machine's accumulated processing time is greater than the time of regular machine maintenance, then the maintenance actions are activated.
- If the accumulated tool usage time is greater than the time of regular tool change, which is less than the average tool life, then the worn tool is replaced.
- If these five checking events occur at the same time, priority is given in the above order.

Once the part processing at the first work station is finished, the part is transported to succeeding work stations in a predetermined sequence. In each work station, the above procedure is repeated. After the part is washed and inspected, it is returned to the load/unload area, removed from the conveyor belt, depalletized, and placed in the finished goods section of the inventory area to satisfy the demand. The same procedure, up to this point, is applied to the other two parts and is repeated during the specified simulation time (here, 960 min). When the simulation time expires, the simulation stops and a FORTRAN subroutine is initiated.

The subroutine utilizes the output data from the simulation and additional data from various departments in the company to calculate the various costs and the performance measures developed earlier. The calculation is rather complex and repetitive in economic analysis. For calculating inventory cost, in particular, the subroutine passes on-hand inventory information between simulation replications. In other words, the subroutine receives the information on on-hand inventory of raw materials and finished goods from the previous replication and leaves the updated information for the next replication. Since each replication represents the daily manufacturing activities, the daily inventory cost changes as the on-hand inventories change.

The simulation time can be any time duration, however, 20 days (a month) was already assumed in this example. The system status at the beginning of each day is not reinitialized since the ending manufacturing condition for a day is used as the initial condition for the next day. The dollar value of total output (O_T) and each cost (C) are accumulated separately for a month to calculate the performance measures. They are, in fact, expected values, and $E[O_T]/E[C]$ does not generally equal $E[O_T/C]$. However, the equality is valid for a renewal-reward process that may represent the sequence of production-monitoring-adjustment with accumulation of costs over a cycle.[23]

The simulation is a sampling experiment for a given set of random numbers, and therefore any result obtained from simulation has to be validated. To accomplish this, the experiment was replicated by re-executing the simulation model with a different set of random numbers. Consequently, an output analysis was performed to find confidence intervals of the true performance measures.

Table 1
Comparison of Performance Between JS and FMS

Measures	JS	CMS
Integral	83	103
Productivity	148	155
Labor	1200	3230
Capital	1480	728
Material	284	287
Overhead	577	793
Quality	627	1060
Process	1010	1130
Product	1640	16700
Flexibility	265	433
Equipment	2030	926
Product	1300	14200
Process	780	2260
Demand	816	1390

Note: Units are indices (\$ output/cost) x100.

Comparison of Performance Between a JS and an FMS. *Table 1* summarizes the JS and FMS performance measures and is the result of the first replication experiment. The values of some partial measures such as capital productivity and equipment flexibility decrease because of a large initial investment and general trend of maintaining overcapacity in IMPS. The values of the remaining partial measures increase significantly, particularly product quality (1018%), product flexibility (992%), process flexibility (290%), and labor productivity (269%). These results demonstrate that an FMS, an IMPS, remarkably reduces product defects, setup, WIP and labor. Since most of the partial measures increase, each total measure also increases: productivity by 5%, quality by 69% and flexibility by 64%. With significant improvement in both quality and flexibility (despite a small productivity increase), the integral performance measure increases by 24%.

The integral performance measure is the most important one since it measures manufacturing performance as a whole. In a JS, for example, using partial or total indexes, which are much greater than 100, misleads management since the integral performance index (= 83) is less than 100, indicating poor performance. This example especially shows the danger of using a conventional performance measure such as a productivity index.

The results in *Table 1* are point estimates based on a single simulation run. By replicating the experiment five times, we obtain the 95% confidence intervals of the true integral performance measures (IMP) as follows:

$$81 < IMP < 85 \text{ for JS,}$$

and

$$101 < IMP < 105 \text{ for FMS.}$$

Since the confidence interval estimate of IMP for the FMS is totally to the right of the upper bound of that of IMP for a JS, there is strong evidence that the FMS provides better total manufacturing performance.

Concluding Remarks

The purpose of this research was to quantify and combine three critical performance measures to one global index, i.e., productivity, quality and flexibility. The conventional productivity measures were improved for use in IMPS. Quality and flexibility measures were defined and quantified. These three measures were combined to evaluate a manufacturing system as a whole. These measures have the following characteristics:

- Integral measure is the primary measure of manufacturing performance. Total and partial measures are surrogates to the integral measure.
- The measures use a simulation technique to collect intangible costs such as parts waiting and equipment idle.
- The measures are closely related to a firm's accounting system. The cost data are easy to access.

A hypothetical case example indicated that the adoption of an FMS improved the integral manufacturing performance, compared with a conventional JS. However, just because an FMS performs better during a specific short period does not mean we abandon the JS. The adoption of IMPS requires a large initial investment under a long-term environment. Economics of IMPS should be considered over the project's life cycle. The performance measures developed here can be useful in strategic planning since they not only evaluate past or current manufacturing performance, but also predict the effect of capital investment on future performance.

Appendix

Derivations of Cost Equations Used in Performance Measures. The cost equations presented in this section are intended to illustrate a possible way to measure each component of total production cost. We certainly recognize that many readers, based on their experience, might consider other parameters to be a more realistic representation in each cost equation. It must also be understood that all cost figures are in *constant dollars* or *deflated dollars*.

Labor Cost

$$C_L = \sum_{d=1}^{L1} c_d N n_d + \sum_{i=1}^{L2} c_i n_i$$

where:

L_1 = number of different jobs using direct labor,
L_2 = number of different jobs using indirect labor,
c_d = wage of job d per unit time,
N = a planning horizon,
c_i = salary of job i during a planning horizon,
n_d = number of direct labor units for job d,
n_i = number of indirect labor units for job i.

In a traditional manufacturing system, wage is usually calculated based on the man-hour.[7] In IMPS, however, a worker can handle many machines and becomes multifunctional. The worker may receive a salary rather than a wage, since the wage calculation for the multifunctional worker is rather difficult.

Depreciation Cost

$$DC = \sum_{y=1}^{Y} \sum_{k=1}^{K4} C_d(k,y)$$

where:

$C_d(k,y)$ = annual depreciation cost of asset k at the end of year y,
$K4$ = number of assets,
Y = number of years during a planning horizon.

Conventional accounting depreciation methods (e.g., straight line) were used for book depreciation purposes. The $K4$ value includes all buildings and manufacturing equipment.

Return on Invested Capital

$$FC = \sum_{y=1}^{Y} \sum_{k=1}^{K4} C_f(k,y)$$

where:

$C_f(k,y)$ = return on invested capital for asset k at year y. Note that the value of $C_f(k,y)$ depends upon the financing mix (equity vs. debt) and the weighted cost of capital.

Raw Material Cost

$$C_R = C_o + \sum_{j=1}^{J} C_d(j) n_d(j) + C_{id}$$

where:

C_o = material ordering cost,
J = number of different parts,
$C_d(j)$ = unit cost of direct material for part j,
$n_d(j)$ = amount of direct material used for part j,
C_{id} = indirect material cost except tools.

Machine Cost

$$C_M = \sum_{k=1}^{K3} \{c_u(k)T_m(k) + c_{mt}(k)T_{mt}(k) + c_r(k) \cdot T_r(k) + aF_k + bF_k\}$$

where:

$c_u(k)$ = utility cost for machine k per unit time,
$c_{mt}(k)$ = maintenance cost for machine k per unit time,
$c_r(k)$ = repair cost of machine k per unit time,
$T_m(k)$ = machine time allocated (or used) for machine k,
$T_{mt}(k)$ = maintenance time for machine k,
$T_r(k)$ = repair time for machine k,
a = insurance premium rate,
b = property tax rate,
F_k = first cost (initial investment) for machine k,
$K3$ = number of machines including machine tools and material handling systems.

In a traditional manufacturing system which uses conventional cost accounting systems, machine cost is an overhead item and calculated based on direct labor hour. Though usually difficult to measure, the machine hour, repair, and maintenance time may be collected through a computer simulation. In IMPS, these types of data are usually available for CNC or DNC machines. Note that $K3 = K4$ - (number of buildings).

Tool Cost

$$C_T = \sum_{m=1}^{M} c_{tl}(m)\{n_w(m) + n_b(m)\}$$

where:

$c_{tl}(m)$ = unit cost of tool type m,
$n_w(m)$ = number of worn tools of type m,
$n_b(m)$ = number of broken tools of type m,
M = number of different tools.

Though the tool cost equation looks simple, it considers the complicated part processing mechanism through a computer simulation.

Floor Space Cost

$$C_S = (C_u + C_m + C_i + C_t)S_M / S_T$$

where:

C_u = total plant utility cost,
C_m = total plant maintenance cost,
C_r = total plant repair cost,
C_i = total plant insurance cost,
C_t = total plant property tax,
S_M = manufacturing floor space,
S_T = total plant space.

In IMPS, the floor space is significantly reduced because of cellular layout, WIP reduction, and so forth.

Computer Software Cost

$$C_{sw} = \sum_{s=1}^{S} c_{ms}(s) n_{sw}(s)$$

where:

$c_{ms}(s)$ = membership fee of software type s,
$n_{sw}(s)$ = number of software type s,
S = number of different software.

Computer software cost may be ignored in a traditional manufacturing system and even in IMPS without computers. The initial or development cost of computer software is relatively large and included in the first cost of manufacturing equipment.

Prevention Cost

$$C_P = \sum_{j=1}^{J} \sum_{k=1}^{K1} c_p(j,k)N$$

where:

$K1$ = number of machines except material handling systems and computers,
N = a planning horizon,
$c_p(j,k)$ = prevention cost of part j for machine k per unit time,
$c_p(j,k)$ = prevention cost of part j for machine k per unit time,
$c_p(j,k) = c_{s1}n_1/t + [c_{cr} + c_{fs}\,\alpha\, exp(-\theta t)/\{1\text{-}exp(-\theta t)\}]/E[T_c] + c_{s2}n_2 + c_{im}/E[T_c]$, if $6\sigma > USL\text{-}LSL$,
$= c_{s1}n_1/t + [c_{cr} + c_{fs}\,\alpha\, exp(-\theta t)/\{1\text{-}exp(-\theta t)\}]/E[T_c] + c_{s2}n_2$, otherwise,

where:

c_{s1} = cost of sampling an item in control chart study,
n_1 = sample size in detecting assignable cause (a.c.),
t = time intervals at which samples are taken,
c_{cr} = cost of correcting an a.c.,
c_{fs} = cost of investigating a false alarm,
α = probability of false alarm,
Θ = mean rate of Poisson process,
c_{s2} = cost of sampling an item,
n_2 = sample size in process capability (PC) study,
c_{im} = cost of improving PC per cycle,
σ = a process standard deviation,
USL = upper specification limit,
LSL = lower specification limit,
$E[T_c] = 1/\Theta + t/(1\text{-}\beta) - \{1 - (1 + \Theta t)exp(-\Theta t)\}/[\Theta\{1\text{-}exp(-\Theta t)\}] + gn_1 + d$

where:

g = time required to process one sample,
d = time required to find and correct the a.c.,
$1\text{-}\beta = 1 - \Phi(q\text{-}\delta\sqrt{n_1}) + \Phi(\text{-}q\text{-}\delta\sqrt{n_1})$,
$\Phi(.)$ = cumulative probability of standardized normal distribution,
q = decision variable for upper/lower control limit,
δ = magnitude of an a.c.

We derived prevention cost by revising the so-called economic design of control charts.[23] In our equation above, the costs of the process capability (PC) study are also considered. Note that the PC study is usually conducted only when the process is in control. In our study, we include the penalty costs due to both out-of-control and out-of-PC in the following failure cost. Note that $K1 = K3$ - (number of MHS and computers).

Failure Cost

$$C_F = \sum_{j=1}^{J} E[c_f(j)]Q_j$$

where:

$c_f(j)$ = failure cost of part j,
Q_j = lot size of part j,
$E[c_f(j)] = e_1c_g + E[p]\{(1\text{-}e_1 - e_2)c_b + (1\text{-}e_2)(1\text{-}w)c_s + e_2c_a\}$

where:

c_g = cost of reworking a good part because of misclassification,
c_b = cost of reworking a bad (defective) part,

c_s = cost of scrapping a defective part that could not be restored,
c_a = cost of accepting a defective part,
$E[p]$ = expected value of proportion defective of a part,
e_1 = error rate of misclassifying a good part into defective (type I error),
e_2 = type II error,
w = rate of restoring a defect to a good part.

We derived failure cost assuming 100% inspection, which is usual in IMPS. In a sampling inspection case which is usual in a traditional manufacturing system, we replace $E[p]$ by $E[p \mid s]$, which is the expected value of proportion defective after observing s defective items in a sample.[24]

Idle Cost

$$C_I = v \sum_{k=1}^{K2} (1-u_k)N$$

where:

u_k = utilization of machine k,
v = profit per unit time,
$K2$ = number of machines excluding computers.

Note that $K2 = K3$ - (number of computers) = $K1$ + (number of MHS), and $K4 > K3 > K2 > K1$.

Setup Cost

$$A = \sum_{k=1}^{K2} C_{su}(k)T_{su}(k)$$

where:

$C_{su}(k)$ = setup cost for machine k per unit time,
$T_{su}(k)$ = setup time for machine k.

In traditional manufacturing, $T_{su}(k)$ value is shown in the routing sheet. In IMPS, we obtain this value using a computer simulation, since the random setup adjustment for different part types after initial setup is usually difficult to measure in analytical means.

Waiting Cost

$$C_W = v \left[\sum_{j=1}^{J} \sum_{k=0}^{K_j+1} T_w(j,k)\{n(j,k-1)-n(j,k)\} + \sum_{j=1}^{J} T_w(j,K_j+1)n(j,K_j+1) \right]$$

where:

K_j = number of processes for part j,
$T_w(j,k)$ = cumulative waiting time of part j up to process k,
$n(j,-1)$ = number of raw materials for part j required in manufacturing,
$n(j,k-1) - n(j,k)$ = WIP between processes k-1 and k,
$n(j,K_j = 1)$ = number of finished part $j = Q_j$.

Note that $k = 0$ indicates the part loading area and $k = K_j + 1$, the part unloading area.

Inventory Cost

$$H = c_{sp}S_I + \sum_{j=1}^{J} [c_{sm}(j)\{I_{Om}(j) + U_j - W_j\} + c_{sf}(j)\{I_{Of}(j) + Q_j - D_j\}], \text{ if } Q_j > D_j,$$

$$= c_{sp}S_I + \sum_{j=1}^{J} [c_{bm}(j)\{W_j - U_j - I_{Om}(j)\} + c_{bf}(j)\{D_j - Q_j - I_{Of}(j)\}], \text{ otherwise,}$$

where:

c_{sp} = space cost per square foot $= (C_u + C_m + C_i + C_t)/S_T$,
S_I = inventory space,
W_j = amount of raw materials of part j used,
Q_j = amount of part j finished,
U_j = raw materials obtained from suppliers,
D_j = demand of part j,
$I_{Om}(j)$ = initial inventory of raw material of part j,
$I_{Of}(j)$ = initial inventory of finished part j,
$c_{sm}(j)$ = cost of service and risk per unit raw material of part j,
$c_{sf}(j)$ = cost of service and risk per unit product of part j,
$c_{bm}(j)$ = back order cost of unit raw material for part j,
$c_{bf}(j)$ = back order cost of unit product of part j.

If $Q_j > D_j$, H value indicates inventory carrying cost; otherwise, back order cost under the assumption that all shortages are back ordered.

References

1. E.P. DeGarmo, J.T. Black, R.A. Kohser. *Materials and Processes in Manufacturing*, Sixth Edition, Macmillan Publishing Company, New York, 1984.
2. R.H. Hayes, S.C. Wheelwright. *Restoring Our Competitive Edge—Competing Through Manufacturing*, John Wiley & Sons, Inc., 1984.
3. R. Kaplan. "Measuring Manufacturing Performance: A New Challenge for Managerial Account Research", *The Accounting Review*, LVIII(4), October 1983, pp. 686-705.
4. P.R. Richardson, J.R.M. Gordon. "Measuring Total Manufacturing Performance", *Sloan Management Review*, Winter 1980, pp. 47-58.
5. S.R. Schneible. "Flexible Manufacturing Systems", *Proceedings of IIE Fall Conference*, 1983, pp. 214-220.

6. R.J. Schonberger. *Japanese Manufacturing Techniques*, Free Press, 1982.
7. C.E. Craig, R.C. Harris. "Total Productivity Measurement at the Firm Level", *Sloan Management Review*, Spring 1973, pp. 13-29.
8. J. Hamlin. "Development in Firm-Level Productivity Measurement", *Proceedings of IIE Spring Conference*, 1979, pp. 194-199.
9. D.S. Sumanth. "Productivity Indicators Used by Major U.S. Manufacturing Companies: The Results of a Survey", *Industrial Engineering*, May 1981, pp. 70-73.
10. D.M. Lundvall, J.M. Juran. "Quality Cost", *Quality Control Handbook*, Third Edition, J.M. Juran, (editor), McGraw-Hill, New York, 1979.
11. H.P. Roth, W.J. Morse. "Let's Help Measure and Report Quality Costs", *Management Accounting*, August 1983, pp. 50-53.
12. A. Masuyama. "Idea and Practice of Flexible Manufacturing System of Toyota", *Proceedings of VIIth International Conference on Production Research*, Volume 1, August 1983, pp. 584-590.
13. J. Browne, D. Dubois, K. Rathmill, S.P. Sethi, K.E. Stecke. "Classification of Flexible Manufacturing Systems", *FMS Magazine*, Volume 2, No. 2, 1984, pp. 114-117.
14. A. Chatterjee, M.A. Cohen, W.C. Maxwell, L.W. Miller. "Manufacturing Flexibility: Models and Measurements", *Proceedings of the First ORSA/TIMS Special Interest Conference*, August 1984, pp. 49-64.
15. G.K. Hutchinson, J.R. Holland. "The Economic Value of Flexible Automation", *Journal of Manufacturing Systems*, Volume 1, No. 2, 1982, pp. 215-227.
16. Sten-Olof Gustavsson. "Flexibility and Productivity in Complex Production Processes", *International Journal of Production Research*, Volume 22, No. 5, 1984, pp. 801-808.
17. D.E. Hegland. "Flexible Manufacturing - Your Balance Between Productivity and Adaptability", *Production Engineering*, May 1981, pp. 38-43.
18. R. Muramatsu, K. Ishii, K. Takahashi. "Some Ways To Increase Flexibility in Manufacturing Systems", *International Journal of Production Research*, Volume 23, No. 4, 1985, pp. 691-703.
19. D.M. Zelenovic. "Flexibility - A Condition for Effective Production Systems", *International Journal of Production Research*, Volume 20, No. 3, 1982, pp. 319-337.
20. Y. Monden. "Toyota's Production Smoothing Methods: Part I", *Industrial Engineering*, August 1981, pp. 42-51.
21. C.S. Park, Y.K. Son. "A Comparative Study of Discounting Effects on Inventory Lot Sizing Models", working paper.
22. Y.K. Son. "An Economic Evaluation Model for Advanced Manufacturing Systems", Ph.D. Dissertation, Department of Industrial Engineering, Auburn University, June 1987.
23. D.C. Montgomery. "The Economic Design of Control Charts: A Review and Literature Survey", *Journal of Quality Technology*, Volume 12, No. 2, 1980, pp. 75-87.
24. M.C. Riew, D.S. Bai. "An Economic Attributes Acceptance Sampling with Three Decision Criteria", *Journal of Quality Technology*, Volume 16, No. 3, July 1984, pp. 136-143.

Author(s) Biography

Young K. Son is an Assistant Professor of Management at Baruch College, the City University of New York. He received a B.S. degree from Seoul National University, Korea, an M.S. and a Ph.D. from Auburn University. Dr. Son has technical experience with IBM in Japan and research experience with KIST and KDI in Korea. His current research areas include economics of advanced manufacturing systems (AMS) and simulation applications to AMS. He is a member of CASA/SME, IIE, ORSA, and DSI.

Chan S. Park is an Associate Professor of Industrial Engineering at Auburn University. He received a B.S. degree from Hanyang University, an M.S.I.E. degree from Purdue University and a Ph.D. from Georgia Institute of Technology. Dr. Park is a senior member of IIE and has served as Division Director of Engineering Economy. He is also on the ASEE Board of Directors and a member of the IEEE Computer Society, TIMS, and Sigma Xi. Dr. Park is the author and coauthor of *Advanced Engineering Economics* (Wiley, 1988), *Interactive Microcomputer Graphics* (Addison Wesley, 1985), and *Inflation and Its Impact on Investment Decision*, and has written numerous technical papers on economic decision analysis and computer applications.

Additional Readings on Control Issues in Systems Integration

Berliner, Callie and James A. Brimson, eds. 1988. *Cost Management for Today's Advanced Manufacturing*. Boston, MA: Harvard Business School Press.

Cooper, Robin and Robert S. Kaplan. 1991. Profit priorities from activity-based costing. *Harvard Business Review*. May-June, pp. 130-135.

Dhavale, Dileep G. 1988. Indirect costs take on greater importance, require new accounting methods with CIM. *Industrial Engineering*. July, pp. 41-43.

Hall, Robert W., H. Thomas Johnson and Peter B. B. Turney. 1991. *Measuring Up: Charting Pathways to Manufacturing Excellence*. Homewood, IL: Business One Irwin.

Hayes, Robert H. and Kim B. Clark. 1986. Why some factories are more productive than others. *Harvard Business Review*. September-October, pp. 66-73.

House, Charles H. and Raymond L. Price. 1991. The return map: Tracking product teams. *Harvard Business Review*. January-February, pp. 92-100.

Kaplan, Robert S. 1988. One cost system isn't enough. *Harvard Business Review*. January-February, pp. 61-66.

________________. 1990. *Measures for Manufacturing Excellence*. Boston, MA: Harvard Business School Press.

VII. ENABLING TECHNOLOGIES

Management systems being designed and upgraded today may take advantage of several technologies in which rapid advances are being made. To be successful, systems integrators must develop a sound understanding of these technologies.

For over a century, industrial engineers have been designing systems that prescribe the flow of material throughout the organization. Even in the earliest days of industrial activity, such systems required information to be generated and maintained. Today, it can be argued that information flow is more critical than material flow. Consequently, systems integrators must deal explicitly with the design of systems that facilitate information flow.

A variety of data communication networks have been devised for conveying data/information among the components of an industrial system. Systems integrators must understand the fundamentals of networks, even though they may enlist the assistance of specialists to deal with technical details.

The many protocol standards associated with electronic data interchange can be a formidable challenge event to electrical engineers who devise communication systems. Nevertheless, systems integrators need to understand the basic issues involved.

Similar issues can be raised relative to data base design and software. There are many issues relative to relational vs. non-relational data bases. Appropriate software specifications and selection is a difficult challenge.

The articles included in this section give the reader an initial exposure to the issues described. Further study will be required to remain competent in these fast moving areas.

NETWORKS: THE BACKBONE OF INTEGRATION

BY ED PASTOR

The profit and loss figures in the annual report on the CEO's desk ultimately derive directly from data originating on the plant floor. Gathering and analyzing these data for the CEO requires the efficient exchange of information among an enterprise's divisions and departments. So how do you achieve efficient information exchange? Through computer networks.

Ideally, a network helps unify a company by linking together all computer-based devices, no matter where they are located within the organization. People, information, and automation can then be integrated to realize the benefits of Computer-Integrated Manufacturing (CIM) which, according to the U.S. National Research Council, include:

- A 30 to 60% reduction in lead times.
- A 15 to 30% reduction in design costs.
- A 30 to 60% reduction in work-inprocess.
- A 5 to 20% reduction in labor costs.
- A 40 to 70% increase in production productivity.
- A 2 to 5-time increase in product quality.
- A 2 to 3-time increase in capital utilization.
- An overall increase in customer satisfaction.

How well an enterprise captures these benefits depends on how well the network is utilized in an enterprise.

Through networks, data are easily collected for process and machine monitoring, fault analysis, and maintenance scheduling. Through networks, Statistical Process Control (SPC) data can be uploaded directly from the plant floor to manufacturing engineering for analysis and process optimization, and statistics from manufacturing engineering to design engineering can be fed to production for product modifications.

Customer requirements for new products can be sent from sales and marketing to design engineering by way of networks. Networks permit CAD-generated bills of materials to be transferred to MRP systems. Product design information can also be transmitted to the manufacturing engineer, ready to be used in process planning. Operation sheets, process plans, and inspection instructions, can be delivered directly to operators on the plant floor as needed. SPC data can even be added to the Electronic Data Interchange (EDI) documentation sent to customers with product.

Achieving full CIM benefits from a network requires a basic framework to sort out the planning, design, and implementation details. Several models describing typical manufacturing hierarchies exist. Each model consists of levels that group the various communication needs. Networks for these levels differ in terms of communication devices, distance, physical transmission media, bandwidth, and communication and protocol functions. Typically, three types of network technologies are required to interconnect the levels: plant floor device-level subnetworks, a plant-wide Local-Area Network (LAN), and an enterprise Wide-Area Network (WAN) that can link separate sites in a corporation and interconnect corporations through EDI.

In the various models, the first level includes machinery and process de-

Characteristic	Enterprise (Level 3 — Level 4)	Plant (Level 2 — Level 3)	Work Cell (Level 1 — Level 2)
Function	Integration of enterprise	Integration of work groups/departments	Automation
Distance	Many miles	10,000+ feet	100 feet
Designed by	MIS/Telecomm	MIS/Manufacturing Engineers	Manufacturing Engineers
Typical media physical connection	WAN: X.25, CCITT, T1 satellite, microwave	LAN: Ethernet, token ring, broadband	Serial Lines PLC LANS Bitbus, MAP
Typical speed required	56 K b/sec to 1.5M b/sec	10M b/sec	< 19.2K bit/sec
Typical protocol	DECnet SNA	DECnet/OSI TCP/IP XNS	Many proprietary
What's connected	Mainframes	Mainframes Minicomputers PCs	Plant floor equipment PLC, Robot, N/C, terminals
Number of nodes	Thousands	Hundreds	< 10
Communication functions	File transfer Program-to-program Virtual terminal connection Remote data file access Network Management Resource Sharing		Read/write data Start/stop device Upload/download control program Device status
Major unmet needs	Standard multivendor protocol: ISO/OSI		Standard protocol for multivendor/ multidevice communications

Different network technologies are required to interconnect equipment operating at various levels of a manufacturing communication-hierarchical model. This table summarizes the characteristics of these network technologies.

Reprinted with permission from *Automation*, March 1990.

Network drives engine assembly

A DECnet/Ethernet network links workers, flexible manufacturing and materials handling systems, and four levels of computerized information systems throughout the BMW engine plant in Steyr, Austria. With this network, BMW/Steyr has reduced costs by reducing machine idle time and work-in-process, improved quality, and created a flexible service system to compete in the worldwide automotive marketplace.

The BMW Moteren Gesellshaft m.b.H builds several types of four and six-cylinder gasoline and diesel engines. The plant employs about 1,900 workers, covers nearly 2 million ft^2 of space, and produces up to 300,000 engines annually, using two shifts.

The plant houses R&D component machining, engine assembly, and a warehouse. On the plant floor, conventional roller conveyors and automatic AGVs transfer machined components and parts from the warehouse to the assembly area. Once assembled, engines, which are scheduled and manufactured in lot sizes of one, are transferred to fully automated test benches for 100% engine functional testing. The tested engines are then sent to the shipping dock, and within 40 hours, they become part of a customer-ordered automobile at another BMW assembly plant.

The backbone behind the Steyr plant is a DECnet/Ethernet network: a 1,650-ft baseband coaxial cable that stretches from the data center in the assembly area, through the warehouse, and into the machining area, connecting plant-floor computers and dedicated monitoring and control devices. A 3,000-ft fiber optic cable connects the data center to R&D; a DECnet/SNA gateway connects the computers to the IBM mainframe; and a DECnet WAN links the data center to BMW in Munich, Germany. These connections support the transfer of CAD data and manufacturing software.

Much of the plant's computers are Digital VAX systems. The DECnet/Ethernet network not only links this equipment together, but also handles time-critical operations such as verifying within two seconds that the right engine component has arrived at each assembly station.

As manufacturing and data processing and control requirements change, network applications are introduced, upgraded, and combined. Likewise, network cable segments, terminal servers, and computer systems are constantly added and modified. Because all the computers run the same VMS operating system, new and existing Digital computers can be freely added or redistributed.

For example, BMW recently acquired the VAX 6220 multiprocessor system to replace four VAX-11/750 minicomputers. The changeover was quick, did not disrupt current manufacturing or data processing operations, and required no costly software conversion.

vices. These devices are hard-wired to single control devices such as robots, programmable controllers (PLCs), numerical control machines, and process controllers. At this level, relatively small amounts of data are involved, communication is usually event-driven, and response times must be fast.

The next level's devices are connected by a variety of subnetworks to upper level cell controllers and factory data-collection systems. Many of these subsystems are proprietary and typically link up to 10 devices. The subnetworks usually extend 100 to 3,000 feet, and provide device communications at speeds less than 19.2 Kbits/s.

These subnetworks read and write data to and from device memory, start and stop devices, upload and download control programs, and monitor the status of devices on the subnetwork. Common subnetwork communication technologies used today include RS-232 serial lines, proprietary PLC LANs, and MAP.

Computers supporting applications at upper levels (cell control, MRP, maintenance management) connect to the plant-wide LAN. These LANs are in turn connected to corporate-level systems used by engineering and marketing, sometimes located at other sites, via a WAN. In this way, all functions are interconnected and information can be exchanged between individuals and departments as required.

The plant LAN and corporate WAN are essentially the same in terms of functionality, types of systems connected, and support responsibility. Their functions include file transfer and accessing remote-files, program-to-program communications, and network management.

However, LANs and WANs differ greatly in speed and distance. A plant-wide LAN typically communicates at 10 Mbits/s over distances of about 10,000 feet. An enterprise-wide WAN typically communicates at speeds from 56 Kbits to 1.5 Mbits/s over many miles. Common LAN and WAN protocols include DECnet, OSI, SNA, XNS, and TCP/IP.

Planning approach. Today, a network is as necessary to the productivity and competitiveness of an enterprise as telephone service. The type of telephone system and service selected typically undergo a rigorous analysis, as should the desired network technology. Factors to consider are:

- Network topology: star, ring, bus, or hybrid.
- Access method: Carrier sense multiple access with collision detection (CSMA/CD) or token passing.
- Communications medium: twisted pair, coax, optical fiber cable.
- Signaling: baseband, carrierband, or broadband.

These technological options must match the enterprise's needs for capacity and growth and fit the organizational structure, finances, and future goals while maximizing network performance at reasonable cost.

Network costs fall into two categories: initial and ongoing. Initial costs include analysis, design, purchase of hardware and software, installation, and start-up expenses. Ongoing costs include hardware and software maintenance, salaries, and the costs associated with expansion and modifications.

The design and type of network directly affect these network costs. For example, a network without redundancy or with minimal network management and monitoring services might have a low initial cost. However, its ongoing cost in performance, maintenance, and troubleshooting could in time exceed any initial cost benefits. On the other hand, an over-designed network may never be cost-effective.

A recent study by the Index Group Inc., Cambridge, Mass., found that approximately 20% of the total costs of a network were in acquiring the network. Operation and incremental changes over a five-year period accounted for the remaining 80%. Moreover, personnel costs constituted over 50% of the total network costs (initial and ongoing) over five years of operation.

Thus to reduce recurring costs, the network designer should select options that allow ease of maintenance, modification, and expansion. Ethernet configurations are well suited for the factory on these points. The bus topology allows easy access for maintenance and updates. New nodes, segments, or subnetworks can be added to a baseband Ethernet network without disrupting network operations. And Ethernet is a proven, reliable technology.

Network justification. The ability to manage and distribute information is essential to reaching an enterprise's corporate goals. In justifying a network, these questions must be answered:

- How valuable is a network that can move SPC data from the plant floor to engineering so that engineering can analyze the data, change the process, and improve yield?
- How much can customer satisfaction and loyalty be improved by providing timely SPC information and electronic invoicing through EDI?
- What reductions in time-to-market can be realized through concurrent engineering, which requires efficient information flow between the design and process engineering departments?

In our own facilities, networks have proved essential. In 1986 we had one year to design, produce, and ship a new circuit board and associated chips for a communications controller for a new industrial VAX computer. That design-to-market cycle was significantly less than that of similar projects.

The group traditionally responsible for this project, along with its CAD tools and design libraries, is located in Massachusetts. However, they were already busy with existing projects.

We solved this problem by making use of our internal EASYnet network. This network links 32,000 computers at almost 500 sites in 33 countries. Project responsibilities were divided and assigned to available groups located at several distant locations.

The network let team members locate available expertise, tools, and resources without relocating anyone. The result: the project was completed in record time. ○

Ed Pastor is Manager of the Industrial Networking Product Marketing Group, CIM Marketing and Product Development, Digital Equipment Corp., Maynard, Mass.

802.3/ETHERNET LANS MEET THE MANUFACTURING ENVIRONMENT REQUIREMENTS

Wayne M. Adams
Industrial Networks Consultant
Digital Equipment Corporation
600 Nickerson Road
Marlboro, MA 01752
U.S.A.

INTRODUCTION

On the plant-floor, 802.3/Ethernet is the Local Area Networking (LAN) technology of choice. Advanced Manufacturing Research [12] surveyed 150 Fortune 500 manufacturing companies for their current usage of LAN technology. The study revealed that 802.3/Ethernet comprised over 61% of the installed LANs in use today. This figure is derived from 44% DECnet/Ethernet installations, 17% Ethernet terminal servers, and an undetermined percentage for Ethernet-based personal computer LANs and intelligent plant-floor device interconnects.

Digital Equipment Corporation has researched the four most important manufacturing environmental requirements for a 802.3/Ethernet LAN. This paper summarizes this LANs capabilities and characteristics for:

1) Electro-magnetic immunity,
2) Network performance,
3) Area coverage, and
4) Cost of ownership.

ELECTRO-MAGNETIC (EM) IMMUNITY

The strength of an EM field is measured in volts per meter (v/m). For compliance with Ethernet V2.0, IEEE 802.3, IEEE 802.4 and IEEE 802.5 specifications, a network must withstand a minimum of 2 v/m up to a frequency of 30 MHz and 5 v/m above 30 MHz while providing a mean undetected bit error rate of less than 10 to the minus 9. 10BASE2 802.3 is specified for 1 v/m across the spectrum [1],[2],[3],[4].

Digital's research measured over 40 pieces of plant-floor equipment in both process and discrete manufacturing industries for radiated EM fields. The fields measured are in the frequency spectrum between 10 KHz to 1 GHz at a 3 meter distance from the source. Figure 1 summarizes the observations. A few EM levels were recorded above 0.1 v/m, but nearly all measured less than 0.01 v/m. None of the recorded EM fields had any disrupting affect on the adjacent 802.3/Ethernet network.

Digital further investigated the correlation between intense EM fields and error-free network operation in an anechoic chamber. Within this chamber, 802.3/Ethernet configurations (10BASE5 and 10BASE2) including transceivers, repeaters, and bridges were radiated with EM fields at levels up to 20 v/m. 10BASE5 and 10BASE2 performed well in these fields, which are beyond the IEEE immunity specifications and an order of magnitude greater than those fields observed in the manufacturing plants.

Reprinted from *1989 IIE Integrated Systems Conference Proceedings*

PERFORMANCE

The well known fact that CSMA/CD (Ethernet/802.3) performs better than token passing methods (802.4/802.5) under light loads (and the opposite is true under heavy loads) is depicted in Figure 2, [6] [7] [9]. Both methods do work under both loading conditions.

Work conducted by the National Bureau of Standards (now National Institute of Standards and Technology) concludes that "CSMA/CD is a reasonable access method for time-critical applications on small plants LANs if loads of less than 40% are anticipated." [5]. Digital's research of installed manufacturing 802.3/Ethernet LANs found average utilization rates of less than 3%. Table 1, Manufacturing Sites Network Utilization, summarizes the findings. (See Table 1 at bottom.)

The maximum peak utilization has been added to Figure 2 to draw attention to the fact that these networks are not heavily loaded and actual response time is excellent. This survey and the networking performance studies reinforce the fact that 802.3/Ethernet has adequate bandwidth to support large distributed applications. This data also dispels the widely accepted belief that the network is the bottleneck in a distributed system.

AREA/DISTANCE COVERAGE

With the definition and introduction of repeater and bridge technology for 802.3/Ethernet networks, these LANs can span distances of 40+ kilometers or cover areas of 11+ square kilometers (4.25 sq miles or 2700+ acres). The world's largest manufacturing facility under one roof spans 204 acres [10]. Even though a few process industry facilities and manufacturing campuses span greater areas, 802.3/Ethernet LAN technology can be easily configured to span these areas [8].

COST OF OWNERSHIP

LAN costs can be broken down into two factors: initial costs and on-going costs. Initial costs include the purchase of hardware and software, the design, the installation, and startup. Ongoing costs include maintaining the hardware and software, people to operate and trouble-shoot the network, and expansion and configuration changes.

The Index Group, Inc. researched the costs of manufacturing LANs over a 5-year period. Their results show that personnel costs constituted over 50% of total network costs (initial and ongoing) over 5 years of operation. LAN equipment costs were 27% of the total. Nearly 80% of the total cost reflects ongoing operating costs, while 20% reflects initial start-up costs. Refer to Figure 3 [13].

To minimize recurring costs, the network designer should select options that allow ease of maintenance, modification, and expansion. Baseband 802.3/Ethernet (10BASE5 and 10BASE2) configurations are particularly suited for the industrial environment on these points. Their bus topology allows easy access for maintenance and updates, and 802.3/Ethernet is a proven, reliable technology. New nodes, segments, or subnetworks can easily be added to a baseband 802.3/-Ethernet without disrupting network operations.

LAN cable media selection does affect network expenses for both equipment purchases and maintenance services. A quick rule-of-thumb cost estimate for installed standard baseband cable is $5 to $7 per linear foot and broadband cable is $11 to $12 per linear foot. Annual LAN maintenance costs for a broadband cable plant can cost 2 to 3 times more than for a baseband cable plant.

SUMMARY

Digital's research has found that manufacturing environments are not electrically harsh, network traffic is light, and 802.3/Ethernet networks are cost effective. These findings provide an in-depth understanding as to why 802.3/Ethernet is the most widely installed plant-wide LAN.

In the 1990s, 802.3/Ethernet LANs will continue to be used. They are easy to expand with new segments without disruptions to manufacturing operations and they have ample bandwidth to accommodate additional network traffic.

REFERENCES

[1] IEEE Standard 802.3, 1985.

[2] IEEE Standard 802.4, 1985.

[3] IEEE Standard 802.5, 1985.

[4] Supplement to IEEE Standard 802.3, 1988.

[5] K. Mills, M. Wheatley, and S. Heatley, "Prediction of Transport Protocol Performance Through Simulation", Sigcom Conference Proceedings, August 1986, National Bureau of Standards.

[6] W. Hawe, M. Kempf, A. Kirby, "The Extended Local Area Network Architecture and LANBridge 100", Digital Technical Journal, September 1986, Digital Press, pages 54-72.

[7] R. Jain, W. Hawe, "Performance Analysis and Modeling of Digital's Networking Architecture", Digital Technical Journal, September 1986, Digital Press, pages 25-34.

[8] Digital's Network and Communications Buyer Guide, Appendix A 802.3/Ethernet cable configuration guidelines ED-32596-42/88.

[9] W. Bux, "Local Area Subnetworks: A Performance Comparison", IEEE Transactions on Communications, COM-29, October 1981, pages 1465-1473.

[10] Guinness Book of World Records, 1987 Edition, Page 242.

[11] Wayne M. Adams, "Manufacturing Network Performance and Environmental Measurements", Enterprise Networking Event 88 Conference Proceedings, June 1988.

[12] The AMR Report, April 1989, Advanced Manufacturing Research, Cambridge, Mass.

[13] Cost of Network Ownership Study, Index Group, Cambridge, MA., available through Digital Equipment Corporation, Part #EJ-32929-42.

BIOGRAPHY

Wayne M. Adams is an Industrial Networks Consultant for Digital Equipment Corporation. He is involved with product and programs marketing for computer LAN technologies and computer communications products for the manufacturing industries. His current research activity encompasses industrial LAN capacity planning for the client/server computing style.

Mr. Adams received a Bachelors of Science degree in Computer Science and Math from the University of Pittsburgh in 1980. He is also a member of the Society of Manufacturing Engineers.

Mr. Adams has published several trade-journal articles and trade-conference papers on manufacturing LAN topics including technologies, performance, environmental characteristics, and ownership costs.

Table 1: Manufacturing Sites Network Utilization [11]

Industry/Plant	Nodes	Average Load %	Peak Load %
Automotive Machine Shop	25	0.2	2.8
Steel Plant	40	2.0	14.0
Heavy Equipment/Engine	80	0.8	3.0
Chemical/Textile	82	0.1	8.1
Automotive/Engine	89	0.8	1.6
Aerospace/Engine	195	1.7	13.9
Aerospace Machining/Fab	650	1.5	4.0

Peak Electromagnetic Field Measurements

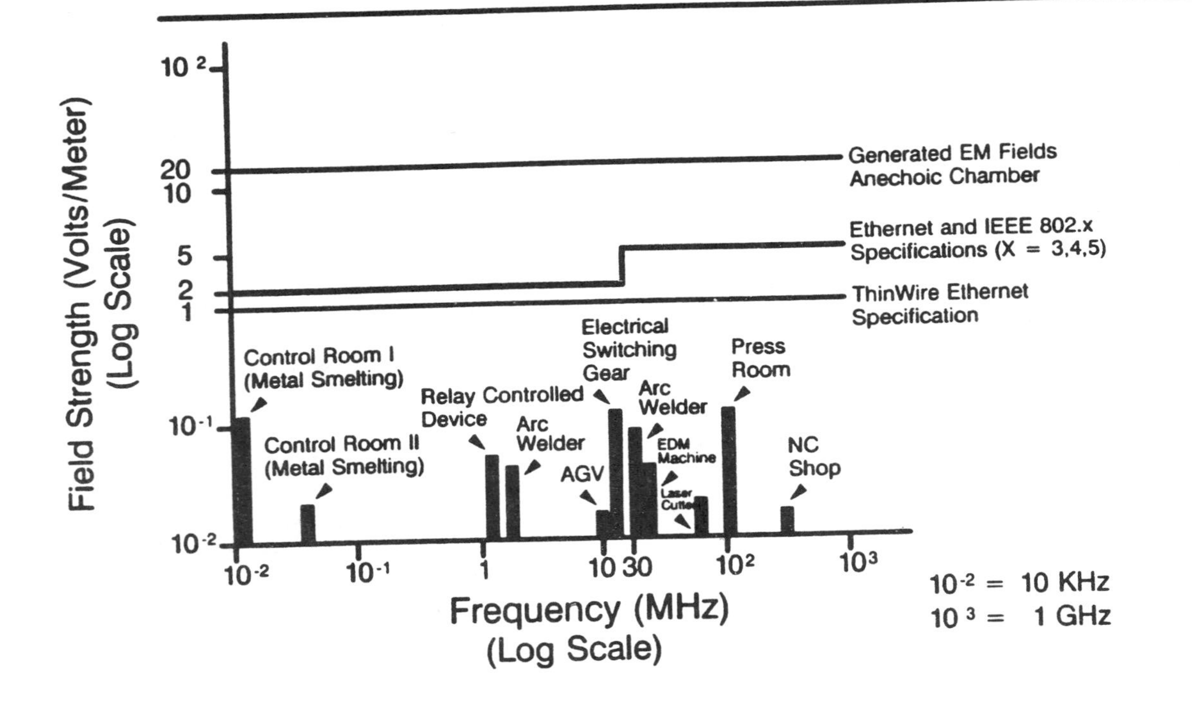

Figure 1

Effects of Loading on Response Time

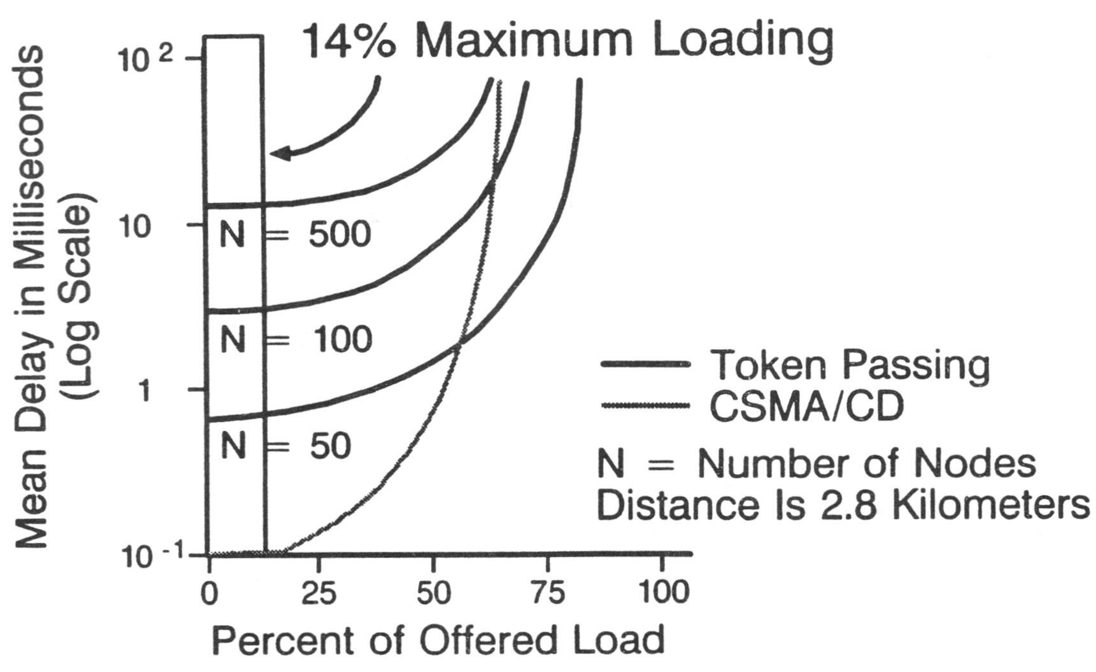

Figure 2

WHERE DO THE MAJOR INDUSTRIAL NETWORK COSTS LIE?

	ACQUISITION	5-YEAR OPERATION	5-YEAR INCREMENTAL CHANGE	
EQUIPMENT	15.1%	12.1%		27.2%
SOFTWARE	2.8%	13.8%		16.6%
PERSONNEL	N/A	36.8%	13.7%	50.5%
COMMUNICATIONS	0.1%	1.4%		1.5%
FACILITIES	3.1%	1.1%		4.2%
	21.1%	78.9%		100.0%

SOURCE

- 98 Terminal, 144 Device Automotive Components Manufacturing Network
- 100 Terminal, 40 Device Automotive Components Manufacturing Network

NOTES

- Network Costs Do Not Include Terminals, Devices, or Plant Computers, Except if Network Processing

Figure 3

Enterprise Networks: One Step At A Time

A new study of IS plans shows creeping networkism in big companies, with manufacturing firms taking the lead. Opening systems up will go hand in hand with cost-reduction measures.

BY MARK SCHLACK

The 1990s is not going to be the decade of the Network Revolution. But it will be the decade of the Network Evolution. That's what a major study of enterprise networking by the Newton, Mass.-based Business Research Group suggests.

"MIS professionals have spent the last five to 10 years building their current backbone networks," says Tom Wood, industry analyst and author of the study. "Vendors who expect them to scrap those networks for 'better, more open' networks will end up getting scrapped themselves. The key words for the '90s networker are *evolution* and *integration*."

BRG, a unit of Cahners Publishing Co., the parent of DATAMATION, questioned 269 IS managers in manufacturing, banking, insurance and trade companies within the *Fortune* 1,000. The survey found little indication that the bedrock systems that have kept companies going for the last decade and more will disappear. For the most part, IS departments are not planning to wheel out their mainframes and minis. Rather, they are planning to wheel in growing numbers of local area networks, along with the hardware and software needed to graft those LANs onto existing backbone networks.

Already A Mixture

Nor are IS managers planning for the Totally Open One Big Network. Proprietary protocols like IBM's System Network Architecture (SNA) and Digital Equipment Corp.'s DECnet remain critical parts of IS plans. They will coexist for the foreseeable future with Transmission Control Protocol/Internet Protocol (TCP/IP) and, later, with the Open Systems Interconnection (OSI) protocol.

If the BRG survey group is typical, today's *Fortune* 1,000 company has an average of 6.5 physically or logically separate backbone networks. Nearly half of these networks use SNA as a primary or secondary protocol, and DECnet accounts for another 16%. Less than 10% of networks use open or standard protocols like TCP/IP or OSI.

Nearly 40% of IS managers agree, to some extent, that the OSI protocol stack will be the foundation of enterprise networks within two years. Yet only 30% of respondents intend to implement new protocols in that time frame, although different industries have different plans. Both manufacturing and insurance are being more aggressive—37% and 34%, respectively, expressing implementation expectations—and the wholesale and retail trade sector lags behind at 15%.

Of those who plan to add protocols to their backbone network, 22% are adding TCP/IP and 18% OSI. TCP/IP's star is rising the fastest in banking, where it will account for 40% of new protocols implemented. OSI shows its greatest strength in manufacturing, accounting for 27% of planned new protocols.

"The enterprise network of the future will be an amalgamation of networks from different vendors," Wood concludes. "Users are waiting for their proprietary networks, like SNA and DECnet, to become OSI compliant so they can integrate them with TCP/IP and other open networks that they are buying now."

Enterprise Cost Cutters

The enterprise networks currently under construction use gateways, bridges and routers to glue together disparate systems. Backbone networks currently have an average of 29 mainframes connected to them, and less than 30% of respondents believe that mainframes, or minis, will diminish in importance. But IS managers showed broad agreement on the importance of integrating LANs into the backbone (83%), putting the pieces in place to support multiple protocols (82%) and using Integrated Services Digital Network (ISDN) or T1 services to bridge remote LANs (69%).

What clearly emerges from BRG's study is that IS managers, in avoiding the trap of technical obsolescence, are not prepared to ignore cost. In fact, enterprise-networking plans are being strongly shaped by one of the great cost-cutting movements in the IS community, data-center consolidation. Among the IS managers polled by BRG, 80% intend to connect large numbers of field offices, 61% will interconnect geographically dispersed networks and 42% will create a shared transmission backbone.

The motivation for these steps is to make the network of the '90s not only more powerful but easier to administer. High on the list of IS managers' consolidation activities are developing applications that span multiple LANs (54%), simplifying management across multiple networks (45%), reducing the number of protocols used (41%) and managing voice and data together (39%).

The tension in IS managers' thinking between reaping the benefits of new technology and controlling the cost of adopting it was evident in their opinions on the problems facing network integrators. When asked what the three biggest problems are, 33% tabbed lack of network management tools, 30% pinpointed the high cost of initial investment and 25% cited the speed of change in technology and products. Close behind (at 23%) were the lack of both application interfaces and technical expertise.

Finally, IS staffs are looking for help. While 97% of the IS managers expect their in-house staffs to carry out product evaluations, they are looking for management and installation assistance from vendors, consultants and systems integrators. And they are not leaving the process to chance: at least 51% of the companies queried have a formal committee investigating the implementation of a unified network strategy.

After the user revolts of the '80s, IS departments are listening more to the user community. Many of the enterprise network committees in industry include people from engineering and manufacturing. In the service companies, sales and marketing are beginning to show up on planning committees. The new style of enterprise computing requires more than just a technical adjustment.

OPEN SYSTEMS

Uniting or Dividing the Organization

By Michael Galane

Computing standards were ballyhooed during the mid '80s as the driving force behind information technology. Islands of automation were quickly becoming continents that could not be bridged by the plethora of proprietary technology. As one colleague put it, "We don't have any problems regarding standards—every organization has its own!"

Today, many vendors are citing the advantages of open systems technology. It seems everyone is claiming to be a leader yet the majority of profits are still being made from decade-old proprietary software and hardware product families. Many users have yet to reap the unfulfilled promises of productivity that these investments were supposed to deliver. Has all of this technology enabled companies to work more effectively across organizational boundaries or have the continents grown larger and the seas become deeper?

It seems everyone is claiming to be a leader yet the majority of profits are still being made from decade-old proprietary software and hardware product families. Has all of this technology enabled companies to work more effectively?

RECENT HISTORY: The first computers to find their way into industry were used by the accounting community. By the late '50s, payroll and cost accounting were the primary functions being performed. By the late '60s, inventory of raw materials and finished goods were being managed by mainframe databases.

As the '70s approached, manufacturing engineers saw the advantages of this power on the factory floor. Mainframe computers were too expensive and unwieldy for this task. Fortunately, a few companies saw the need for a different type of computer — the minicomputer. Mainframe companies did not see these products as legitimate computers. In fact, one of these companies refused to acknowledge the term "minicomputer" and later referred to its product line as "small computers." A new subculture rose to power by solving real industrial problems with these "illegitimate" machines. The tectonic plates of the information age were forming.

The engineering community, not to be

ABOUT THE AUTHOR:

Michael Galane is the Director of Stategic Consulting at Hewlett-Packard.

outdone, also became frustrated by the timeshare/mainframe concept. They too wanted to be the masters of their information destiny. The workstation burst onto the computing scene in the '80s and wrought more diastrophic consequences. With vendors now touting the advantages of moving toward open systems in the '90s, the user community is more focused on improving productivity than ever. Will this move act as the catalyst that actually allows information to move freely across the company or will the continents of automation move farther apart?

WHY AUTOMATE? A company is in business to generate a profit by delivering a product or service. The only reason to invest in automation is to improve this capability. A customer measures the value of a product or service by its cost, quality, availability, and features.

The ability to meet customer expectations and make money has never been easy and is becoming more difficult every day. Proprietary systems met the needs of most companies until recently. This is because different parts of the organization acted autonomously.

A company is in business to generate a profit by delivering a product or service. The only reason to invest in automation is to improve this capability. A customer measures the value of a product or service by its cost, quality, availability, and features.

This paradigm required each functional unit to perform its tasks or processes more quickly. Engineering, marketing, and manufacturing were encouraged to segregate their activity. Indeed, each fiefdom could generate more activity when the chain of command had complete control of specific processes. Proprietary systems allowed each unit to "optimize" on a specific technology without regard for other groups. The accounting system rewarded those who improved the productivity of isolated operations. Computer purchases could be justified on mere return on investment analyses of bounded processes. Computer "isolated" manufacturing was the result.

This was good enough to show improvement through the '60s and '70s. Networking, albeit proprietary, permitted some sharing of information. Decisions regarding technology selection that were once viewed as one-time buys were now seen as strategic.

The decision process for purchasing computers was indicative of the decision structure for the entire company — large companies formed committees that took forever to decide on strategic directions while the operating entities approved funds for tactical buys. Productivity improvement was inconsistent and only appeared within the walls of each fiefdom. Automation did pay dividends but the cost of supporting all this technology was becoming a drag on the organization. The pockets-of -technology information system approach reflected deeper problems in the company.

Instead of focusing on competitors as the competition, these business enterprises acted as if other units of the company were the real competition. As one person put it, "If our company was an airplane, we would be a bunch of parts flying in close formation!" No doubt they would also be computer controlled.

WHY OPEN SYSTEMS? Most companies now realize that much more teamwork across the organization is required to achieve business breakthroughs. This requires information to be shared at all levels in the organization. The proprietary cocoons that contributed to success in the old paradigm are a handicap in the new paradigm.

This new paradigm requires information be shared much earlier in the innovation cycle across functional organizations. The single hand-offs from each functional organization are now replaced by nifty passing of information among the functions. Just as today's organization must be agile to respond to the needs of customers, so too must be the information infrastructure.

Open systems will enable individual entities to make optimal "buy" decisions without mortgaging the future. These systems, based on standards, will ensure that growth will not be stifled by the limitations of any single vendor. Competition will fuel innovation that will directly benefit users of systems technology.

There are some companies that will not see an open system strategy as supporting competitive advantage. These companies are either content to operate in the traditional environment of functional fiefdoms or are meeting their information needs through the proprietary products of single vendors. The former scenario hinders quick product development. The latter casts the user as hostage to the availability and price premiums associated with proprietary products.

THE NEW WORLD ORDER OF COMPUTING — Although most users agree open systems are good, they are threatened by the changes they may bring. The MIS community has been the information broker for decades. Open systems thinking may threaten their empire. If all computing components can "plug-and-play," why will we need such a large organization to manage them?

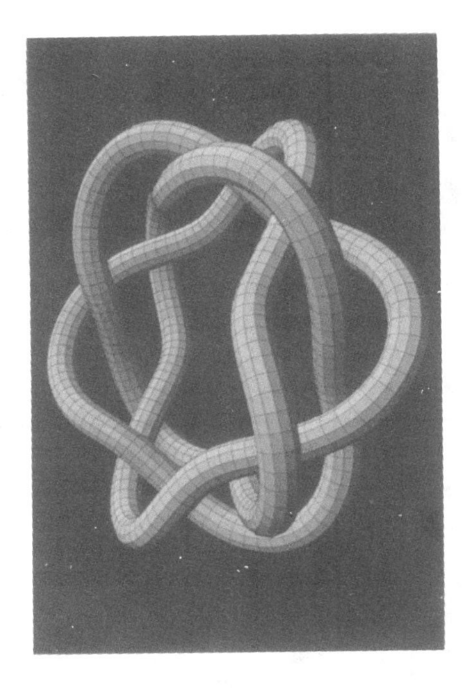

Power plays based on information availability will also be threatened. In the past, the "owner" of the system decided how, who, and when information would be distributed. This allowed one organization to exert dominance over others. If information is allowed to be accessed by all those having a legitimate need, the "owner" will not be able to use the system as a political weapon.

Open systems are bad news for the bullies of the old paradigm. These power bases took many years to establish. Just as computers facilitated the flow of information to reduce layers of management, open systems will eliminate the gate keepers of proprietary technology and democratize the use of information.

The result has sometimes been ugly. Fiefdoms built on the independence of the old paradigm have used the open system thrust to polarize the system strategy. They argue these standards will create near term turmoil. Incremental investments to the tried and proven proprietary technologies are in the best interest of the company.

Organizations lacking the fortitude to make such a bold open system decision and the leadership to execute this strategy will fall back into the proprietary comfort zone. The subsequent inability to capitalize on the accelerating price/performance advantages of new technology will put them at a serious competitive disadvantage in the long run.

WHAT TO DO? Open systems technology will enable a company to share information more cost effectively than ever before. This will provide significant advantages if the sharing of information has been identified as a critical success factor.

Many businesses have already recognized the need to reduce impediments to information sharing at all layers of the organization. Open systems technology will allow a company's information needs to adapt quickly to changing needs.

Companies that have not yet realized how effective business processes span functional organizations will not reap all the benefits that open systems provide. For them, evolutionary investment in existing technology will probably yield a satisfactory return on investment in the short term. Similar investment analyses were probably done for slide rule and mechanical cash register manufacturers in the early '70s. **IC**

Instead of focusing on competitors as the competition, these business enterprises acted as if other units of the company were the real competition. As one person put it, "If our company was an airplane, we would be a bunch of parts flying in close formation!" No doubt they would also be computer controlled.

THE STATE-OF-THE-ART DESKTOP

So, how do we get down to brass tacks and really take advantage of these open systems? The cover of this issue shows the screen of a PC that illustrates some of what can be done today. The photo and the description below are not hypothetical, they are operating every day in our company.

PC hardware—The four screens on the cover are being driven by a single PC. In this case, the PC is an Intel 33 MHz 486, with 8 MB of RAM and a 200 MB hard drive. All four screens are addressed as if they were one large 1280×960 (1,228,800 bit) field, using two VGA+ boards and Microsoft MS-windows software drivers from Colorgraphics Communications. This provides a "megapixel display," which until very recently was found only on very expensive workstations. The PC was ordered as separate parts (AMI motherboard, power supply, disk drives, case, etc.) and assembled in house for about $5,000. No difficulty was experienced in getting these standard parts to function together — the PC is itself an open system.

PC software—The PC is running MS-DOS and MS-windows 3.0, which provides a graphical user interface, multitasking, and the ability to cut and paste data and graphics amongst a wide mix of low-cost personal productivity applications software. Here, each individual user will customize his or her personal working "desktop" with favorite word processing, spreadsheet, graphics, and other software. WIN/TCP (TCP/IP) and Pathway (NFS) from the Wollongong Group and HCL-eXceed Plus (X-windows) from Hummingbird Communications software provide workstation connectivity.

Companywide network—A standard thin coax ethernet provides a low-cost, high speed (10 Mbit/sec) connection to an in-house network of 40 or so computers. These are a mix of more DOS PCs, Apple Macintosh PCs, and workstations. A Novell file server on the network provides large shared disk storage, and connectivity between the ethernet and an Appletalk network. The file network and server hosts PC-based financial software, DBase-compatible databases which can be simultaneously accessed by all PCs on the network, electronic mail, bulletin boards, a FAX server, and dial-in capability for off-site access.

Engineering workstations—On the same ethernet are X-windows based UNIX and VMS workstations — DECstation, VAX, IBM RS-6000, Sun Sparcstation, Concurrent, HP 9000, and Apollo 3500. These provide far greater compute and graphics capability than any of the PCs in the company.

What is remarkably powerful about this is that from a single desktop one can integrate and display data and graphics from all the information systems in the company. This really illustrates the extent to which open systems have arrived. From the Program Manager window, a kind of control panel for this PC, one can log on to any one of six workstations by clicking the appropriate X-windows icon. Files can be transferred or mailed, and one can cut and paste from any window to any other.

The software shown on the cover photo shows MS-windows software running on the PC and X-windows software running on an IBM RISC-6000 UNIX system. The X-windows software can either put each X-window in a separate MS-window or create a single large MS-window that contains all the X windows. In this case, for clarity (and to show off the power of this system), we have set up the X server in single window mode. This window (title "HCL-eXceed/W") looks exactly the same as the display on the RS-6000 itself. This illustrates the client-server model of computing implemented by X. All the software inside the HCL window is executing on the UNIX system, but is displayed on the PC.

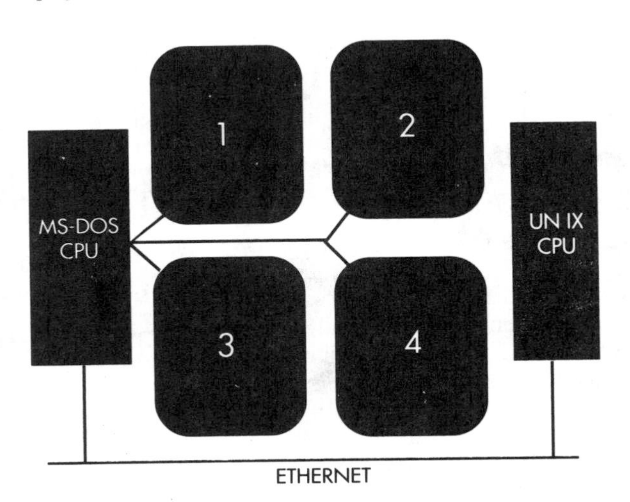

In the UNIX window, the OSF/Motif window manager and the X desktop software are showing. Note that one manipulates the Motif windows in exactly the same way as the MS-windows. Also, X desktop provides a similar iconic interface as does MS-windows. One very important component of open systems is that they be open to the user, by conforming to the same user interface conventions. This has now been achieved — on the Apple Macintosh, MS-windows, and X-windows, one can start an application by a double-clicking on its icon.

Finally, three applications software packages from our own company are shown. These packages are offered in versions for all the major operating systems — DOS, MS-windows, and X-windows UNIX (DEC, IBM, HP, Sun, Concurrent, Modcomp) and VMS (DEC). The LT CONTROL process mimic diagram (blue background) is running locally on the PC under MS-windows. The LABTECH NOTEBOOK data acquisition (green background) is running on UNIX, and the Chrom/RT chromatography analysis is running on MS-windows. This is another aspect of open systems standards — they greatly facilitate the porting and support of software across many different hardware platforms, which allows the best hardware for each job to be selected. — *Reported by Fred Putnam as a user at LABTECH.*

MODULAR SOFTWARE FOR INTEGRATED FACTORY MANAGEMENT SYSTEMS

Steven A. Vogel
President, AME Associates
2642 Cedarvue Drive
Pittsburgh, Pa. 15241
(412) 831-7644

ABSTRACT

This paper will present a definition of a Factory Management and Control System and describe several examples of FMCS systems being implemented today. It will also discuss a methodology to evaluate the requirements of an FMCS, based on a company's manufacturing characteristics, business objectives, and technical capabilities.

DEFINITION OF FMCS

The term Factory Management and Control Systems is being used increasingly to define a layer of software within the overall Computer Integrated Manufacturing systems architecture that resides below the level of an MRP (Manufacturing Resource Planning) system and above the Process Control level (see Figure 1). This layer of software functionality has gone by many different names in the past, including Shop Floor Control, Shop Floor Management, Production Control, and others.

Because of the diverse characteristics of factory management and factory control disciplines across industries, the standardization of this layer of software has been difficult to achieve. Over the past several years, however, there is a growing commonality in both the functions and the terminology being applied to the FMCS domain.

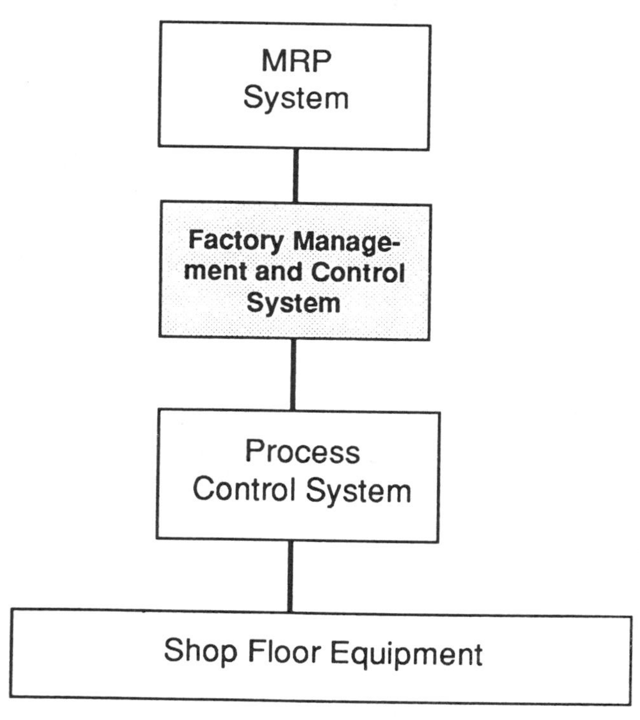

Figure 1 - FMCS Software Layer

Reprinted from *1991 IIE International Industrial Engineering Conference Proceedings*

To understand FMCS, one must first understand MRP. MRP is both a set of procedures and a software solution. Its goal is to make sure an adequate supply of raw material and purchased parts are available to support the production schedule. At the heart of the MRP system is the master schedule, the bill of materials, and the process plan. MRP does not create these item. They are created by Scheduling, Design, and Process Planning Systems or they are created ad-hoc as part of the input to the MRP system.

Based on these three items, material requirements are developed for several time frames, normally ranging from several months out to several days out. Obviously, the resolution and accuracy of the material requirements gets better and better as time gets closer to present.

The most important aspect of MRP in relation to a total CIM architecture is not so much what it does, which most people understand, but what it does not do, which most people misunderstand. MRP developed as a financial application to control the ordering, receiving, and payment of materials from suppliers. It has evolved through several generations of enhancements, and now MRP is used extensively throughout the manufacturing community, from small job-shop oriented manufacturing sites to multi-location repetitive manufacturing corporations.

Its primary emphasis, however, remains the financial management of materials. Improvements have been made to minimize inventory levels and support JIT material management practices. However, extensions beyond this level of functionality are really outside of the intended purpose of the MRP system. And generally, the underlying structure of the MRP system is not well suited to perform lower level factory management functions that require a more real-time messaged-based distributed processing environment.

Next, let's turn to the layer of software below the FMCS layer. We refer to this in Figure 1 as the Process Control layer. This layer of software focuses on the machinery and processes that transform raw materials and purchased parts into subassemblies and end items. There are a wide variety of Process Controllers, including PLC's, CNC's, robot controllers, cell controllers, vision system controllers, and so forth. The common element of this layer of software is that it addresses all of the control needs of the process.

Some typical functions found in Process Control includes data collection, graphical process monitoring, analog and discrete I/O control, program upload and download, SPC, and low-level process reporting.

Again, it is important to understand not only what Process Control does, but also what it does not do. Generally, Process Control technology comes with the machine or process being controlled.

Because the solution is typically supplied by the machine or process vendor, the functionality, the hardware and software architecture, and the communications capabilities are tightly tied to the specific process. Because of this, even though similar functions may exist in two different Process Control capabilities within a facility, they are most likely implemented differently and are not capable of being merged into one solution.

A good example of this is SPC (Statistical Process Control). Many processes now have SPC as a standard or optional feature. However, most SPC solutions that come with a process have their own algorithms, their own data base, their own output capabilities, and so forth.

This problem is driving many CIM architects into taking functions that may logically belong at the Process Control level and moving them to the FMCS level. The advantage is that the function can be performed with a common set of software across all processes, enhancing maintainability as well as reducing training time of shop floor personnel to use these applications.

In summary, then, the Process Control level should be thought of as those unique applications that directly interact with the process of changing raw materials and purchased parts into subassemblies or finished goods. Applications which are common across multiple processes should be included in the FMCS layer.

Now let's turn our attention to the Factory Management and Control System Layer. Again, FMCS does what MRP and Process Control are not intended to do. That is, it deals with the overall logistics on the shop floor. The main audience of FMCS is the manufacturing management and shop floor production workers.

In general, FMCS addresses the three key elements of the shop floor, that is, production, product, and process. It is convenient and helpful to think of the applications of an FMCS system along these catagories. Depending on whether the facility is work-order oriented, flow-line oriented, or a combination of the two, an FMCS will have a set of applications unique to the manufacturing environment. The master set of applications one could expect in an FMCS are as follows:

- Production Management

    Short Term Scheduling
    Material Dispatching
    Work Order Management

- Product Tracking

    Serialized Tracking
    Failure Reporting and Tracking
    Component Traceability

- Process Monitoring

    Machine Fault Monitoring
    Statistical Process Control
    Production Monitoring
    Work in Process Monitoring

Briefly, we shall discuss each of these areas of capability both in terms of functionality and in terms of the business objectives they address.

Production Management

The "highest" layer of software in a Factory Management and Control System deals with the management of overall production resources. This most closely matches what many call Shop Floor Control. It is also the closest set of functions to MRP.

Production Management takes the master schedule and resource requirements generated by the MRP system, and manages its execution on the factory floor.

The first step of this process is Short Term Scheduling, where the daily and hourly schedules are developed for each area and process of the facility. In most plants, the Short Term Scheduling logic is too complicated to allow this function to be fully automated. Therefore, a compromise is struck. Rather than generating the short term schedule automatically, a set of tools are provided which determine the effect of theoretical schedules and schedule changes. This is called "what-if" scheduling.

Or even more simply, many plants cannot justify the investment in any shop floor scheduling technology, and simply maintain the short term schedule and resources manually in a data base. Whether it is automatically, semi-automatically, or manually entered, the short term schedule is a key element of the FMCS data base and drives the lower level applications.

Another key element of Production Management is Material Dispatching. This can be a very simple "pull" signal generated when a material queue is below a certain lower limit or it could be a very sophistocated application that anticipates shortages, manages changeovers, and drives automated material handling systems. Material Dispatching generates material reports and material movement authorizations to keep production running with the least amount of material on the shop floor. Typically, Material Dispatching is found in facilities or areas of a facility that are set up as continuous flow lines or repetitive manufacturing lines.

Work Order Management is a key application of Production Management in work-order based manufacturing facilities. It is to a job shop environment what Material Dispatching is to a repetitive environment. There are a number of standard solutions on the marketplace that do a fine job and have been enhanced over a period of years.

Product Tracking

In high volume repetitive manufacturing, individual products may not be tracked on a unit basis. However, most facilities that are either work-order driven or that do repair operations track individual units of production. For these facilities, a set of applications that track individual units is often beneficial.

The core application for product tracking is called Serialized Tracking. This of course requires that each unit of production, from sub-assemly to final assembly, have a unique serial number during its life in the factory. With this degree of product tracking, specific unit history can be maintained. This makes it easy to enforce proper process routings, to determine test, repair, and rework history, and to relate individual purchased part component lot numbers to final assemblies.

In some cases, the tracking of each individual unit by serial number cannot be justified, so a compromise is struck. Only units that have failed a test or inspection process are given serial numbers. In this manner, the unit can be tracked through the test and repair operations to ensure that they re-enter the production process as repaired units. Also, if there are subsequent failures on the unit further downstream, a decision can be made to scrap the item after a certain number of failed tests or inspections.

Failure Reporting and Tracking is the application that monitors and reports on the activities that occur to a product after it has failed a test and before it is re-introduced to production. This application generates reports, both on a unit product basis, as well as across a number of test processes. For example, it may be revealing to report on the failure rate for a particular attribute of a product amoung a group of testers. If one tester is generating more failures than the others for that attribute, perhaps the tester is at fault.

Component Traceability makes the final link in the product tracking area. It relates specific purchased parts by lot code to the subassembly and/or end item that they go on.

This application can be extremely useful for purging parts from a lot that has been found defective after it was used in production. It can also support recalls in the field after items have left the plant. Finally, it provides quality reports to parts suppliers that go beyond the receiving inspection area, allowing the supplier to monitor the fall-out of his parts throughout the production process.

Process Monitoring

This final set of applications forms the "lower level" of the Factory Management and Control System. Often, the applications in this layer overlap to some degree with the applications we discussed earlier in the Production Control layer that is closely linked with the process equipment itself.

The advantage to having as many process-related applications within the FMCS layer as opposed to the Process Control layer is that it promotes a standardized approach to the applications. As stated earlier, if similar applications are developed and supplied by the equipment suppliers, the chances of those applications having consistency in terms of functionality, look and feel, data base structure, and communications interfaces is very low. By taking these applications up one layer, the same applications can perform similar functions across a variety of production processes.

Machine Fault Monitoring is a module that generates alarms and reports related to the machines and processes. The purpose of this application is to detect as soon as possible any performance that is below standard for a process or group of processes.

Statistical Process Control is an application that is often bundled with the process control system. This application has two components, a real-time alarming capability that detects out of control conditions and signals when one has occurred, and a higher-level reporting capability that generates long-term SPC reports.

The Production Monitoring application collects information from the Process Control layer and generates production reports, such as end of shift counts by product type, process efficiency, yields, and process status. It also generates alarms in real time when any process characteristic is not performing to standard.

Finally, Work in Process Monitoring tracks the WIP queues between processes. It generates pull signals, material reports, and alarms related to these material queues.

FMCS EXAMPLES

Not so long ago, the installed base of Factory Management and Control Systems was limited to a small number of companies that invested in the development of custom software solutions for their factories. However, over the past five to ten years, a handful of companies have emerged which specialize in the design, development, and installation of FMCS solutions.

The type of solution offered by these vendors falls into one of several catagories. First, there is the custom solution that has very little software product content, but may use previous project experience and technology as a baseline for the new system.

There are few vendors who provide totally custom solutions for FMCS, primarily because it is very difficult to make money at it. However, many internal development groups of larger companies will create custom solutions that are highly specific to their manufacturing practices and requirements.

Next is the software product solution. Several companies have developed rather complete software modules that very adequately perform the functions we have outlined for an FMCS solution. These companies tend to focus on particular market niches, such as assembly, electronics, injection molding, defense contractors, etc. This is because a product solution by nature must be constrained to a set of customers that all do business in a similar way.

A good example of an FMCS software product is the DELTA system from Cimflex Teknowledge. This system consists of a set of nine standard modules that operate off of a common data base and communications backbone. This solution is best suited for repetitive assembly processes such as high volume electronics.

Finally, there is a solution that is a compromise between the product solution and the custom solution. This is refered to as a "tool set" solution. The supplier provides a set of tools to the customer that can be used to create a custom solution. The customer then is left with the task of creating an integrated solution using these building block tools.

A good example of a building block solution is the FactoryLink Application Enabler from U.S. Data.Corporation. This solution provides a set of library routines that can be programmed in several languages to build FMCS applications. It supports a variety of machine interfaces that are commonly found on the shop floor. It also provides basic FMCS utilities such as data base management, communications, alarming, graphical monitoring, and reporting.

IMPLEMENTING AN FMCS SOLUTION

We have covered the definition of FMCS and have reviewed several approaches to implementing this technology, from custom solutions to product-based solution to software building blocks. How do you know what functions you need and which implementation approach makes the most sense for you?

Developing Functional Specifications

The first step toward a successful FMCS solution is to develop a set of functional specifications that address your key business requirements. This step is always the most difficult, because it forces a business to ask questions that have no clear cut answers. For example, how much is an improvement in quality worth in terms of increased market share. Or what tangible benefit can be realized by better performance to schedule?

This is one area where an outside consultant may be of benefit. Some consulting firms have specialists in the development of cost justifications for FMCS technology. And they can often bring to bear case studies from related industries.

The first level of detail for the FMCS functional specification should be very high level and should clearly relate functions to business objectives. This is extremely important. As the project moves to implementation, these initial assumptions and objectives must be constantly reviewed so tradeoffs can be made. Most likely, some functions will be too expensive to implement, at least in the initial installation. Without a solid foundation of functionality vs benefit, the wrong decisions will be made in these tradeoffs.

The first level functional specification should be developed by a small team representing the major organizations that the FMCS will serve. This generally includes manufacturing, design engineering, materials, purchasing, MIS, and manufacturing engineering. Again, it is extremely important that the specification has the full support of this team, because each organization will invest heavily in the following stages of design and implementation.

Next, each major function should be assigned to an individual team member or several team members for detailed specification. The objective is to have sufficient detail to use as a basis for evaluating FMCS solutions in the marketplace. If the specification is taken to too great a level of detail at this stage, it may prematurely lock you into a custom solution. Allow some freedom to accomodate the variations in solutions that you will find from FMCS suppliers.

FMCS ARCHITECTURE

After the functional specification has been developed, the next step is to determine what, IF ANY, restrictions you must specify for the "architecture" of the FMCS solution. This must cover at a minimum the computer platform. Is the solution to be implemented on a particular brand of computer (eg HP,DEC,IBM) and with a particular operating system (eg Unix, MS-DOS, VMS, etc)? Are there specific models that must be used (PC vs AS400, etc)?

These decisions will restrict the potential third party solutions much more than decisions about functionality, so be careful not to put constraints on the architecture unless it is really an issue. And if possible, allow for several choices. For example, if it must be a DEC solution, could you still go with a VMS or Unix operating system?

Some companies will also want to develop architectural requirements for such items as data base, communications, and programming languages. Just remember that you are making huge tradeoffs when you do this. The more specific you are in these areas, the more likely you are going to have a highly customized solution. And the more likely the price tag is going to skyrocket.

EVALUATING SOLUTIONS

Once the functional specifications have been completed and the architecture constraints are understood, it is time to start evaluating solutions that fit your requirements. Again, this may be a step that could benefit from a specialist from a consulting firm who is familiar with the many solutions that are being implemented in your industry. Because the FMCS business is still in its infancy, there is no comprehensive reference material that details all of the potential suppliers for a particular industry. One must rely on the network of FMCS experts that have been exposed to the many solutions offered in various industries.

At this point, you must be prepared to invest some time in thoroughly researching the FMCS prospects. Not only must you understand their technical solutions, you must also understand their delivery capability. Can the vendor support the up-front planning and specification activities? Can they develop custom applications and custom communication drivers? Can they provide training and maintenance? Often, these are the issues that differentiate one vendor from another. And depending on the expertise you have in house, you may only desire a pure product-based solution, or you may want the supplier to do a full "turn-key" system. Most likely, you will be somewhere between the two.

FINAL SELECTION

The final selection of a vendor should be made based on a competitive bid process. You will probably find two or three vendors that have passed your previous screenings. It is money well spent to invest in a detailed bid package that goes to the finalists. It is also advisable to select as your finalists several alternative solutions. For example, you may want to have bids from a product-oriented provider, from a "tool kit" provider, and from a "turn-key" provider. This will help greatly in quantifying the total costs of the project, both internal and external.

CONCLUSIONS

The ideas presented in this paper are not revolutionary, but it is sometimes surprising how often these simple procedures are not followed. And the results often show it. Implementing an FMCS should not be the painful experience it may have been five years ago. There is a better understanding of the functions that are cost effective to support the shop floor operations in various industries. And there are suppliers who are well qualified that can reduce the risk and cost of implementation.

BIOGRAPHY

Mr. Steven Vogel is President of AME Associates, a consulting firm specializing in advanced manufacturing engineering services. Mr. Vogel has extensive experience in planning, designing, and implementing Factory Management and Control Systems for manufactureres such as Ford, Whirlpool, Rockwell, TRW, Chrysler, and others. Before founding AME, Mr. Vogel held executive positions with Cimflex Teknowledge in Pittsburgh and General Electric's Factory Automation Group in Charlottesville, Virginia. He has written many papers on CIM, founded the Central Virginia CASA chapter, and has been a senior member of SME/CASA. Mr. Vogel has lectured internationally on CIM, process planning, and FMCS and has published numerous papers and articles in these fields.

Getting to Know You—Manufacturing and MIS

Cultural differences are the norm between manufacturing and the management information systems department staffs. When they need to be working partners, a look at the world through each other's eyes is needed.

Larry M. Singer
Contributing Editor
Columbus, OH

Manufacturing in the 1990s requires computers, and automation ideas often lead to joint projects with the folks in Management Information Systems, or MIS. Yet too many potentially rewarding endeavors flounder, not because of poor planning, indifferent management, or complex technical issues, but because of the cultural differences between the people in manufacturing and MIS.

Obviously, different areas of the same organization will have different attitudes, but a problem seldom develops until two groups are forced to work together to accomplish a common goal. The most sophisticated technology in the world still needs old-fashioned cooperation between diverse groups. And cooperation starts with understanding.

The differences are more than terminology, education, training and function. There are profound cultural differences in terms of outlook, procedures, attitudes, responses and inclination. If not understood by both sides, such differences can lead to conflict. The project itself may suffer.

Every article and book on manufacturing automation projects invariably demands that the MIS technicians learn the manufacturing side of the business, and that the manufacturing staff become computer-literate. Hardware and software are only the surface of the issue. The articles and books fail to emphasize the importance of understanding the people behind the jobs.

Know thy cohort

How can such problems be avoided? First, by recognizing that manufacturing employees can be and often are different than computer types. Second, by accepting the need for both groups to work toward a common goal. And third, by acknowledging specific differences. It is the responsibility of those in the manufacturing side of the business to understand those in the MIS department, just as the MIS employees have an equal obligation to learn about their counterparts in manufacturing.

MIS has its own philosophy and history. While the old saying that one should know one's enemies is still true, it is more important to know one's friends. The people in MIS are more than computer wizards, just as the manufacturing people are more than magicians who turn raw materials into finished products.

Four MIS characteristics seem to trigger the most misunderstandings with manufacturing people.

- MIS typically follows a general-purpose methodology that may seem frustratingly slow to manufacturing people.
- MIS employees are generally oriented toward other areas of the company, and do not "think like plant people."
- MIS personnel are concerned with matters that seem irrelevant to manufacturing managers.
- MIS sends a battery of specialists rather than one data processing expert who can do it all.

Systems development

MIS creates computer systems in the same way engineers create bridges or highways, and most departments follow standardized steps to create a computer system that meets the needs of the users. In the early days of data processing (DP), programming itself was the novelty, and DP employees rushed into its mysteries before the needs were fully understood.

Forced by embarrassing mistakes, in the 1980s MIS departments began applying consistent rules for creating software. The term "software engineering" became popular. Management realized that programming was only one ingredient of a successful project, and that planning, design, and attention to users were equally important. Creating computerized systems turned into more science than black art. However, these System Development Life Cycles (SDLCs) have both benefits and drawbacks.

Methodologies come in many forms and flavors, and some, quite frankly, are better suited to manufacturing projects. Generally they have some combination of the following steps or phases:

- define the project scope and objectives
- document detailed requirements
- design the hardware and/or software solution
- define the programming specs
- program

- test the individual programs
- test the complete system
- install
- convert to new system.

The most effective methodologies are a series of optional steps that cover all key ingredients in a computer development project. Most projects require only a subset of the steps, and specialized areas such as manufacturing usually need additional activities. But the methodology is the starting point, and manufacturing managers normally see MIS professionals trying to plan the project in terms of phases.

The crux

How can such a logical, professional and impressive operational tool as a methodology cause a conflict between manufacturing and MIS? After all, one would certainly prefer a surgeon who does things "by the book" rather than one who decides to experiment a little with his trusty scalpel.

Manufacturing people may nod approvingly when the MIS approach is initially presented over coffee, doughnuts and goodwill during the first team meeting, but they become highly frustrated when later sequential progression is viewed as too slow.

Formal methodologies take time. To carry out each step in sequential order, MIS prepares documents, requests users to review each one, and plans the next step. Such a progression frustrates the manufacturing line manager waiting anxiously for his new computer software. A strict methodology forces everyone to approve the results of the preceding step or phase before the next one begins. The rationale is that both users and team members share a common understanding of the design. A stubborn adherence to a step-by-step progression often prevents embarrassing mistakes.

But following a strict methodology can be too slow in time-critical situations.

MIS management is usually willing to modify its rules for approval if manufacturing management accepts the possibility that rework may be necessary. The amount of rework—due to misunderstandings—is proportional to the sequential progression of the project. If the project plan calls for deliberate approvals of each step in the cycle, the chance for rework is low. If, however, the team agrees to proceed with one step before all details of the preceding phase have been approved, the risk is higher.

The strictly sequential approach may be safer, but both MIS and company management are justifiably concerned about the time required to implement software under those constraints. As a result, organizations do overlap phases, which allows project teams to move faster by performing steps concurrently. Quality can still be maintained and rework minimized, if all team members remain flexible and cooperative.

This recent development is good news for those involved in manufacturing automation. Automation team members should understand that MIS needs an SDLC as much as engineers need consistent methods to design and construct a bridge. They should also understand the effects on the project schedule. Everything has a price.

Changing tiger stripes

Those in the relatively isolated manufacturing world may forget that even highly competent MIS employees find a plant to be a strange and unique environment. This is usually exacerbated when MIS staffers are thrust onto a project team with manufacturing representatives who talk about using PLCs to track robotic milling machine downtime for a statistical quality control analysis system.

But the problem is more than a lack of knowledge. In the 1990s, most MIS programmers, analysts and managers are products of other business applications. The lack of manufacturing knowledge is usually obvious and generally easily rectified with training and a little patience. What they also lack is a manufacturing orientation.

The typical MIS team member not only knows the vocabulary of accounting, finance, payroll, order entry, distribution, logistics, decision support systems or pricing, but has also absorbed the philosophy and approaches of those areas. While they may not be experts, their experience with other business applications has certainly taught them the nuances or styles of those disciplines.

For example, a good MIS professional learns to think like an accountant when designing financial systems. This may be wonderful for a project involving a new accounts receivable system, but not so great when designing a factory data collection system. Accounting personnel do not think like manufacturing people, and those who work with accountants ultimately absorb some of the financial orientation to information processing.

Multiple faces

Manufacturing personnel must appreciate the mental orientation of their new MIS team member who spent five months working on a payroll system, three months implementing a decision support system and nine months developing an order entry system. In discussions, the staff member may show the precision of a payroll clerk, the overview of a senior executive, and the detail of an order entry manager. The most competent MIS professionals eagerly learn the philosophy of the particular business unit they are working with, and this attitude will definitely carry over into their next project.

While they will eventually adjust to the manufacturing lifestyle, their previous experiences will always show. It is a mistake to expect people accustomed to working with other departments to immediately

follow a true manufacturing orientation. A little understanding and assistance from the plant staff will do wonders.

The big picture

Manufacturing employees are often puzzled when MIS professionals bring up issues that seem to have no direct relationship to the specific project needs. A frequent complaint is, "All we want is a simple data collection system, and the computer guys want to make a big deal out of it."

Are MIS team members looking at extraneous matters? The success of a computerization project may depend on factors not readily apparent to those outside of MIS. This is one area where the experience of data processors can be helpful. Questions that seem unrelated to the specific job may provide answers that will make the completed system a success or a failure. Remember that MIS professionals attempt not only to build a computer system, but a human/machine interface that meets the user needs. MIS departments have universally been criticized for completed software projects that do not help the end users. By asking questions, MIS professionals try to avoid that sad result.

A good systems analyst will always discover how the information is to be used. Will the fabrication manager use the daily defect count to investigate potential problems on the line? If so, he may need date- and time-stamped records showing the location of each defect. These additional data elements may not have been initially considered. Will the engineering supervisor need more than a simple downtime listing for each machine? Is the information requested of value by itself, or does the end user really need other facts to create a complete picture? Collecting information for the sake of information is a clerical exercise—collecting useful information makes the project a success.

Integrating systems

The term "integration" is a rallying cry for MIS managers who have suffered through the pains of connecting incompatible equipment and software. Computers can and do talk to each other, and experienced MIS professionals will look carefully at the prospects of integrating computer systems even if the specific project is only a single PC-based system.

Management will eventually ask that a bar code reader capturing lot numbers be connected to another computer that produces work-order statistics. The trend in factory automation is for computers to share information. Incompatible computer systems will invariably cause everyone headaches. MIS experts will wrinkle their foreheads when the request is for a "simple standalone computer."

Defining real time

Another controversy that puzzles manufacturing people is the endless discussion on real-time versus intermittent reporting. There is a fundamental difference between the need to control industrial processes in a real-time mode and the need to display information real-time. Process control must be real-time, if the functions are truly driven by timing considerations. But what about showing information as it is gathered? Should it be displayed real-time?

Everyone is impressed by the ability to display real-time information. A screen that displays overall line speeds, rate per minute, and reject rates per workstation at the press of a function key is certainly exciting. Such screens dazzle even the most skeptical.

Yet, MIS knows that the price of displaying real-time information is high and its true value can be overestimated. Does anyone really use such information? Will someone be sitting in front of a terminal

checking line speeds, rate per minute and reject rate per workstation? Will such fascinating technology assist the line supervisor in managing his operation?

The alternative is the older but often highly satisfactory interval reporting, often called "batch reporting." Daily shift reports may give the line supervisor the information he needs to take corrective action and monitor his staff. Indeed, many manufacturing projects have delivered exciting real-time displays, only to have the manufacturing users later ask for summary reports to "get the real work done."

Real-time data collection and quality control monitoring are generally more expensive than periodic reporting. The situation and user requirements must determine the needs. It is a mistake to automatically assume that all information must be collected in real time.

The MIS team

In the early, less-complicated days of factory automation projects, the MIS Director would send over a systems analyst and a programmer. In the 1990s, with the growing complexity of manufacturing systems, manufacturing automation projects can require an interdisciplinary team. MIS has become a department of specialists!

The plant manager holding an organizational meeting for his factory automation project may find not one lonely MIS representative, but a systems development manager, project manager, systems analyst, minicomputer programmer, mainframe programmer, data base designer, software engineering supervisor, telecommunications analyst, network maintenance technician and quality assurance analyst.

The project manager serves as the conductor of an orchestra of specialists. Many of the specialists will perform specific tasks or serve as reviewers, but all are necessary in a complex automation project.

The problems of group communication are multiplied because these specialists are assigned part-time rather than full-time to the project. Documentation—and lots of it—is the tool that allows individual contributors to perform their activities, and the more competent MIS professionals consistently document every meeting, discussion and decision. A design for a data collection system will be described in detail. The order-filling priority of an automated storage/retrieval system will be listed in laborious detail.

Armed with a basic understanding of who, why and how MIS professionals approach a plant automation project, manufacturing management can build the foundation of a team committed to success. MIS employees should be viewed as computer experts who can provide valuable assistance to the manufacturing business. But the manufacturing manager and his employees still have to take responsibility for the project. MIS staff will come and go as they are assigned to specific projects, but the manufacturing staff must use whatever the project delivers. Understanding one's team members is the first critical step toward success. MS

A Standards Framework for the Computer-Integrated Enterprise

Albert J. Gibbons

The objective of an enterprise is to provide products and services at a profit. Related objectives include assurance of customer satisfaction and the increase in the value of the enterprise to its owners. All product life cycle activities are concerned with value enhancement.

In pursuit of increased productivity, many organizations have applied computer technology to various operations and activities. Reports of results, however, indicate that though improvements have been achieved in individual, isolated applications, the overall benefits have been limited. It is not sufficient to use computerized tools (software) and systems in a piecemeal fashion. All parts of the enterprise, the people as well as the computerized portions, must operate in harmony to reach the maximum effectiveness level.

Albert J. Gibbons is a principal engineer in the engineering systems department, Product and Process Technology Division, at Westinghouse Electric Corp's Productivity and Quality Center. He is chairman of the Westinghouse Special Interest Group for Sharing and Exchange of Graphics Data and Product Data. His responsibilities include participation in several US and international projects for integrating computerized tools and systems. He is a member of IEEE, the IEEE Computer Society, the Software Engineering and Design Automation Technical Committees, the Design Automation Standards Subcommittee, and the US Technical Advisory Group for ISO Technical Committee 184, Subcommittee 4. He holds a BS degree in mathematics from Carnegie Mellon University.

Configuration of the computer-integrated enterprise (CIE) as a substantial generator and user of information results in a computer system architecture that is hierarchical, distributed, and heterogeneous. According to the US National Research Council, CIM (or CIE) occurs when:

- All of the processing functions and related managerial functions are expressed in the form of data.
- This data is in a form that may be generated, transformed, used, moved, and stored by computer technology.
- This data moves freely between functions in the system throughout the life of the product.

The primary objective is that the enterprise as a whole has the information needed to operate at optimum effectiveness.[1]

Islands of Automation

Computer-aided technology applied to standalone activities such as CAD, numerical control (NC), and robotic work cells is an example of an island of automation. The designation recognizes that each island was developed by a vendor to be independent of any other tools or systems used elsewhere in the enterprise. As originally applied within a substantially paper-based environment, the islands provided significant improvement over previous methods. Acquisition of additional tools and systems continued the trend of separation and independence. Bob Ivey of Westinghouse observes that:

> Vendors of these systems and software make every effort to focus the attention on local capabilities . . . because it is the local features

that distinguish one vendor from another and become the deciding factors in the selection process.

Often, vendors and users initially overlook the need to use the tools and systems other than in an isolated mode. The islands are functionally oriented with little or no means for communication or data transfer with other islands. Moreover, the internal data bases associated with the islands contain "redundant and conflicting data in incompatible formats."[2] Paper output and data reentry—a detriment to efficiency, timeliness, and accuracy—are typically followed by direct translator interfaces, data exchanges through nonproprietary intermediate standard formats, and eventually, data integration.

The Activity Center Concept

Advancement of computer hardware and software technology advancements in the moderate price range now provide support for grouping application tools such as computer-aided engineering (CAE) workstations. It is convenient to consider the enterprise as consisting of several activity centers, engineering workstations, or manufacturing cells. The enterprise should configure each activity center so that it provides effective support of the organization, customer requirements, and products.

The incompatibility assumption

Conventional wisdom assumes that tools and systems selected for various product life cycle activities are inherently incompatible; they were developed by different vendors for somewhat different purposes and they provide somewhat different features with proprietary internal workings. In 1986, the National Research Council indicated that:

> The most serious immediate barrier to the integration of manufacturing data is the incompatibility of CAD data with information needed by CAE and process planning.

Furthermore, because the tools and systems are incompatible, there is a presumed need to use data exchanges (transfers) to interface any tools or systems. These incompatibilities are unnecessary and even detrimental to enterprise effectiveness.

Enterprise data propositions

Data is an asset of the enterprise and should be reusable and shareable—it must be complete, consistent, and accurate. Redundant and conflicting copies of data are undesirable. Product data belongs to the enterprise, not to the tools and systems. Evolutionary changes in product technology and performance, customer requirements, features of tools and systems, and other factors generate an evolutionary need to modify enterprise or activity center configuration of tools and systems. According to Robert E. Young and Richard Mayer:

The most serious and immediate barrier to the integration of manufacturing data is the incompatibility of CAD data with information needed by CAE and process planning

> The technology implementations most likely to make manufacturing more efficient are those which can make major improvements in the effectiveness of information creation and use.[3]

Thomas Smith believes that:

> An integrated design and manufacturing process composed of an ever-changing set of independently developed and rapidly evolving tools, configured in different combinations by different organizations, places severe demands on the design of the data structures and transfer mechanisms used to glue the process together.[4]

Computerized tools and systems must interact with an evolving product model, potentially consisting of multiple representations of the product. Each representation describes the product at a particular level of abstraction and at a particular point in the life cycle and may include data specific to one or more life cycle activities or project participants (i.e., graphical, functional, logical, and physical representations). Dependent representations are those that can be derived completely from independent representations without manual intervention.

Data as hostage

An interactive session using a computerized tool generates or enhances data which is then stored within the data structure of the tool. Extraction of the data for other enterprise use can be very difficult, is often incomplete, and may not preserve design intent.

In the computer-integrated enterprise, however, it is essential for the enterprise data to be independent of the tools and systems. That is, the data describing the product must be external but readily accessible to the tools.

Direct translation for data transfer

Although direct translators have proven satisfactory in a number of limited applications, they are not supportive of the CIE. And in the face of changes in the configuration of the computerized tools and systems or even in the installation of the latest version of the same tool, they can require a large commitment of resources to maintenance and revisions. Direct translators are sometimes used as a first response to an interfacing need.

Data exchange

It is reasonable and appropriate to use data exchange where a complete and consistent data packet can be defined for transfer between an organization and a subcontractor, supplier, project partner, customer, or geographically separate facility. Data exchange is not, however, an enterprise-integrating mechanism—it proliferates pieces of the product data throughout the enterprise. An exchange from one system to another results in four copies, not necessarily identical or consistent, of the data packet:

- System A native form.
- System A outgoing intermediate form.
- System B incoming intermediate form.
- System B native form.

Data exchange propagates multiple copies of pieces of product data redundantly and inconsistently. Such a data packet is a static representation of specific information, a snapshot from one perspective at an instant in time. Many enterprise activities, however, are dynamic, iterative, and interrelated.

Integration, Beyond Data Exchange

There are several difficulties associated with the use of data exchange among tools and systems:

- The definition of the required data (content and form) is a function of the receiving system.
- The definition of the form and content of the data packet to be transferred is a function of the sending system.
- Transfers are incomplete because of differences in the function of sending and receiving systems and in the mappings used between the internal representations of each system and the constructs available with the transfer mechanism.
- The data required by the receiving system has more than one source.

Many enterprise activities are dynamic, iterative, and interrelated

Transfer mechanisms move packages of data of undefined content, meaning, and function from one isolated black box to another for an unspecified purpose—transfers are not based on a determination of the associated data requirements.

The 1984 National Research Council defined integration as:

> Having an information structure supporting a free flow of all information resident in the system to any part of the system as needed. The information stream can continually grow and be enriched while supporting all required functions throughout the organization and during the life of the product.

It is correct that the integrating factor within an enterprise is the information (generation, management, and use), but this definition should be modified to de-emphasize the free flow of information in favor of ready accessibility. A free flow of information could be accomplished through data transfers and exchanges among the computerized tools and systems, but true integration requires a substantially different approach. Data exchange is not the solution to the data integration requirement—it can be the antithesis of design and configuration management.

Requirements for integration

It is essential to integrate information to provide consistency and integrity of the results of design activities and the definitions that drive manufacturing, testing, and other enterprise activities. Without integration, the information is located in several places and represents more than one version.

Achievement of the CIE requires:

- Serious commitment by the organization.
- Supportive data base, physical communication, and logical communication technology.
- Integrated design of products and processes.
- Information models of all enterprise activities.

Propositions for data integration

Achievement of the desired levels of automation and integration depends on the ability to manage the information resource effectively. Such effective management is dependent on the development and application of standards for the environment of the tools and systems to be used. To provide for current integration and evolutionary reconfiguration, an open systems approach to the data is needed.

All enterprise tools and systems must have access to the data required to perform the intended activities. The change must be made to a data-driven environment from a process-driven environment, standardizing the meanings of the data. It is necessary to use a single data, multidiscipline approach rather than a single discipline, multiple data approach. Input and output information requirements for each enterprise activity must be determined:

> Axiom 1—Business survival in the future will require functioning in environments containing heterogeneous data. Axiom 2—Standardizing data appears to be the only feasible solution to the problems that will be thereby posed. Corollary—Enterprise survival is likely to depend on data standardization efforts.[5]

Requirements for Computerized Tools and Systems

Every computerized tool and system must include:

- The definition of functional purpose and result (the transform).
- Satisfactory application software that provides the functions.
- The definition and existence of a satisfactory computer environment (i.e., hardware, operating system, and utilities).

Each activity center must have an appropriate data structure that supports the separation of tools and product data and the sharing of data among the tools. The enterprise needs a hierarchy of data structures for data management and sharing among multiple activity centers over the complete product life cycle. All data elements, characteristics, and relationships associated with all stages of the life cycle must be defined, and all generators, enhancers, and users of data (i.e., tools and systems) must be identified. Data exchange, which preserves the meaning and design intent of the data packet, is necessary for support of physically or organizationally separate facilities.

Standards Requirements

Industry standards are consensus solutions to common problems; they represent substantial effort and technical expertise. Technical excellence and a worldwide perspective are important. John Rankine, past chairman of ANSI, indicates:

> More and more this is a single world in several closely related fields—telecommunications, computer and space technology, and industrial automation. Since the application of technology is increasingly international, the standards must also be international.
>
> Standards accepted internationally facilitate world trade and the application of high technology. They are of great importance to the human race. They can provide the means for more economic use of our shrinking world

resources. By unifying market requirements, standards can provide competitive options to users and consumers.

Needed are the best international standards, no matter what their source—ISO, IEC, national standards bodies such as ANSI or Deutches Institut fur Normung (DIN), regional standards organizations, or any national professional, trade, or technical troup whose work has attained worldwide respect. They are acceptable so long as they are the best solution, so long as they meet world requirements.

A standards framework

A standards framework, or skeletal structure, supports orderly coordination and development of data integration standards and related enabling information technologies. Identification of the types of standards required, with appropriate partitioning and interrelationships, will allow for the efficient and effective application of resources by the various technical committees. The accompanying inset identifies various organizations active in the standards area.

By unifying market requirements, standards can provide competitive options to users and consumers

Identifying requirements

S. Jeane Ford suggests more attention to the information requirements of the enterprise:

> Data base technology must define how information is identified and manipulated within the computing environment . . . before data base technology can be effectively advanced to support CAD/CAM, the requirements must be thoroughly analyzed . . . Additionally, the data environment must be identified with respect to the total data flow, relationship of computerized and manually processed information, optimum user interface requirements, organizational accountability, and validation procedures. The establishment of an overall functional diagram of the design/manufacturing cycle which can be decomposed to the lowest data element level is needed.[6]

In 1984, the Commission of the European Communities established ESPRIT—the European Strategic Program for Research and Development in Information Technologies. Its long term objective is to increase European productivity and competitiveness in the development and application of information technologies, both for CIM vendors and the manufacturing industry in general. The description of R & D area 5.1, Integrated System Architecture, includes:

> A wide variety of CIM systems will be implemented within manufacturing industry. Each system has to serve its specific application in an optimum manner. In order to be competitive and preserve flexibility by taking advantage of new technologies as they become available, effective design methods for CIM must be developed. At present there are no guidelines or international standards available for the design of system architecture for CIM. To enable the development of integrated system structures it is necessary to include the requirements, specification, and possibly the development of data management techniques. The efficient structuring and manipulation of the stored data is critical to both storage needs and execution speed.[7]

According to the work of the ANSI/X3/SPARC Database Architecture Framework Task Group (DAFTG), a reference model is a conceptual framework for use in defining manageable pieces of a larger effort and for illustration of the interrelationships of the pieces. One of the best examples is the ISO reference model for the Open Systems Interconnection (OSI) layered architecture for communication. The OSI reference model is the basis for the Manufacturing Automation Protocol (MAP) effort and has become a "major tool for the study and organization of standards activities relating to interprocess communications."[8]

Related projects

Examples of existing projects involved in portions of necessary development include the Electronics Automation Program of Computer Aided Manufacturing-International Inc (CAM-I); International Organization for Standardization (ISO) Technical Committee 184, Subcommittee 4;

Organizations for Standards

For more information on the organizations mentioned in this article, please contact the following. However, keep in mind that these are only a few of the organizations that are pursuing standards within the CIE.

American National Standards Institute
1430 Broadway
New York NY 10018
Daniel Smith

International Organization for Standardization (ISO)
Case Postale 56
CH-1211 Geneve 20
Switzerland
ISO Technical Committee 184/Subcommittee 4 (External Representation of Product Definition Data)
Secretariat: United States (ANSI/NBS)
Chairman: Bradford M. Smith, National Bureau of Standards, Gaithersburg MD 20899
Convenor, Working Group 1: Jerry A. Weiss, McDonnell Douglas Corporation, St Louis

IGES/PDES Organization
Room A101, Sound Building
National Bureau of Standards
Gaithersburg MD 20899
Chairman: Bradford M. Smith

Product Definition Data Interface (PDDI)
United States Air Force
Materials Laboratory
Air Force Wright Aeronautical Laboratories
Air Force Systems Command
Wright-Patterson Air Force Base OH 45433

Computer Aided Manufacturing-International Inc
Suite 1107
611 Ryan Plaza Drive
Arlington TX 76011

CIM-OSA/AMICE (ESPRIT)
489 Avenue Louise
B14-B1050 Brussels
Belgium

---

the ESPRIT Computer Integrated Manufacturing Open Systems Architecture (CIM-OSA) project; and the US Air Force Product Definition Data Interface (PDDI) projects.

The Electronics Automation Program (EAP) of CAM-I, a joint industry research and development activity, sponsored a 1986–1987 project—Product and Process Data Definition. Its intent was to identify the required contents and preferred form for product and process data needed for automated manufacture of electronic products. The project objective, outlined in the Statement of Work, is:

> To identify the type and content of input data required for, and the output data required from each electronic design, manufacturing, and test work cell.
>
> A great variety of CAD . . . and CAM . . . tools are presently in use in the electronics industry. The desirability of having a common data base which contains all data defining a product has been widely advocated. This data would ideally be nonredundant, easy to access and secure, and would possess configuration integrity. Having such a common data base and agreed standards for data representation and transmission could enable the utilization of alternate CAD or CAM tools to perform equivalent tasks with a minimum amount of perturbation.[9]

In 1984, ISO Technical Committee 184 (Industrial Automation Systems), Subcommittee 4 (External Representation of Product Definition Data), began work on a new draft International Standard, the Standard for Exchange of Product Model Data (STEP). US participation in the ISO activity is provided through the Product Data Exchange Specification (PDES) effort, administered by the National Bureau of Standards. Over 600 companies and organizations are involved in the US and 10 other participating countries.

A stated objective of the PDES/STEP effort is to develop and apply the technology necessary to communicate digital product definitions within a heterogeneous computing system environment. Although one of the deliverables of the project is a data exchange capability, the three-schema development methodology separates identification of the information requirements from the designation of the specific format and the set of constructs to be used for the data exchange. The conceptual models produced for representation of the enterprise information requirements, for a variety of disciplines and products over the product life cycle, form the basis for the CIE.

In a modern enterprise there should be freedom to follow an optimal evolution path, to select the best supplier, and to interact with the environment in an unrestricted way

ESPRIT Project 688, the CIM-OSA, is one of the activities sponsored by the Commission of the European Communities. The project brochure states:

> Manufacturing systems containing "Islands of Automation" are not an adequate answer . . . As long as these islands remain isolated, as long as communication remains burdensome, there is no hope. Progressive integration, with a fully integrated information system covering the whole enterprise, is the ultimate target and the key to the future.
>
> Yet this integrated environment should never become a constraint. In a modern enterprise there should be freedom to follow an optimal evolution path, freedom to select the best supplier, freedom to interact with the environment in an unrestricted way, unhampered by limitations inherited from the past.
>
> . . . CIM-OSA aims at an all-embracing conceptual framework, adaptable to any concrete situation, so as to allow CIM users to evolve in an open and decentralized way; suppliers to aim for a well-defined target in developing new products; systems houses to establish migration paths for existing applications toward more and more integrated systems.[10]

The CIM-OSA project will develop a reference architecture consisting of three generic models:

- A manufacturing enterprise reference model.
- An information reference model.
- A CIM implementation reference model.

This architecture will serve as input to standardization activities, to vendor projects for new CIM tools and systems, and to enterprises to derive individual reference models.

Each model will be used by the enterprise to guide implementation of its desired CIM system, specifically suited to the individual enterprise products and business rules. The vendors will apply the CIM-OSA generic models and the appropriate CIM-OSA integration standards for development of the computerized tools and systems available for selection by the enterprise.

Several projects, conducted under contract to the US Air Force, carry the designation or are related to PDDI. The PDDI projects study data requirements for replacement of engineering drawings by digital data sets.

The Prescription for Success

It is necessary that a consensus be reached on the need for data integration. Achievement of the desired result will require effective application of available technical expertise and related resources through the following: definition of the standards framework, coordination and partitioning of the standards development effort, and interaction with the vendor community to ensure that the evolving data structure definitions are satisfactory. Vendors and users must support an evolutionary path from independent and isolated tools and systems (each with a proprietary internal data structure which limits access to enterprise information) to tools and systems interacting with external enterprise and activity center data structures, using query and update transactions.

Many of the steps currently needed to update all enterprise data to reflect design changes or support the iterative design stages are cumbersome and costly. Integration of enterprise data will provide a consistent representation of the product during the entire life cycle and will support controlled revisions to the design.

Concluding Comments

If it is believed that incompatibility of computerized tools and systems is inevitable and that data exchange is the solution to the data integration requirement, CIM will fail. However, if managers resolve to identify and subdue the appropriate problem, US manufacturing will gain immeasurably. According to Edward de Bono:

> There are two opposite ways of improving a process. The first is to try to improve it directly. The second is to recognize, and then remove, those influences that inhibit the process. . . . It is not possible to dig a hole in a different place by digging the same hole deeper. . . . It is not possible to look in a different direction by looking harder in the same direction.[11]

The justification of an enterprise's investment is based on the projection of a favorable cost to benefits ratio and return on investment and on an assessment of business risk factors. Measurement criteria include the rate of return on safe investment instruments, such as the money market. A decision to commit substantial resources to new products, technology, facilities, or organizational structure must be financially and technically sound. It is essential that the sound technical foundation for the CIE be established—the standards framework, the enabling technologies, and the appropriate development of industry standards.

The enterprise, as computer representions, must learn how to understand its products and associated processes, including the portions of the enterprise outside the engineering and manufacturing realm. Development of industry standards and related enabling technologies must have a suitable framework or structure for partitioning and coordinating the overall problem.

For world class industry there must be world class standards and the associated automation and integration architecture. Cooperative optimal application of technical expertise and other resources is essential. ▲

Notes

1. Commission on Engineering and Technical Systems, *Computer Integration of Engineering Design and Production: A National Opportunity* (National Academy Press: Washington DC, 1984).
2. R.B. Kurtz, *Toward a New Era in U.S. Manufacturing: The Need for a National Vision* (Washington DC: National Academy Press, 1986).
3. R.E. Young, and R. Mayer, "The Information Dilemma: To Conceptualize Manufacturing as Information Process," *IE* (September 1984).
4. T.R. Smith, "A Data Architecture for an Uncertain Design and Manufacturing Environment," *22nd Design Automation Conference Proceedings*, IEEE Computer Society (Las Vegas NV, 1985).
5. C.W. Klomp, presentation to the PDDI Executive Review (September 1985).
6. S.J. Ford, "CIM's Bridge from CADD to CAM—Data Management Requirements for Manufacturing Engineering, IPAD II, Advances in Distributed Data Base Management for CAD/CAM," *IPAD II Proceedings*, The National Aeronautics and Space Administration, the US Naval Material Command, and the Industry Technical Advisory Board (Denver CO, 1984).
7. "Information and Notices, Draft Council Decision adopting the 1984 work programme for the European strategic programme for research and development in information technologies (ESPRIT), Subprogramme 5, Computer Integrated Manufacture," *Official Journal of the European Communities* C 47, vol 27 (February 20, 1984).
8. Database Architecture Framework Task Group (DAFTG) of the ANSI/X3/SPARC Database System Study Group, *Reference Model for DBMS Standardization*, ed D.K. Jefferson and E.N. Fong, National Bureau of Standards NBSIR-85/3173 (May 1985).
9. Computer Aided Manufacturing-Inc, "Data Definition—Identification of Required Contents and Preferred Format of Product and Process Data Needed for Automated Manufacture of Electronic Products" (Statement of Work for the Electronics Automation Program, 1986).
10. ESPRIT, *ESPRIT Project 688, CIM-OSA/AMICE* (Brochure, Brussels, Belgium).
11. E. de Bono, *New Think* (Avon Books: New York, 1971).

ARE YOU READY FOR EDI?

BY LESLIE C. JASANY, SENIOR EDITOR

> THE LATEST ACRONYM TO GAIN INDUSTRY ATTENTION PROMISES TO DO MORE THAN JUST ELIMINATE PAPER BETWEEN AND WITHIN COMPANIES. IT WILL FUNDAMENTALLY CHANGE THE WAY YOU DO BUSINESS, . . . INCLUDING MANUFACTURING.

Future Freight's computer sends a freight bill to the computer at New Customer. New Customer's computer checks the information fields on the bill for correct data and finds an error. It sends the bill back to Future Freight's computer along with information on the error. Future Freight's computer checks that information, finds the error, corrects it, and electronically resends the bill to New Customer's computer. It accepts the bill this time and passes it to the accounts-payable computer, which checks it, and then issues instructions for Very Rich Bank to transfer payment to Future Freight. It has taken only a few minutes for these transactions. And no human has seen the bill, keypunched data from that bill into a computer, or in any other way intervened in handling that bill. It was all done electronically.

Computers automatically handling business transactions between companies without people intervening—the visions of futurists are not that far from reality. LTV Steel in Cleveland, for example, is working on a project close to this scenario. "We are testing right now, with some of our freight carriers, a system that automatically generates an EDI transaction set [(a message)] that tells the freight carrier what's wrong with a freight bill as we see it," says Philip Kozsey, EDI Client Services, Information Systems Group, LTV Steel Co. "We're looking to get to a complete automation of the freight-bill activity. If there are problems [with data accuracy] we want the system to generate the error notification to the carrier [and reduce manual involvement]." Of the 71% of the EDI-transmitted freight bills LTV receives, 80% of these are automatically paid through the EDI system.

The technology that makes the futuristic scenario a future probability is called Electronic Data Interchange (EDI). It is formally defined as the computer-to-computer electronic interchange of structured business documents between trading partners. It promises to eliminate data errors, because data are not manually reentered into a computer; it promises to reduce paper processing costs, because data are transmitted electronically to computers rather than mailed and processed manually; and it promises to change the way a company does business, partly because it can eliminate paper transactions.

The number of EDI implementations is rapidly growing. In 1990, this market, which includes networks, software, and professional services, is projected to grow to $1 billion, according to Input Research & Analysts Group. The EDI software market alone is expected to grow at an annual average rate of 44% to $230 million by 1993. As many as 75% of the Fortune 100 and 39% of the Fortune 500 companies have implemented EDI systems. The Yankee Group expects the high growth rates to continue through 1991 when prices will begin to drop as the number of new users declines. Industry analysts expect that almost all companies will eventually implement EDI.

LTV Steel began its implementation in the early 1980s, even though EDI has been available for some time. Texas Instruments has been working with it since the late 60s and early 70s. But it has taken many years for enough companies to have EDI for it to become a more prevalent method of transacting business. In fact, it has only been within the last six to eight months that activity has jumped. "A lot of implementations have gotten off the ground and they're rolling," says Kozsey. "It might be a year or two before you get that critical mass where there's no stopping EDI any more."

EDI was initially implemented to reduce and speed up the processing of invoices, purchase orders, freight bills, and so on between companies. In its present form, EDI does not directly affect those machine-dependent processes that add value to material; that change or shape material into something more than what it was. Most users though, foresee, in a not too distant future, that EDI will be used to send machining instructions directly into a machine

Transmitting more than paper documents

"The EDI that exists today is the first generation of exchanging data," says Roger Willis, partner and product executive for Aerospace & Defense Software Products, Andersen Consulting. "When it was conceived, . . . the only electronic data people were familiar with were orders, pieces of paper. Today, we're talking about much more complex data—all the CAD type information before it's committed to paper."

EDI cannot be used to transmit graphic data yet, primarily because it would take days in some cases to complete the transmission. But that is an area several companies are looking into.

The Department of Defense is working on establishing and promoting implementation of Computer-aided Acquisition and Logistics Support (CALS). CALS address the generation, access, management, maintenance, distribution, and use of technical data in digital form in the design, manufacture, and support primarily of weapon systems, ships, and equipment. But this standard can be used for any graphic data. Several companies from the private sector have joined the DoD in promoting acceptance of CALS. Six parts of CALS covering technical data, CAD and graphic applications, technical journals, and engineering drawings and text in technical manuals have been released. For more information on CALS, write to Mr. David Bettwy, National Institute of Standards & Technology, Room B 146, Building 233, Gaithersburg, MD 20899.

control.

Presently, EDI impacts those departments that interface with manufacturing, particularly purchasing, order entry, and accounting. But even now, EDI can be more than just automated paper handling. According to a report from the Yankee Group, user understanding of the broader definition of EDI is increasing. Users are realizing substantial benefits when EDI is integrated with a company's business applications; when transmitted data go directly into the business computers and software that need them, without people keying or rekeying in the data.

"We're already seeing a trend toward people wanting to look out beyond just that business transaction area. . . ," says Paul Iverson, manager, corporate manufacturing, EDI Project Office, International Business Machines Corp. "We approached it as a total process rather than a piece of it. When we adopted our manufacturing strategy to implement EDI . . . we defined it in the broad sense, that EDI is electronic communication. And whether it's a free-form note, a formatted business transaction, a schedule, the transfer of a complicated CAD-type file, or electronic funds transfer of payment, all of this is under the definition of EDI in our strategy. . . ."

Says Alex Beavers, partner in charge of Operations for the Center for Manufacturing Technology, Coopers & Lybrand, "EDI really should be part of an enterprise strategy."

Digital Equipment Corp. and LTV Steel are just two of the companies that have implemented EDI as part of an enterprise strategy. "On our first pilot in our Augusta, Maine plant," says Curt Anderson, EDI marketing manager, Digital Equipment Corp., "EDI was put in as part of our manufacturing control software, part of our MRP environment. As a result, with one trading partner we were able to reduce the purchasing order cycle from 3 weeks to 3 days, . . . reduce the paper handling and processing time of an order from between 5 to 10 hours a week to 30 minutes a week, . . . and reduce our inventory from $800,000 to about $43,000."

Digital also reduced costs in what it calls their manufacturing acquisition costs—the overall costs associated with acquiring goods and services that go into the manufacturing process. "In our internal program," continues Anderson, "our costs were running at about 8%. The company looked at EDI as a vehicle to reduce these costs just one percentage point. Now one percentage point for Digital on a $2 billion business represented about $20 million of potential savings using EDI. So far, the program is working."

"If implemented correctly, there's some real paybacks [with EDI]," says Anderson. "In our approach to EDI, it isn't a separate technology such as, 'This is my MRP system. This is my payroll system. And this is my EDI system.' Rather, EDI is looked at as an integrated component. . . . It's a component in an MRP system, a component in a purchasing system, and a component in a logistics and distribution system."

"[For LTV Steel] EDI is a strategic application of computer and telecommunications technology," says Kozsey. "This is how we expect to be doing business for the foreseeable future. I don't see anything, even in the far distant horizon, that's going to replace EDI."

Communicating with distant processors. Some of LTV's manufacturing is done by processors located throughout the U.S. Says Kozsey, "We have an EDI initiative underway with our outside processors and warehouses. We outsource some of our production requirements to processors who cut the steel, slit it into smaller coils, coat it, pickle it, or add value to our product. Our processing services department . . . deals with these processors as if they were extensions of our production line. We've targeted 85 of them to participate in this program with us."

A lot of communication must go back and forth between LTV corporate, its processors, and its mills to keep track of the steel. For example, each time the processors receive a shipment of steel, begin some activity on that steel, and finish that activity, they notify LTV corporate offices. All of this information is taken by corporate and passed onto the mills so that they know at what process stage the steel is in, what to expect, and when to expect it to schedule their processes. The only way to get all this information to flow fast enough is through EDI.

"The direction for this EDI initiative is to manage and track inventory. That's the business reason for doing this," says Kozsey. "Of the processors we've got trained, they cover 91% of the business we have out there. We're up to 52% [of those processors] that have passed the 'good data rating' [and have implemented EDI]. So, of our inventory that is out in processor sites, 52% of it is being managed by the EDI system. Over one-half of our inventory. We want to get it all that way."

True JIT. Because EDI increases the speed at which supplies and orders are processed, material can be ordered weekly, even daily, making Just-In-Time (JIT) manufacturing a reality.

In fact, some people say that it is because of JIT manufacturing that EDI has emerged. "The single most important fact that's pushing EDI to the forefront is JIT inventory. There's no way to do JIT without EDI," says Kozsey.

Paul DiBono, vice president, marketing, American Software, agrees, "Everybody got excited about JIT because they were going to reduce inventories, but nobody looked at what kind of demands JIT put on the supporting systems.

"What you're really trying to do with JIT is minimize safety stock. . . . A significant piece of the lead time component [in inventory control] is made up of how long it takes me to receive an order, because I have to plan on having that many days of inventory as a buffer."

As DiBono points out, it does not do any good to order on a daily basis if it takes several days or weeks to send and process an order. Thus, JIT is making those departments that interface with manufacturing increase their productivity. "Now," continues DiBono, " [companies] are beginning to recognize that they have to go back into their purchase-order systems and reevaluate those systems, their order-entry systems, their account-receivables systems, . . . because with even weekly orders, they have four times the volume for the same revenue. They have four times more orders to process, four times more orders to pick, four times more orders going into the manufacturing plant. This is placing a tremendous burden on all the supporting systems. . . ."

Iverson points out that timely information is vital to the manufacturing process. A speedy, flexible manufacturing process should not be burdened with "an antiquated, dated method of data movement."

EDI reduces that data-handling burden and speeds the flow of information. In fact, LTV uses EDI to schedule material shipment to a specific hour. Several of its processors do not have storage facilities. "When coil is brought in," says Kozsey, "it's brought in right into the [manufacturing] line."

Changing the competitive playing field. As the majority of companies achieve quality—that is, 100% first-pass yield on their products—competition will shift from providing a quality product to providing quality service. "When you get into a commodity like steel," says Kozsey, "where quality is a given, . . . and cost is pretty much a given, how do you distinguish yourself from your competitors? What's left? The level and quality of service."

Adds Beavers, "If you define quality to mean that you deliver to a customer something that makes him happy, then EDI can play a significant role. Many times you order something and what's delivered isn't what you ordered. This is a source of frustration, expense, cost, and delay. To the extent that EDI can insure the accuracy of orders, . . . then that will be a contributor to quality *and* reduced costs."

Achieving this level of quality service often means a company must fundamentally change its relationships with its suppliers. Says Anderson, "We realized that we were really extending the enterprise; we were starting to build a relationship with our supplier as another part of the company, actually extending the four walls of the company to include our suppliers. We've gotten to the point where we're sending advanced forecasting data so that the supplier could better plan his manufacturing to support our needs."

As these new relationships are forged, the competitive playing field will change. The competitive advantage will go to those companies that can use EDI better, faster, more accurately, and for less cost than competitors. Adds Kozsey, "Even though the rules of the game are being standardized [by everyone eventually using EDI], it's going to be who can play the game the best." ○

Getting started

EDI basically requires computers, software (referred to as translation software) that translates your formatted data into the formats of your trading partners, and a network between trading partners.

Many EDI implementations are made of custom software. But with more companies working to implement EDI, standards are being formed. ANSI ASC X12 is the accredited standards committee for EDI. Their charter is to develop the formats, or transaction sets, to get computers "talking" to each other. This group also houses the North American EDIFACT board for international standards. Several industries are working with the X12 group to standardize their formats. For example, the electrical industry is working on EDIX, and the automotive industry is standardizing on ODETTE.

The networks most companies use are third-party networks, called Value-Added Networks (VANs). In a VAN, each trading partner has a "mailbox" into which data are entered. Periodically partners dial into their mailbox to retrieve the data. Major third-party VANs include General Electric Information Services (GEIS), McDonnell Douglas Tymnet, and Supply Technology Infonet. IBM, CompuServe, and Control Data also offer VANs.

There are also over 100 suppliers of EDI software. Many systems integrators and consultants are offering software as well as their consulting services. Included in this list are:

ACS Network Systems *CIRCLE 430*
Andersen Consulting *CIRCLE 431*
American Software Inc. *CIRCLE 432*
Coopers & Lybrand *CIRCLE 433*
Digital Equipment Corp. *CIRCLE 434*
Future Three Software Inc. *CIRCLE 435*
GE Information Services *CIRCLE 436*
International Business Machines Corp. *CIRCLE 437*
McDonnell Douglas *CIRCLE 438*

Reprinted from *Computer-Aided Engineering*, February 1990. Copyright 1990, by Penton Publishing Inc., Cleveland, Ohio.

PDES Shapes Data Exchange Technology

An emerging standard will make it easy to transfer engineering data between different applications and computers.

BY BARBARA WARTHEN

The Product Data Exchange Specification, or PDES, does not yet exist as a standard. But a volunteer organization comprised of vendors, users, consultants, educators, and government representatives are hard at work to make PDES a reality. The first version of PDES may become a stable draft standard by late 1990 and an international standard by 1991. Vendors will soon be writing translator products, and perhaps as early as 1992, users will be utilizing PDES for true, intelligent data exchange.

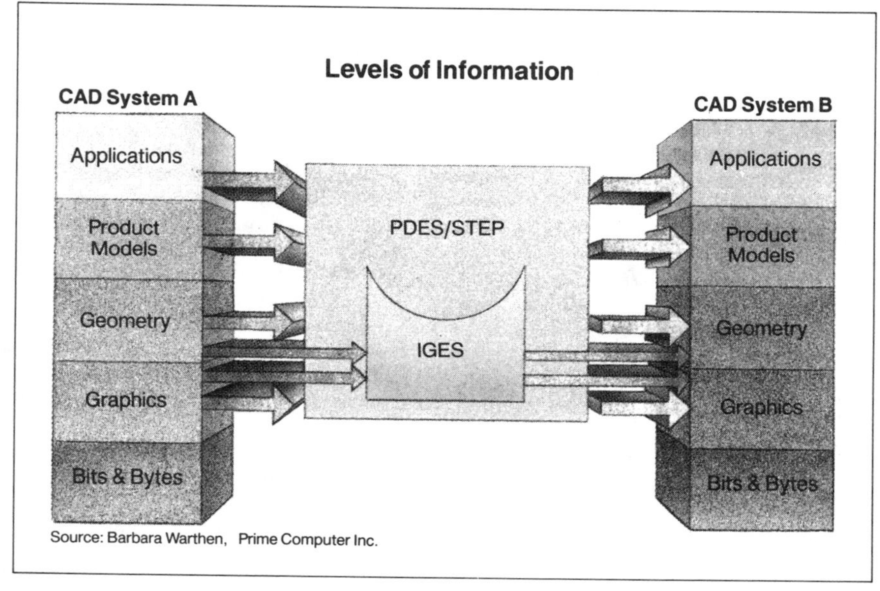

Source: Barbara Warthen, Prime Computer Inc.

A Technology Driver

Known informally as STEP in international standard circles, PDES is an outgrowth of IGES, a 10-year-old industry standard used for basic graphic and geometric data exchange. PDES will go beyond IGES, providing powerful data exchange capabilities not available now. With PDES, for example, users will be able to exchange intelligent objects—items such as steel beams or bolts with all their associated attributes (size, material, and properties)—not just their single line or solid representations.

The goals of PDES are multiple. This standard will provide a foundation for static and dynamic data exchange as well as for distributed databases and knowledge-based systems. It addresses exchange capabilities not just for basic graphics and geometry but also for user-level information in three major CAD/CAM areas—mechanical engineering, electrical/electronics, and AEC.

These areas contain a number of applications with enormous amounts of data that must be defined. For example, mechanical engineering and manufacturing include tolerancing, form-feature modeling, finite-element analysis, product structure, and NC. AEC includes plant design (encompassing conceptual design, final design, construction models, and operations models), building design, facilities management, road and bridge construction, and shipbuilding. Electrical/electronics entails printed wiring boards, integrated circuits, and hybrid circuits.

Developers are taking a top-down approach in working on these areas, defining information they require, not just information that currently exists in their CAD systems. Hence, PDES is proactive and will drive technology, not just react to it, like most standards. PDES is aiming at systems of the future, and in the process is developing

Whereas IGES focuses on graphics and geometric data exchange, PDES adds capabilities for product model and application data exchange.

technological tools to do so. The PDES effort involves a significant amount of true research, which is taking place worldwide.

The Global Picture

The first version of PDES is expected to include capabilities for basic geometory, topology, form features, tolerances, layered electrical products, ship structures, and more. Future versions will include any postponed Version 1 capabilities; AEC distribution systems such as piping, HVAC, electrical, and raceway (cable tray); and additional functions.

As developers see it, PDES will consist of a dynamic set of objects called the Generic Product Data Model (GPDM). This set will contain the objects mechanical engineers, NC experts, geometric modelers, shipbuilders, architects, plant design engineers, and others define as necessary for their applications. By putting objects into GPDM, entities can be shared. Integration—the recognition of similarities and the linking together of similar objects and concepts—is a key part of PDES. A major goal is to develop a minimal set of entities to satisfy all applications.

GPDM will serve as the core of PDES. But just as users would not directly utilize an entire database for one transaction, they will not exchange this entire global pool of information. Rather, application protocols (APs) will allow users to "grab" the specific set of objects they require from the GPDM for a particular application or project. APs now under development include ones for engineering drawings, 2D building design, 3D piping, and ship structures. Overall, APs define user subsets along with GPDM user requirements and the mappings to GPDM objects. Supporting documentation and test cases are also part of each AP.

Developers are utilizing a three-layer approach in building PDES. At the top level, user experts define conceptual models for their universes of information. At the second level, information models from the various universes are integrated. Finally, level-two information is converted into a physical file format.

IGES to PDES—A Parallel Challenge

Since 1980, a committed, international, volunteer consortium comprised of many resources has been working on IGES; in 1985, this group initiated work on PDES. Today, in the U.S. alone, there are nearly 1,000 professionals involved in IGES/PDES efforts; approximately 250 are active members who meet quarterly. Input from participants contributes to the broad database since business, industry, universities, and government agencies are represented.

In 1988, PDES Inc., an adjunct organization to the PDES development group, was established by several U.S. companies that wanted to get involved with PDES software development and testing. Today, there are over 20 members, including companies such as Prime Computer Inc., Boeing, General Electric, Digital Equipment Corp., Hewlett-Packard, McDonnell Douglas, and General Motors.

Positive results from PDES implementation projects at McDonnell Douglas, Allied Signal, and SDRC are encouraging, especially since the standards community anticipates that PDES/STEP will be an international standard by 1991. The National Institute of Standards and Technology (NIST—formerly the National Bureau of Standards) also has a testbed facility which performs and monitors various functions, identifies problems, and ensures compliances.

A Multi-Level Methodology

Developers are utilizing a three-layer approach in building PDES. At the top level, user experts define their universes of information in one of two information modeling languages. These structured graphic languages let experts clearly define information that can be passed on to the second group of experts.

At this second level, the information models from the various universes are integrated. Commonalities are identified, duplications are removed, and links are established. Integration leads to a minimal set of information in the global pool.

For example, one type of integration may involve a bolt used in different applications. In a chemical plant application, only the bolt's ID and location are important. In a manufacturing application, however, more data on the bolt may be required such as specific shape, materials, and manufacturing specifications. At this second level, the two

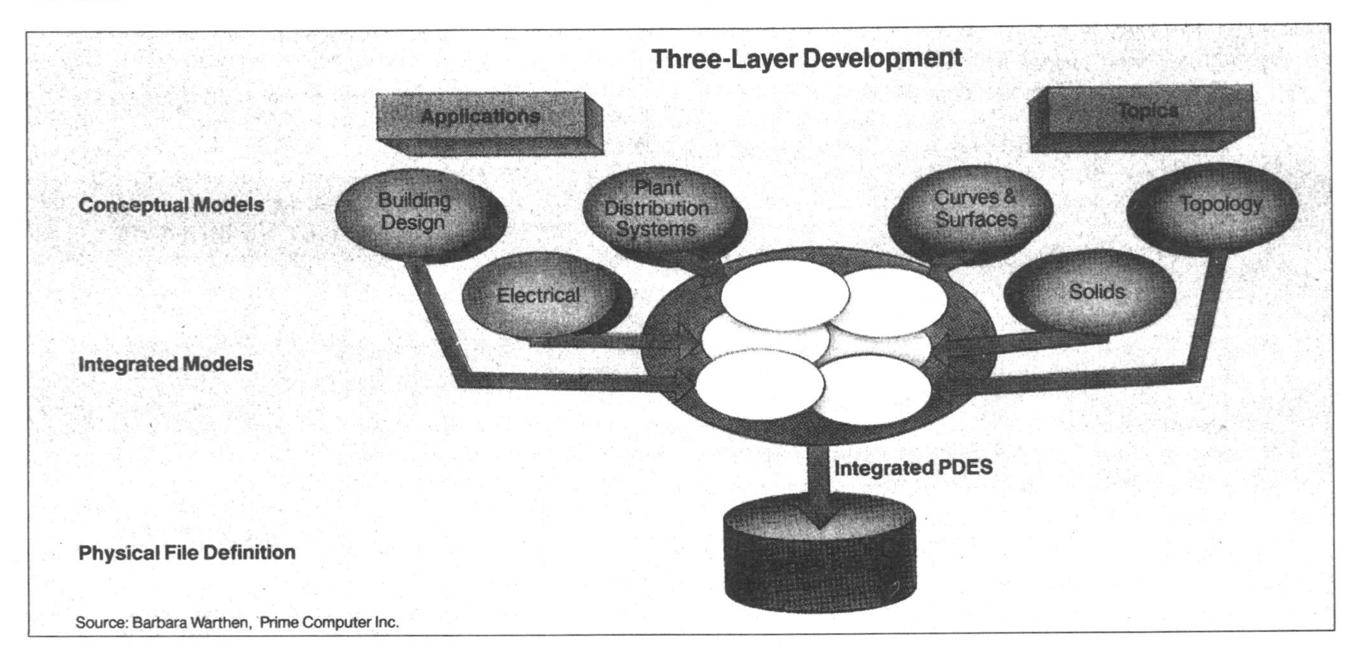

Source: Barbara Warthen, Prime Computer Inc.

views of the bolt must be merged into one. In the real world, the bolt is one bolt, whether manufactured or placed in a building.

Another type of integration recognizes common concepts. For example, the elevator of a building, important to an architect, is similar in concept to the passageway that conntects two layers of a printed-circuit board. These two share some characteristics and so will be integrated.

The third layer is the one that allows PDES implementations. It is the conversion of the second-level information into a physical file format. In this multi-stage process, the user view of real-world objects will be tranformed into an integrated computer representation for implementation in any computer.

Not a CAD System

PDES will not be a CAD System, although PDES *may* provide the structure for a CAD system's database. PDES defines the underlying information required for various applications as well as for displaying drawings and models. It will not, however, contain the additional information that makes each CAD system unique.

This up-and-coming standard will provide underlying data. Vendors using PDES in their core databases will continue to provide product pieces, such as the user interfaces. Vendors also will continue to write applications to utilize the database information, whether it is structured in PDES or another format. Such applications allow users to create and link pipes, check for interferences, run structural analyses, and generate reports showing the results.

Targeting the Future

An established standard such as IGES is nearly a decade old in its underlying concepts. It cannot, therefore, always take advantage of current technologies. Efforts such as the Department of Defense's Computer-Aided Acquisition Logistics Support (CALS) program and the European Commission's plans for European economic and trade harmonization in 1992 require more powerful standards.

In looking at the capabilities of each, it is easy to see how IGES and PDES compare:

- **Objectives:** IGES has been designed for simple graphic and geometric data exchange. PDES not only handles these but adds capabilities for full product and project data exchange and databases.
- **Basis:** IGES is entity-based while PDES is based both on intelligent applications and topics.
- **Approach:** Developers have used a bottom-up, or entity-by-entity approach, in designing IGES. PDES, on the other hand, is built from the top down, using a three-layer approach—from applications to the data definitions.
- **Target:** IGES reacts to existing technology and systems. PDES is proactive, geared to future technology and systems.
- **Goals:** IGES was developed to exchange basic graphics and data between different systems. PDES will do this and more. The PDES group has defined four levels of data exchange.

The first level deals with the transfer of product data from one application to another in a batch process, similar to how it is done in IGES. The second level allows data to be accessed from a local work form model using common subroutines. The third level supports the organization of information across distributed databases. Finally, the fourth level taps knowledge-based technology, allowing product data and methods operating on that data to be stored together as an object, perhaps in an object-oriented database.

- **Status**: IGES is a U.S. ANSI standard. PDES will be an ISO standard (IS 10303).

PDES is forward-looking, a state-of-the-art exchange and database standard to serve today's users and tomorrow's computer systems. The PDES effort is moving as rapidly as it can, considering its ambitious international technical and political goals. ■

Barbara Warthen is chair of the AEC Committee of IGES/PDES/STEP and is a Principal Analyst for Prime Computer, San Diego, CA.

Look to the 1990's...

1989 in the PDES history book was a landmark year because the first round of voting for PDES/STEP Version 1 Draft Proposal took place. This document reflected the coordinated efforts of the various PDES/STEP committees. Fourteen nations are ISO members, but only 9 participated in the voting (8 disapproved, 1 abstained). None of the developers expected acceptance of the Proposal; however, the more than 1,600 comments which accompanied the ballots provided valuable technical input for PDES developers to take back to their drawing boards.

Since then, the committees have rewritten the document, addressing such issues as functionality, status, cross-referencing, and consistency as they relate to the PDES/STEP model. These issues, in turn, have been re-organized to include Overview, Express (PDES/STEP language), Physical File, Conformance Testing Framework, Shape, Presentation, Ship Structures, and at least one Application Protocol. These proposal considerations will probably be submitted to ISO members for their votes in two separate sessions during 1990. The two sets of ballots are likely to lead to a Draft International Standard (DIS) by the end of 1990 and, by 1991, an International Standard.

Though PDES/STEP is expected to eventually replace IGES, it won't happen overnight. Developing and refining standards are definitely complex and time-consuming. Version 1 of PDES is still being reviewed by ISO. So, even considering a problem-free schedule of events, the first PDES translators will probably not be available commercially until late 1991.

There are several things users can do while waiting for PDES to become a reality. If their requirements are limited, IGES may be adequate. Or, they can develop more "intelligent" direct translators or neutral file format translators. Although these latter efforts can be costly, they solve immediate data exchange problems until PDES products are available.

Users may also consider becoming active participants in the PDES/STEP volunteer group. Getting involved will not only keep users up to date on standards information but also provide an opportunity to contribute to standard development in their particular application areas. To find out when the next PDES meeting will be held, contact Barbara Warthen at Prime Computer Inc., 9805 Scranton Rd., San Diego, CA 92121 or call (619) 587-3000.

CAD/CAM Data Exchange

Philip Smith

The use of CAD/CAM is widespread, and its benefits are now being achieved by many organizations. Most of these benefits are in-house improvements and as yet do not significantly affect customer and supplier relationships. This article discusses the benefits of CAD/CAM in relation to productivity gains, improvements in quality, reduced lead times, and improved presentation to customers.

Philip Smith is a manager at CADDETC, Leeds, UK.

There is usually a significant volume of drawing traffic between an organization and its customers and suppliers. Few organizations make no use of purchased components, and equally few have no need to produce engineering drawings for the customer. Often an organization adds its own part number to a supplier's drawing or reworks the drawing to include in its own assemblies. Organizations near the end of the manufacturing chain may have very few parts of their own in the final product and spend a large part of their time copying suppliers' drawings. These drawings are likely to be in paper form, even when both parties have CAD/CAM equipment installed. This is one of the main areas in which product data exchange can improve productivity. For example, if company A with CAD/CAM system X can supply a part description of its product in a form that company B with CAD/CAM system Y can understand, the savings to both parties could be considerable.

The second main area of interest in product data exchange standards is that of archiving. There is no guarantee that files written using 1980s technology will be readable in 20 or 30 years. Even though CAD/CAM vendors now promise future compatibility, a change in the technology could occur. A positive vendor-user relationship could also change, and a different, noncompatible system could be used in the future. If, however, the user has the capability to store drawings in a system-independent format, these possible concerns are negated or at least lessened.

A system-independent format also gives users a larger number of systems to select from. There is often a compromise between the best system for mechanical design and drafting and the best system for manufacture. A neutral-format data base allows a more open system approach.

Such a format would also allow users to transfer information from a CAD system to other functional areas within the organization—for example, a bill of materials (BOM) to a manufacturing resource planning (MRP) system.

Data Exchange Standards and Specifications

Although various technical and commercial reasons for the use of product data exchange in computer format exist, there are many problems of compatibility. Each of the 300 or more CAD/CAM systems available has its own data base structure and representation for the different entities that it handles. For example, the data base of one system cannot handle data from another.

The two main solutions to this incompatibility problem are the use of a direct translator and a neutral format.

Direct translators. A direct translator is software that reads a specific CAD/CAM system data base format and converts it to another specific CAD/CAM system data base format. Direct translators have the advantage of being fast; they need to deal only with the entities that the two systems have in common. One disadvantage is that for every other system with which an exchange is needed, two programs are required. This is no problem if exchange with only one other system is desired, but the software proliferates when exchange with many other systems is required (see Exhibit 1). For example, for n systems, $n(n-1)$ translators are required, with an additional $2n$ for each additional system. Updating can become very complex; each pair of translators may need to be updated for each new release of CAD/CAM system software. Translators also require access to the data base formats of each CAD/CAM system, which are certainly not widely available and are likely to be proprietary.

The neutral format. A neutral format specifies a data structure that is defined for the public domain and attempts to be independent of all CAD/CAM internal formats but capable of carrying all of the information that might be contained in any format. Each system requires only two programs: a postprocessor that converts from the neutral format to the system's own internal data base and a preprocessor that converts from the system's own data base into the neutral format. Therefore, for n systems, only $2n$ programs are required, with 2 additional programs needed for each additional system. Vendors are responsible for updating and converting their own data formats to neutral formats.

Current standards and specifications

IGES. Version 1.0 of the Initial Graphics Exchange Specification (IGES) was introduced in 1980. Version 4.0 has entities ranging from simple geometry (e.g., lines, arcs, points) to constructive solid geometry and can handle such applications as finite-element data. It is directed primarily at the mechanical design market but has some coverage of electrical entities. Most of the major mechanical CAD/CAM vendors have at least some implementation of IGES, and it is the most widely accepted data exchange format.

Exhibit 1. ***The Direct Translator Solution***

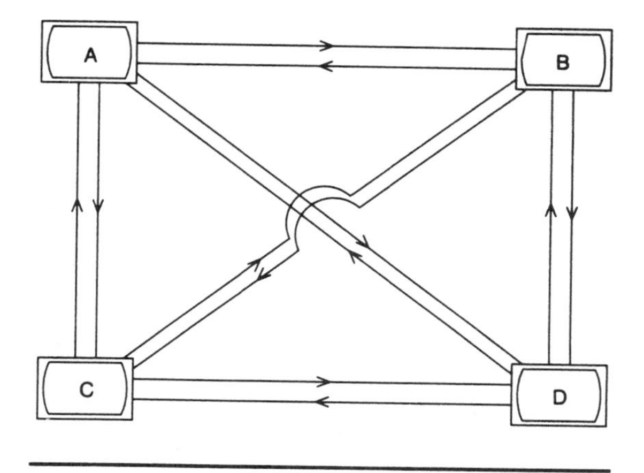

VDA-FS. VDA-FS (Verband des Automobilindustrie FlachenSchnittstelle) is a standard developed by the German Car Manufacturers Association that covers points and parametric polynomial curves and patches. It has been designed to keep the interface simple and handles essential elements only for surface transfer. It is intended to supplement IGES in an area that some feel IGES falls short in. It is in use both in Germany (as German national standard DIN 66301) and in the UK, mainly in the automotive industry. Most of the vendors involved in that industry have a processor available.

SET. SET (Standard d'Echange et de Transfer) was developed as a more compact data exchange form than IGES and has facilities for transferring geometry (including rational polynomial surfaces, annotation, and structure), but it is not more functionally complete than IGES. It has become French national standard z68300, and future versions will include finite elements, boundary representations, constructive solid geometry, and numerical control (NC) toolpaths. It is especially popular in the aerospace industry, which is using it in collaborative work on the European Airbus projects.

EDIF. The Electronic Design Interchange Format (EDIF) was initially conceived to transfer gate array and standard cell design approaches; it currently handles integrated circuit transfers, and future versions will have printed circuit board facilities. It has a LISP-like language structure that is easily extended. EDIF is supported at a prototype level by several vendors in the electronic and electrical CAD/CAM sector. Although there has been little interaction with the IGES standard until now, there is pressure, especially in the UK, for the two standards bodies to consult with each other.

Other formats. Several other formats are in use for data exchange that are not organized by standards bodies. These include DXF, which is a format used by AutoDESK, and the Intergraph Standard Interchange Format as well as others used by various automotive manufacturers. DXF is the most widely used, but it is different from such standards as IGES in that it is a proprietary product.

GKS. The Graphics Kernel System (GKS) is not a data exchange format but a graphics standard that transfers two-dimensional graphics through a standard subroutine interface for the graphics programmers. The associated computer graphics metafile is a format for transfer of graphics (not product) data.

Future developments

Work is currently being undertaken by the International Standards Organization (ISO) Standard for the Exchange of Product Data (STEP) project to develop a three-layer model. The top application layer describes the structure and specification of data of the various applications to be covered by the standard. The second layer is the logical layer, which defines and classifies entities. The third layer is the actual physical file specification. Formal data modeling techniques are used—the EXPRESS language is used for the logical layer, and the IDEF1X data model is used for the various application models. Application areas that currently exist in an integrated model are geometry, including wireframe, surfaces, and solid models; constructive solid geometry and boundary representation; and tolerances and finite-element modeling. Areas currently being integrated are presentation, drafting, mechanical products, and architecture and engineering construction. Much research is required to understand the information content of the applications and the appropriate data models and to find the best way of formally achieving the specification. Whether this process will lead to the definition of a completely universal data model has yet to be seen. STEP has amalgamated with the Product Data Exchange Specification (PDES) project in the US, which should result in a single, international standard (see Exhibit 2).

Current Data Exchange Experience

In general, many CAD/CAM users have exchanged data with other interested parties on a test basis, but there are few who are transferring large amounts of data. This should change in the near future as more users become aware of the need for product data exchange and as vendor implementations of standards achieve higher quality. For example, there are several reasons for the slow implementation of IGES. First, users have concentrated on getting their CAD/CAM installation working correctly and efficiently within their own organizations and are only now looking for the advantages of product data exchange. Although several major companies (notably the automotive industry) have said that they would insist on CAD/CAM data transfer as a condition of trade, the time frames for this compliance have moved back.

A second cause of delay has been the poor quality of IGES processors, which is only now beginning to improve. There is a considerable time lag between a vendor improving its processor and customers in other countries receiving the new

Exhibit 2. ***The Convergence of Data Exchange Standards***

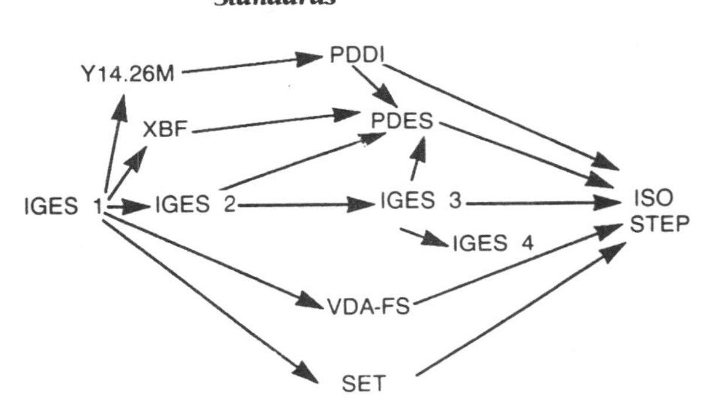

software release. Because there are several standards, the effort has been divided among several processors. This is where the STEP project can help; however, the possibility that STEP will become the standard has made some vendors question the effort that should be put into supporting IGES. Because IGES processors are not specified by standards writers, the IGES standard is open to different interpretations. IGES itself is a verbose file format with lots of wasted space in the file and is difficult to understand. It is also difficult for users to identify how a transfer has failed; this process may require the cooperation of all vendors and users involved in the transfer, with a possible adverse impact on customer-supplier relationships.

The absence of neutral expertise in IGES has led to problems in resolving these issues. The verbose nature of IGES has not posed serious problems but may do so as the volume of transfers increases; the size of IGES files are often four or five times the size of the original part in the native CAD system.

Solutions

To address these problems, the CAD/CAM Data Exchange Technical Center was formed in the UK. This center was initially funded by the Department of Trade and Industry (DTI) and is now financed through membership fees, consultant work, software sales, and income from validation services. The center exists to offer advice on all aspects of data exchange and also has a major input into work on both new and existing standards. It has considerable expertise in data exchange formats and is currently undertaking a major testing program for IGES processors, which will lead to a formal validation service.

The testing program helps solve the problems previously discussed in several ways. Experience with a large number of IGES files has resulted in the publication of guidelines for using IGES that describe the most effective ways of obtaining a successful data exchange with much practical advice. The guidelines are available from the CAD/CAM Data Exchange Technical Center (171 Woodhouse Lane, Leeds LS2 3AR UK) and include sections on the initial liaison between parties, drawing office practice, creating and receiving drawings, and reading and writing IGES files. The center also has various software tools that allow accurate syntax checking of IGES files; this has allowed any errors in IGES processors to be reported to the vendors, many of whom are members of the CAD/CAM Data Exchange Technical Center.

The center also has data bases and software available that can be used to predict the expected level of success of a transfer. This information can be used to choose a limited subset of entities that should transfer successfully. These data bases are based on vendor documentation and are being updated as the testing program progresses.

Summary

If STEP/PDES is rushed, the new standard will not be significantly better than IGES. The STEP standard has much wider coverage than existing standards, and current CAD/CAM systems will be unable to use it fully for some time. IGES is quite capable of handling engineering drawings and simple models and is likely to be around for some time. The users must insist that vendors continue to enhance exchanges.

There are considerable technical and commercial benefits to being able to freely transfer product data between CAD/CAM systems, and the ability to exchange CAD/CAM product data must be supported by senior management. The state of the art is constantly progressing as vendors improve their processors and as users learn methods for effective transfer. The availability of validated processors will also help to take the uncertainty out of transfers.

With careful planning and agreement between the two parties, successful transfers of three-dimensional wireframe models, surface models, and detail drawings (including dimensions and annotation) can be obtained. In the future, users will be able to transfer more and more entities with greater accuracy and less planning. The long-term goal must be to transfer data completely from one CAD/CAM system to another in a manner that is transparent to the user. ▲

Additional Readings on Enabling Technologies

Aranow, Eric. 1991. Modeling exercises shape up enterprises. *Software Magazine*. January, pp. 36-46.

Frank, Howard. 1991. Building a management system for tomorrow's enterprise network. *Networking Management*. March, pp. 32-34.

George, Joey F. and John L. King. 1991. Examining the computing and decentralization debate. *Communications of the ACM*. July, Vol. 34, No. 7.

Korzeniowski, Paul. 1990. Everything to everything in a network of networks. *Software Magazine*. June, pp. 69-74.

Scheer, A. W. 1989. *Enterprise-wide Data Modelling*. New York, NY: Springer-Verlag.

Stusser, Daniel I. 1991. Eleven approaches to tailoring a system to your needs. *Networking Management*. March, pp. 42-48.

VIII. TOOLS AND TECHNIQUES FOR THE SYSTEMS INTEGRATOR

Methodologies for performing systems integration are fragmented and incomplete. Much research is currently being devoted to the development of better and more appropriate tools and techniques. For the purposes of this study guide, four groupings of tools are presented.

Tools for systems analysis and design include various quantitative methods and simulation systems. Queueing models can be very helpful in high-level scoping of a system design, with simulation tools available for more detailed analysis. A variety of graphical presentation schemes are also available. Data flow models appear to be particularly helpful for information system analysis and design.

Tools for the justification of advanced technologies include those that focus primarily on cost, as well as others that attempt to incorporate *all* costs and benefits, even those that are intangible. There is growing recognition of the existence of critical strategic factors that are not dealt with explicitly in traditional approaches to justification.

Every successful systems integrator must master the concepts and mechanics of **project management tools**. A large system integration program will typically consist of several dozen discrete projects, may of which are being pursued simultaneously. In a typical company, a particular technical person may be assigned to two or more of these projects at the same time. It is important to coordinate all these diverse activities through the use of multi-project management practices.

Systems integrators spend much of their time conducting meetings of specialists in various areas. Specified groups or task forces are studying various aspects of the overall system. These "think tank," or "brainstorming" sessions are absolutely necessary. If left unstructured, however, the process can become clumsy, inefficient, and discouraging to participants. Tools and techniques have been developed to assist the systems integrator in **facilitating group interactions**.

The articles in this section are organized according to the four groupings described above.

Effective Analysis of Manufacturing Systems Using Appropriate Modeling and Simulation Tools

S. Wali Haider
IBM Corp.
M.S. R27B
1600 River Edge Pkwy
Atlanta, GA 30328
(404) 956-5782

and

Rajan Suri
University of Wisconsin
Dept of Industrial Engineering
1513 University Avenue
Madison, WI 53706
(608) 262-5536

ABSTRACT

Sometimes it seems that analysis studies take too long. At other times the studies may not be accurate enough or address the right issues. Why is this the case? Effective performance analysis of manufacturing systems requires answering the right questions, in a timely fashion, by using the appropriate modeling and simulation tools. In this paper, we first develop a framework for understanding what questions need to be addressed. Based on this, we present guidelines to assist manufacturing analysts in the choice of tool for their studies. It is hoped that by using this structured approach, analysis can be performed in a timely manner, addressing the right issues, and at the level of accuracy that is appropriate to the situation at hand.

1.0 INTRODUCTION

While there has always been a need for some form of analysis of manufacturing systems since the advent of the industrial era, simple and informal analysis usually sufficed for early manufacturing systems. The reason was that these systems were themselves relatively simple and consisted of almost autonomous manufacturing facilities. Modern systems, in contrast, require much more sophisticated analysis techniques. This is due to several reasons ([1],[2]), the main ones being:

- Greater complexity of manufacturing processes and process flows. To a large extent this is due to automation which enables high speed production lines and greater integration of facilities and resources.
- Larger enterprises and bigger manufacturing facilities.
- Increased world wide competition requiring highly efficient use of manufacturing resources.
- New manufacturing approaches (e.g. Just-in-Time or JIT) and, as a result, more responsive facilities. This requires short lead times and minimal work-in-process (WIP) in the system.

In response to the demand for increasingly sophisticated analysis techniques, significant advances have been made in manufacturing modeling and simulation (M&S) technologies over the last decade. Today, well over 50 M&S software products are available on a wide variety of computers, from micros to mainframes [3]. These products offer a variety of features and range from a few hundred dollars for a microcomputer based system to well over $100,000 for a mainframe product.

Does one need a single M&S software package to meet a variety of analysis needs that one often has during the life cycle of a manufacturing system? How does one select an appropriate M&S software product category in a marketplace that provides products with such widely differing capabilities and prices?

Proper answers to the above questions cannot be provided without an appropriate framework and analysis strategy. Here we develop such a framework and strategy. This then provides a basis for recommending the appropriate M&S product categories for the various types of analysis.

The focus here is on evaluative tools rather than generative tools. Basically, the former enable one to analyze the performance of a system, while the latter help to generate decisions. Thus a generative tool might provide a potential production schedule while an evaluative tool would predict the performance of the system if such a schedule were implemented. More on this categorization, along with pros and cons of each type of tool, can be found in [4]. We recognize that the two types of analysis are not entirely independent. Still, a good understanding of, and proper use of, evaluative tools forms the foundation of effective manufacturing analysis, and this paper is restricted to such tools.

In order to develop a framework and analysis strategy we will follow a number of steps. First, by using the

Reprinted from *1990 IIE Integrated Systems Conference Proceedings*

perspective of manufacturing system life cycle, a classification scheme is developed for the analysis categories. Next, M&S products are categorized. On the basis of the two categories so generated, a performance analysis strategy is developed. Finally, the benefits obtained from using a good strategy for selecting appropriate M&S products from appropriate categories are described.

2.0 MANUFACTURING SYSTEM LIFE CYCLE

During its life cycle, a manufacturing system goes through a number of stages. First, during the business planning stage, when a decision is made to go forward with a new manufacturing system, such a system goes through various design stages. Following this, installation and testing is performed and the system is ramped up to a target production rate. A manufacturing system performs around its target capacity over a period of time. During this time it may be improved or modified in various ways. Finally, when the system is no longer competitive for the product it manufactures and it is not economically feasible to re-tool it and make it competitive for another set of products, or for a new product set, a decision is made to terminate such a system. These major phases during the life cycle of a manufacturing system are briefly described below (more details can be found in [1]).

System Planning

The objective during this stage is to establish the economic attractiveness of the system size and the layout alternatives which will satisfy the strategic corporate objectives. During this stage a number of candidate systems may be identified.

Initial Design

Candidate systems from the system planning stage are studied further, and a handful of alternatives, which are both operationally and functionally attractive, are identified. The study is at a high level and details such as material handling and storage requirements are considered in a limited way. Typical issues included in the study are:

- What processing capacities are needed based on:
  - -Workstation reliability and yield?
  - -Current and projected throughputs and product mixes?
- What are the implications of various lot sizes on lead times?

Detailed Design

In this stage most of the details of the system are decided. Specific characteristics of the processing, material handling, and storage equipment requirements are considered explicitly for the first time. The possible production planning and operating procedures that will be used are also identified and studied at this stage.

Installation and Testing

Once the system design has been approved a set of vendors is selected. The processing equipment and the supporting facilities are put in place and unit tested. Finally, the portions are integrated and tested as one system.

Start-up

At this stage of the system life cycle, the manufacturing system starts production with a limited set of products with small volumes. The product variety, if it exists, and the volume are gradually increased. At each stage of the ramp-up, any hardware/software problems that crop up are reviewed and corrected, as are operating procedures and other system level problems.

On-going Operations

This is the major stage of the life cycle. During this stage, the product variety and the volume, for which the system was designed, is manufactured. At this stage, one is interested in making sure that the system continues to operate most efficiently at the required production level and, in the event of minor disruptions, appropriate adjustments are made. To support this objective during the on-going operations stage, the following activities are performed on a continuing basis:

- Long Range Planning: The ability of the system to support changes in product volumes and mixes over a period of time greater than a year is ensured, plans for any system modifications are developed, and long range production targets are set.
- Tactical Planning: Potential problems in meeting the medium term production goals are identified and recommendations made to correct the problems. Additionally, ways of improving and enhancing the system are constantly sought out. Examples of recommendations could be: adding a second shift for the next quarter; or the need to reduce setup times and lot sizes in an area with long lead times.
- Operations Analysis: Short term operating procedures are developed and effective short term solutions are found to counter to disruptions on the factory floor. Examples of such disruptions are machine breakdowns and unavailability of component parts. Short term solutions might involve routing a product to another work center, or changing the priorities of production for various parts.

Termination

When the manufacturing system is no longer cost-effective for the products it produces, it is ramped-down and, finally, closed. There can be several reasons for its inability to compete. They are:

- The processing technology used is obsolete resulting in products that are too expensive or of inferior quality.
- Lack of demand for the products manufactured by the system.
- The inability of the system to be re-tooled economically for new products.

3.0 CATEGORIES OF PERFORMANCE ANALYSIS

We now propose three categories of performance analysis and illustrate how they support the various stages of the manufacturing system life cycle.

Aggregate Analysis

This level of analysis does not include the details of the operating and control policies. Typical issues studied include:

- Changes in product volumes and mixes
- Equipment layout alternatives
- Different equipment and material handling choices (type and capacity)
- Lot size choices
- Setup reduction programs
- Buffer size variations
- Pull vs. push environments

while the main performance measures analyzed are equipment utilization and queues, product lead time, WIP, and throughput rates.

Detailed Analysis

This type of analysis evaluates operating alternatives, details of material handling and storage, and fine tunes system performance.

On-line Analysis

Here historical data and current product and resource information are used to support short term decision making. This type of analysis is confined to very short time horizons like a shift or a day. It may or may not be performed in real time. It requires additional real-time information such as the current location of products and the status of various machines.

4.0 MAPPING OF ANALYSIS TO LIFE CYCLE

Table 1 shows the mapping of the performance analysis categories to the manufacturing system life cycle. As can be seen in the table, performance analysis techniques are not relevant during the installation/startup (debugging) phases.

The on-going operations phase is the longest phase in a manufacturing system life cycle and a variety of activities are performed during this phase on a continuing basis. To support these different types of activities all three performance analysis classes are needed. In situations where more than one analysis category is indicated, more details will be provided below to distinguish the various cases.

TABLE 1
Mapping of Performance Analysis Categories to Manufacturing System Life Cycle

Manufacturing System Life Cycle	Performance Analysis Category
System Planning	Aggregate Analysis
Initial Design	Aggregate Analysis
Detailed Design	Detailed Analysis
Installation and Testing	-
Start-up	-
On-going Operations:	
Long-range Planning	Aggregate Analysis
Tactical Planning	Aggregate Analysis, Detailed Analysis
Operations Analysis	Detailed Analysis, On-line Analysis
Termination	-

5. CATEGORIES OF MODELING AND SIMULATION PRODUCTS

The vast variety of M&S products available in the marketplace today can be placed into four categories.

Analytic Models

Products in this category perform aggregate dynamic analysis of discrete manufacturing systems using mathematical models based on queueing theory [2]. The models take into account the dynamics of the manufacturing system and provide estimates of various performance measures. Such products provide rapid model development and fast execution of "what-ifs", but do not model certain details as explained later.

An example of one such product whose use in industry has been widely documented is MANUPLAN.

Manufacturing Simulators

These are computer packages that allow one to simulate a system contained in a specified class of manufacturing systems, using menus and graphics, with little or no programming. While model development and analysis is easy and rapid, a drawback of many simulators is that they are limited to modeling only those configurations allowed by their standard features. This difficulty can be overcome if a simulator allows one to drop down to a lower level language (Fortran, C, etc.).

Examples of manufacturing simulators available in the market today are PROMOD, SIMFACTORY, STARCELL, WITNESS, and XCELL+.

General Purpose Simulation Languages

Products in this category are general, flexible, and powerful enough to model any type of manufacturing system at any level of detail. A model is developed by writing a program using the language's modeling constructs. However, model development and analysis is much more involved and time consuming, and considerable programming expertise is needed.

Examples of such products are GPSS (including GPSS/H and GPSS/PC), SIMAN, SIMSCRIPT II.5, and SLAM II. Most of these packages also allow the simulations to be animated, which can have many benefits [3].

On-line Simulators

Products in this category are designed to support the day to day decision making processes on the manufacturing floor. They often provide utilities to access and use on-line MRP and shop floor status data for model execution. Due to the amount of detail, customization, and fine tuning required, the model development cycle time is long. Examples of such products are FACTOR and InterFaSE.

6. ANALYSIS STRATEGY

Insight gained from a good performance analysis strategy provides the basis for good management decisions. An effective performance analysis strategy allows initial exploration of a wide range of alternatives easily and quickly, during which the less desirable alternatives are eliminated; and performing in-depth analysis of a selected few alternatives later on. A poor analysis strategy might spend too much time investigating just one or two alternatives, and is likely to miss a highly desirable alternative.

Products using analytic models provide the most convenient vehicle to develop and explore a wide range of alternatives very quickly. They are therefore also known as "rapid modeling" tools, and can be very responsive to the time-sensitive needs of competitive manufacturing [5]. In certain cases manufacturing simulators may be used instead of analytic models. Two such instances are:

- When analytic models cannot model certain peculiarities of the required process flow or equipment characteristics (e.g. an oven with very special loading requirements).
- When some animation is desired even at the high level analysis.

After using analytic models or simulators to generate the main options, the alternatives are narrowed down to one or two for detailed analysis using simulation languages. Finally, for the day to day operations, on-line simulators are used to control and fine tune a specific alternative. Figure 1 illustrates these points.

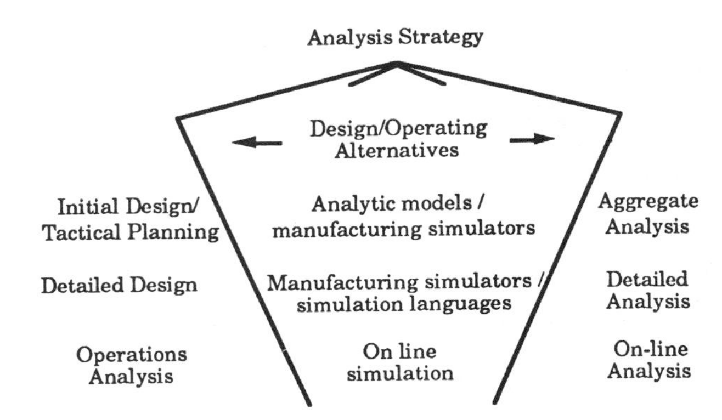

Fig. 1 Desirable Performance Analysis Strategy

Figure 2 gives another perspective through the users. Manufacturing and planning engineers typically engage in aggregate analysis. Often, these engineers do not have a strong background or interest in modeling or programming. Their primary responsibilities are to evaluate product and facilities planning alternatives. These plans often span a horizon from several months to years. Analytic models or manufacturing simulators should be used for this. Wherever possible, the use of analytic models should be preferred to manufacturing simulators for the following reasons:

- Rapid model development
- Very fast execution speed (as little as a minute on a PC, compared to possibly hours)
- Steady state predictions eliminating the need to handle statistical issues like run length, warm-up period, replications, etc.

However, to balance the above it should be noted that analytic models do not allow one to model certain details of resource sharing, finite buffer sizes, sequencing logic, etc. Also it is not possible to provide animation through analytic modeling.

Detailed analysis is performed by people who are experienced in modeling activities and have some aptitude in programming. Such people come from various sources like consulting groups or industrial engineering. Simulation languages are appropriate for this user group.

Users Analysis Type	Mfg. and planning Engineers	Plant/Corporate/External consultants, Industrial engineers	Manufacturing supervisors
Aggregate Analysis	Analytic Models Manufacturing simulators		
Detailed Analysis		Pre and post processors General purpose simulation languages	
On-line Analysis			On-line simulators

Fig. 2. Relationships between users, analysis type, and M&S software categories

Line or sector supervisors who have little or no background in M&S are the primary users of on-line analysis. Models used in this category are developed by experienced modelers and systems analysts, so, in this case, model developers and users are two very different groups with different backgrounds.

One variation here is to write flexible pre- and post-processors for the general purpose languages to make them look like a manufacturing simulator, thereby, making it available to the manufacturing and planning engineers (see the arrows in Figure 2).

It is desirable to be able to carry forward as much of the modeling effort invested during the aggregate analysis to the subsequent detailed as well as on-line analysis. In other words, it is not desirable to build models from scratch when moving from aggregate analysis to detailed and then to on-line analysis. Given that no single M&S package currently exists to support the wide variety of analysis that one needs to engage in to design and operate manufacturing systems effectively and efficiently, the proposed strategy also calls for integrating the different analysis products that one must use during the various design and operating stages.

Consider a firm that wishes to investigate whether it makes sense to put a family of products in a dedicated group technology (GT) cell. A manager may only want a quick study to indicate whether or not it is worth pursuing the idea. If this initial study shows that there may be substantial benefits, the manager could then allocate more resources to a detailed study. In such instances it may become necessary to switch analysis products when going from initial to detailed analysis. This is where an integrated set of products shows its potential; one can move from the simpler to a more complex one with minimal duplication of modeling effort, thereby reducing the total analysis cycle significantly.

An example of such an integrated set of modeling products is in Figure 3. This set of products provides analysis capabilities for a broad range of situations. Lotus 1-2-3 enables a database of basic data to be built, and allows simple engineering and financial calculations to be performed quickly and easily. Next, this data is supplied directly to MANUPLAN II using its Lotus 1-2-3 interface, and aggregate analysis is performed rapidly. This analysis considers the dynamics of the system and provides information regarding capacity, flow time, and work in process. SIMSTARTER allows almost instant conversion of MANUPLAN II models into SIMAN or SLAMSYSTEM simulation code. The simulation language allows one to extend the original model and capture more complex features of the manufacturing system (scheduling strategies, material handling control policies, etc.). Finally, the CINEMA or TESS animation packages give the analyst a greater ability to observe and communicate the results of the analysis.

At this point, the reader may be concerned about the time requirements for using a number of different M&S products on one project. While there is a clear advantage in having several different tools available, the time requirements for creating a number of different non-integrated models would make such an approach infeasible. The set of integrated M&S tools mentioned above allows the user to build, in effect, only one base model, enhancing it as it becomes appropriate.

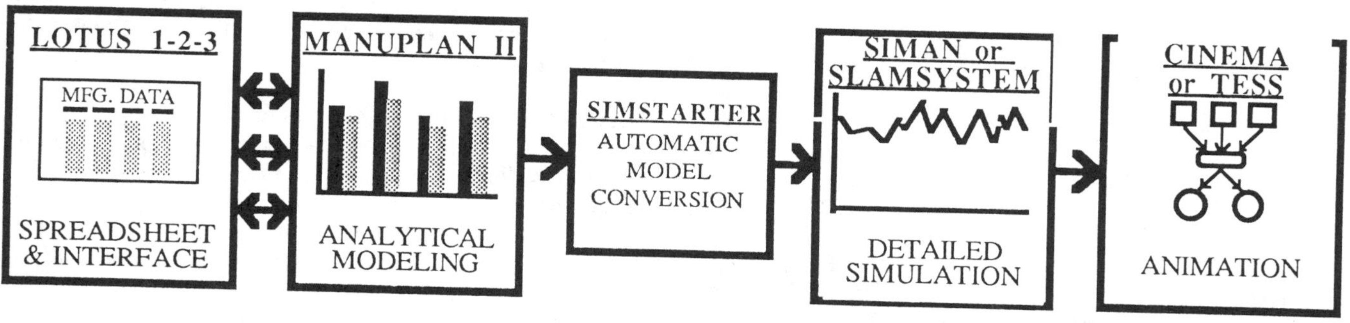

FIGURE 3: INTEGRATED TOOLKIT FOR EFFECTIVE ANALYSIS

7. BENEFITS OF PROPOSED APPROACH

The approach proposed here leads to timely and effective analysis of manufacturing systems. As an example, Nymon [6] describes the use of rapid modeling, simulation, and on-line analysis tools, at various stages of a factory design project. Application of the right tool provided timely feedback at each stage of the project. Haider et al. [7] discuss the use of rapid modeling and detailed simulation in the design of a factory of the future. Some of the benefits they derived from using this sequence of tools included early identification of critical process constraints, and judicious choice of material handling strategy. Use of an integrated toolkit is described by Shimizu and Van Zoest [8]. In addition to the benefits just discussed, they also illustrate the gains attained from having an integrated set of tools.

In summary, analysts using the sequence of tools described here can provide rapid and meaningful feedback to decision makers, leading to better utilization of resources, improved performance, and as a result, a more competitive manufacturing enterprise.

REFERENCES

[1] R. Suri, "A New Perspective on Manufacturing Systems Analysis", in *Design and Analysis of Integrated Manufacturing Systems,* W.D. Compton (Ed.), National Academy Press, 1988.

[2] Y.T. Leung and R. Suri, "Performance Evaluation of Discrete Manufacturing Systems", *IEEE Control Systems,* Vol.10, No.4, 1990, 77-86.

[3] Law, A.M. and S.W. Haider, "Selecting Simulation Software for Manufacturing Applications: Practical Guidelines and Software Survey", *Industrial Engineering,* May 1989, 33-46.

[4] R. Suri, "An Overview of Evaluative Models for Flexible Manufacturing Systems", *Annals of Operations Research,* Vol. 3, 13-21, 1985.

[5] R. Suri, "Lead Time Reduction Through Rapid Modeling", *Manufacturing Systems,* July 1989, 66-68.

[6] Nymon, J. (1987), "Using Analytical and Simulation Modeling for Early Factory Prototyping," In *Proc. 1987 Winter Simulation Conference*, 721-724.

[7] Haider, S.W., D.G. Noller and T.B. Robey, "Experiences with Analytic and Simulation Modeling for a Factory of the Future Project at IBM", In *Proc. 1986 Winter Simulation Conference*, 641-648.

[8] Shimizu, M. and D. Van Zoest (1988), "Analysis of a Factory of the Future Using an Integrated Set of Software for Manufacturing Systems Modeling," *Proc. 1988 Winter Simulation Conference*, 671-677.

BIOGRAPHICAL SKETCHES

S. Wali Haider is currently in Atlanta, Georgia with the Metals Industry Application Development group in the Industrial Sector Division of IBM Corp. He is an application specialist in Computer Integrated Manufacturing with a focus on Production Planning, Scheduling and Modeling. Prior to moving to Atlanta, he was responsible for the Center of Competence for Manufacturing Modeling in the CIMS group in Boca Raton, Florida.

Before joining IBM in January 1984, he was a faculty member in the Department of Industrial Engineering at Texas A&M University. Dr. Haider has also worked for Bethlehem Steel Corp., where he was involved in the applications of simulation and mathematical modeling techniques. He participated in the development of two software packages, one using linear programming and the other using the SLAM simulation language.

Wali Haider received both his Ph.D. and M.S.I.E. from Purdue University in Industrial Engineering. He also received a graduate research award sponsored by IBM in 1973, and is a senior member of IIE.

Rajan Suri is Professor of Industrial Engineering at the University of Wisconsin-Madison, where he is also one of the faculty responsible for the Manufacturing Systems Engineering Program. He received his Bachelors degree from Cambridge University (England) and his M.S. and Ph.D. from Harvard University.

He has been instrumental in extending the theories of queueing networks and perturbation analysis for manufacturing applications, and is the author of over 50 technical publications, several books and edited volumes. He is Editor-in-Chief of the *Journal of Manufacturing Systems,* Associate Editor of the *International Journal of Flexible Manufacturing Systems,* and Area Editor of the *Journal of Discrete Event Dynamic Systems.*

Dr. Suri combines his academic credentials with considerable practical experience. He has consulted in this area for leading firms including 3M, Alcoa, AT&T, DEC, FIAT, Ford, Hewlett Packard, McDonnell-Douglas, IBM, Pratt & Whitney and Siemens. He is also a principal of Network Dynamics Inc., a firm specializing in software for manufacturing systems. In 1981 Dr. Suri received the Eckman Award from the American Automatic Control Council for outstanding contributions in his field.

Graphical process description—views and diagrams

KOSTIA MANDEL

Abstract. Process design (and description) is largely an *ad hoc* discipline composed of several diagramming techniques with no clear definition of the process characteristics described by any one, usually with no definition of the relationship between the diagrams, and in most cases with no definition of the process design stages. This is in sharp contrast with product description and design, where the design stages and the various formats of the product definition information capture (diagrams and others) are well defined.

In most cases today, when a process is designed, a process diagram is selected arbitrarily and is used to determine the process characteristics inherent in that diagram. Other characteristics not presented in that particular diagramming technique, are simply ignored, or in the best case, vaguely addressed. Whatever method or diagramming technique for process description is selected, it will not specify all the essential process characteristics needed for a full and complete specification of the process.

This paper addresses the above problems. It discusses the need for process description and the process characteristics that need to be described. It presents a structure to categorize process diagrams and the concept of process views. In this context, the application of the $IDEF_0$ method for modelling the functionality of systems/environments for process description is presented, and some modifications to this method are discussed.

1. Introduction

The phrase 'one picture is worth more than a thousand words' is as true in the technical world as anywhere else. Sketches and drawings have been used in almost any technical discipline: from mechanical drawing to electrical schematics, from landscape diagrams to power plant schematics, from computer programs' flow charts to project-tracking diagrams. Every traditional, well-defined discipline has several graphical representation techniques or forms, each emphasizing a different aspect of that discipline. In each graphical representation it is very clear what aspect of the discipline is presented and its relationship to other representations. For example, when a mechanical drawing defining the dimensions of a moving mechanism is viewed, it is not expected that the kinematics of the mechanism will be shown.

The major impact that the process used to produce a product has on the quality, cost and timely delivery of the product is slowly being realized by managers and engineers alike. The challenge this realization offers to researchers and developers is to devise formal methods for process design and description, as well as means to collect, store and display the knowledge about a process—this is the Process Description discipline. The large 'bandwidth' of information transfer a picture has (Levine 1988) lends itself easily to process description needs, and is the reason why the vast majority of process description methods are graphical (Malcolm *et al.* 1959, Gane 1977, ICAM 1982, Leclair 1982, Harrington 1984, Alford 1985, Mandel *et al.* 1986, Harel *et al.* 1988).

Process description, not being a well-defined discipline, lacks the clear distinction between the various graphical representations being used, as product description disciplines do. Many process description techniques exist. Some techniques are based on solid theoretical grounds like Petri Nets (Peterson 1977), but most not. Some are very common like PERT (Malcolm *et al.* 1959) and Gantt diagrams, but most are specialized for a specific discipline. In addition, the relationship between the various graphical representation formats used to describe processes, as well as the content of some of these representations, are vague. The purpose of this paper's discussion is specifically to address two issues:

1. to gain understanding about the process phenomenon through discussion of some process description issues and formal definitions; and
2. to describe a single aspect (functionality) of a process through the deployment of the Modified $IDEF_0$ modelling method.

Author: Kostia Mandel, PAMAT Engineering Ltd, Netanya, Israel.

Reprinted courtesy of Taylor & Francis from *International Journal of Computer-Integrated Manufacturing*, vol. 3, no. 5.

2. Process phenomenon overview

The dictionary (Webster 1985) defines a process as 'a natural phenomenon marked by gradual changes that lead toward a particular result'. The term 'process' is not bound by size, complexity or structure; it can be either a stand-alone 'thing' or it can be part in a larger process in any discipline. If a process is to be described formally, the description methodology should be able to capture and present the various aspects of the process to a variety of users. In this section some of the process characteristics are identified and potential uses and users of process description are discussed.

2.1. Process description issues

When describing a process formally, attention should be given to the 'three dimensions of process description' as follows.

A. Depth (level of detail). The process should be described in a level of detail that will make it useful. It is doubtful that describing product life cycle as 'design–test–manufacture–market' is of any use; however, continued decomposition of these high-level phases, will reveal the real relations between the modules. The more detailed the decomposition, the more accurate and reliable the description is. The more detailed and explicit the description, the less likely assumptions will be made by one activity in the process about another.

B. Scope (portions of the process). Although somewhat beneficial, describing one group's portion of a multi-divisional project is not very useful. It is commonly acknowledged that most problems in a process occur when organizational boundaries are crossed. Therefore, in order to increase the potential for harvesting benefits from describing the process formally, the involvement of as many possible organizations from those participating in the process should be described.

C. Breadth (level of completeness). For example, scheduling is an important aspect of process description and design, but describing only that aspect of a process through a PERT or Gantt diagram does not do justice to the complexity of a process. Issues like information flow, process management, and material management need to be described formally as well.

In addition to the 'three dimensions of process description', attention should be given to the following.

D. Separation of concerns (Wallace 1987). Formal process description can and should address the needs of many uses and users: project managers, planners, MIS, etc., each having a different set of concerns. A 'good' process description methodology should address the specific needs and concerns of one type of user at a time, while hiding others. In a different representation, another set of concerns should be addressed while the first ones are hidden.

E. Global responsibility vs local autonomy. These two concepts are subsets of the separation of concerns principle. The Global Responsibility and Local Autonomy concepts provide decoupling between the global view of a process and the local view of an activity. The global view of a process (sometimes referred to as 'process knowledge') concentrates on the interfaces between activities, the relationship between activities and aids in determining how change is propagated through the process. The local view of an activity concentrates on the procedures and algorithms used by the activity to convert its input to output. *Global Responsibility* is the process wide (global) definition of the activity's output. It allows identification of the unique source of every information item or material used in the process. When a modification of an item is required, e.g. an iteration occurs, the global responsibility can be used to determine the activity that can modify this item, and the activities that should be notified about the change. *Local Autonomy* is the knowledge and authority of an activity to generate the required output from the provided input. Local autonomy is defined for one hierarchical level of the described process. Local Autonomy is, among other things, the authority to: allocate the activity's resources; to define internal procedures for performing the task; to decompose the task into sub-activities (and knowledge about the interactions between those sub-activities). As long as the activity's interfaces are unchanged, the local autonomy is the activity's ability to modify its internal structure at will.

F. Conceptual clarity. It should be very clear what aspect of the process is being described and by what method. For example, what are the building blocks of the description: activities, organization or states? Does an arrow represent an information exchange, material flow or a 'this after that' relationship?

2.1.1. Essential process characteristics. A process has a finite set of essential characteristics which defines 'all there is to know about a process'. These essential characteristics should be planned for and specified during the project or process planning phase in exactly the same manner product characteristics are determined during the product design phase. It is not the purpose of this paper to discuss the essential process characteristics, but for illustration purposes some are listed below.

- The set of activities or states which compose the process—both the activities or states and the hierarchical relations between them.
- All required input—for every activity or state at any level of abstraction.

- Required resources—for every activity, specification of the required resources.
- Logical relations among the input—if an activity can be performed under multiple conditions, requiring different sets of input each time, the relations between the input sets need to be specified.
- Activation constraints—all the constraints specifying values of external parameters required before an activity can start or a state can be reached. For example, 'the activity can start if the temperature is between 40 and 55°C, the relative humidity is between 40 and 60%, and the wind velocity is less than 20 km/h'. Or 'the state will be reached when all the inputs are present'.
- Termination conditions—specification of all the possible causes of activity termination. For example, production of all output, exceeding time limit, exceeding some other parameter limit (e.g. temperature), value of a particular information output, etc.

Since a process is a universal phenomenon and is not tied to any description methodology or application domain, the essential characteristics are common to all processes. In addition to these, there are additional characteristics particular to a specific discipline that may be captured by a special description methodology.

2.1.2. Iterations in a process. One of the more important features of many processes is their iterative nature. Many processes are not executed in one direction—from the start to the finish, without portions of the process being revisited. This phenomenon of repeating a portion of a process, called iteration, is now addressed by most, if not all, available process description methods and tools.

Iterations can be expected, as in a design review that results in re-execution of portions of the design, or unexpected as in changes of customer's specifications by the customer. Iterations can be initiated inside or outside the process. Externally caused iteration is the result of changes outside the process' boundaries, for example, new regulations that prohibit the use of a certain component. Internally caused iteration is triggered by a feedback message from an activity within the process.

2.1.3. Feedback. The concept of feedback, as used in the context of process description, can be understood by relating it to the notion of state in time of the process' progress. A state in time is a 'wave front' of the process' progress at a given instant. The heavy line in Fig. 1(*a*) illustrates the state of the process' progress at time t_1: activities to the left of the line were completed while activities C and D are being executed. Between t_1 and t_2, activity C detected a problem in one of activity A's output, a feedback message was sent to A and A has to be re-executed in order to correct the faulty output (an iteration occurred). The state of the process progress at time t_2 (just as the iteration started) is illustrated in Fig. 1(*b*).

Examples of feedback include a message from a review or an audit detailing the results of a check, or a message from an activity explaining why the information provided to it was found to be incorrect or incomplete, or even just a message from an activity stating that it cannot do its task for no apparent reason.

Even if the verification was performed only on a portion of the items produced at one state, the resulting iteration causes the state in time of the entire process to be different. If, in the light of new constraints imposed by successive activities, a subset of the items produced at the previous state in time of the process is found to be correct, the feedback is positive. If a subset of the items produced at the previous state in time of the process is found to be

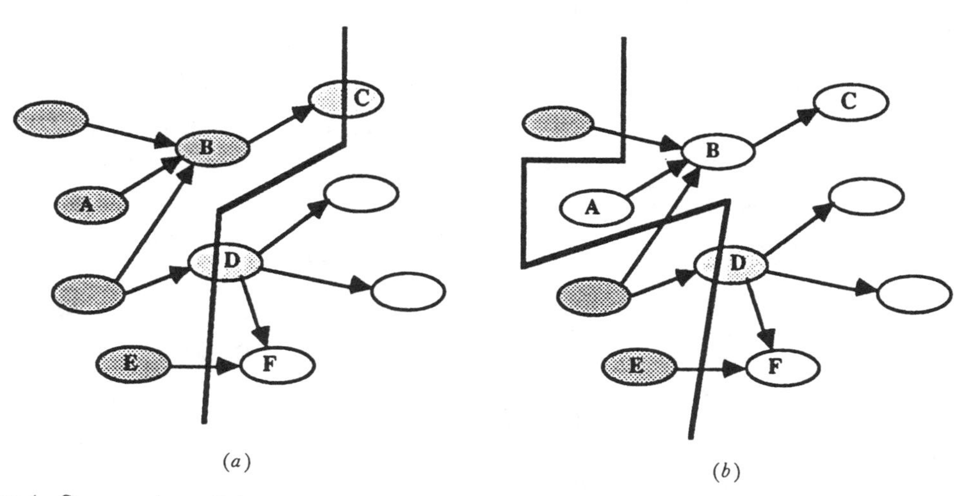

Figure 1. State at time of the process progress. (*a*) State at time t_1; (*b*) state at time t_2 (where $t_2 > t_1$).

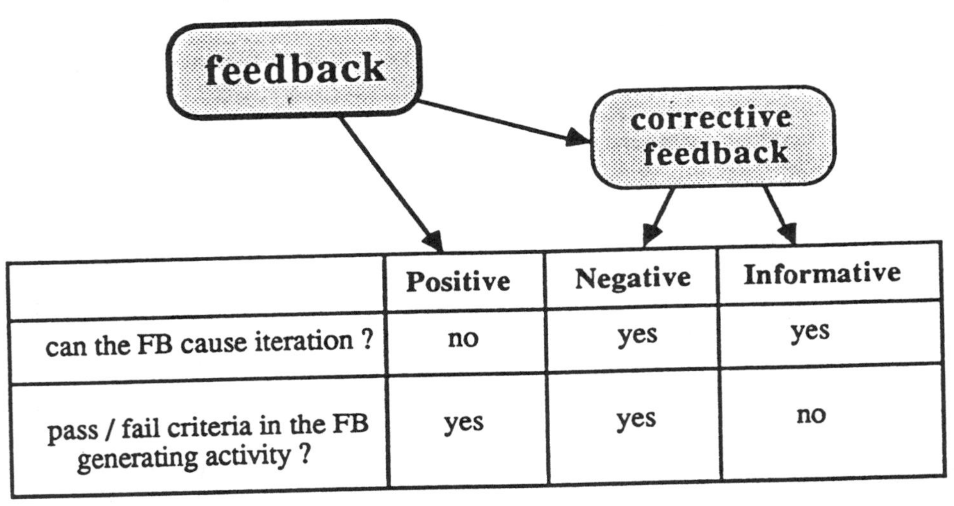

	Positive	Negative	Informative
can the FB cause iteration ?	no	yes	yes
pass / fail criteria in the FB generating activity ?	yes	yes	no

Figure 2. Properties of feedback.

incorrect, an iteration occurs—the current state is aborted and the process is brought back to a state before the previous one.

There are two kinds of feedback: positive and corrective. *Positive feedback* is an approval of the output of a previously executed activity and cannot cause an iteration. *Corrective feedback* is any other message sent from one activity to an activity that was previously performed. Corrective feedback can be further divided to negative feedback and informative feedback. *Negative feedback* is a rejection of the output of a previously executed activity. Negative feedback causes the activity whose output was rejected to be re-executed (fully or partially) and change the rejected item it had generated. Both positive and negative feedback types can be generated only by activities that know the pass/fail criteria about the item on which the feedback is provided. *Informative feedback* contains review results, comments, problems or any other type of information that should be analysed by the activity where the item was generated. The pass/fail criteria, in the case of informative feedback, does not reside in the feedback generating activity. These properties of feedback are summarized in Fig. 2.

The expected internally caused feedbacks should be shown distinctly from information flow, as the two convey different meanings. Information flow is unidirectional—it describes flow from one state in time to another state in time, where every successive state in time is closer to generating the process output than the preceding one. Feedback, of all kinds, flows in the opposite direction. It flows from activity that is being executed to activity that had been previously performed. An iteration triggered by corrective feedback causes the successive *state in time* of the process' progress to be further away from producing the process output than the preceding one.

2.2. Process description's potential uses

Formal process description can serve a wide range of users needing either *descriptive* or *executional* information about the process. The information that should be provided about a process for descriptive purposes is different than that for executional purposes. Managers may wish to understand the behaviour of the process they manage, or they may need support for their process control decision making. Practitioners may want to understand the context of their activity in the process or may want to determine a source of a problem they have. System engineers may want to analyse and improve a process. Generally, a formal process description can be used to achieve four basic purposes as follows.

- *Understand an existing process*—many existing processes are not fully defined and/or understood. A formal process description can go a long way in helping to show how activities in the process are related, the expectations from every activity in the process, the requirements of every activity in the process, and how the process generally behaves. It can surface hidden information about responsibilities of parties to actions and output.
- *Improve an existing process*—when every action, piece of information, or product, can be traced to a responsible party, feedback sending and receiving is facilitated. Feedback is a major element of a continuous process improvement, as the receiving activities can improve their performance after every feedback. Another potential for process improvement is the examination of existing organizational structure versus the process flow. Existing organizations can be modified in such a way that will

minimize crossing organizational boundaries. Yet another area of potential improvement is defining the interfaces between activities. This will eliminate much of 'I thought they should have done it ...'.

- *Design a new process*—the ability to determine as many characteristics of a process as possible in the planning phases will reduce the number of surprises and subsequently the number of mid-course corrections. A process description methodology should help design a new process throughout its phases: starting with a vague notion of what the process should be, through functional design and scheduling, to process control and resource allocation. It should have simulation capability to develop response policies for various scenarios while in the planning phase, which can save precious production time later.
- *Control a process*—status reports regarding process' progress and resource load and utilization are important elements in process control that should be provided for. In addition, the ability to run 'what if' scenarios off-line will enhance the quality of process control decision making. For example, if a problem occurs and there are several possible solutions, the ability to determine *formally* the best solution based on a specified criteria, like least impact on deadlines or lowest capital investment, may be very helpful in selecting the solution.

3. Process description methods

Many process description methods have evolved over the years, ranging from textual to graphical, from formal to informal, from hierarchical to 'flat'. One reason for the ever-growing arsenal of process description methods is the complexity of the phenomenon called process. An important, and often forgotten, step in defining or deploying a process description method is the definition of the process characteristics described by the method. Once it is understood which process characteristics are included in a description and more importantly which are *excluded*, defining or understanding a process becomes much easier. In this section, the concept of process view is introduced and, in the context of the functional view of a process, the IDEF$_0$ method for constructing a functional model of a process is presented.

3.1. The multiple views of a process

The concept of a view as a sub-set of a whole is not new. In database terminology a view is defined (Ullman 1982) as 'an abstract model of a portion of the conceptual database'. Views have also been one of the fundamental concepts in mechanical drawings. Any mechanical object has a finite set of features, and every view of the object (like a top view or side view) shows a subset of these features. The choice of view to show a certain feature depends on which view shows the particular feature best. The number of views needed to describe all the object's features is determined by the object's complexity. Similarly, a process view is the portion of the process characteristics, related to a particular aspect of the process, like dependencies, resource allocation, information flow, control, etc.

The notion of a process view is best illustrated by an example of an actual process. Building a house is a complex process that requires rigorous planning and coordination among numerous professionals under the constraints of a schedule. For each professional, the house-building process is different. For a carpenter, the house-building process is composed of the rough carpentry of erecting walls and ceilings, to the fine carpentry of window frames and wooden floors. For the plumber, the house-building process is composed of installing the water pipes, the sewer pipes and the fixtures. For each of these and other professionals, building a house means something different—each has a different *view* of the house-building process. However, it is very well understood that the complete house-building process is a union of the views of all the professionals that build the house.

Everything said about describing a mechanical object or building a house can be said about describing a process. Any process can also be described by a finite set of characteristics, a small subset of which was discussed in Section 2.1.1. Any view of a process should contain, or emphasize, a subset of these process characteristics. Several diagramming techniques may be used to represent the set of characteristics described by a single view. Figure 3 illustrates the relationship between views, diagrams and the comprehensive set of all the essential process characteristics. These relations and views are not unique and other opinions regarding the views and diagrams required to describe a process do exist [ICAM 1982, Alford 1985, Harel *et al.* 1988].

The Distributed Computing Design System (DCDS) uses the System Specification Language (SSL) (Alford 1985) to describe a system through the use of two views, or structures in SSL terminology. One structure (I-net) describes the flow on input and output items between activities, while the second structure (F-net) describes the temporal relations between the activities.

STATEMATE® (Harel *et al.* 1988), is an environment for development of complex reactive systems. STATEMATE uses three views to define a process: structural view, functional view and behavioural view. The structural view describes the hierarchical decomposition

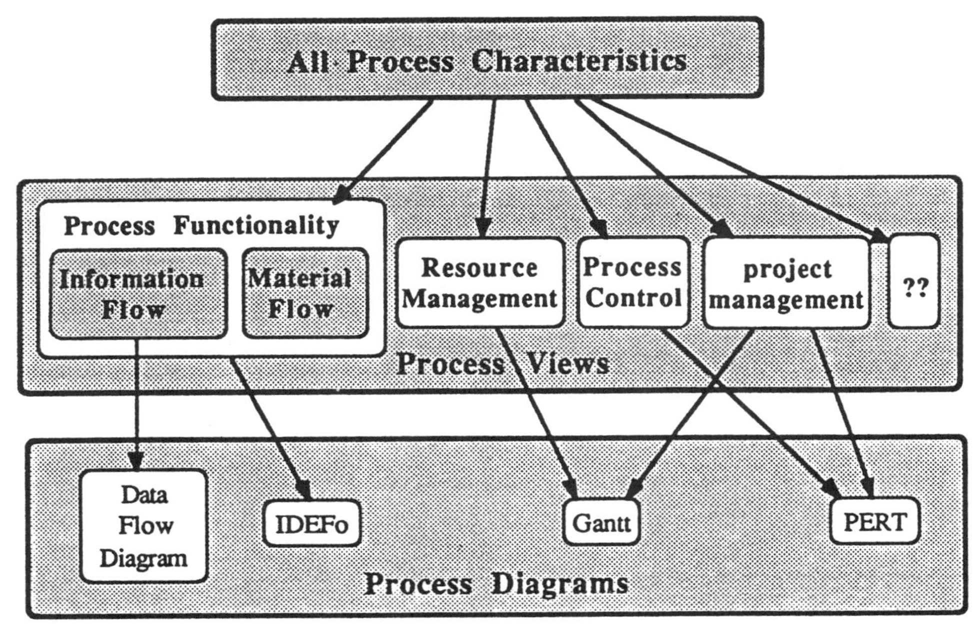

Figure 3. The relationship between process characteristics, views and diagrams.

of the system into its physical components, which in the case of a process will be the organizational structure of the entity executing the process. The functional view depicts the hierarchy of activities and the possible information exchange between them. The behavioural view describes the control of the system: it specifies when, how and what will happen with the activities in the process. The three views use three specifically defined diagrammatic languages: module charts, activity charts and state charts, respectively.

The US Air Force program for Integrated Computer Aided Manufacturing (ICAM) (ICAM 1982) has developed a system definition method called IDEF (Icam DEFinition) that consists of three graphical languages to describe three models (views) of a system. $IDEF_0$ is a functional model of the system which represents the activities and the exchange between them. $IDEF_1$ is an information model of the system describing the common, shared and discrete information needed to support the activities of the system. $IDEF_2$ is a dynamic model of the system that shows the time-varying behaviour of the activities, information and resources of the system.

In either the mechanical drawing or the house building examples, the views were derived from a previously created, well-defined and complete set of characteristics. The various drawings handled by the house-building professionals describe all the house characteristics which the architect had defined, as the mechanical drawings show the various characteristics of a part fully defined by the designing engineer. This is usually not the case with process design. In most cases today, when a process is designed, a process diagram is selected (usually arbitrarily) and used to determine the process characteristics inherent in that diagram. Other characteristics not presented in that particular diagramming technique are simply ignored, or in the best case, vaguely addressed. It is of utmost importance to realize that whatever method or diagramming technique for process description is selected, it will not specify all the essential characteristics of a process needed for a full and complete specification of a process. This realization has some implications:

- how are the various process views related?
- should one view be specified before another?
- what about maintaining consistency among the various process views?
- how should this multi-view process specification be carried out?

Although these issues are fundamental, discussion of them is beyond the scope of this paper.

3.1.1. Process functionality view. When the full list of essential process characteristics is analysed, it can be intuitively divided into two parts. In the first part, process functionality is described—what is converted to what by an activity and who is using the results—this is the *what* portion of process description. In the second part, control and timing issues like dependency, scheduling, resources, activation/termination conditions, sequence/parallelism, etc. are addressed—this is the *how* portion. The first part can be further divided into information flow and material flow. The information flow diagram of a process shows the exchange of information between the activities, using information items (Mandel *et al.* 1986), while the material flow diagram shows the flow of materials and tangibles throughout the process. To aid in the understanding of the described process' functionality, the two diagrams can be superimposed to form a process functionality diagram or view. The activities in this view, as well as the information and materials items (indifferently referred to as interface items), can be decomposed and/or aggregated. In addition to identifying the source and destination(s) of every interface item and its hierarchical structure, process functionality should show the feedbacks in the process.

3.2. *IDEF$_0$ view of a process*

One diagramming technique mentioned in Fig. 3 is IDEF$_0$ (pronounced as 'I-def-zero'). 'IDEF$_0$ is a technique whose main goal is to capture a description of a system/environment in a form that is independent of implementation' (IDEF$_0$ 1981). It is the result of the work done by the ICAM task force chartered by the US Air Force, and is based on SofTech's Structured Analysis and Design Technique (SADT). The ICAM task force's goal was to '...develop structured methods for applying computer technology to manufacturing processes and to use those methods to better understand how best to improve them' (IDEF$_0$ 1981). The task force started working in September 1978 and produced its results in May 1981. The following organizations were among those who participated in the ICAM project: US Air Force, SofTech, MIT, Rockwell International, Control Data Corporation, Hughes Aircraft Company, Northrop Corporation, Boeing Computer Services, Boeing Commercial Airplane Company, Pritsker & Associates, Dan Appelton Company, Higher Order Software, and Vought Corporation. The results of the ICAM task force are available in 11 volumes. One of those results is the Icam DEFinition methodology for describing a system's functionality is gaining popularity and recognition both in the USA and in Europe. In the USA the Department of Defence recommends its use in an increasing number of defence-related programs, and the European research programme ESPRIT uses it to design CIM system's architectures.

3.2.1. Original IDEF$_0$. 'IDEF$_0$ is a functional model which is a structured representation of the functions of a system or environment and of the information and objects which interrelate those functions. ... The IDEF$_0$ methodology may be used to model a wide variety of systems, where systems may include any combination of hardware, software and people. For new systems, IDEF$_0$ may be used first to specify the requirements and functions and then to design an implementation that meets the requirements and performs the functions. For existing systems, IDEF$_0$ can be used to analyse the functions the system performs and to record the mechanisms by which these are done.' (IDEF$_0$ 1981).

The IDEF$_0$ method, as defined by the ICAM publication (IDEF$_0$ 1981) and described in others (LeClair 1982, Harrington 1984), has four major characteristics: simple graphics, data abstraction, preciseness, and differentiation between organization and function. The separation between organization and function is a guideline that focuses the IDEF$_0$ method on *process functionality* (what is happening) and not on *process organizational structure* (who is doing it). The other three characteristics are explained below.

The simple graphical characteristic of the IDEF$_0$ diagram is achieved by composing the diagram from only two basic elements: boxes and arrows. The boxes (or nodes) represent activities while the arrows represent interfaces. There are four kinds of interfaces determined by the side of the box they are attached to: input, output, control and mechanisms, as illustrated in Fig. 4.

Input arrows represent anything that will be processed by the activity to produce its output. Output arrows represent anything produced by the activity. Control arrows represent conditions, circumstances or information that govern the activity and/or are needed by the activity to perform its task. Mechanism arrows represent people or equipment that the activity utilizes in its conversion of the input to the output, i.e. resources. The input and output show *what* is done by the activity. The controls show *why* and *how* it is done, while the mechanisms show *who* is doing it. The terms input and output convey the notion that a box, or activity, represents a transition from a 'before' to an 'after' state of the system or environment.

Data abstraction is achieved in the IDEF$_0$ diagram through a hierarchical decomposition of the described system/environment. A one-box diagram, the top level, provides the context for the entire model. The box, or activity, in this diagram is labelled A0 and it represents

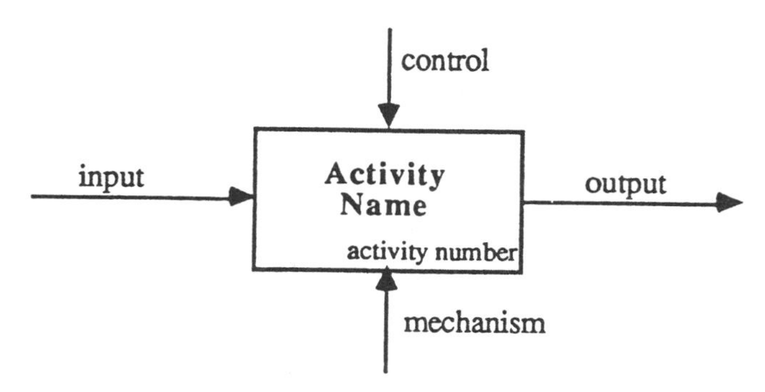

Figure 4. IDEF$_0$ basic elements.

the entire system. In the second level, the box describing the system being analysed (A0) is 'opened' and its major sub-systems are shown as boxes, this time numbered as A1, A2, A3, etc. In the third level, each major sub-system is decomposed into its activities: A1 into A11, A12 ...; A2 into A21, A22 ...; A3 into A31, A32 ... etc., and so on until the desired level of detail is reached.

The fourth characteristic of the $IDEF_0$ diagramming technique, preciseness, is reached by the many rules and conventions that govern the building of the diagram and aid in its understanding. Some of the major ones are as follows:

- Every decomposed activity (parent), is always divided into no less than three and no more than six sub-activities (children). The upper limit of six forces the use of hierarchy to describe complex subjects, while the lower limit of three insures that enough detail is introduced to make the decomposition of interest. The upper limit of six was chosen since psychological experiments have shown that it is difficult for humans to grasp 5–7 distinct concepts at one time.
- A child activity is restricted to have only those interfaces that are defined in the parent activity. Further, the child activity can not add or omit any of those interfaces. Thus the parent activity provides a context for the child activity.
- Child activities inherit the number of the parent activity and add a number, e.g. when activity number 3 (A3) is decomposed into three child activities, their numbers will be A31, A32, A33.
- An $IDEF_0$ diagram does not contain any *explicit* sequence or duration information. It may convey some notion of sequence from the direction of the arrows, but this notion is undefined.
- An $IDEF_0$ diagram may contain arrows that have both source and destination on that diagram and arrows that have only the source or destination on that diagram—boundary arrows. The source or destination of a boundary arrow can only be found by examining the parent diagram.

These are only some of the major rules and conventions of the original $IDEF_0$ diagram. A full detailed definition of the $IDEF_0$ model can be found in $IDEF_0$ (1981), while additional discussion about the method and its implementation can be found in LeClair (1982) and Harrington (1984). An example of a process described through the $IDEF_0$ method is given in Figs 5 and 6.

3.2.2. Modified $IDEF_0$. Throughout the author's utilization of the $IDEF_0$ methodology for describing both design and manufacturing processes (Mandel *et al.* 1986, 1988), several major limitations of the $IDEF_0$ method were encountered.

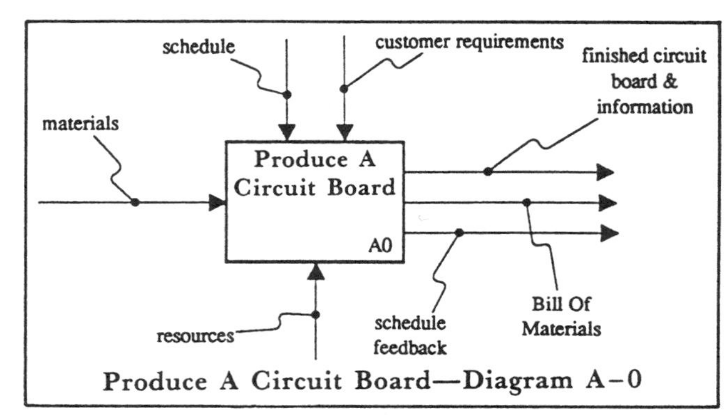

Figure 5. Context diagram of a process in the original $IDEF_0$ method.

1. When a detailed description of a complex process is conducted, the description pages become very rapidly cluttered, making them illegible by casual users. This problem is caused by the capture and representation of too many process characteristics (information, mechanism and control) in one $IDEF_0$ diagram. In order to solve this problem, the process characteristics shown in the diagram were limited to only information and material flow.
2. The $IDEF_0$ methodology lacks the ability to correlate an interface item with its originating activity directly. Explicit and direct identification of every interface item's source is important throughout the process life cycle, and becomes invaluable in cases of problem solving, iteration and conflict resolution.
3. The $IDEF_0$ methodology lacks representation clarity in several areas including the hierarchical relations between parent–child pages, distinction between the various types of interfaces, and others. Representation improvements were made to enhance the readability of the $IDEF_0$ diagram in those areas and others.

The modifications that were devised to the original $IDEF_0$ method, which include the abovementioned among others, are discussed in more detail in the following two sub-sections. In the first sub-section, the conceptual modifications are discussed. In the second sub-section, the syntax modifications that support the conceptual ones and other modifications are discussed.

3.2.2.1. Concepts refinement: The focus on process functionality (*what* happens in the process and not *who* or *how* does it happen) in the Modified $IDEF_0$ method, results in focusing the diagram on input/output interfaces. The

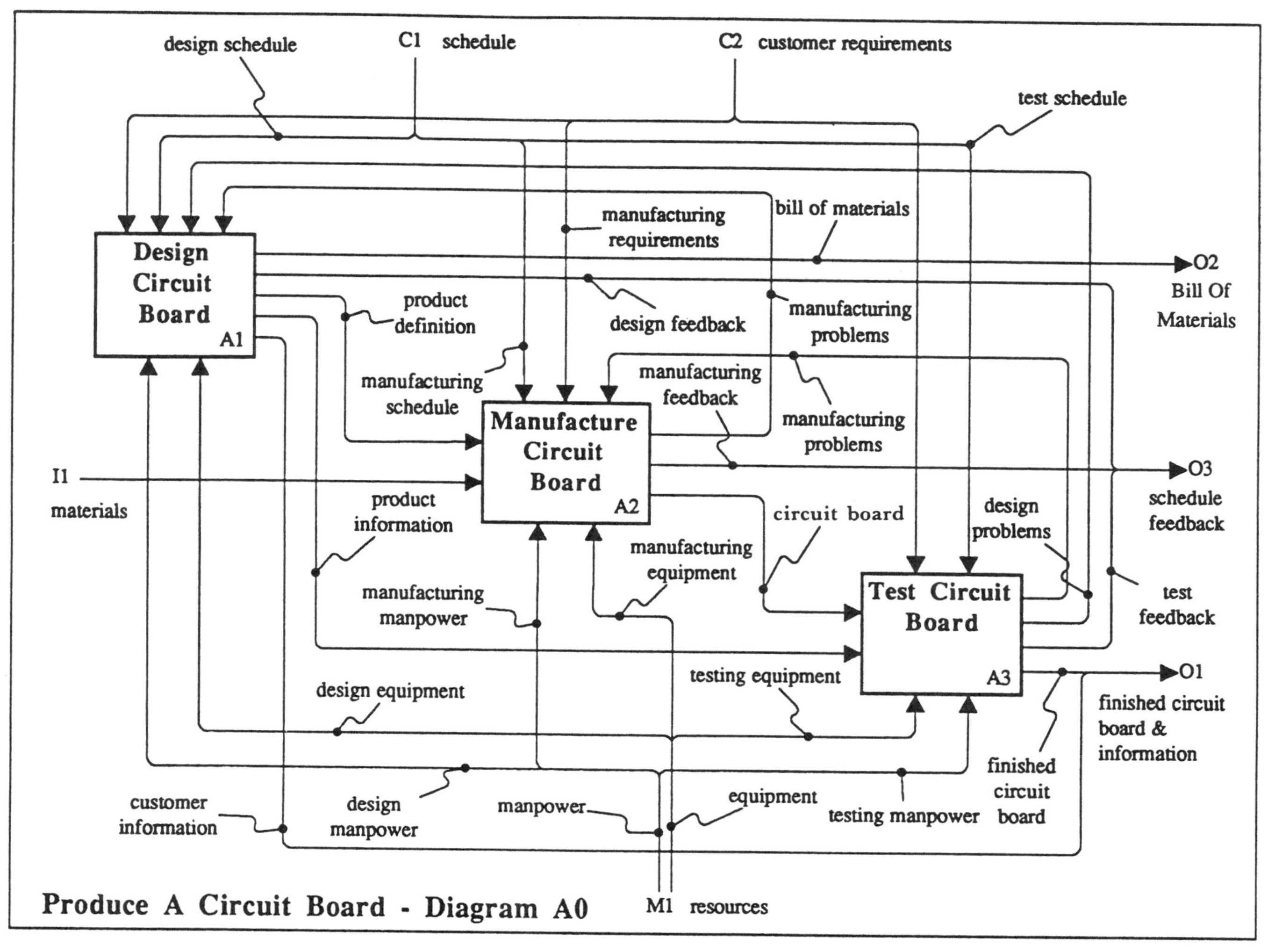

Figure 6. Diagram A0 in the original $IDEF_0$ method.

implications of this emphasis on the mechanism and control aspects of the original $IDEF_0$ method are discussed below.

A mechanism is defined ($IDEF_0$ 1981) as '...the person or device which carries out the function'. In the context of describing process functionality, as depicted by the information and material flows, specifying or understanding resource requirements does not play a major role. It is not essential to have the mechanism specified in order to understand what is converted to what, which is the goal of the process functionality view. Thus mechanisms are not shown in the Modified $IDEF_0$ diagram.

Control is defined ($IDEF_0$ 1981) as 'the conditions or circumstances that govern the function', while input is defined there as '...converted by the function into output', The assumption in the original $IDEF_0$ is that '*every incoming arrow is control unless it obviously serves as an input' (consumed by the activity)*. Based on these definitions, customer requirements are not an input to the design activity but are control; detailed drawing is not an input to the manufacturing activity—it is control. It was observed that most people associate control with execution issues like schedule, activation conditions, termination conditions, logical relations among multiple sets of input, etc. Input is associated with the information and material needed (consumed, converted, used, or otherwise) to perform an activity. Based on these observations it is assumed, in the Modified $IDEF_0$ method, that *every incoming arrow is input unless it obviously serves as control*, where control is the executional issue like those mentioned above. In the context of the process functionality view that depicts the input/output relations between activities, control (executional) issues are not addressed.

In the Modified $IDEF_0$ method, interfaces that represent control, i.e. schedule, execution conditions, termination conditions, etc., are not shown. However, even in the process functionality view's context, there is one element that governs the transformations of input to out-

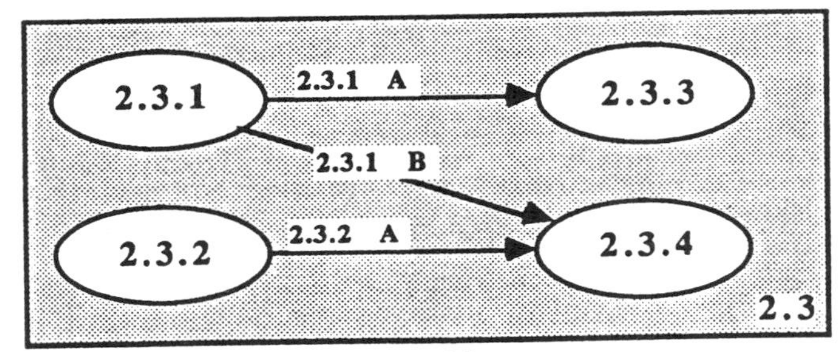

flow between activities on one page

(*a*)

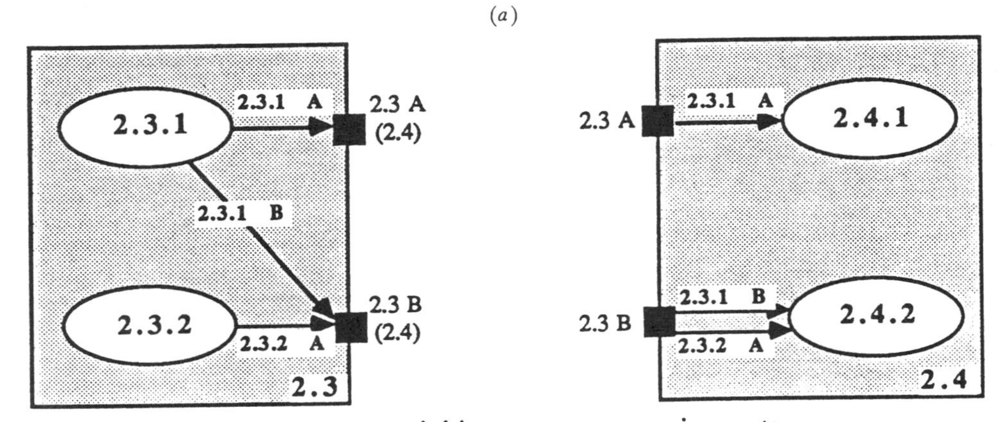

flow between activities on two pages using ports

(*b*)

Figure 7. Example for the use of ports in the Modified $IDEF_0$ diagram.

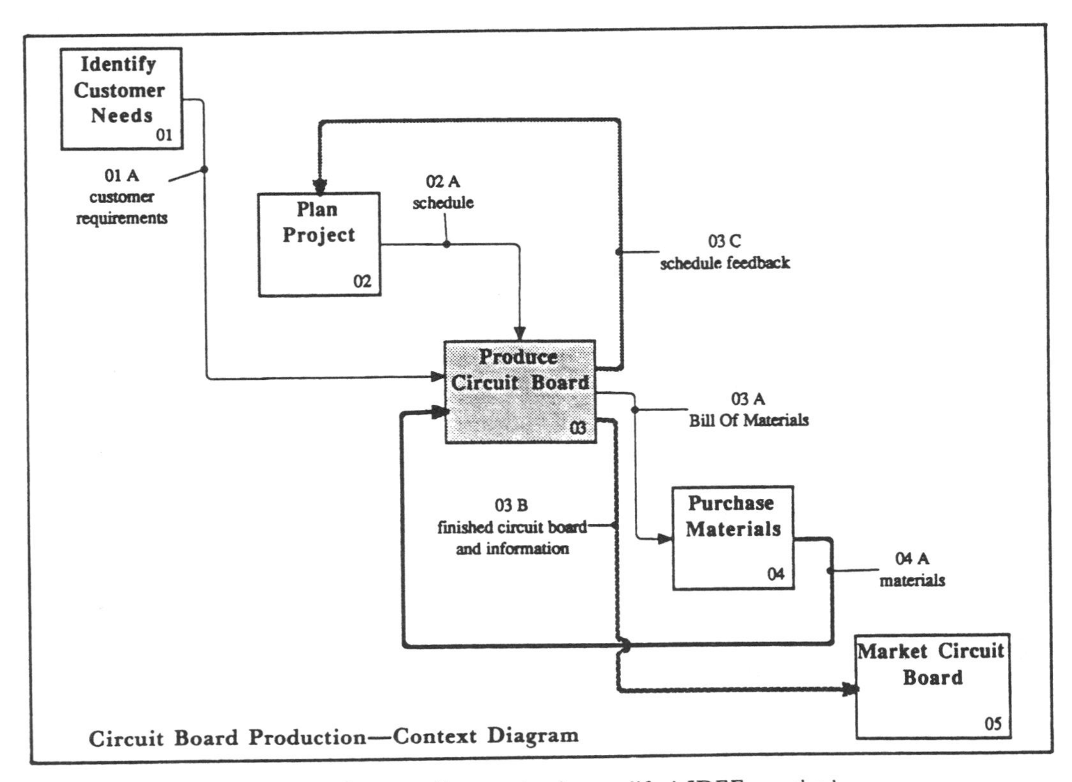

Figure 8. Context diagram in the modified $IDEF_0$ method.

put—feedback. As long as there is no need to modify any of the interface items (all feedbacks are positive), the flow in the process is unidirectional. When a problem does occur, it shows up as corrective feedback from the activity that identified the problem to the activity that was identified as the source of the problem. Based on that corrective feedback, an interface item will be modified by the activity that originally generated it. Therefore feedback controls the circumstances under which an activity is performed—it adds new information that was not known the first time the activity was performed and the interface item was produced. Hence feedback, in the context of process functionality view, modifies the transformation of input to output and thus is designated as control.

In addition to the mechanism and control issues there are other conceptual modifications that were made to the original $IDEF_0$ method.

The context diagram in the Modified $IDEF_0$ method contains, in addition to the one box representation of the entire system being analysed as in the original $IDEF_0$ method, all the activities that have interfaces with the analysed system.

Supplementary textual documents containing indexed information about the activities and interfaces in the $IDEF_0$ diagram are provided. For process functionality representation there are two such documents, or dictionaries: activity and interface. The activity dictionary contains the activity's number, name, input, output, controls, textual description and indexes of sub-activities. The interface dictionary contains the interface item's number, name, activities that use it, textual description and indexes of the interface items it is composed of. Other qualifiers of activity and/or interface can be added as needed.

Every interface item is uniquely identified using an indexing scheme that enables a direct and explict correla-

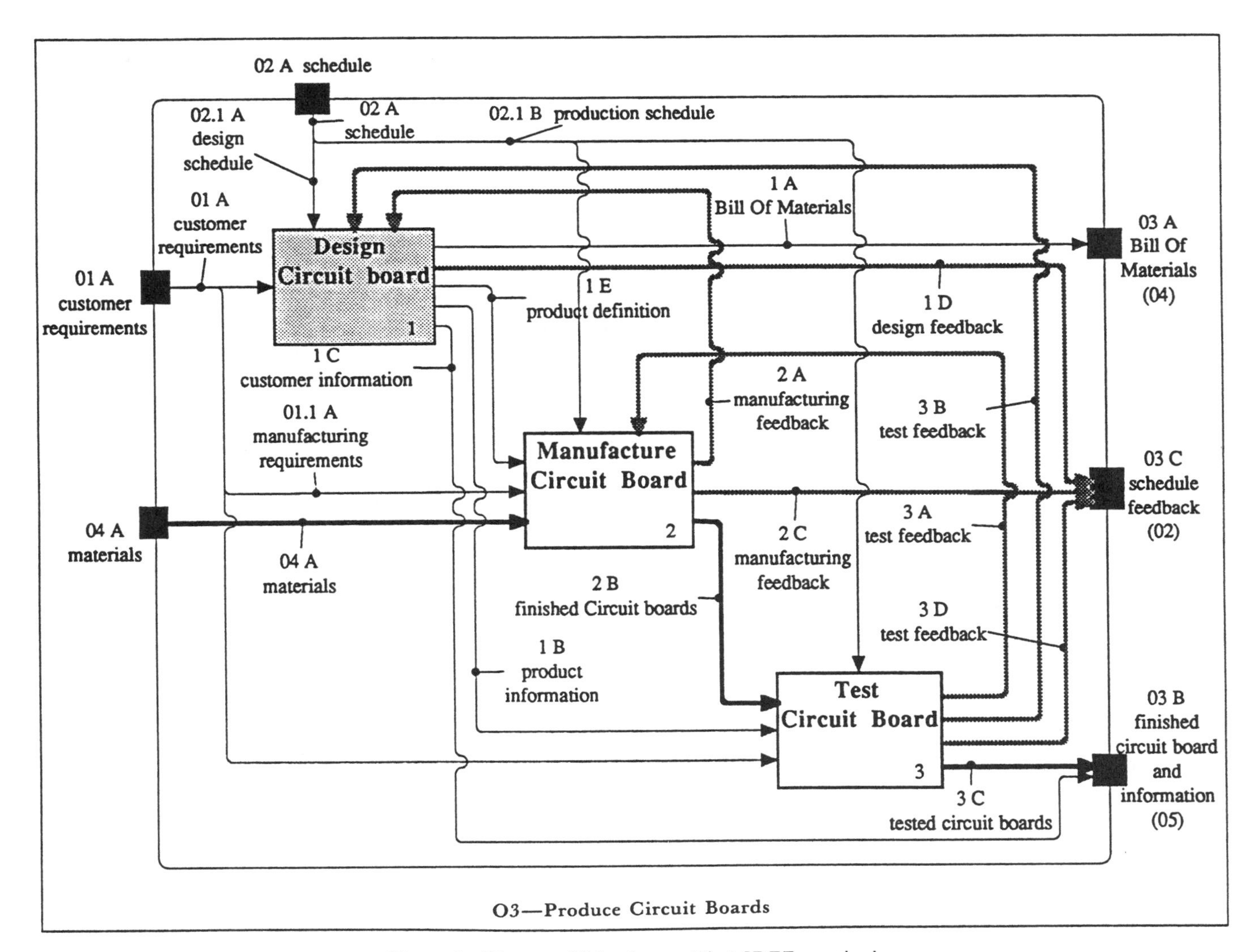

Figure 9. Diagram 03 in the modified $IDEF_0$ method.

tion of an interface item with its originating activity. The syntax of the scheme is explained in the next section.

Joining two arrows is allowed in $IDEF_0$ and implies that '...the same data may be produced by more than one activity' ($IDEF_0$ 1981). Due to the need to correlate every interface item with one and only one generating activity, joining two or more interface items into one is not allowed. Aggregation of interface items is permitted, however, as it maintains the identity of individual items.

Ports are used to designate interfaces to and from activities on different pages. Every port has a label corresponding to the interface item it represents. This is done to eliminate the need to view the context of the child activity on the parent page, as in the original $IDEF_0$ method.

3.2.2.2. Syntax modifications: Syntax enhancements were made to the original $IDEF_0$ method in order to make the Modified $IDEF_0$ diagram more readable. The major goals that were targeted by the syntax enhancements are:

A. elimination of the need to flip between the parent and child diagrams when reading the child diagram (in other words, provide all the information needed to read a diagram on that diagram);
B. support the conceptual modifications; and
C. add graphical distinction between the various types of elements in the diagram.

The following are the major syntax enhancements in the Modified $IDEF_0$ method.

- Information flow is shown as a thin solid line while material flow is shown as a thick solid line.
- Feedback is shown as a thick shaded line to highlight the distinction between it and the regular interface.
- Thick hatched lines are used to represent interfaces that are aggregates of more than one type, e.g. information and material.
- Decomposed activities (parents of other activities) are shaded.
- Activities are numbered in an outline format. For example, the children of activity 3 are numbered 3.1, 3.2, etc.
- The interface items are numbered based on the

Table 1. Portion of the activity dictionary.

1. Design circuit board
Input: 01 A customer requirements
Output: 1 A bill of materials, 1 B product information, 1 C customer information, 1 D design feedback, 1 E product definition.
Control: 02.1 A design schedule, 2 A manufacturing feedback, 3 B test feedback.
Sub-Activities: 1.1 architectural design, 1.2 custom device design, 1.3 circuit board development.
Description: This activity encompasses all the product definition activities.
Hierarchical level: context (– 0)
Appears in drawings: 1
Performing organization(s): Development division
Authorizing person: Development VP.
Required resources: development division personnel.
Duration: 15 men years
Activation conditions: arrival of customer requirements *and* design schedule.
Termination conditions: Product development terminates with beginning of mass production.
Execution stages: Start, start and end of sub-activities, end.

2. Manufacture circuit board
Input: 01.1 A manufacturing requirements, 04 A materials, 1 E product definition.
Output: 2 A manufacturing feedback, 2 B finished circuit boards, 2 C manufacturing feedback
Control: 02.1 B production schedule, 3 A test feedback.
Sub-Activities (None).
Description: All the production and repair-related activities that are performed in order to produce the circuit boards and repair them.
Hierarchical level: context (– 0)
Appears in drawings: 1
Performing organization(s): production planning and manufacturing divisions.
Authorizing person: Production VP.
Required resources: production planning and manufacturing divisions personnel.
Duration: Entire project.
Activation conditions: Initiation of design activities.
Termination conditions: Management decisions.
Execution stages: Start, start and end of sub-activities, end.

activity that generated them. For example, the interface items generated by activity 2.4.2 are numbered as 2.4.2A, 2.4.2B, 2.4.2C, etc.

- The interface items have hierarchy similar to that of the activities. For example, interface item 4.2.3A is composed of any number of interface items generated by the children of activity 4.2.3 (4.2.3.1, 4.2.3.2, etc.).
- Ports are used to designate interfaces with the activity: ports on the LHS represent input interfaces, ports on the RHS represent output interfaces and ports on the top represent control. Each port is designated by the code of the interface it represents; in addition, output ports show parenthetically the activity that receives the interface item. The use of ports is illustrated in Fig. 7. Figure 7(a) shows a flow between four activities on one page, while Fig. 7(b) shows how the same flow looks when the activities are on two pages and ports used.

The Modified $IDEF_0$ diagramming technique, as described in this section, has been used intensively to

Table 2. Portion of the interface dictionary.

Activity 1. Design circuit board
Inputs
01 A customer requirements—see activity 01

Outputs
1 A Bill of materials
Interface type: Information
Format: Document
Description: A listing of all the parts, sub-assemblies and raw materials that go into the circuit board showing the quantity of each one in a single board.
Generated on drawing: 2
Consuming activities: 04 Purchase Materials
Generating organization: Design Division.
Authorizing person: project manager.
Quality Metrics: Interface item is too broad for specific metrics, see sub-interfaces.
Sub-interfaces: None.
1 B Product information
Interface type: Information
Format: Document, drawings.
Description: The circuit's schematic diagram and functional description that will enable the test engineers to develop test plans and strategies.
Generated on drawing: 2
Consuming activities: 3 Test Circuit Board.
Generating organization: Design Division.
Authorizing person: Schematic Design manager.
Quality Metrics: Interface item is too broad for specific metrics, see sub-interfaces.
Sub-interfaces: None.
1 C Customer Information
Interface type: Information
Format: Document and drawings
Description: functional description and parameters, diagnostics and operating instructions for the customer.
Generated on drawing: 2
Consuming activities: 05 Market Circuit Board
Generating organization: Design Division.
Authorizing person: project manager.
Quality Metrics: Interface item is too broad for specific metrics, see sub-interfaces.
Sub-interfaces: None.
1 D design feedback
Interface type: feedback
Format: Document, oral.
Description: Comments and suggested modifications to the design schedule made by the Design Division.
Generated on drawing: 2
Consuming activities: 02 Plan Project
Generating organization: Design Division.
Authorizing person: Design manager.
Quality Metrics: Interface item is too broad for specific metrics, see sub-interfaces.
Sub-interfaces: None.

describe the functionality of a complex process—design through manufacturing of a product that contains both software and hardware (Mandel *et al.* 1988).

The same process used to illustrate the original $IDEF_0$ method in Figs 5 and 6 is used to illustrate the Modified $IDEF_0$ method in Figs 8 and 9. Table 1 contains part of the Activity Dictionary and Table 2 contains part of the Interface Dictionary corresponding to this example.

4. Summary

Process description can be a very powerful tool. It can be used to understand and/or improve an existing process, to design a new process, or to aid in controlling a process. Regardless of the use, any process description should describe large portions of the process, at sufficient level of detail and completeness to make it useful. Any process should be described or specified through multiple views and diagrams.

The major purpose of this paper is to discuss some fundamental issues in the process description area in general, and graphical process description in particular.

In this paper, the concept of a process view was substantiated and the process functionality view, along with some of the essential process characteristics, were presented and discussed. One diagramming technique that can be used to describe process functionality, $IDEF_0$, was analysed and some of its shortcomings were discussed. These shortcomings are addressed by the Modified $IDEF_0$ method. This method is different from the original $IDEF_0$ in the following major points:

- restriction of the process characteristics shown in the diagram to only information and material flows (no mechanism and control);
- an indexing scheme that uniquely correlates every interface item (information or material) in the process with its originating activity; and
- the use of ports to show interfaces between activities on different pages, thus providing the context for every page in the description on the page itself.

The Modified $IDEF_0$ method was implemented to describe process types: large and small scale, engineering (design and manufacturing) and financial, and software development processes. The method was judged to be very readable and facilitated numerous activities by the process participants ranging from eduction to improvement, from accumulating knowledge to conflict resolution.

Acknowledgment

I would like to thank my colleague Doyte L. Perry whose attention to detail and thoughtful scrutiny helped shape some of the ideas presented in the first part of the paper.

References

ALFORD, M., 1985, SREM at the age of eight; the distributed computing design, *Computer*, **18** (4).

GANE, C., and SARNSON, T., 1977, *Structured System Analysis: Tools and Techniques* (Improved System Technologies Inc.).

HAREL, D. *et al.* 1988, STATEMATE: a working environment for the development of complex reactive systems. *Proceedings of the 10th IEEE International Conference on Software Engineering*, Singapore, April.

HARRINGTON, J., 1984, *Understanding The Manufacturing Process* (New York: Marcel Dekker).

ICAM, 1982, Integrated Computer Aided Manufacturing (ICAM) Architecture, Part II, Volume VIII—Technology Transfer, AFWAL-TR-82-4063, October.

$IDEF_0$, 1981, Integrated Computer Aided Manufacturing (ICAM) Architecture, Part II, Volume IV—Functional Modeling Manual ($IDEF_0$), AFWAL-TR-81-4023, June.

LECLAIR, S. R., 1982, IDEF the method, architecture and means to improved manufacturing productivity, SME Technical Paper MS82-902.

LEVINE, R., 1988, Visualization barriers. *Computer Graphics World*, August, p. 28.

MALCOLM, D. G., *et al.*, 1959, Application of a technique for research and development of program evaluation. *Operations Research*, September–October.

MANDEL, K., 1988, Functional description of the IMS product realization process, AT&T Bell Laboratories Internal Document, June.

MANDEL, K., BEN-ARIEH, D., and VENUGOPAL, R., 1986, Information flow from design to manufacturing in the electronics industry. *International Journal of Computers and Industrial Engineering*, **11**, (3).

PETERSON, J. L., Petri nets. *Computing Surveys*, **9**, September.

ULLMAN, J. D., 1982, *Principles of Database Systems*, 2nd edn, (Computer Science Press), p. 7.

WALLACE, R. H., STOCKENBERG, J. E., and CHARETTE, R. N., 1987, *A Unified Methodology for Developing Systems* (Intertext Publications), pp. 172–173.

WEBSTER, 1985, *Webster's Ninth New Collegiate Dictionary* (Merriam-Webster Inc.).

Making the Most of IDEF Modeling—The Triple-Diagonal Concept

Dan Shunk, Bill Sullivan, and Jerry Cahill

Implementing computer-integrated manufacturing (CIM) systems is not easy. Some of the challenges include timely, cost-effective software creation, user education and cooperation, and computer systems interfaces. A valuable aid in meeting these challenges is a user-friendly tool for visually defining integrated systems. Such a tool helps system definers and builders communicate more effectively among themselves and with users. This article presents a proven technique for system definition that is rigorously defined yet easy to create.

Dan Shunk is director of the Center for Automated Engineering and Robotics and associate professor of industrial engineering at Arizona State University. Previously, Dr. Shunk was vice-president and general manager at GCA Corporation, manager of Group Technology at International Harvester, and manager of Manufacturing Systems at Rockwell International. He began his career as co-founder of the USAF Integrated Computer Aided Manufacturing (ICAM) program. He is an experienced practitioner and consultant in integrated system design, group technology, and automated systems.

William G. Sullivan, P.E., is professor of Industrial Engineering at the University of Tennessee-Knoxville. Dr. Sullivan is also director of the Center for Computer Integrated Engineering and Manufacturing in the College of Engineering. He received his Ph.D. from the Georgia Institute of Technology and is author of three books and over 50 articles.

Jerry Cahill is a senior associate with GCA and, in that capacity, has worked on modernization projects at Hughes Aircraft and McDonnell Douglas. He previously worked in manufacturing planning for Emerson Electric, where he played a key role in its implementation of group technology.

The first step in developing successful CIM strategies is to understand the basic integration requirements. As we review the levels within the factory hierarchy, a few integration principles become evident. First, information must tie systems together and flow between systems with ease and absolute accuracy. Second, control functions must link software systems with lower-level material movement and process control. Third, material flow must be integrated so that parts are handled in known locations. These three principles lead us to three main aspects to systems integration:

- Information integration—Information must be available in the form needed for optimal planning, tracking, and control. This often requires that various pieces of computer hardware have data base access without sacrificing data integrity.
- Controls integration—Control strategies and systems must be integrated to achieve the closed-loop process control that links processors and operates each process. To achieve this integration, the control system must know what process parameters are involved in making a good part and monitor these parameters to ensure that a good part is made. The control system must also track part location.
- Material Flow Integration—Parts must be handled in known orientations through known moves to achieve transfer line speed with job shop flexibility.

For a CIM system to be successful, integration must be accomplished in each of these three areas (information, controls, and material flow). If the most productive processes are then used at bottleneck operations, the CIM system will meet the fundamental goal of producing the best product in the least time. The techniques described in this article are intended to help CIM system designers meet these goals.

A Modeling Methodology

To begin the creation of a CIM system, we must define the architecture of the existing system (i.e., the as-is model). Exhibit 1 illustrates a functional manufacturing systems model that allows top-down decomposition. The benefits of creating this type of as-is model include:

- The project team gets an overview of the operation.
- Users have an opportunity to participate in a contributory yet terse manner.
- The formal decomposition of all functions provides the groundwork for all future systems definition.

The methodology for the as-is approach can be homegrown, but we have found that a formal, rigorous technique is best for integrated system definition. The syntax for the suggested modeling methodology is shown in Exhibit 2. The methodology is known as USAF ICAM *Definition* Language ($IDEF_0$). (Level 0 addresses function modeling, IDEF level 1 ($IDEF_1$) provides a method for information modeling, and $IDEF_2$ supports dynamic modeling.) Because this method provides a means for capturing functions, input, controls, output, and mechanisms in a consistent manner, the utility of the resulting document will endure the test of time.

Using $IDEF_0$

The objective of the $IDEF_0$ function modeling technique is to decompose each function into its simplest level, and then quantify the cost and time drivers. By interviewing those involved in the various functions and analyzing the information

Exhibit 1. ***Manufacturing System Architecture***

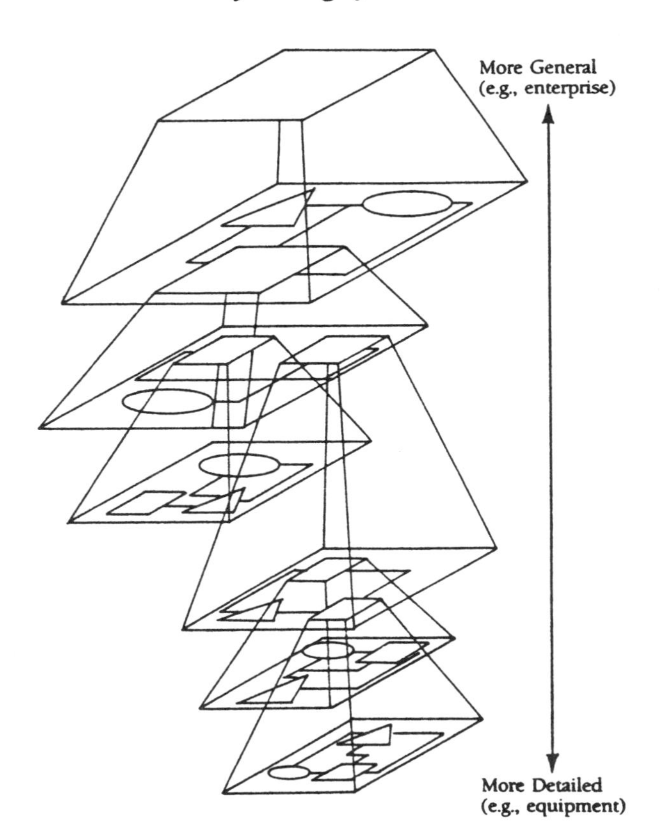

Exhibit 2. *Syntax of the IDEF Technique*

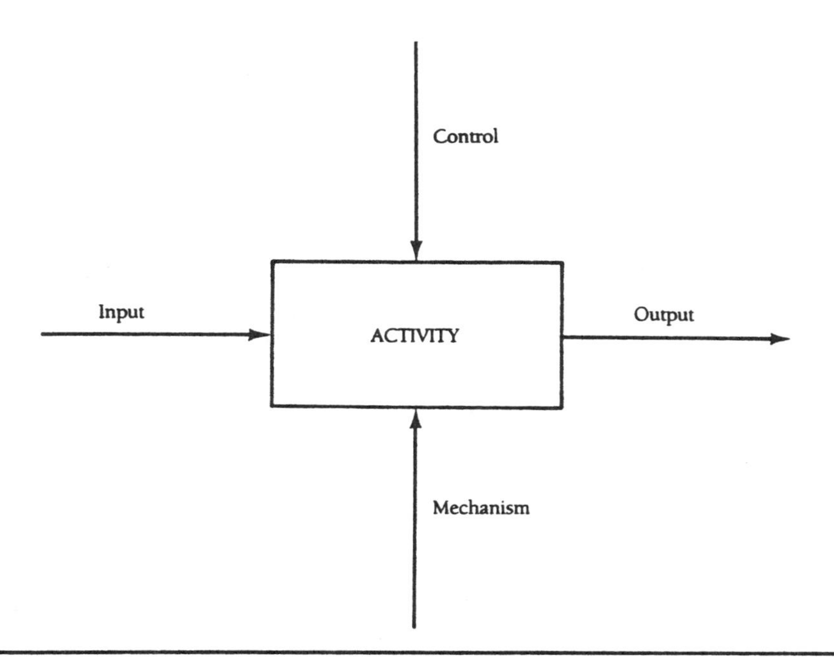

gathered, we can define the resources required in each function (e.g., direct and indirect labor requirements, material expense, overhead support, reliability).

Often, IDEF$_0$ diagrams go unused

These resources can be assigned to functions in units (e.g., number of people, dollars). The objective is to highlight and graphically portray major cost drivers. In theory, the IDEF$_0$ functional diagramming tool appears to be an efficient and effective way to seek remedies for manufacturing integration problems. However, far too often the diagrams go unused.

Obstacles to IDEF$_0$ application

Why do completed IDEF$_0$ kits sit on the shelf? Even though the IDEF$_0$ model does a thorough job of breaking down each high-level function into its most basic components, a tremendous amount of data is generated that defies the sorting and assimilation efforts of many system developers. Thus, in the process of looking closely at individual functions, the IDEF$_0$ author can identify input, output, mechanisms, and controls, yet may not see this process as part of the larger CIM system.

Consider the analogy of a biologist studying a cell. The biologist can isolate a cell, carefully label all of its components, and even describe the internal and external dynamics that affect it. However, to consider a cell in isolation is to ignore its more complex purpose—through organization and association, cells acquire larger, more meaningful functions. The same notion applies to the IDEF$_0$ modeling tool. It is precisely at the broader functional levels that there is the greatest potential for improvement through integration. Unfortunately, this opportunity can be overlooked in the maze of details presented by top-down IDEF$_0$ functional diagrams, which are typically bound in one-inch-thick documents with three to six functional activities shown on each page.

A Remedy: The Triple-Diagonal Technique

The IDEF$_0$ triple-diagonal (IDEF$_0$-TD) technique is a bottom-up method of diagramming a system prototype. Its three main components are diagrams of information, control, and material flow systems. These components are developed sequentially, approved, layered together, and quantified to compose a complete overview of the manufacturing architecture.

The basic goals of the triple-diagonal technique are to shorten the lengthy $IDEF_0$ modeling process (to make it workable) and to show the total integration of the factory. Thus, the triple-diagonal concept is an extension of $IDEF_0$ that addresses both of its potential problem areas—the generation of superfluous information and the obstruction of the big picture. For example, CIM's essential information, control, and material flow systems are depicted on one diagram. Imposing this superstructure on the detailed breakdown of the $IDEF_0$ model illustrates the ways in which the various organizational units relate to the processes being linked, and depicts their relationship with other units sharing the same function. This innovation counteracts the tendency toward organizational fragmentation because it forces everyone to look at the forest as well as the trees. Thus, so-called functional foxholes appear as a single coordinated line of defense in the battle for survival.

The triple-diagonal modeling method makes $IDEF_0$ results manageable

Another advantage of structuring the functional model around the information, control, and material flow systems is that they also provide a practical framework for organizing the data generated during $IDEF_0$ diagramming exercises. The triple-diagonal model helps management compile the $IDEF_0$ information into a form that is applicable to its immediate concerns. For example, with this enhanced integration model, management will find it easier to pinpoint areas that require software development and to ensure an effective information flow. In essence, the triple-diagonal method provides management with a tool to shape the results of the $IDEF_0$ method into a logical, managable base of data.

Triple-diagonal modeling procedures

The first step in developing a triple-diagonal model is to create a high-level, top-down diagonal of functions, using basic $IDEF_0$ procedures to identify cost and time drivers. Next, identify key projects that can substantially reduce cost and time. After the projects have been identified, create a bottom-up implementation plan for each project by using the triple-diagonal technique. Position the functions into which each project has been decomposed along the appropriate diagonal (i.e., integrated information system, integrated control system, or integrated material flow system).

The next step is to get expert feedback from all possible sources. At the information system level, ask central management, "Is this how information feedback flows?" At the control system level, ask control management, "Is this how the controls operate?" At the material flow system level, ask the shop floor manager and workers, "Is this the way material flows through your department?" Finally, quantify each project for strategic and tactical input to the company.

The steps involved in constructing a triple diagonal can be summarized by the words *gather, bound, control,* and *quantify.* To show how a triple-diagonal model is built, we will consider an opportunity identified in product assembly. The following example uses a generic assembly model that is an extract of two actual uses of the $IDEF_0$ triple-diagonal modeling procedure.

To construct the triple-diagonal diagram, we begin with the potential productivity improvement, which is based on the assembly analysis. The assembly model, which is identified from opportunity studies, begins with a bottom-up look at material flow and processes. Breaking down the material flow system creates a process-flow diagonal that shows the basic process flow without explaining materials movement and information flow. Basic material handling functions are then added as inputs and outputs are incorporated to show the flow of materials or parts through the shop. This produces the results shown in Exhibit 3. This first diagonal, the material-flow diagonal, is actually a process-flow diagram.

Exhibit 4 constitutes the second level of the triple diagonal, or control diagonal. This level describes the controls and feedback mechanisms that provide control information to the process, or station, level. Exhibit 4 shows how feedback from process and material flow can be shown within the control-system level.

Exhibit 3. *The Material-Flow Diagonal*

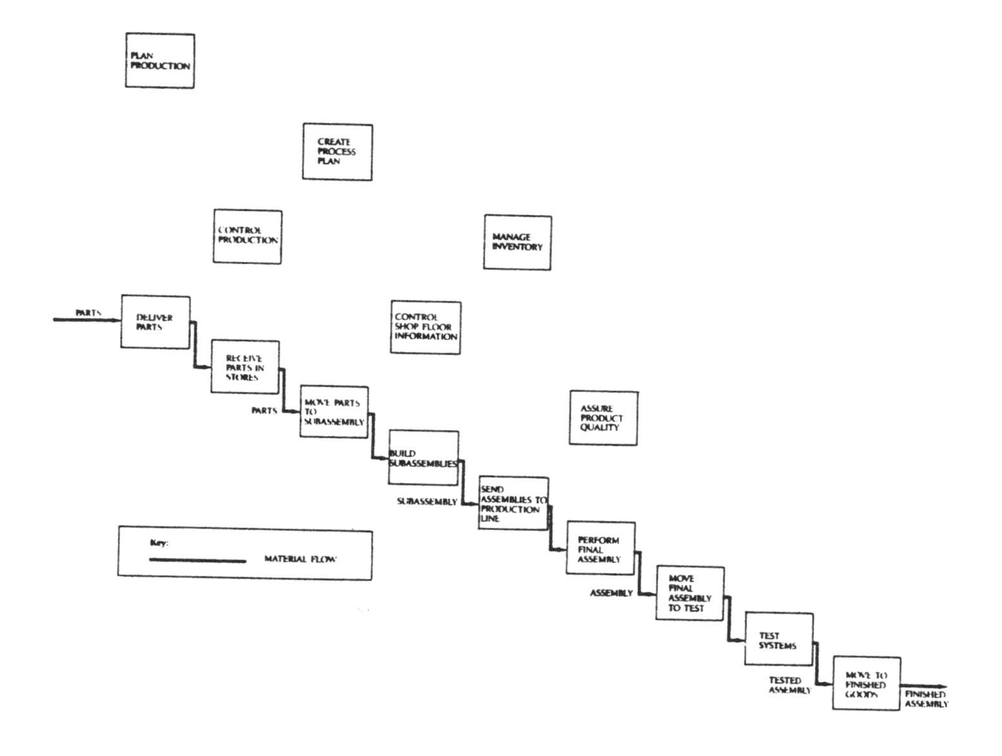

Exhibit 4. *The Control Diagonal*

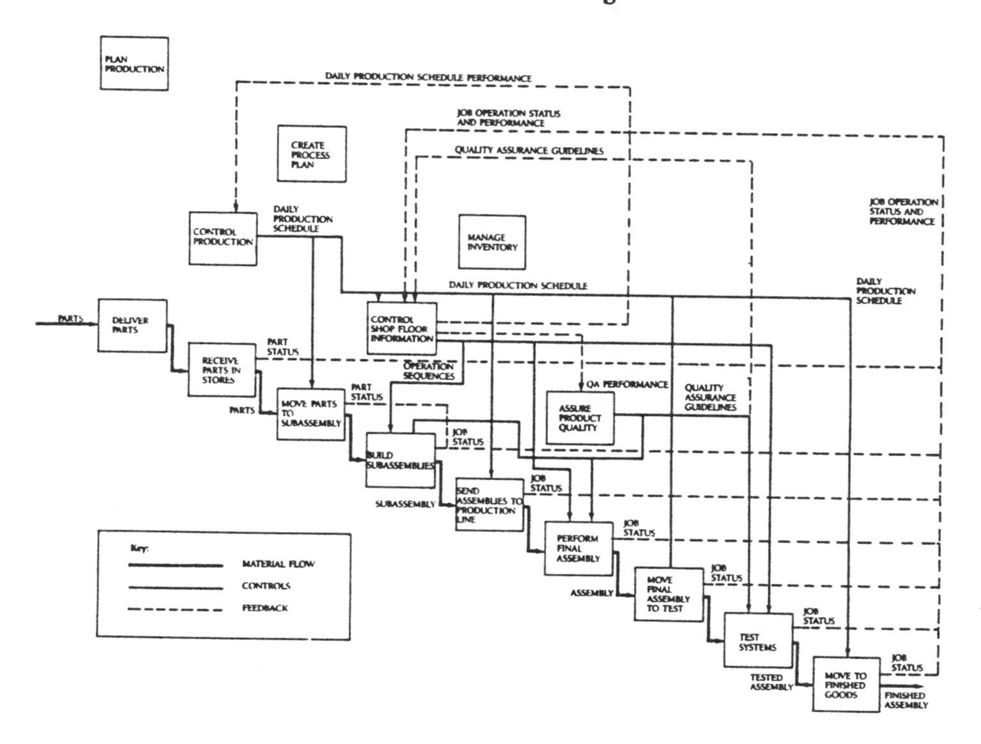

Exhibit 5. *The Information Integration Diagonal*

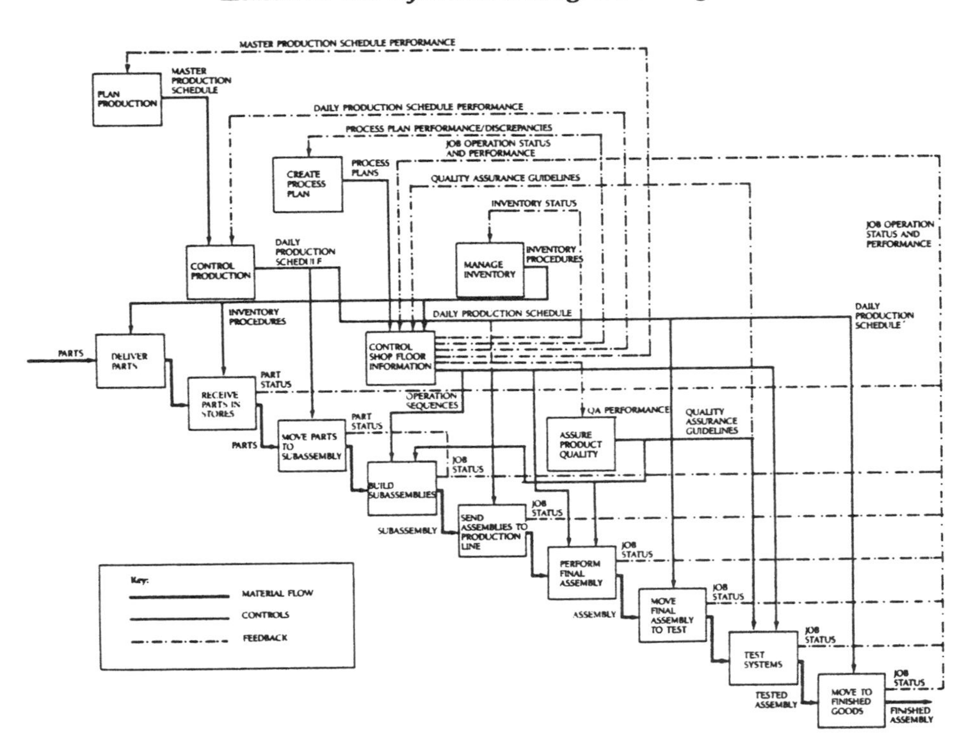

The top tier of the three-tiered diagonal diagrams the integrated information systems. These are the planning tools that fulfill management's needs. As illustrated in Exhibit 5, the top level of the triple diagonal provides controls to its own components and to those of the control-level diagonal. Adding feedback completes the definition.

Obviously, changes resulting from the improved, integrated factory concept can also be scrutinized easily by using the IDEF$_0$-TD enhancement. For example, when local area networks (LANs) are analyzed, all data flows are shown in the IDEF$_0$-TD for proper LAN analysis.

The final steps of constructing the triple diagonal are quantifying material and information flow. Quantification at the material-flow level includes measurements from the current production level and various statistics, including mean time between failure, mean time to repair, mean time between human assistance, and actual speed of the processes. Quantification of information flow addresses points at which information hang-ups typically occur. At this point, timing is crucial and the distribution of the material or parts-per-time unit can be presented in a distribution curve. Simulation models can then be constructed using the IDEF$_0$ TD model as a basis.

User Acceptance

When presenting the IDEF$_0$-TD model, progressive disclosure is the key to describing the results. Beginning with the process and material-flow diagonal, direct labor and floor supervisors can be convinced to accept the diagonal as diagrammed with inherent boundary influences from upper levels. Progressing upward, top management can also be presented with the overall picture to persuade it to buy into the system; management can also recommend further changes (the to-be condition). This approach is different from the IDEF$_0$ diagram approach, which would begin with top-level approval and subdivide from that point. The triple diagonal provides a bottom-up route to planning for an integrated factory of the future and produces a workable extension to the customary top-down IDEF$_0$ models. ▲

DATA FLOW DIAGRAMS
A TOOL FOR THE MODERN INDUSTRIAL ENGINEER

Donald L. Byrkett and David C. Haddad
Miami University
Oxford, Ohio 45056

INTRODUCTION

In today's information age many industrial engineers find themselves developing computer systems incorporating operations research models. Most industrial engineers are well prepared for their role of choosing or developing a reasonable mathematical model to convert data into useful information to facilitate informed decision making. However, this is only part, albeit an important one, of the task expected of today's industrial engineer. The remainder of the task is to develop a computer information system that makes the model a useable tool.

Why does the industrial engineer find him or herself in the role of a systems analyst, designer, and programmer? Each business or corporation has computer systems analysts trained in developing computer information systems. Why don't they take the engineer's model and develop the computer system around it? Why don't they simply use the industrial engineer as a consultant when the need for a mathematical model becomes apparent? The answer lies in the fact that most computer analysts are "business" systems analysts. Most have little or no background in mathematics, statistics, and model building. While they are totally competent and comfortable in developing an accounts payable computer system, many do not even recognize that there is a need for a mathematical model in many systems, such as inventory, scheduling, preventative maintenance, or environmental systems. Many think in terms of tracking inventory rather than seeing an opportunity to save money by optimizing the inventory policy. Even if a mathematical model is developed by an engineer for inclusion into the computer system, many analysts become uncomfortable with developing a system containing a mathematical model. They may find the model intimidating or incomprehensible and thus are reluctant to build a system whose essential component is a mystery. In addition the implementation of the model in the system may require a "scientific" computer language or package with which they are unfamiliar. One of the authors has seen corporations where the MIS group has turned away projects once they became aware that a mathematical model was involved or that a language other than COBOL should be used. In these cases the industrial engineer must become the systems analyst, designer, and programmer as well as the modeler.

The purpose of this paper is to show how a tool used by systems analysts -- the data flow diagram -- can be used to design and model the complete computer system. The mathematical model must be placed in the context of a larger computer information or decision support system. Data flow diagrams can clarify where the mathematical model fits into the computerized system, where the data to support the model will come from and how it is to be checked, what programs are needed to implement the model, and how the people involved in decision making will interact with the computer system. Data flow diagrams serve a number of useful functions; they can at times help identify the need for a mathematical model, they are extremely useful to communicate to the user and the systems analyst what the envisioned computer system will look like, and they help eliminate conceptual errors and ambiguity in the specification of the system. Most industrial engineers have little background in computer systems development beyond a couple courses in FORTRAN programming. Most are familiar with flow charts but few are knowledgeable with current analysis and design methods including data flow diagrams.

SYSTEM LIFE CYCLE

Before getting into the details of data flow diagrams, it is useful to take a bird's eye view of the systems development effort and place data flow diagrams in this larger context. The development of an information system is inherently an engineering activity since design is a major and important aspect of systems development. Systems analysts have bor-

Reprinted from *1988 IIE Integrated Systems Conference Proceedings*

rowed the phased approach used in engineering and applied it to systems development. This phased approach is called the systems development life cycle. There are a number of variations of the systems development life cycle, but most have the following elements: initial analysis, feasibility study, functional requirements definition, functional design, computer system design, detailed design, system construction and testing (including programming), and evaluation. For smaller systems several of these components may be combined into one phase. For some systems where either the requirements are not well defined or the method of solution is not well known, the systems analyst may iteratively cycle through a number of these phases developing successive approximations to the desired system.

Data flow diagrams accompanied by a data dictionary and structured english are the principal products of the functional design. They describe what the new, desired system will do and how it will do this from the perspective of the user of the system. Data flow diagrams are also used in the earlier phases of the systems development life cycle to model the existing (perhaps manual) system that is to be replaced. The data flow diagrams produced in the functional design are the primary input to the computer systems design. It is in this latter phase where design decisions are made about files and programs. Programs are written and tested in the system construction phase.

EXAMPLE DATA FLOW DIAGRAM

In this section of the paper we describe an example of a system an industrial engineer might be asked to develop and give examples of the resulting data flow diagrams, data dictionary, and structured english.

A manufacturing plant maintains a spare parts inventory under the control of its materials management group. Spare parts are issued to production units as they are needed. Periodically the part is reordered to bring the inventory up to a specified target level. The parts manager receives a periodic inventory report, makes decisions on targets, and initiates the reordering of parts. The inventory system is currently a manual one whose successful operation depends heavily on the parts manager's experience and expertise. The parts manager would like the system computerized to make it more efficient and have it function more effectively in her absence.

This is a rather vague description of the problem, but rather typical of the type of description that is initially given. A high level model of the current manual

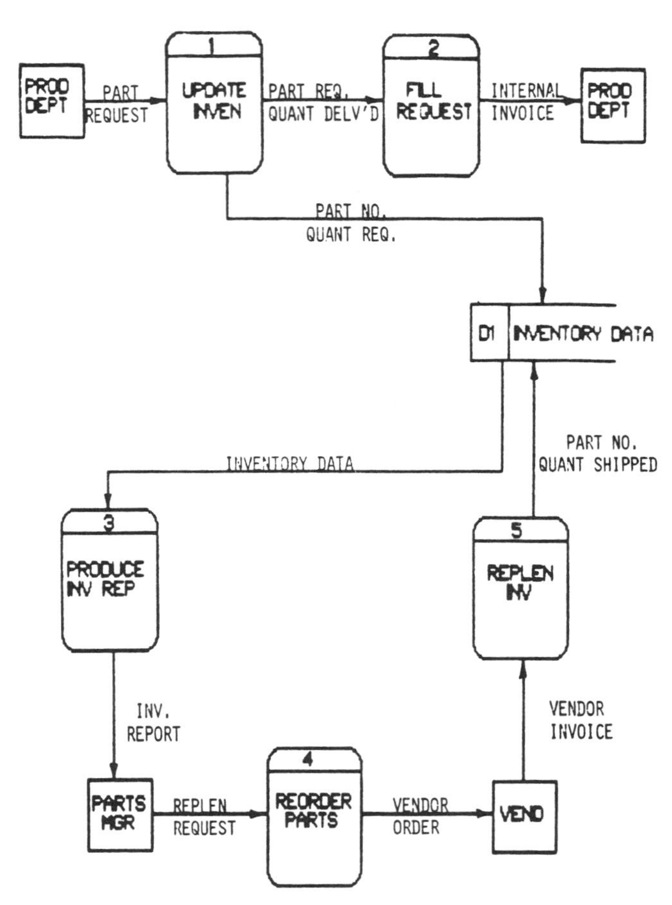

Fig. 1 High level data flow diagram for old manual system

system is given by the data flow diagram in figure 1. A number of interviews with the parts manager, others in the materials management group and with production unit heads will yield a wealth of detailed information on how the current system operates and its deficiencies. The successive levels of details obtained can be built into the model of the current system by "exploding" or providing a detailed close up of each of the entities in the diagram.

Once the current system is understood, requirements for the new system are specified and a high level data flow diagram for the the new system is produced (see figure 2). A comparison of figures 1 and 2 shows that figure 2 is nearly identical to figure 1 except for the addition of 2 entities entitled FORECAST DATA and UPDATE FORECASTS/ TARGETS. These additional entities provide for the efficient operation and optimization of the inventory system.

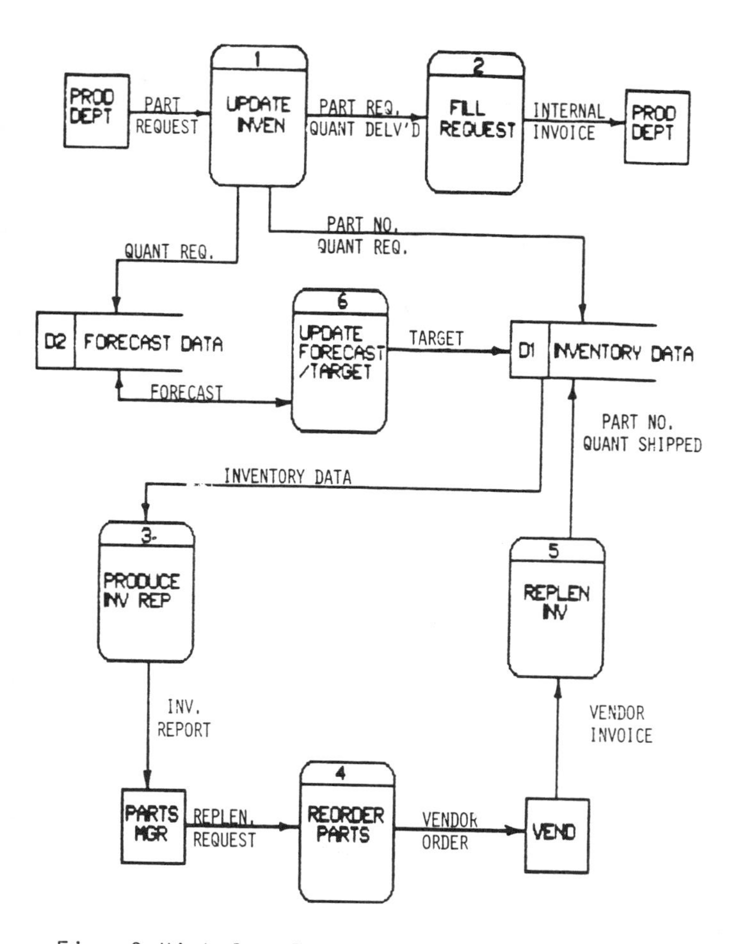

Fig. 2 High level data flow diagram for new computerized system

However, to really understand how the new system is to operate more detail is needed. This is accomplished by exploding and expanding the processes in the high level data flow diagram. For example, Figure 3 contains an explosion of process 1, UPDATE INVENTORY, and Figure 4 contains an explosion of process 6, UPDATE FORECASTS/TARGETS. These explosions provide more detail concerning the handling of shortages (backorders), the tracking of demand, the periodic revising of forecasts, and the calculation and managerial review of new targets. The functional design is not a "one shot" affair, but is arrived at after a number of interviews and trial designs are shared with the users of the system.

Even further detail is provided by the data dictionary in Table 1 and the structured english process descriptions in Table 2. The data dictionary simply provides an explanation of the data depicted in the data flow diagram and the structured english provides a structured explanation of the algorithms used in each process in the data flow diagram. The data dictionary and the structured english process descriptions clearly define the position and implementation of mathematical models in the computer system. Normally, a complete system specification would include explosions of each process, a complete data dictionary, and structured english process descriptions for each process. For the sake of brevity, we have included only two process explosions, only part of the data elements, and only two process descriptions.

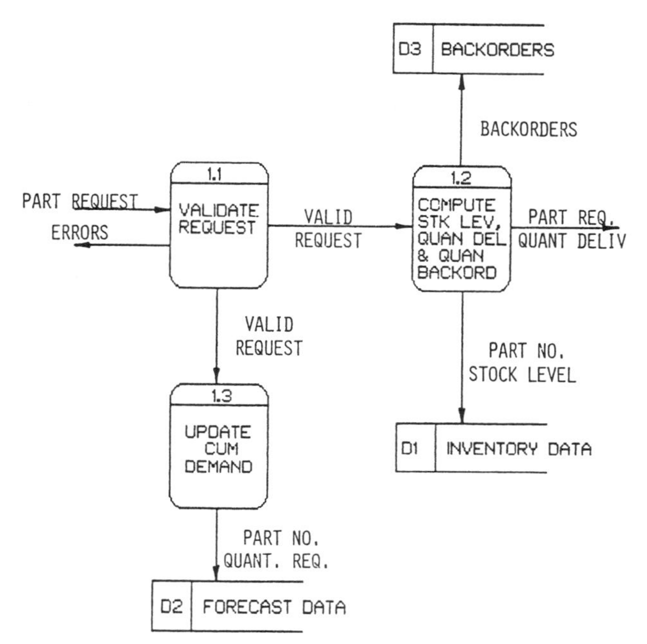

Fig. 3 Explosion of process 1, Update Inventory

DATA FLOW DIAGRAMS

We shall now give a brief explanation of data flow diagrams and their development. The reader is encouraged to see DeMarco [1] and Gane and Sarson [2] for a thorough explanation of data flow diagrams. In addition, Davis [3] has a brief overview of data flow diagrams that is more complete than that given in this article.

A data flow diagram (DFD) is a logical model of a system. It contains no implications on the physical design of programs, files or hardware. It is a pictorial representation of the sources and destinations of data, the processes required to transform the data, the stores or collection of data that will be maintained, and flow of the data through the system. A source or destination (square) is usually a person or organization. A process (rounded rectangle) can depict the manual effort of a clerk, a computer program or part of one, or a combination of manual

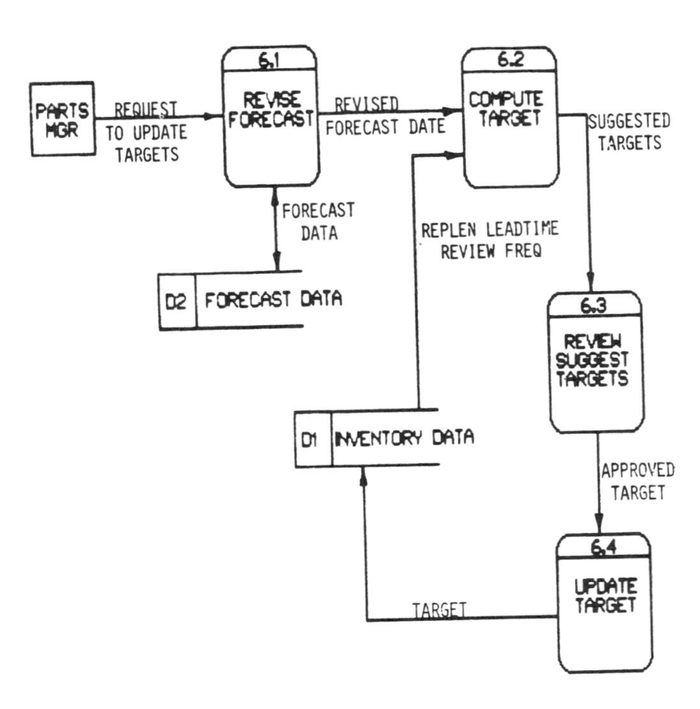

Fig. 4 Explosion of process 6, Update Forecasts/Targets

and computerized effort that transforms data. A data flow (arrow) can represent any means of passing data from one entity to another, such as an invoice, a memo, a phone call, a report, or passing of data between two computer programs. A store of data (open-ended rectangle) can depict any means by which data is stored including a folder in a filing cabinet, a box of index cards, a computer file or part of it. This is why a DFD is said to describe the logical view of the system; it gives no information about the physical manner in which the system is implemented.

So how do we begin drawing a DFD to model a system? Below are some steps that are helpful in drawing a DFD [2].

1. Identify external entities involved, for example, the production department, the parts manager, and the inventory data store. These will be data sources, destinations, and stores.

2. Identify routine and scheduled inputs and outputs expected in the normal conduct of business, for example, a spare part request, an inventory report, and the parts replenishment order. These will be data flows. At this point in the process put aside any input or output related to error handling or exception conditions; these will be handled at a lower level, later. Identify any inquiries or on-demand reports. These will also be data flows.

3. Identify any obvious processes, for example, order parts. All processes should be described with a strong active verb and direct object. It should be noted that it is not normally possible to identify all data sources, destinations, stores, flows, and processes at this time. But we have a start that can be expanded upon later.

4. Our first drawing should be a very high level depiction of the system showing the data flowing from the sources to their destinations and showing major processes. Try to make the flow go from left to right, and, if necessary, duplicate data sources, sinks or stores to prevent the sketch from becoming a hopeless tangle. It will take several drafts to get even this high level diagram to look reasonable.

5. Check the diagram for reasonableness.

a. There should not be direct flows between sources, destinations or stores without passing through an intermediate process.

b. Flows of materials are not modeled in the diagram but any accompanying flows of data are.

c. For every data flow depicted by an arrow, be sure to indicate the data flowing by a descriptive word or two.

d. Data flowing in and out of a data store must be contained in that store.

e. Data flowing out of a process must either flow into the process or be derived from the the data flowing into the process.

f. No decisions should be depicted in the diagram. Decisions will be modelled with structured english later when the process is explained in detail.

g. Data stores representing something in the real world will have to be created and maintained.

6. Begin creating a data dictionary for the data stores and data flows in your high level diagram. A data dictionary describes the data structure (that is, the collection of related data elements) in each data store and data flow. In the example above see the data structures for PART REQUEST and INVENTORY DATA. Data stores can be thought of as stationary data structures that have to be maintained; data flows are data structures in motion.

7. Explode the processes contained in the high level diagram into their functional parts by drawing a separate DFD for each process. The explosion depicts the proc-

TABLE 1 - DATA DICTIONARY

```
PART REQUEST                BACKORDERS
  DEPARTMENT NUMBER           DEPARTMENT NUMBER
  PART NUMBER                 PART NUMBER
  PART NAME                   QUANTITY REQUESTED
  QUANTITY REQUESTED          QUANTITY DELIVERED
                              QUANTITY BACKORDERED

INVENTORY DATA
  PART NUMBER
  PART DESCRIPTION
  STOCK LEVEL
  TARGET
  QUANTITY ON ORDER
  REPLENISHMENT LEADTIME (months)
  REVIEW FREQUENCY (Weekly or Monthly)

FORECAST DATA AND REVISED FORECAST DATA
  PART NUMBER
  FORECAST (monthly)
  MEAN ABSOLUTE FORECAST ERROR
  CUMULATIVE DEMAND SINCE LAST REVISION
```

esses, data flows, stores, sources and destinations contained in the process in the original high level diagram. In our example, see Figure 3 depicting the explosion of the process UPDATE INVENTORY and Figure 4 depicting the explosion of the process UPDATE FORECASTS/TARGETS. Error handling and exception conditions are modelled in this second or lower levels of explosions. It may be necessary to add additional data stores (see BACKORDERS in Figure 3) and data sources or destinations (see PARTS MGR in Figure 4). Flows in and out of the original process being exploded should also be present in the explosion. Check explosion diagrams for reasonableness, and develop data dictionary entries for data stores and flows.

8. Continue exploding processes into their functional parts as long as the functional parts can be described by a strong verb direct object combinations with data flowing between them. Stop exploding a process once further subdivision requires describing how the process is to be implemented. Describing how the process is to be implemented is done with structured english and not by further explosions. Again, check diagrams for reasonableness and develop and refine data dictionary entries for data stores and flows.

9. Make data stores as simple as possible. Examine them for overlaps in data and to determine whether simplification is possible.

10. Remember that it will take multiple drafts of a data flow diagram just to make it readable. As you refine the data stores and the exploded processes, you will gain a better understanding of the system and this may necessitate making changes to higher level diagrams.

11. Finally describe each process in the lowest level diagrams using structured english.

TABLE 2 - ALGORITHMS IN STRUCTURED ENGLISH

```
COMPUTE STOCK LEVEL, QUANTITY DELIVERED,
AND QUANTITY BACKORDERED
Procedure (Process 1.2)

Read STOCK LEVEL from INVENTORY DATA store
If QUANTITY REQUESTED is greater than
STOCK LEVEL Then
   Set QUANTITY DELIVERED = STOCK LEVEL
   Set QUANTITY BACKORDERED = QUANTITY
      REQUESTED - STOCK LEVEL
   Set STOCK LEVEL = 0
   Save STOCK LEVEL in INVENTORY data store
   Save QUANTITY REQUESTED, QUANTITY
      DELIVERED, and QUANTITY BACKORDERED
      in BACKORDER data store
Else
   Set STOCK LEVEL = old STOCK LEVEL -
      QUANTITY REQUESTED
   Save STOCK LEVEL in INVENTORY DATA store
   Set QUANTITY DELIVERED = QUANTITY
      REQUESTED
End if

REVISE FORECAST Procedure (Process 6.1)

Set ALPHA = .25
Repeat for each PART NUMBER
     Read FORECAST, MEAN ABSOLUTE FORECAST
        ERROR, and CUMULATIVE DEMAND SINCE
        LAST REVISION from FORECAST
        DATA store
     Set FORECAST ERROR = CUMULATIVE
        DEMAND SINCE LAST REVISION -
        FORECAST
     Set NEW FORECAST = FORECAST + ALPHA *
        FORECAST ERROR
     Set NEW FORECAST ERROR = ALPHA *
        absolute value of FORECAST ERROR +
        (1-ALPHA) * MEAN ABSOLUTE FORECAST
        ERROR
     Save NEW FORECAST and NEW FORECAST
        ERROR in FORECAST DATA store
     Set CUMULATIVE DEMAND SINCE LAST
        REVISION to zero
End repetition
```

STRUCTURED ENGLISH

The purpose of structured english is to describe a process in clear unambiguous english that is easy to follow. It is useful to imagine that you are describing how to perform the process to a clerk. Since all logic is a combination of a sequence of simple instructions, decisions, and a repetition of simple instructions, we can describe the logic in any process using combinations of these three "structures". You may recognize sequence, decision, and repetition as the three basic structures of structured programming. This makes structured english not only unambiguous and easy to read but also easy to transform into a computer program.

Table 2 contains process descriptions of two processes written using the structured english style. We purposely selected a simple computational procedure (COMPUTE STOCK LEVEL, QUANTITY DELIVERED, AND QUANTITY BACKORDERED) and a mathematical procedure (REVISE FORECAST). Each instruction is stated as a short imperative sentence and all three structures are illustrated. Simple instructions typically begin with the verb Set in order to assign a particular value to a data element. Decisions are represented by the If/then/else structure illustrated in the first procedures. Generally, if a particular condition is true then one group of instructions are performed or else if the condition is false the other group of instructions are performed. Finally, the repetition structure is illustrated in the second procedure using the verb Repeat to indicate a group of instructions that are to be repeated for each part number.

Notice the indentation to highlight the structures being used and the use of capital letters to designate data elements from the data dictionary.

DATA FLOW DIAGRAMS AND MATHEMATICAL MODELS

Notice that each process in the data flow diagram requires an algorithm to transform the data inputs to the process into useful information. Many of the processes require rather simple algorithms to update data elements, calculate totals, or perform simple checks. These processes may be handled by the systems analyst. However, some processes require sophisticated algorithms or mathematical models to efficiently control or optimize an information decision system. This is where the expertise of the industrial engineer can be of assistance in defining the algorithms used for these processes. The authors believe that the data flow diagram provides a key tool for communication between the industrial engineer and the user and also between the industrial engineer and the systems analyst.

It should be noted that the definition of algorithms is not a one way street with the data given and the model to be specified. Rather, there is a tightly woven relationship between the data and the model. The processes in the data flow diagram specify the need for models and the models specify the need for data. The industrial engineer must take care to insure consistency between the model specfied and the data requirements necessitated by the model. For example, the reader may have noticed that the forecasts and forecast errors were revised using a simple exponential smoothing model. The industrial engineer may find that the demand for spare parts is seasonal and want to use a seasonal model. This would change the elements of the FORECAST DATA store to include seasonal indices and would change the structured english process description to include revision of the seasonal indices.

In summary, there is often a tightly woven relationship between the modelling efforts of the industrial engineer and the specification of information systems. The industrial engineer must be a competent systems analyst who can build information systems as well as specify models needed in those systems. The data flow diagram provides a useful tool for specifying and designing information systems and putting the mathematical model in context of the entire system.

REFERENCES

1. T. DeMarco. Structured Analysis and System Specification, Yourdon Press, New York (1978).

2. C. Gane & T. Sarson. Structured Systems Analysis: Tools and Techniques, Prentice-Hall, Englewood Cliffs, NJ (1979).

3. W. Davis. Systems Analysis and Design: A Structured Approach, Addison-Wesley, Reading, MA (1983).

BIOGRAPHICAL SKETCH

Donald L. Byrkett is an associate professor in the Systems Analysis Department at Miami University. He received his Ph.D. in Industrial and Systems Engineering from The Ohio State University. His current teaching and research interests are in simulation and manufacturing information systems.

David C. Haddad is a professor in the Systems Analysis Department at Miami University and is serving as Acting Dean of the School of Applied Science. He received his Ph.D. in Mathematics from Purdue University. In addition to teaching and administration at several universities, he spent five years developing information systems for Union Carbide.

"Managers need not – and should not – abandon the effort to justify computer-integrated manufacturing on financial grounds. Instead, they need ways to apply the DCF approach more appropriately."

Must CIM be justified by faith alone?

Robert S. Kaplan

When the Yamazaki Machinery Company in Japan installed an $18 million flexible manufacturing system, the results were truly startling: a reduction in machines from 68 to 18, in employees from 215 to 12, in the floor space needed for production from 103,000 square feet to 30,000, and in average processing time from 35 days to 1.5.[1] After two years, however, total savings came to only $6.9 million, $3.9 million of which had flowed from a one-time cut in inventory. Even if the system continued to produce annual labor savings of $1.5 million for 20 years, the project's return would be less than 10% per year. Since many U.S. companies use hurdle rates of 15% or higher and payback periods of five years or less, they would find it hard to justify this investment in new technology – despite its enormous savings in number of employees, floor space, inventory, and throughput times.

The apparent inability of traditional modes of financial analysis like discounted cash flow to justify investments in computer-integrated manufacturing (CIM) has led a growing number of managers and observers to propose abandoning such criteria for CIM-related investments. "Let's be more practical," runs one such opinion. "DCF is not the only gospel. Many managers have become too absorbed with DCF to the extent that practical strategic directional considerations have been overlooked."[2]

Faced with outdated and inappropriate procedures of investment analysis, all that responsible executives can do is cast them aside in a bold leap of strategic faith. "Beyond all else," they have come to believe, "capital investment represents an act of faith, a belief that the future will be as promising as the present, together with a commitment to making the future happen."[3]

But must there be a fundamental conflict between the financial and the strategic justifications for CIM? It is unlikely that the theory of discounting future cash flow is either faulty or unimportant: receiving $1 in the future is worth less than receiving $1 today. If a company, even for good strategic reasons, consistently invests in projects whose financial returns are below its cost of capital, it will be on the road to insolvency. Whatever the special values of CIM technology, they cannot reverse the logic of the time value of money.

Surely, therefore, the trouble must not lie in some unbreachable gulf between the logic of DCF and the nature of CIM but in the poor application of DCF to these investment proposals. Managers need not – and should not – abandon the effort to justify CIM on financial grounds. Instead, they need ways to apply the DCF approach more appropriately and to be more sensitive to the realities and special attributes of CIM.

Mr. Kaplan is Arthur Lowes Dickinson Professor of Accounting at the Harvard Business School and a professor of industrial administration at Carnegie-Mellon University, where for six years he was dean of the business school. His first article for HBR, "Yesterday's Accounting Undermines Production" (July-August 1984), was a McKinsey Award winner.

Technical issues

The DCF approach most often goes wrong when companies set arbitrarily high hurdle rates for evaluating new investment projects. Perhaps they believe that high-return projects can be created by setting high rates rather than by making innovations in product and process technology or by cleverly building and exploiting a competitive advantage in the marketplace. In fact, the discounting function serves only to make cash flows received in the future equivalent to cash flows received now. For this narrow purpose – the only purpose, really, of discounting future cash flows – companies should use a discount rate based on the project's opportunity cost of capital (that is, the return available in the capital markets for investments of the same risk).

It may surprise managers to know that their real cost of capital can be in the neighborhood of 8%. (See Part I of the *Appendix* at the end of the article.) Double-digit hurdle rates that, in part, reflect assumptions of much higher capital costs are considerably wide of the mark. Their discouraging effect on CIM-type investments is not only unfortunate but also unfounded.

Companies also commonly underinvest in CIM and other new process technologies because they fail to evaluate properly all the relevant alternatives. Most of the capital expenditure requests I have seen measure new investments against a status quo alternative of making no new investments – an alternative that usually assumes a continuation of current market share, selling price, and costs. Experience shows, however, that the status quo rarely lasts. Business as usual does not continue undisturbed.

In fact, the correct alternative to new CIM investment should assume a situation of declining cash flows, market share, and profit margins. Once a valuable new process technology becomes available, even if one company decides not to invest in it, the likelihood is that some of its competitors will. As Henry Ford claimed, "If you need a new machine and don't buy it, you pay for it without getting it."[4] (For a more realistic approach to the evaluation of alternatives, see Part II of the *Appendix* at the end of the article.)

A related problem with current practice is its bias toward incremental rather than revolutionary projects. In many companies, the capital approval process specifies different levels of authorization depending on the size of the request. Small investments (under $100,000, say) may need only the approval of the plant manager; expenditures in excess of several million dollars may require the board of directors' approval. This apparently sensible procedure, however, creates an incentive for managers to propose small projects that fall just below the cut-off point where higher level approval would be needed. Over time, a host of little investments, each of which delivers savings in labor, material, or overhead cost, can add up to a less-than-optimal pattern of material flow and to obsolete process technology. (Part III of the *Appendix* shows the consequences of this incremental bias in more detail.)

"I still think 'Buyout' is not a proper name for a dog."

Introducing CIM process technology is not, of course, without its costs. Out-of-pocket equipment expense is only the beginning. Less obvious are the associated software costs that are necessary for CIM equipment to operate effectively. Managers should not be misled by the expensing of these costs for tax and financial reporting purposes into thinking them operating expenses rather than investments. For internal management purposes, software development is as much a part of the investment in CIM equipment as the physical hardware itself. Indeed, in some installations, the programming, debugging, and prototype development may cost more than the hardware.

There are still other initial costs: site preparation, conveyors, transfer devices, feeders, parts orientation, and spare parts for the CIM equipment. Operating and maintenance personnel must be retrained and new operating procedures developed. Like software development, these tax-deductible training and education costs are part of the investment in CIM, not an expense of the periods in which they happen to be incurred.

Further, as some current research has shown, noteworthy declines in productivity often accompany the introduction of new process technology.[5] These productivity declines can last up to a year, even longer when a radical new technology like CIM is installed. Apparently, the new equipment introduces severe and unanticipated process disruptions, which lead to equipment breakdowns that are higher than expected; to operating, repair, and maintenance problems; to scheduling and coordination difficulties; to revised materials standards; and to old-fashioned confusion on the factory floor.

We do not yet know how much of the disruption is caused by inadequate planning. After investing considerable effort and anguish in the equipment acquisition decision, some companies no doubt revert to business as usual while waiting for the new equipment to arrive.

Whatever the cause, the productivity decline is particularly ill timed since it occurs just when a company is likely to conduct a postaudit on whether it is realizing the anticipated savings from the new equipment. Far from achieving anticipated savings, the postaudit will undoubtedly reveal lower output and higher costs than predicted.

Tangible benefits

The usual difficulties in carrying out DCF analysis—choosing an appropriate discount rate and evaluating correctly all relevant investment alternatives—apply with special force to the consideration of investments in CIM process technology. The greater flexibility of CIM technology, which allows it to be used for successive generations of products, gives it a longer useful life than traditional process investments. Because its benefits are likely to persist longer, overestimating the relevant discount rate will penalize CIM investments disproportionately more than shorter lived investments. The compounding effect of excessively high annual interest rates causes future cash flows to be discounted much too severely. Further, if executives arbitrarily specify short payback periods for new investments, the effect will be to curtail more CIM investments than traditional bottleneck-relief projects.

But beyond a longer useful life, CIM technology provides many additional benefits—better quality, greater flexibility, reduced inventory and floor space, lower throughput times, experience with new technology—that a typical capital justification process does not quantify. Financial analyses that focus too narrowly on easily quantified savings in labor, materials, or energy will miss important benefits from CIM technology.

Inventory savings

Some of these omissions can be easily remedied. The process flexibility, more orderly product flow, higher quality, and better scheduling that are typical of properly used CIM equipment will drastically cut both work-in-process (WIP) and finished goods inventory levels. This reduction in average inventory levels represents a large cash inflow at the time the new process equipment becomes operational. This, of course, is a cash savings that DCF analysis can easily capture.

Consider a product line for which the anticipated monthly cost of sales is $500,000. Using existing equipment and technology, the producing division carries about three months of sales in inventory. After investing in flexible automation, the division heads find that reduced waste, scrap, and rework, greater predictability, and faster throughput permit a two-thirds reduction in average inventory levels. (This is not an unrealistic assumption: Murata Machinery Ltd. has reported that its FMS installation permitted a two-thirds reduction in workers, a 450% increase in output, and a 75% cut in inventory levels.[6])

Pruning inventory from three months to one month of sales produces a cash inflow of $1 million in the first year the system becomes operational. If sales increase 10% per year, the company will enjoy increased cash flows from the inventory reductions in all future years too—that is, if the cost of sales rises to $550,000 in the next year, a two-month reduction

Example of an FMS justification analysis

With the following analysis, one U.S. manufacturer of air-handling equipment justified its investment in an FMS installation for producing a key component:

1
Internal manufacture of the component is essential for the division's long-term strategy to maintain its capability to design and manufacture a proprietary product.

2
The component has been manufactured on mostly conventional equipment – some numerically controlled – with an average age of 23 years. To manufacture a product in conformance with current quality specifications, the company must replace this equipment with new conventional equipment or advanced technology.

3
The alternatives are:
Conventional or numerically controlled stand-alone.
Transfer line.
Machining cells.
FMS.

4
FMS compares with conventional technology as Table A shows.

5
Intangible benefits include virtually unlimited flexibility for FMS to modify mix of component models to the exact requirements of the assembly department.

6
The financial analysis for a project life of ten years compares the FMS with conventional technology (static sales assumptions, constant, or base-year, dollars) as Table B shows.

7
With dynamic sales assumptions showing expected increases in production volume, the annual operating savings will double in future years and the financial yield (still using constant, base-year, dollars) will increase to more than 17% per year.

On the basis of this analysis and recognizing the value of the intangible item (5), which had not been incorporated formally, the company selected the FMS option.

Table A

	Conventional equipment	FMS
Utilization	30 %-40 %	80 %-90 %
Number of employees needed (including indirect workers, such as those who do materials handling, inspection, and rework)*	52	14
Reduced scrap and rework	–	$ 60,000 annually
Inventory	$ 2,000,000	$ 1,100,000†
Incremental investment	–	$ 9,200,000

*Each employee costs $36,000 a year in wages and fringe benefits.
†Inventory reductions because of shorter lead times and flexibility.

Table B

Year	Investment	Operating savings	Tax savings ITC and ACRS depreciation	After-tax cash flow 50 %
0	$ 9,200	$ 900‡	$ 920	$ –7,380
1		1,428§	1,311	1,370¶
2		1,428	1,923	1,675
3		1,428	1,835	1,632
4		1,428	1,835	1,632
5		1,428	1,835	1,632
6		1,428		714
7		1,428		714
8		1,428		714
9		1,428		714
10		1,428		714

After-tax yield: 11.1 %.
Payback period: during year 5.

‡$ 900 = Inventory reduction at start of project.

§$ 1,428 = 38 fewer employees at $36,000/year + $60,000 scrap and rework savings.

¶$ 1,370 = (1,428) (1 – 0.50) + (1,311) (0.50).

in inventory saves an additional $100,000 that year, $110,000 the year after, and $121,000 the year after that.

Less floor space

CIM also cuts floor-space requirements. It takes fewer computer-controlled machines to do the same job as a larger number of conventional machines. Also, the factory floor will no longer be used to store inventory. Recall the example of the Japanese plant that installed a flexible manufacturing system and reduced space requirements from 103,000 to 30,000 square feet. These space savings are real, but conventional financial accounting systems do not measure their value well–especially if the building is almost fully depreciated or was purchased years before when price levels were lower. Do not, therefore, look to financial accounting systems for a good estimate of the cost or value of space. Instead, compute the estimate in terms of the opportunity cost of new space: either its square-foot rental value or the annualized cost of new construction.

Many companies that have installed CIM technology have discovered a new factory inside their old one. This new "factory within a factory" occupies the space where excessive WIP inventory and infrequently used special-purpose machines used to sit. Eliminating WIP inventory and rationalizing machine layout can easily lead to savings of more than 50% in floor space. In practice, these savings have enabled some companies to curtail plant and office expansion programs and, on occasion, to fold the operations of a second factory (which could then be sold off at current market prices) into the reorganized original factory.

Higher quality

Greatly improved quality, defined here as conformance to specifications, is a third tangible benefit from investment in CIM technology. Automated process equipment leads directly to more uniform production and, frequently, to an order-of-magnitude decline in defects. These benefits are easy to quantify and should be part of any cash flow analysis. Some managers have seen five- to tenfold reductions in waste, scrap, and rework when they replaced manual operations with automated equipment.

Further, as production uniformity increases, fewer inspection stations and fewer inspectors are required. If automatic gauging is included in the CIM installation, virtually all manual inspection of parts can be eliminated. Also, with 100% continuous automated inspection, out-of-tolerance parts are detected immediately. With manual systems, the entire lot of parts to be produced before a problem is detected would need to be reworked or scrapped.

These capabilities lead, in turn, to significant reductions in warranty expense. When General Electric automated its dishwasher operation, for example, its service call rate fell 50%. Designing manufacturability into products, making the production process more reliable and uniform, and improving automated inspection can all contribute to major cash flow savings. Although it may be hard to estimate these savings out to four or five significant digits, it would be grossly wrong to assume that the benefits are zero. We must overcome the preference of accountants for precision over accuracy, which causes them to ignore benefits they cannot quantify beyond one or two digits of accuracy.

We can estimate still other tangible benefits from CIM. John Shewchuk of General Electric claims that accounts receivable can be reduced by eliminating the incidence of customers who defer payment until quality problems are resolved.[7] Consider too that because improved materials flow can reduce the need for forklift trucks and operators, factories will enjoy a large cash flow saving from not having to acquire, maintain, repair, and operate so many trucks. All these calculations belong in a company's capital justification process.

Intangible benefits

Other benefits of CIM include increased flexibility, faster response to market shifts, and greatly reduced throughput and lead times. These benefits are as important as those just discussed but much harder to quantify. We may not be sure how many zeros should be in our benefits estimate (are they to be measured in thousands or millions of dollars?) much less which digit should be first. The difficulty arises in large part because these benefits represent revenue enhancements rather than cost savings. It is fairly easy to get a ballpark estimate for percentage reductions in costs already being incurred. It is much harder to quantify the magnitude of revenue enhancement expected from features that are not already in place.

Greater flexibility

The flexibility that CIM technology offers takes several forms. The benefits of economies of scope–that is, the potential for low-cost production of high-variety, low-volume goods – are just beginning to flow from FMS environments as early adopters of the technology start to service after-market sales for discontinued models on the same equipment used to produce current high-volume models. We are also beginning to see some customized production on the same lines used for standard products.

Beyond these economy-of-scope applications, CIM's reprogramming capabilities make it possible for machines to serve as backups for each other. Even if a machine is dedicated to a narrow product line, it can still replace lost production during a second or a third shift when a similar piece of equipment, producing quite a different product, breaks down.

Further, by easily accommodating engineering change orders and product redesigns, CIM technology allows for product changes over time. And, if the mix of products demanded by the market changes, a CIM-based process can respond with no increase in costs. The body shop of one automobile assembly plant, for example, quickly adjusted its flexible, programmed spot-welding robots to a shift in consumer preference from the two-door to the four-door version of a certain car model. Had the line been equipped with nonprogrammable welding equipment, the adjustment would have been far more costly.

CIM's flexibility also gives it usefulness beyond the life cycle of the product for which it was purchased. True, in the short run, CIM may perform the same functions as less expensive, inflexible equipment. Many benefits of its flexibility will show up only over time. Therefore, it is difficult to estimate how much this flexibility will be worth. Nonetheless, as we shall see, even an order-of-magnitude estimate may be sufficient.

Shorter throughput & lead time

Another seemingly intangible benefit of CIM is the great reductions it makes possible in throughput and lead time. At the Yamazaki factory described at the beginning of this article, average processing time per work piece fell from 35 to 1.5 days. Other installations, including Yamazaki's Mazak plant in Florence, Kentucky, have reported similar savings, ranging from a low of 50% reduction in processing time to a maximum of nearly 95%. To be sure, some of the benefits from greatly reduced throughput times have already been incorporated in our estimate of savings from inventory reductions. But there is also a notable marketing advantage in being able to meet customer demands with shorter lead times and to respond quickly to changes in market demand.

Increased learning

Some investments in new process technology have important learning characteristics. Thus, even if calculations of the net present value of their cash flows turn up negative, the investments can still be quite valuable by permitting managers to gain experience with the technology, test the market for new products, and keep a close watch on major process advances.

These learning effects have characteristics similar to buying options in financial markets. Buying options may not at first seem like a favorable investment, but quite small initial outlays may yield huge benefits down the line. Similarly, were a company to invest in a risky CIM-related project, it could reap big gains should the technology provide unexpected competitive advantages in the future. Moreover, given the rapid pace of technological change and the advantages of being an early market participant, companies that defer process investments until the new technology is well established will find themselves far behind the market leaders. In this context, the decision to defer investment is often a decision not to be a principal player in the next round of product or process innovation.

The companies that in the mid-1970s invested in automatic and electronically controlled machine tools were well positioned to exploit the microprocessor-based revolution in capabilities – much higher performance at much lower cost – that hit during the early 1980s. Because operators, maintenance personnel, and process engineers were already comfortable with electronic technology, it was relatively simple to retrofit existing machines with powerful microelectronics. Companies that had earlier deferred investment in electronically controlled machine tools fell behind: they had acquired no option on these new process technologies.

The bottom line

Although intangible benefits may be difficult to quantify, there is no reason to value them at zero in a capital expenditure analysis. Zero is, after all, no less arbitrary than any other number. Conservative accountants who assign zero values to many intangible benefits prefer being precisely wrong to being vaguely right. Managers need not follow their example.

Author's note: Especially helpful comments on the preliminary draft were made by Robin Cooper and Robert Hayes (Harvard Business School), Alan Kantrow (*Harvard Business Review*), George Kuper (Manufacturing Studies Board), and Scott Richard and Jeff Williams (Carnegie-Mellon).

One way to combine difficult-to-measure benefits with those more easily quantified is, first, to estimate the annual cash flows about which there is the greatest confidence: the cost of the new process equipment and the benefits expected from labor, inventory, floor space, and cost-of-quality savings. If at this point a discounted cash flow analysis – done with a sensible discount rate and a consideration of all relevant alternatives – shows a CIM investment to have a positive net present value, well and good. Even without accounting for the value of intangible benefits, the analysis will have gotten the project over its financial hurdle. If the DCF is negative, however, then it becomes necessary to estimate how much the annual cash flows must increase before the investment does have a positive net present value.

Suppose, for example, that an extra $100,000 per year over the life of the investment is sufficient to give the project the desired return. Then management can decide whether it expects heightened flexibility, reduced throughput and lead times, and faster market response to be worth at least $100,000 per year. Should the company be willing to pay $100,000 annually to enjoy these benefits? If so, it can accept the project with confidence. If, however, the additional cash flows needed to justify the investment turn out to be quite large – say $3 million per year – and management decides the intangible benefits of CIM are not worth that sum, then it is perfectly sensible to turn the investment down.

Rather than attempt to put a dollar tag on benefits that by their nature are difficult to quantify, managers should reverse the process and estimate first how large these benefits must be in order to justify the proposed investment. Senior executives can be expected to judge that improved flexibility, rapid customer service, market adaptability, and options on new process technology may be worth $300,000 to $500,000 per year but not, say, $1 million. This may not be exact mathematics, but it does help put a meaningful price on CIM's intangible benefits.

As manufacturers make critical decisions about whether to acquire CIM equipment, they must avoid claims that such investments have to be made on faith alone because financial analysis is too limiting. Successful process investments must yield returns in excess of the cost of capital invested. That is only common sense. Thus the challenge for managers is to improve their ability to estimate the costs and benefits of CIM, not to take the easy way out and discard the necessary discipline of financial analysis.

References

1 This example has appeared in several articles on strategic justification for flexible automation projects. Clifford Young of Arthur D. Little has traced the example to *American Market/Metalworking News*, October 26, 1981. Other examples of the labor, machinery, and throughput savings from flexible manufacturing system installations are presented in Anderson Ashburn and Joseph Jablonowski, "Japan's Builders Embrace FMS," *American Machinist*, February 1985, p. 83.

2 John P. Van Blois, "Economic Models: The Future of Robotic Justification," Thirteenth ISIR/Robots 7 Conference, April 17-21, 1983 (available from Society of Manufacturing Engineers, Dearborn, Michigan).

3 Robert H. Hayes and David A. Garvin, "Managing As If Tomorrow Mattered," HBR May-June 1982, p. 70.

4 Quoted in John Shewchuk, "Justifying Flexible Automation," *American Machinist*, October 1984, p. 93.

5 See Robert H. Hayes and Kim B. Clark, "Exploring the Sources of Productivity Differences at the Factory Level," in *The Uneasy Alliance: Managing the Productivity-Technology Dilemma*, ed. Kim B. Clark, Robert H. Hayes, and Christopher Lorenz (Boston: Harvard Business School Press, 1985), and Bruce Chew, "Productivity and Change: Understanding Productivity at the Factory Level," Harvard Business School Working Paper (1985).

6 "Japan's Builders Embrace FMS," *American Machinist*, February 1985, p. 83.

7 John Shewchuk, "Justifying Flexible Automation."

Appendix

Getting the numbers right

Part I
The cost of capital

A company always has the option of repurchasing its common shares or retiring its debt. Therefore, managers can estimate the cost of capital for a project by taking a weighted average of the current cost of equity and debt at the mix of capital financing typical in the industry. Extensive studies of the returns to investors in equity and fixed-income markets during the past 60 years show that from 1926 to 1984 the average total return (dividends plus price appreciation) from holding a diversified portfolio of common stocks was 11.7% per year. This return already includes the effects of rising price levels. Removing the effects of inflation puts the real (after-inflation) return from investments in common stocks at about 8.5% per year (see *Table A*).*

These historical estimates of 8.5% real (or about 12% nominal) are, however, overestimates of the total cost of capital. From 1926 to 1984, fixed-income securities averaged nominal before-tax returns of less than 5% per year. Taking out inflation reduces the real return (or cost) of high-grade corporate debt securities to about 1.5% per year. Even with recent increases in the real interest rate, a mixture of debt and equity financing produces a total real cost of capital of less than 8%.

Many corporate executives will, no doubt, be highly skeptical that their real cost of capital could be 8% or less. Their disbelief probably comes from making one of two conceptual errors, perhaps both. First, executives often attempt to estimate their current cost of capital by looking at their accounting return on investment – that is, the net income divided by the net invested capital – of their divisions or corporations. For many companies this figure can be in the 15% to 25% range.

There are several reasons, however, why an accounting ROI is a poor estimate of a company's real cost of capital. The accounting ROI figure is distorted by financial accounting conventions such as depreciation method and a variety of capitalization and expense decisions. The ROI figure is also distorted by management's failure to adjust both the net income and the invested capital figures for the effects of inflation, an omission that biases the accounting ROI well above the company's actual real return on investment.

The second conceptual error that makes an 8% real cost of capital sound too low is implicitly to compare it with today's market interest rates and returns on common stocks. These rates incorporate expectations of current and future inflation, but the 8.5% historical return on common stocks and the less than 2% return on fixed-income securities are *real* returns, after the effects of inflation have been netted out.

Now it is possible, of course, to do a DCF analysis by using nominal market returns as a way of estimating a company's cost of capital. In fact, this may even be desirable when you are doing an after-tax cash flow analysis since one of the important cash flows being discounted is the nominal tax depreciation shield from new investments. I have, however, seen many a company go seriously wrong by using a nominal discount rate (say in excess of 15%) while it was assuming level cash flows over the life of their investments.

Consider, for example, the data in *Table B*, which is excerpted from an actual capital authorization request. Notice that all the cash flows during the ten years of the project's expected life are expressed in 1977 dollars, even though the company used a 20% discount rate on the cash flows of the several investment alternatives. This assumption of a 20% cost of capital most likely arose from a prior assumption of a real cost of capital of about 10% and an expected inflation rate of 10% per year. But if it believed that inflation would average 10% annually over the life of the project, the company should also have raised the assumed selling price and the unit costs of labor, material, and overhead by their expected price increases over the life of the project.

It is inconsistent to assume a high rate of inflation for the interest rate used in a DCF calculation but a zero rate of price change when you are estimating future net cash flows from an investment. Naturally, this inconsistency – using double-digit discount rates but level cash flows – biases the analysis toward the rejection of new investments, especially those yielding benefits five to ten years into the future. Compounding excessively high interest rates will place a low value on cash flows in these later years: a 20% interest rate, for example, discounts $1.00 to $.40 in five years and to $.16 in ten years. If companies use discount rates derived from current market rates of return, then they must also estimate rates of price and cost changes for all future cash flows.

Table A **Annual return series**
1926-1984

Mean annual returns

Series	**1926-1984**	**1950-1984**	**1975-1984**
Common stocks	11.7 %	12.8 %	14.7 %
Long-term corporate bonds	4.7	4.5	8.4
U.S. Treasury bills	3.4	5.1	9.0
Inflation (CPI)	3.2	4.4	7.4

Real annual returns
net of inflation

Series	**1926-1984**	**1950-1984**	**1975-1984**
Common stocks	8.5 %	8.4 %	7.3 %
Long-term corporate bonds	1.5	0.1	1.0
U.S. Treasury bills	0.2	0.6	1.6

Part II
Measuring alternatives

Look again at the capital authorization request in *Table B.* The cash flows from alternative 1 assume a constant level of sales during the next ten years; the cash flows from alternative 5 show a somewhat higher level of sales based on a small increase in market share. The difference in sales revenue as currently projected, however, is not all that great. Only if managers anticipate a steady decrease in market share and sales revenue for alternative 1, a decrease occasioned by domestic or international competitors adopting the new production technology, would alternative 5 show a major improvement over the status quo.

Obviously, not all investments in new process technology are investments that should be made. Even if competitors adopt new technology and profits erode over time, a company may still find that the benefits from investing would not compensate for its costs. But either way, the company should rest its decision on a correct reading of what is likely to happen to cash flows when it rejects a new technology investment.

Table B **Example of a capital authorization request***

Alternative 1 **Rebuild present machines**

Year	1977	1978	1979	1980	1981	...	1986
Sales	$ 6,404	$ 6,404	$ 6,404	$ 6,404	$ 6,404	...	$ 6,404
Cost of sales:							
Labor	168	168	168	168	168	...	168
Material	312	312	312	312	312	...	312
Overhead	1,557	1,557	1,557	1,557	1,557	...	1,557

Alternative 5 **Purchase all new machines**

Year	1977	1978	1979	1980	1981	...	1986
Sales	$ 6,404	$ 6,724	$ 7,060	$ 7,413	$ 7,784	...	$ 7,784
Cost of sales:							
Labor	167	154	148	152	152	...	152
Material	312	328	344	361	380	...	380
Overhead	1,557	1,440	1,390	1,423	1,423	...	1,423

*Adapted from Robert S. Kaplan and Glen Bingham, *Wilmington Tap and Die,* Case 185-124 (Boston: Harvard Business School, 1985).

Part III
Piecemeal investment

Each year, a company or a division may undertake a series of small improvements in its production process–to alleviate bottlenecks, to add capacity where needed, or to introduce islands of automation based on immediate and easily quantified labor savings. Each of these projects, taken by itself, may have a positive net present value. By investing on a piecemeal basis, however, the company or division will never get the full benefit of completely redesigning and rebuilding its plant. Yet the pressures to go forward on a piecemeal basis are nearly irresistible. At any point in time, there are many annual, incremental projects scattered about from which the investment has yet to be recovered. Thus, were management to scrap the plant, its past incremental investments would be shown to be incorrect.

One alternative to this piecemeal approach is to forecast the remaining technological life of the plant and then to enforce a policy of accepting no process improvements that will not be repaid within this period. Managers can treat the money that otherwise would have been invested as if it accrued interest at the company's cost of capital. At the end of the specified period, they could abandon the old facility and build a new one with the latest relevant technology.

Although none of the usual incremental process investments may have been incorrect, the collection of incremental decisions could have a lower net present value than the alternative of deferring most investment during a terminal period, earning interest on the unexpended funds, and then replacing the plant. Again, the failure to evaluate such global investment is not a limitation of DCF analysis. It is a failure of not applying DCF analysis to all the feasible alternatives to annual, incremental investment proposals.

*Roger G. Ibbotson and Rex A. Sinquefield, *Stocks, Bonds, Bills and Inflation: The Past and the Future* (Charlottesville, Va.: Financial Analysts Research Foundation, 1982). The author has updated this study for returns earned during 1982-1984.

This estimate should be adjusted up or down, depending on whether the project's risk is above or below the risk of the average project in the market. A detailed discussion of appropriate risk adjustments is beyond the scope of this article. Good treatments can be found in David W. Mullins, Jr., "Does the Capital Asset Pricing Model Work?" HBR January-February 1982, p. 105, and in chap. 7-9 in Richard Brealey and Stewart Myers, *Principles of Corporate Finance*, 2d ed. (New York: McGraw-Hill, 1984).

Reprinted courtesy of Thomas Publishing Co. from *Managing Automation*, August 1986.

Cost Justification Is Possible

But it's going to take a change in culture—a change from the top down.

By Tony Baer

Like motherhood and apple pie, the benefits of automation are easy to describe but hard to quantify. The problem of justifying automation is well-known: the traditional cost accounting criteria of industry—direct labor, materials and indirect overhead—simply cannot place fair dollar values on benefits such as quality and responsiveness. Just about the only modern innovation in manufacturing costing has been the introduction of discounted cash flows to reflect inflation. Otherwise, industry is still justifying 1980s computer integrated manufacturing technology with 1950s criteria.

To rectify this situation, accounting practices are going to have to change. In particular, accountants are going to have to learn to communicate with engineers and manufacturing personnel in ways they haven't before. It's probably going to take a change in corporate culture, and it likely won't take place without encouragement from the most senior managers.

Traditional costing methods were fine when manufacturing was a high

Illustration: David Klein

volume, labor intensive business. Direct labor once accounted for up to a third of a product's cost. But labor's share of the pie today has dropped to 10 percent, while materials and overhead have climbed to 55 and 35 percent, respectively, according to manufacturing audit partner Steven Hronec of Arthur Andersen & Company.

"If you look at a standard cost sheet, many of them have the labor component quantified to the fourth decimal place," says Hronec. "Manufacturers are analyzing numbers based on a very small component of the product."

The "other" numbers can make a difference, maintains Robert Kaplan, the Arthur Lowes Dickenson Professor of Accounting at the Harvard Business School and professor of industrial administration at Carnegie-Mellon University. "We're spending money on inventory, bad quality and floor space. If we can forecast the savings, then we can have at least a baseline. It won't be to three or four digits, but at least it will get us within ten percent."

The benefits of automation and computer integration are as well known as the problems of justifying them. Higher quality products can open up new markets, and faster cycle times can deliver the product to the customer ahead of the competition. Linkups between shop floor and computer aided design systems can improve design consistency and rationalize product lines. Advanced manufacturing systems typically use state-of-the-art sensors and instrumentation, which improve data collection and status reporting. That in turn leads to better job flow and tracking, reducing the need for heavy inventories.

Then there's the question of what to do with the additional floor space freed up by the reduced inventory. It can be sold, rented or dedicated to new product lines.

All of these advantages should be considered together, advises James Brimson, vice president of Computer Aided Manufacturing International, Inc., (CAM I), Arlington, Texas. For instance, a flexible manufacturing system, in conjunction with an effective MRP II system, will reduce work in process. But a group technology (GT) system to reduce parts routings and tooling changes may only be as good as the CAD/CAM systems with which it is linked. However, together or apart, virtually all of these costs and benefits still tend to fall through the cracks of traditional accounting systems, which tend to lump the more quantifiable ones into overhead and the others as "intangible."

Just because these costs have not been tracked before, the task is not impossible. The amount of homework will vary depending on the type and scope of the data desired. For instance, tracking inventories might require only the part-time services of a single person or some reprogramming of the MRP system. Analyzing the costs of quality might require a more ambitious beneath-the-surface investigation involving a company-wide effort ranging from the shop stewards and design engineers up to the accountants.

Manufacturers are taking intangibles into account to justify automation.

Such a comparison of the quality costs of manual and automated processes would have to document and evaluate the additional steps required whenever a bad piece is produced. Defective parts or products will likely necessitate additional inspection time (especially if they have to go through twice) and rework, or they might be tossed out altogether. If the defective items slipped through the quality control department, they could later wind up in warranty service. Each of these activities incurs materials and labor costs, not to mention the additional paperwork generated.

Less quantifiable, but just as real, is the cost of lower market share when a company gets a consistent reputation for poor quality, as Detroit learned the hard way. These costs are usually aggregated under overhead. However, the effort to disaggregate the data might encounter resistance from plant personnel who "just want to do their jobs."

The other side of the coin is the "funny money" syndrome, where the figures aren't taken seriously. In one instance, a comptroller for a Midwestern manufacturer of machined parts ran the numbers on a new $1 million FMS cell containing six lathe and milling machines which reduced cycle times from 12 to 8 weeks. Even with very conservative marketing figures, predicting a seven to eight percent sales gain if the product reached customers a month sooner, return on investment for the new cell still hit an astronomical 100 percent. The comptroller knew that figure would never sell to his superiors, so he readjusted the market projections downward to one percent, deriving a more believable 25 percent return. The cell was eventually approved and went on to cut lead times as predicted, posting a 40 percent return after its first year.

Yet no matter how modest or reasonable the presentation, there will always be those in the head office who will reject anything based on projections. Faced with a choice between rough estimates and zeros, says Harvard's Kaplan, conservative accountants will invariably go for zero.

"Yet zero is no less arbitrary than any other number," he maintains. Robert Chovanec, manufacturing consultant with Ernst & Whinney, Chicago, finds running the numbers in multiple best, worst and base case scenarios often makes these projections somewhat more palatable to top managers.

Manufacturers are beginning to taking these "intangibles" into account to justify automation, reports a 1985 survey by the National Electrical Manufacturers Association. Not surprisingly, few were quantifying their criteria. Most companies were justifying advanced manufacturing technologies based on strategic objectives to remain in a given market or to get their feet wet in CIM or FMS.

Ironically, cost reduction may be the lowest priority for automation. "Cost reduction is often the last thing we do," reports a top industrial financial analyst. "If we've done the strate-

Tony Baer is a science and technology writer based in New York.

gic planning correctly, achieving the plan should automatically get the costs down."

General Electric Company made a strategic decision to stay at the top of the dishwasher market when it modernized its much-heralded Louisville, Ky., appliance complex several years ago. The cost reductions were almost incidental: scrap and rework dropped 40 percent, product service calls declined by 50 percent, and original equipment manufacturer private label orders jumped. Such figures would have been difficult to predict during cost justification.

The purpose at GE wasn't necessarily to build the cheapest dishwasher on the market, but to build the best. Evidently, the strategy worked. According to John Shewchuck, GE's manager of automation projects, the payback for the Louisville plant came well before the projected 3.4 years.

In fact, a growing school of thought questions whether cost reduction in itself should be the real goal of automation. "No longer does the least expensive product on the market necessarily command the dominant market share," says manufacturing consultant John Kabbes of Touche Ross & Company. Quality and reliability rank high, he says, pointing to Japanese cars as one example.

If automation always guaranteed a better product and market share, there would be no need to justify CIM or FMS investments. But the world isn't perfect. A poorly planned CIM system can be worse than none at all. So the question of how to justify automation is very much alive.

Few proven solutions have emerged, but the dilemma has spawned its share of theories and speeches on the rubber chicken circuit. Most of the theories revolve around strategic planning, delineating lump-sum overheads to specific product lines, isolating cost centers through functional approaches, factoring in opportunity costs of getting a better quality product to market sooner or freeing up floor space for other activities or product lines.

Some cost justification theories even conjure up visions of future shock. "Why not pay the robots?" ventures Keith McKee, director of the Manufacturing Productivity Center at the Illinois Institute of Technology. His colleague, Martin Bariff, associate professor of information resource systems at the IIT's Stewart School of Business, recently developed an unorthodox concept that shifts direct costs from labor to machine hours. Bariff admits that the manufacturers he visited reacted quizzically to his concept.

In justifying an FMS cell, return on investment can be so high that the numbers have to be adjusted downward.

Systems integrators, particularly from the Big Eight accounting firms, have become especially active in formulating and applying new justification techniques. Of course, that's not surprising given the nature of their business to promote new technology, and the receptiveness of their market shouldn't be that surprising, either. A client seeking the services of an integrator is likely to be more predisposed to modernization and innovative thinking.

And who are these clients? Not your typical company, says John Senatore, a partner at Arthur Young and Company, Boston. Most of his clients come from opposite ends of the spectrum: either they are already at the

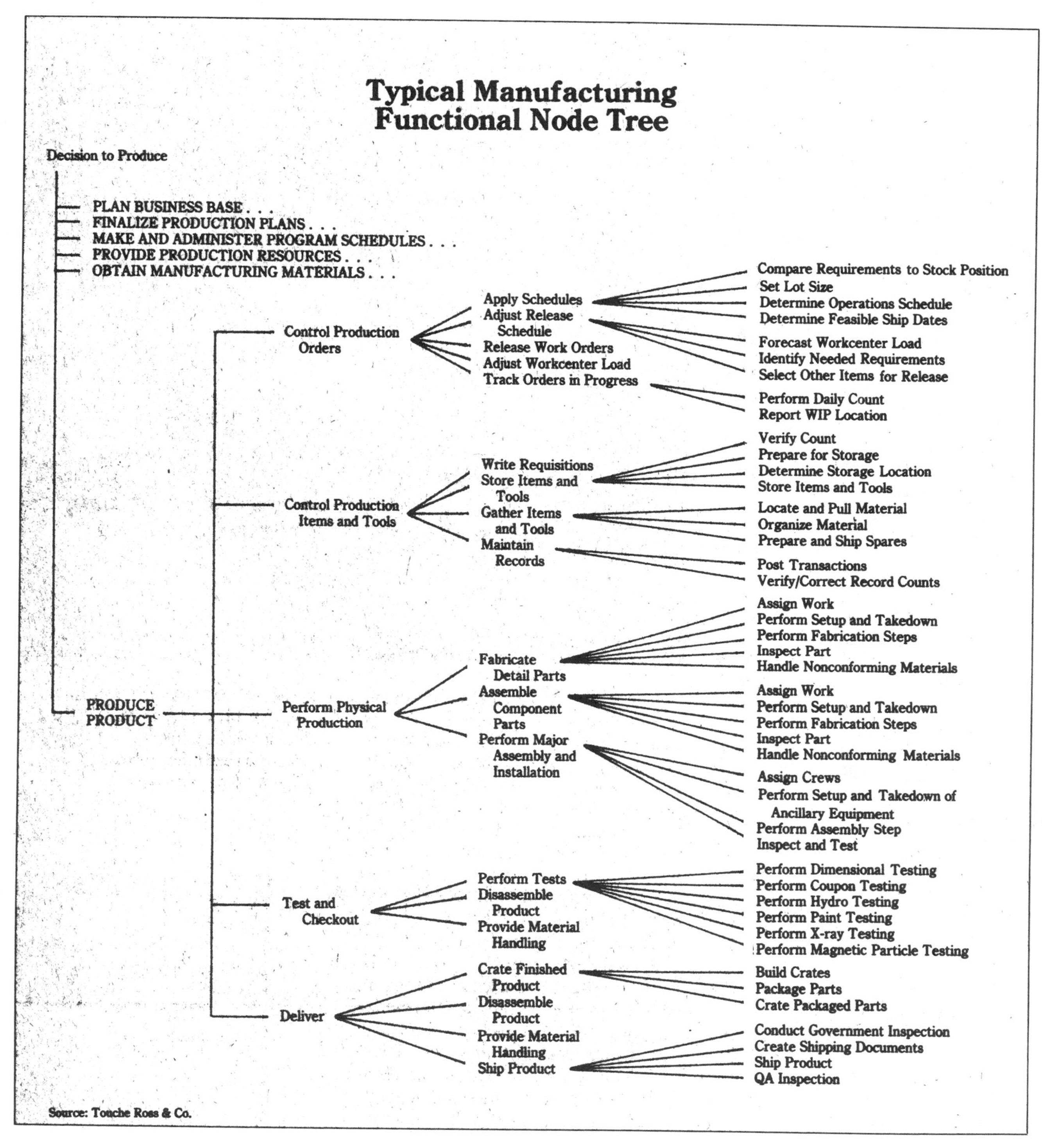

top of their markets and want to maintain their positions or they are struggling just to stay in business.

Joining the movers and shakers from the Big Eight is a small cadre of in-house analytical accountants and financial analysts from major corporations and manufacturing trade associations, with a few business school professors sprinkled in. Just last year, CAM I came into being for the sole purpose of updating industry cost management practices. A not-for-profit consortium backed by three dozen members drawing from the automotive and aerospace industries, the Big Eight accounting firms and the military, CAM I says that its initial goal is to produce a rudimentary CIM cost management system by the end of the year.

In an ideal world, it would be nice to factor out overhead and identify the cost bottlenecks or break down scrap and rework categories to see which products or processes are causing the most problems. In the real world, it is usually more expedient to work with the data that are already there, according to Ernst & Whinney's

Chovanec. Improving plant data collection systems, he says, should evolve with time as new production systems go on line.

Although overhead may hold the buried treasure of data, Chovanec tends to target inventories and work in process first, going through relatively obscure data such as stockroom move tickets, which record all new raw and finished goods inventory additions. Chovanec looks for variances in labor, performance and materials data to see if anything is falling through the cracks. The discrepancies may provide a clue to scrap and rework levels. Work in process, admits Chovanec, is harder to track because it is a moving target.

Naturally, he also looks at MRP data, but checks the procedures by which data are entered to ascertain their timeliness. Even the most detailed MRP systems, says Chovanec, can suffer the age-old "garbage in, garbage out" problems.

He often compares the data by source. Accounting departments, says Chovanec, tend to project budgets based on original engineering data, and often aren't as up to date as the manufacturing end, which carries data on the latest product specification or configuration changes. These may force the revaluing of inventories. On the other hand, Chovanec has found some in-house accounting departments that have entered the 1980s, using software tools such as Lotus 1-2-3 or other spreadsheet programs to stay current with inventory revaluations.

Though some automation consultants are resourceful enough to reinterpret obscure data, others take even more ambitious approaches. They dissect companies from top to bottom to determine the true functional relationships among different activities. Then all the activities required to produce a product can be factored in, and the true cost centers emerge.

A node tree approach, which traces such relationships, was used by Touche Ross to identify the cost drivers at a defense contractor that was already using automated equipment (see opposite page). The computerized numerical control cutters were turning out pieces that regularly exceeded their tolerances. Traditional analyses might have fallen down here because the cutting machines themselves worked fine. By building a tree examining floor shop activities based on throughput, it was found that related activities were causing the cutting center problems.

For instance, parts emerging from the CNC cutters went on to a quality control area where the temperatures were much cooler. This caused the parts to contract and fall out of tolerance. Other problems included inconsistent tooling procedures used by different shifts and the overall low level of training given to the operators. The solutions, which cost $1 million to implement, included stabilizing temperatures in the CNC cutter area to match those of the quality control area, adopting GT practices and CAD system links to limit retooling as well as improving operator training.

Is it worth $100,000 per year to reduce work in process, lead times and inventories and get products to the customer faster?

Many systems integrators now have tools that can perform rough-cut justifications on personal computers. Ernst & Whinney has developed a menu-driven manufacturing cash-flow model that compares base case and new options. The user enters baseline sales and cost data. Costs include administrative, marketing and other expenses not categorized.

The model also asks for the typical calendar duration of most accounts receivable, both for the base case and for any new system. Here the user can specify how much sooner accounts would be paid if the system issued invoices faster.

The system also asks the user to project future sales and inventory levels as well as the costs of operations and maintenance, including manufacturing equipment, computer hardware and software. Labor, training and indirect overhead costs are also requested. The operator enters the discount rate, which the model computes for all expenses and savings. What comes out is a series of three-year spreadsheets detailing cash flows attributable for the base case and new manufacturing systems.

One such factor that can emerge from these analyses is opportunity costs. If, after examining the cash flows attributable to direct costs and overhead, a new CIM system still incurs a negative cash flow, Kaplan suggests a value judgment be made. For instance, is it worth $100,000 per year to reduce work in process, lead times and inventories and get products to the customer faster? Or is it worth that much to stay in or enter a particular market?

Virtually all of the proposals to reform cost justification revolve around one tenet: long-term corporate strategy. A company should decide where it wants to be and what it wants to make, then implement whatever is necessary to achieve that goal. Tools such as Touche Ross's node tree, says Kabbes, can translate those strategies into the context of day-to-day activities.

The flip side of considering opportunity costs is deciding whether it is worthwhile to sacrifice that strategy because a project falls a couple of points short of the ROI goal, or, as CAM I's Brimson puts it, whether the company is willing to sacrifice its place in the market because its technology is outdated.

When it comes to corporate strategy, automation often looks like a white knight. But when it comes to justifying it with numbers, it is guilty until proven innocent. The alternatives to traditional cost accounting systems are still embryonic. Most corporate accountants view with skepticism anything that gets away from quantifying direct costs to those four or five decimal points.

"It's going to take a change in culture," says Kaplan. "They are going to have to get out and talk with engineering and manufacturing and marketing people, and get closer to the underlying process of the product." That change will likely have to come from the top down. ■

Reprinted courtesy of Taylor & Francis from *International Journal of Computer-Integrated Manufacturing*, vol. 2, no. 6.

Calculating investments for integrated manufacturing: looking at the overall costs and benefits

DIETER BOELZING and HERBERT SCHULZ

Abstract. Traditional investment calculations focus on singular aspects and functions of the investment. Integrated information technologies affect more than the small area of the actual investment; most of their real positive economic effects are generated in other areas of the enterprise by using primarily generated data, and giving a higher reliability to the information basis. This leads to a justification method, which includes the overall aspects of these kinds of investment, by systematically analysing the cost and benefit impacts on major cost categories in every functional unit of the enterprise. The result is the potential for cost reduction, since it shows the most efficient amount of investment to be made.

1. Why traditional methods have to fail in evaluating integrated manufacturing

Investments in computerized manufacturing automation are looked at differently by technical and by business investment planners. Engineers tend to focus on the technical advantages and necessities of these technologies; business people tend to prefer a short payback period to justify the high cost, which is usually not realistic for these investments. The necessity to act is often recognized, but—and this is the point—there are major stumbling blocks: which systems to choose, in which to extend automation, and with which systems to start. The variety is wide: for design there are computer aided design systems (CAD), process planning might be supported by computer aided planning systems (CAP), in manufacturing the computerized controlled machines and devices of computer aided manufacturing (CAM) promise to be much more effective than the old, conventional machines and practices. Computer based systems for production planning and scheduling (PPS, MRP) are used in some companies, but are often driven by organizational necessities to harness the otherwise inevitable chaos, and without estimating the real benefits.

The major feature of these investments is the long-term nature of the introduction process and economic advantages. The strategic character of the decisions and investments is no longer a point of discussion (Wildemann 1988, Hayes *et al.* 1988, Gerelle and Stark 1988, Kaplan 1986). This creates the need to use hurdles other than just the payback-period to evaluate investments in computer based factory automation (CBFA). In common economic evaluation calculations the future costs of a system or machine might be quite well known (Cooper and Kaplan 1988). But the flow of revenues caused by this investment is always based on estimates. The assumed accuracy of these methods usually suggests that one should not invest in high-risk, high-technology investments with long payback periods.

But all these methods have in common that they are oriented towards financial flows and financial issues which are based on single technical features of the investment (see, for example, Gupta *et al.* (1988), and Baudin (1985)). versus the existing one, the focus is on higher output per unit time, or lower costs per unit. But the impact on other functional units, the changed interaction with the surrounding machines, and the future capabilities for a higher integrated concept are not included in those calculations. Engineers could argue with those aspects by using cost–benefit matrices, but these are often neglected by controlling interests if the short payback is not visible.

Besides the long-term issue connected with investments for CBFA, there is another aspect of vital importance. When looking at a 'traditional' investment, it is often enough to concentrate on the functional area in which the investment is placed. Effects on other functional units of the enterprise are not look at (see, for example, Herroelen *et al.* (1986) and Lederer and Singhal (1988)). But—and

Authors: Dieter Boelzing and Professor Dr.-Ing. Herbert Schulz, Technical University of Darmstadt, Department of Mechanical Engineering, Institute for Cutting Technologies and Machine Tools, Petersenstrasse 30, D-6100 Darmstadt, F.R. Germany.

this is the major difference of the 'new' investments—most computer based systems are aimed at an integration of functions, of information flows, and of data generation and storage. It is therefore not enough to take only the 'direct' effects into consideration; the 'indirect' effects are usually much more important. Savage (1988) pointed out that about 80% of the tasks and challenges connected with the introduction of CIM are organizational. Perhaps the difference between the two effects should be explained briefly. *Direct effects* are generated inside the functional unit where the investment is placed. For example, when purchasing and installing a CAD system, the direct effects are within the design department. *Indirect effects* are generated in other functional units, which can benefit from the former digitalization of data, from higher accuracy of transmitted data, from faster receiving of data, or by avoiding multiple data-entry. In our example, the process planning department can benefit from CAD data by directly taking the geometric information in the form of the CAD data-set for generating the NC program, or by using the drawing information to create an explosion chart which shows the order of assembly for the production workers. By avoiding the new data entry, this department benefits from the investment placed in another department and therefore receives an indirect benefit. Comparable effects take place in other departments, when drawings are received earlier, or with higher accuracy, or cost information can be given faster to the accounting department.

New approaches for justifying new manufacturing technologies usually focus on just one point: to justify the enormous expense of the investment it is suggested that more and more of the qualitative aspects be taken into consideration, which often means that the financial risks are neglected (see for example Meredith (1986, 1987), Sullivan and Liggett (1988), Canada and Sullivan (1988). Or, as Fine (1988) shows, one can concentrate on a narrow, model based analysis (Lederer and Singhal 1988, Gupta *et al.* 1988, Fine and Freund 1988) and neglect to view the company holistically, which is absolutely necessary when starting to evaluate manufacturing technologies.

The method discussed in the next section enables one to take these integrating effects of advanced manufacturing technologies explicitly into consideration on a monetary basis, and so to integrate the direct and the indirect effects of investments in CBFA with the economic analysis in order to justify—or cancel—the investment.

2. Cost reduction and performance enhancement dominate the evaluation process

The basic principle of the method presented is that all cost, benefit, and performance effects are translated into their impact on the total costs of the company, here called the overall cost. This also means it is easier to show the impact on the profitability of the company. The overall cost is focused on costs for generating products; costs for financial transactions or other 'non-relevant' costs are excluded. This will be explained later.

Costs and output can each be described by a volume component and a value component. A cost reduction therefore might be reached by a reduction in consumption of a raw material (for constant ouput), or by an enhancement in output to higher volume (while costs are constant). So the fundamental effects can be described by

(a1) same output for less input *volume*
(a2) same output for less input *value*

(b1) same input for higher output *volume*
(b2) same input for higher output *value*

Usually it is quite difficult to distinguish these effects clearly when analysing a factory. They are recognized as superimposed on four major categories:

(c1) higher output value for less input volume;
(c2) higher output value for less input value;
(c3) higher output volume for less input volume; and
(c4) higher output volume for less input value.

It is very important consciously to realize these different types of effects when evaluating the effects of the analysed systems. For this analysis cases (a1)–(b2) are most relevant for showing the impacts of investments in integrated technologies more clearly. But in practical use, cases (c1)–(c4) can also be used to evaluate the investment's impact.

As shown in Table 1, most cost and performance effects of CBFA can be broken down into four major categories: productivity, quality, flexibility, and capacity utilization. Productivity is defined by the relation of production output to the necessary input. This input might be human work time, capital, or material, with related forms of productivity (productivity of work, capital, or material). Flexibility is described by the capability of the production system to adapt to a change in input variables per unit of time. This includes the costs of the adaption measure, and the speed of reaction measured in throughput time. Quality can refer to the product-quality (quality in using the product), and the manufacturing quality (costs for scrap, rework, and quality prevention). Capacity utilization uses the time-based utilization of machines and production equipment for manufacturing the products; it is calculated by the machine cost per produced unit.

For the evaluation of the overall costs and benefits of

Table 1. Cost and performance implications of computer based factory automation (CBFA), (separated for the impact on the volume and the value components of cost and performance).

Change in cost/benefit	Volume/value component	Measure of computer based factory automation	Main effect	Group of effect
Benefit = 0 Cost < 0	Volume < 0	Less used labour time through CAD, CAP, CAM, PPS:		throughput time
		CAD: Faster drawings, faster changes	Productivity	
		CAP: Faster process planning and NC programming	Productivity	
		CAM: Less set-up time in manufacturing	Flexibility,	
		Less double work/multiple fixtures Less rework and scrap	Productivity	
		PPS: Less planning time, higher transparency Higher accuracy of planning	Flexibility	
	Value < 0	Better use of values by CAD, CAP, CAM, PPS:		productivity
		CAD: Less material input by optimized design	Productivity	
		CAP: Less material input by more accurate process plans and better NC programs	Productivity	
		CAM: Less material input by optimized use of machinery	Productivity	
		PPS: Less bound capital, less loss by over-maturity of material	Productivity	
Benefit > 0 Cost = 0	Volume > 0	Higher volume output by CAD, CAP, CAM, PPS:		throughput time
		CAD: More drawings by same staff volume	Productivity	
		CAP: More process plans/NC programs by same staff volume	Productivity	
		CAM: Higher capacity utilization, more produced goods by same capacity	Productivity,	
		PPS: More orders to be planned by same staff volume	Productivity	
	Value > 0	Higher value output by CAD, CAP, CAM, PPS:		quality
		CAD: More accurate and unequivocal drawings	quality	
		CAP: More accurate process plans, NC programs with fewer mistakes	quality	
		CAM: Higher manufacturing quality, less rework, Higher on-time delivery rate, and therefore higher prices	Quality	
		PPS: Execution of short-term changes possible	Flexibility	

integrated manufacturing technologies it is sufficient to concentrate on the four major cost categories: direct labour, material, manufacturing equipment cost (i.e. depreciation) and interest (for work-in-process, etc.), and other costs (Schulz and Boelzing 1989). All are effected by CBFA investments, but in different ways, according to the special kind of investment.

3. Overall perception of the company

As pointed out before, it is not only the cost and benefit impacts within the investing unit which are of importance. Much more important is the impact on all the other functional units in the company. A functional unit is defined as one which fulfils a complete task in the creation process of the products, and which can be analysed for cost independently of the others. Functional units are, for example, the design department, the process planning department, finance and administration, mechanical processing, or assembly. In more detailed analysis the mechanical processing department can be further divided into a drilling, a turning, and a milling unit (in companies with shop-oriented layouts) or into product-oriented cells.

A major distinguishing feature is the orientation of this method to the targets of factory automation, but not only in cost categories. It is the performance which is enhanced by application of the technologies to which these calculations are directed, and therefore the impacts of an enhanced peformance according to the special targets of the factory automation project are the centre of interest. Thus some costs may fall while others may rise. It is not single aspects but the overall balance which provides the basis for the final decision process.

The steps one must go through in the evaluation process are, in short, the following:

1. Determine the *cost structure* of the company for functional units, and for the major cost elements (those affected by automation, see above).
2. Define the *targets* for the investment, according to the overall strategy of the company.
3. Investigate the *tasks* which *might* be supported by a computer based system (CBS), in every functional unit. If possible, distinguish between tasks which can be fulfilled with the CBS, and those resulting from the use of the system.

4. Determine the number of *tasks* which *should* be done by application of the CBS.
5. Take an assumption of a possible *configuration* of the CBS, to make a mental model on which to orientate the following estimates.
6. Revise the *targets* for the factor automation project, and then analyse the impacts of enhanced performance and different costs in every functional unit of the company for the four major cost categories, according to every target of the project.
7. *Sum up* the cost effects, once for every automation target, and once for every functional unit, and the overall sum.
8. *Compare* the overall sum with the necessary investment, and perform a *break-even-analysis* to get the impact on the fixed and variable costs, and thus an indicator for the risk structure.

This leads to thinking in terms of overall cost, as in logistics, which is a cross-sectional function, and affects various other functions.

Most of these steps do not need further explanations, and will become obvious with the further descriptions and pictures. But the heart of this method, the calculation of the profitability of the investment, deserves some further description. It is based on the arithmetical connection of overall costs, cost share in the functional unit, and the factor for reduction of the functional unit. This reduction factor (RF) is given for a certain period of time in which the investment should be implemented, settled, and be put to its intended use (i.e. 5 years).

To figure out the potential for cost reduction, and its cost impacts, we use a trick: all costs are regarded as variable. The reason is simple: although many costs are fixed (even labour costs become more and more fixed, in some cases only 5% of the labour costs are really variable), we want to concentrate on the productivity effects. This means, for example, that more things can be done in the same time, or the same things can be done in shorter time, although the worker's wage is fixed and doesn't vary with the output achieved. But in our methodology it is necessary to grasp this important difference, and therefore all costs are virtually assumed to be variable. The real impact on fixed and variable cost is determined later by performing a break-even analysis, where fixed and variable costs are explicitly considered.

Let 'i' represent the functional unit, and 'j' the technology. The reduction factor for the functional unit is then given by the cost share, which is influenced by the measure and its impact on this cost share:

$$RF_{ij} = (\text{influenceable cost share})_{ij} \times (\text{change of cost share})_{ij} \quad (1)$$

This is the more theoretically correct description. When applying this methodology, an estimate of this factor can be made for easier application. The overall costs of the enterprise can be summed up by the costs of every functional unit:

$$\text{overall cost} = \sum_i (\text{cost share factor}_i \times \text{overall cost}) \quad (2)$$

The effect of the cost of a particular investment within a functional unit can be a cost reduction or a cost increase. By recognizing this effect, the main characteristic of investments in CBFA can be taken into consideration: often some costs *within* the functional unit, where the investment is placed, might rise, i.e. when the defining time for a CAD-drawing for a new design is longer than with conventional means. But all following functional units can use the generated data set and other consequences, and their costs might be reduced. When only focusing on the design department, the investment would not promise to have an adequate payback period, but by taking all effects systematically into consideration it is possible to determine the real benefits, and a better payback determination.

So the total cost effect of a functional unit is not necessarily directed towards a cost reduction, and therefore can be described according to the real causes:

$$\text{cost effect}_{ij} = \text{cost}_i \times RF_{ij} \quad (3)$$

or, when the single cost categories relevant to the automation project (indexed by 'k') are analysed, by

$$\text{cost effect}_{ijk} = \text{cost}_{ijk} \times RF_{ijk} \quad (4)$$

Thus the overall cost effect is given by

$$\text{overall cost effect}_j = \text{overall cost} \times \sum_i (\text{cost share}_i \times RF_{ij}) \quad (5)$$

This more formal explanation is quite easy to apply by making a chart, such as Table 2, which guides one along the targets of factory automation via all the relevant points. This follows three basic steps. In the first step the overall costs of the enterprise are to be ascertained. As in a cost centre analysis, the cost share of every single functional unit in the overall cost is determined. This is done first for the functional unit as a whole, and later, if more detail should be necessary, for the relevant types of cost. These shares of the functional units appear at the top of Table 2. In this company with about 500 employees and with revenues of $45 million, the design department has a cost share of 9% of the overall cost, the operations scheduling and process planning 5%, production 37%, assembly 14%, and so on. 17% of the overall cost is due to other activities of the company, i.e. finance transactions which are not linked to the manufacturing activities.

In the second step the evaluation criteria for the

Table 2. Estimates of cost and benefit impacts, given by functional units and targets of factory automation, shown by the example of CAD in a company with 500 employees and annual revenues of $45 million.

	CAD					Status: 0%		Plan: 40%	
Target of factory automation	Design	Planning and scheduling	Production	Assembly	Quality control	Offering department	Material planning	Accounting	Cost impact of target (1000 $)
Share of overall cost	9%	5%	37%	14%	2%	6%	5%	5%	45000
Shorter through-put time	- 2.80%	- 1.20%	- 0.10%		- 0.10%	- 2.00%	- 0.50%		- 223
Reduced material cost			- 1.50%				- 0.10%		- 252
Higher on-time delivery	- 0.20%	- 2.00%	- 0.50%	- 0.70%		- 0.90%	- 0.15%		- 208
Reduced expense for changes	- 2.50%	- 1.30%	- 0.30%	- 0.26%	- 0.25%		- 0.03%		- 200
Higher quality, reduced scrap		- 0.03%	- 0.90%	- 0.20%	- 2.60%				- 187
Higher standardization	- 0.80%	- 1.20%	- 0.30%	- 0.20%	- 0.40%	- 2.30%	- 0.18%		- 192
One-time data entry	3.00%	- 0.90%			- 0.08%	- 0.13%		- 0.80%	79
Optimized use of material			- 0.70%						- 117
Higher transparency of processes		- 0.02%				0.05%			1
Cost impact within functional unit	- $134k - 3.3%	- $150k - 6.7%	- $716k - 4.3%	- $86k - 1.4%	- $31k - 3.4%	- $143k - 5.3%	- $22k - 1.0%	- $18k - 0.8%	- $1298 - 2.88%
Overall cost impact	- $1298k - 2.88%								

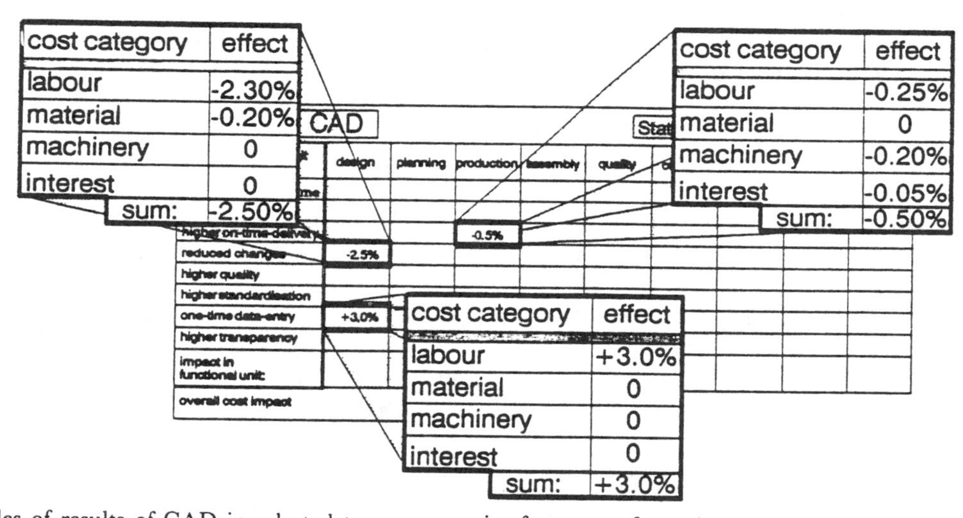

Figure 1. Examples of results of CAD in selected target categories for types of cost in design and manufacturing. Compare with Table 2.

impacts of the investment are put into the left-hand column of the table. Here the targets of factory automation or other specific targets can appear. According to the impacts of the technology shown in Table 1, for every target the impact in every functional unit has to be estimated for the regarded period of time. This can be calculated for each of the four types of cost (direct labour, material, machine cost, interest), or globally for the functional unit. Every result is fixed in the matrix. Therefore the impact on a percentage of cost effect is fixed; this might be a cost reduction (' – '), or a cost increase (' + '). The investment itself is not included in this analysis! This is done later when comparing the amount of the investment with the potential for cost reduction. Only the effects of the desired measure are sought.

For instance, a higher on-time delivery rate of drawings and manufacturing data by use of CAD leads to cost reductions of 0.5% of the cost in the manufacturing department (Fig. 1). This is the result of a 0.25% reduction in labour cost, because of less waiting or trouble-shooting time, a 0.2% reduction of machine cost, due to less idle time while waiting for orders, and a 0.05% reduction in interest cost because of less work-in-process (WIP) and less bound capital. For this example, material costs are not affected.

Reduced effort in changes leads in this example to a 2.5% reduced cost in the design department; nearly all from less labour cost (− 2.3%), but also less material cost (− 0.2%). The higher efforts and longer time for the first data entry, or data generation, leads to 3% higher costs in the design department, all due to labour cost.

In the final step, all cost-effect values are summed up by multiplying the percentage value with the costs of the appropriate functional unit. The sum along the lines is used to figure out the contribution of every target of the automation project, and the sum of the columns is used to get the cost effect for every functional unit. Thus the major and the minor targets become obvious; respectively their contribution to the project, and the structure of cost and benefit implications in all functions of the enterprise become apparent. This shows, in supplement to the basic analysis of the company for the investment, which areas to focus on, or where other means can or should be taken into consideration, along with the investment. This might even suggest changes to the desired actions.

In the example reduced throughput time resulting from the use of CAD leads to an overall cost reduction of $223 000, or − 0.49%. Reduced changing efforts lead to an overall cost reduction of $200 000, or − 0.44%. The total cost effect inside the design department adds up to − $134 000, which is 3.3% of the total cost of the design department, and 0.30% of the overall cost. The improvements for the manufacturing department add up to —$716 000 (− 4.3% of their total cost), 1.6% of the overall cost.

When comparing the direct cost impacts within the CAD-using design department with the total cost effects, the difference between direct and indirect effects become obvious. The overall effects add up to − $1 298 000—only 15% are within the design department, and the other 85% are results from improved performance in the other areas of the company. The major impact can be seen in the manufacturing department: 55% of the overall cost effect is generated there. But that is no wonder, because the quality of planning functions is gauged by the quality with which the following departments can perform by using the information obtained.

4. Which investments?

The last step described above supplies the gross potential for cost reduction of the analysed technology or measure. The economically determined investment can be found by subtracting the risk premium, special interest, or desired profit that results from the net potential or the economically maximum investment budget. As shown later in the break-even analysis, only the net potential should be considered for the investment, the influence of the changed fixed costs on the break-even point and therefore on the risk structure of the investment.

The financing of the payments, which mostly occur during the beginning of the project, must be assured for long-term. The positive cost effects will be set free only slowly and long-term, and with an additional demand for extra costs in the starting phase. So a restriction on figuring out an investment from the potential is given by the possibly financeable amount:

$$\text{feasibility}_j = \text{financeable amount}_j/\text{investment}_j \quad (6)$$

Small businesses in particular often don't know where to focus their investment activities. On one hand there are various technical necessities or opportunities, on the other hand there are rapid changes in price and performance of the new computer based systems. Small companies cannot test different technologies in small units like big companies can. Therefore they have to guide their small investment budgets into the most efficient areas. By applying the described method for basic investments in design, planning or manufacturing the potential for single investments can be found and compared with the necessary investment, or even a kind of return on investment can be figured out:

$$\text{priority}_j = \text{net potential}_j/\text{investment}_j \quad (7)$$

So it is quite easy to determine the most lucrative investment by the best ratio of potential to investment. By ranking the measures according to this index, the technical presuppositions and peculiarities also have to be taken into consideration. That is, before generating an NC program using the CAD/CAM system, it is necessary to gain experience with CNC machines. Therefore the financial hierarchy has to be superimposed on the technical hierarchy, to develop a long term investment plan to guide the way for the 'factory with a future'.

This comparison of overall potential for cost reduction with the necessary investments yields a new way of evaluation for this investment. The more traditional explicit cost comparison of old and new technology is removed. That standard of comparison would only bias the evaluation process, because the basis for judging the investment is not known, or is significantly different. Investments for CBFA have a much higher sphere of impact than 'traditional' investments.

Another remarkable sign of CBFA is the impact on revenues. Many authors claim automatically higher revenues and profits by introducing and applying these technologies. But that is often simple minded or even dangerous (as will be shown later). When comparing the long term potential for cost reduction with revenues, those should be considered as constant (if desired, adapted by inflation rate). One reason is that increasing revenues caused by the investment would imply a growth that may or may not occur. Some technological leaders might take advantage of this to increase their market share, or to participate on an overall market growth. But for most users of these technologies they are not to gain a higher relative market power, but to keep up with an overall increase in performance capabilities, and to maintain the relative market power. Implying higher revenues or higher profits might turn out to be quite dangerous when the expectations are not fulfilled. One major reason for this effect is the influence of CBFA on the fixed cost.

5. Influence of fixed cost on strategic investment planning

While performing the calculation method presented here, all costs were considered as variable in order to get a quantitative index for the possible overall productivity enhancements. But for the final investment decision process, the real influence of this investment on fixed and variable cost must become clear. First, the investment in CBFA tends to increase the share of fixed cost of the company. This results from purchasing hardware and software, installation and adaptation of the software, and education of the users. In addition to this increase there is a significant remanence of cost: when increasing the capacities to produce higher volumes, the costs rise. But they do not fall by the same amount when these additional capacities are utilized less, following reductions in the production volume. But, on the other hand, within a certain boundary these technologies have a high degree of flexibility to adapt to changing production tasks. So in the long run short-term adaptations to production needs can be fulfilled with lower costs than with traditional equipment. Therefore it is extremely important to differentiate between the quantitative and the qualitative advantages and risks in regard to the flexibility of the computer based equipment.

Further aspects of designing the optimized investment policy and budget are shown by an analysis of the cost functions. With the higher fixed costs of introducing an integrated computer based structure the gradient of the marginal cost function is smaller (Fig. 2), and one result is the creeping break-even point phenomenon.

In case when the break-even point should not move

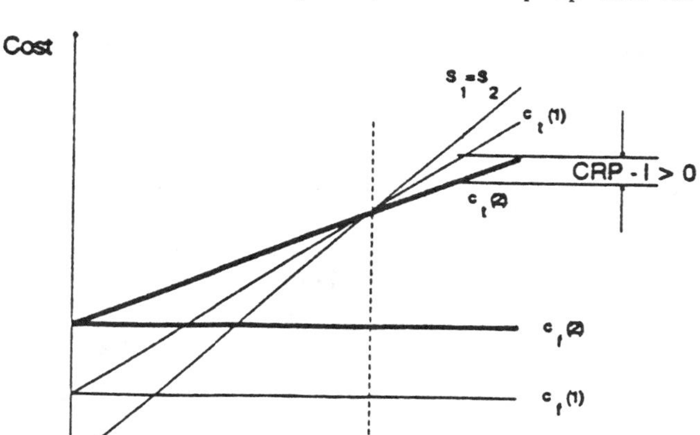

(*a*)

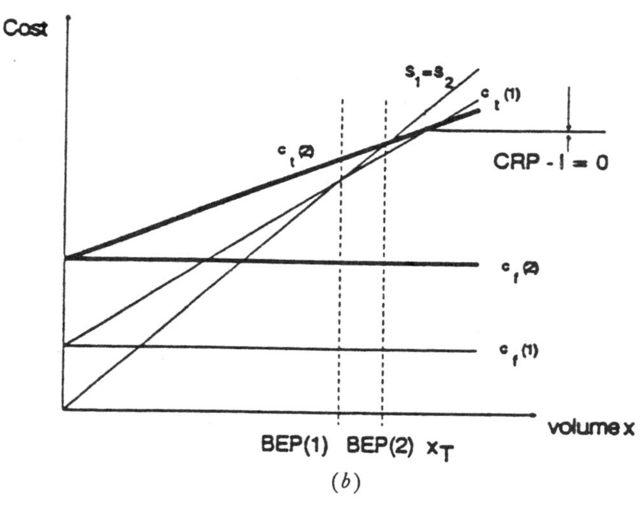

(*b*)

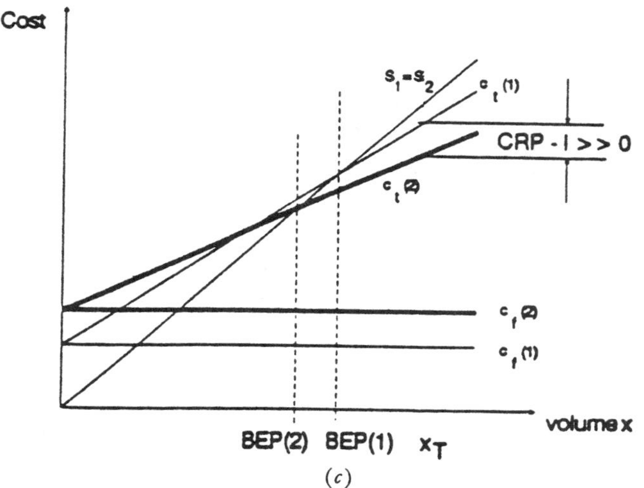

(*c*)

Figure 2. Course of cost functions before (1) and after (2) the investment. (*a*) 'Optimized' investment budget: the potential for cost reduction is not scooped by the investment and the break-even point doesn't change. (*b*) Investment at the same amount as the potential for cost reduction; the break-even point moves towards higher volumes, dangerous when sales or production volume diminishes. (*c*) The potential for cost reduction is much higher than the necessary investment, the break-even points moves towards lower volumes, and lower volumes can be produced at economic cost. CRP is the cost reduction potential, I the investment, Cf the fixed cost, Ct the total cost, S the sales (revenue), BEP the break-even point and xT the volume for planning. The index '1' indicates before investing and '2' indicates 'after investing'.

towards higher cost at the planning volume, the investment has to be of a kind to reduce the variable cost while operating within the amount of the potential for cost reduction. If the production volume, which is the basis for the evaluation, is higher than the break-even point, the investment *must* be smaller than the potential (Fig. 2(*a*)).

If the technological and/or organizational necessary investment is at, or even higher than, the potential, it is only advisable to invest in the case of certain expectations for a constant or increasing sales volume, caused by a more straightforward total cost function (Fig. 2(*b*)). The break-even point moves towards higher volumes.

Another example explains the reason for the capability to produce in smaller lot sizes by use of flexible production equipment. If the ratio between the potential for cost reduction and the necessary investment can be modelled as shown in Fig. 2(*c*), there is no need to step up the volume, because the break-even point moves towards smaller volumes.

Most of the evaluated investments will be of type 2(*a*) or 2(*b*). These cases provide the chance for higher production volumes with the same total cost, caused by lower variable cost. The average cost per piece starts to diminish when the break-even point is passed. Therefore investments in CBFA only have a manageable risk if size, steps, and timing of the investment can be harmonized with the requirements of the market. To perform these investments in periods of low demand (and thus low volumes) can lead to severe financing problems, especially concerning the short-term liquidity (the major cause of bankruptcy). On the other hand, making sensible use of the potential by investing into CBFA provides the capability to operate on a constant, or even higher level of customer-oriented performance in stagnant markets.

Additionally, Fig. 4(*b*) shows clearly why early investors could stand the high investments for CBFA when this technology was starting to emerge: although the necessary investments were much higher than the expected savings, there is a profitable use of these systems. One reason is that the actual potential was higher than analysed, because the total extent of the indirect benefits was not appreciated. The other reason results from the general way these companies are managed. They usually operate highly successfully in growing markets or with growing market shares—even without the use of CBFA. In adding the powers of CBFA systems to their basic capabilities they could realize a higher sales volume, higher revenues, and thus take those advantages which are usually claimed by new production technologies. Only due to this increase in the relative market position and the combined enhancement in sales were the higher costs financeable, while with deteriorating revenues these projects, and the whole enterprise, would have been bound to fail.

6. System integration influences types and structure of cost

The change in fixed and variable cost is the result of changing the types of cost. These are due to the individual introduced systems; generalizations about the size are very difficult to make. For the German machines and equipment building industry a recent analysis (Schulz and Boelzing 1989) has shown overall cost reductions by CAD of about 2.5%, by CAP of about 2%, by CAM of about 3%, and by PPS of about 2.8% The contribution of technologies to the targets of factory automation and the direction of changes in cost types, according to this in-depth analysis are shown in Tables 3–6. These results might help to analyse and estimate the individual impacts of CBFA.

One of the major reasons why shortening throughput time or product development time cannot be explicitly

Table 3. Types of cost which are influenced by CBFA.

	Direct functional cost influence by				
Type of cost	CAD	CAP	CAM	PPS	Integration
Wages and salary	–	–	–	–	– –
Auxiliary wages	–	–	0	–	0
Fringe benefits	–	–	–	–	–
Material	–	0	–	0	–
Tools and devices	0	–	+	0	0
Maintenance	+	+	+ +	0	0
Energy, heating	0	0	+	0	0
Taxes, fees, insurance	0	0	+	0	0
Calculated depreciation	+	+	+ +	+	+
Calculated interest	+	+	–	– –	–
Other indirect expenses	0	0	0	0	0

Table 4. Types of cost which are influenced by integrating computer based technologies.

Type of cost	CIM linking CAD–NC	CAM/DNC	CAD–PPS	CAP–PPS	CAM–PDA
Calculated depreciation	+/–†	+/–†	0	0	0
Calculated interest	+/–†	+/–†	0	0	0
Material	0	0	–	–	0
Labour cost:					
Direct wages	0	–	0	0	–
Assembly	0	0	0	0	–
Indirect wages	–	–	–	–	–
Design	0	0	0	0	0
Administration and sales	0	0	0	0	0
Cost for capital (WIP)	– (TPT)	– (TPT)	– (Parts)	– (TPT)	– (Info)

†Cost increase for additional equipment, cost reduction via better capacity utilization (no additional capacity necessary).

Table 5. Cost (C) and performance (P) impact of the targets of factory automation on the relevant types of cost (+ +, intensive cost-/performance increase; +, moderate cost-/performance-increase; 0, no or small cost-/performance-increase; –, moderate cost-/performance-decrease; – –, intensive cost-/performance-decrease).

Type of cost	Shorter throughput time		Reduced material		Higher on-time delivery		Reduced cost for changes		Higher quality, less scrap		Standardization		One-time data entry		Better capacity utilization		Changing flexibility		Optimized use of equipment	
	C	P	C	P	C	P	C	P	C	P	C	P	C	P	C	P	C	P	C	P
Direct labour cost	0	+	0	0	–	+	– –	+	–	0	0	0	–	0	+	+	–	0	0	0
Material	0	0	–	+	0	0	0	0	–	+	–	0	0	0	0	0	0	0	–	+
Calculated depreciation, interest	0	0	0	0	–	0	0	0	0	0	0	0	0	0	0	+	0	+	0	0
Bound capital (WIP)	–	0	–	+	–	0	0	+	0	+	–	+	0	0	0	0	–	+	0	0
Tools and devices	0	+	0	0	0	0	–	0	–	0	–	0	0	0	+	0	0	0	–	+
Indirect labour cost	+	+	0	0	–	+	–	+	–	0	0	+	–	+	+	0	–	+	0	0

Table 6. Contribution of CIM technologies and their integration for reaching the targets of factory automation (+ +, intensive positive contribution; +, moderate positive contribution; 0, no or small contribution; –, moderate negative contribution; – –, intensive negative contribution).

Technology	Shorter throughput time	Reduced material	Higher on-time delivery	Reduced cost for changes	Higher quality, less scrap	Standardization	One-time data entry	Better capacity utilization	Changing flexibility	Optimized use of equipment
CAD	+	+	+	+	+	+ +	+	+	0	+
CAP	+	+	0	+	+	+	+	+	0	+
CAM	+	+	0	0	+ +	0	0	+ +	0	0
PPS	+ +	+	+ +	+	0	+	0	+ +	+ +	0
CAQ	0	+	+	0	+ +	+	+	0	0	+
CAD-NC	+ +	0	+	+	0	+	+ +	0	+	0
DNC	+	0	0	0	0	0	+	+ +	+	0
CAD-PPS	+	+	0	+	0	+	+	0	0	0
CAP-PPS	+	0	0	+	+	+	+	0	+	0
CAM-PDA	+	0	+	0	0	0	+ +	+	+	0
CAM-CAQ	0	+	0	0	+	0	+	0	0	+
CAD-CAQ	+	+	0	0	+	+	+	0	0	+

included into this scheme is that 'Delays in commercializing innovative products not only increase the expenditures for research and development, but simultaneously they often shorten the yield potential, especially if there is a decay in prices during the remaining product life cycle' (Sommerlatte 1988). The method presented here is a method for the evaluation and decision process for new manufacturing technologies. With the holistic view of the company, in regards to the overall costs and benefits, major economic and technical effects can be shown, but it can't replace long-term oriented entrepreneurial decisions.

References

BAUDIN, M., 1985, Experience curve theory: a technique for quantifying CIM benefits. *CIM Review*, Summer, 51–58.

CANADA, J. R., and SULLIVAN, W. G., 1989, *Economic and Multiattribute Evaluation of Advanced Manufacturing Systems*, (Englewood Cliffs, NJ: Prentice Hall).

COOPER, R., and KAPLAN, R. S., 1988, Measure costs right: make the right decisions. *Harvard Business Review*, Sept./Oct., 96–103.

FINE, C. H., and FREUND, R. M., 1988, Optimal Investment in Product-Flexible Manufacturing Capacity. Working paper, Sloan School of Management, MIT, Cambridge MA.

FINE, C., 1989, Developments in Manufacturing Technology and Economic Evaluation Models. Working paper, Sloan School of Management, MIT, Cambridge MA.

GERELLE, E. G. R., and STARK, J., 1988, *Integrated Manufacturing—Strategy, Planning, and Implementation* (New York: McGraw-Hill).

GUPTA, D., BUZACOTT, J. A., and GERCHAK, Y., 1988, Economic Analysis of Investment Decisions in Flexible Manufacturing Systems. Working paper, Department of Management Sciences, University of Waterloo.

HAYES, R. H., WHEELWRIGHT, S. C., and CLARK, K. B., 1988, *Dynamic Manufacturing—Creating the Learning Organization* (New York: The Free Press).

HERROELEN, W., DEGRAEVE, Z., and LAMBRECHT, M., 1986, Justifying CIM: quantitative analysis tool. *CIM Review*, Fall, 33–43.

KAPLAN, R. S., 1986, Must CIM be justified by faith alone? *Harvard Business Review*, March/April, 87–95.

LEDERER, P. J., and SINGHAL, V. R., 1988, The Effect of Cost Structure and Demand Risk In The Justification of New Technologies. Working paper, Simon School of Business, University of Rochester NY.

MEREDITH, J., 1986, *Justifying New Manufacturing Technology*, Norcross Ga.: Institute of Industrial Engineers).

MEREDITH, J., 1987, New justification approaches for CIM technologies. *CIM-Review*, Spring, 37–42.

SAVAGE, C. M., 1988, The challenge of CIM is 80% organizational. *CIM-Review*, Spring, 54–58.

SCHULZ, H., and BÖLZING, D., 1989, *Rechnergestützte Fabrikautomatisierung—Kosten senken, Leistungen steigern*, (Frankfurt, Main: Maschinenbau-Verlag).

SOMMERLATTE, T., 1988, Innovationsfähigkeit und betriebswirtschaftliche Steuerung—lässt sich das vereinbaren? *Die Betriebswirtschaft*, **48**, (2), 161–169.

SULLIVAN, W. G., and LIGGETT, H. R., 1988, A decision support system for evaluation investments in manufacturing local area networks. *Manufacturing Review*, **1**, (3), 151–157.

WILDEMANN, H., 1988, *Strategische Investitionsplanung—Methoden zur Bewertung neuer Produktionstechnologien* (Wiesbaden: Gabler Verlag).

The Fresh-Start Approach to CIM Investment Justification

David J. Storm and Steven J. Sullivan

Decisions involving CIM implementation present managers with a fundamental dichotomy. On the one hand, managers are being overwhelmed with information suggesting that CIM is the ultimate weapon for attaining competitive advantage. They are warned to invest now or be buried by far-sighted competitors. On the other hand, managers are faced with pressures to improve bottom-line results as measured by return on investment (ROI).

Managers are also handcuffed by outmoded cost-management and investment-justification approaches. To many of these managers, therefore, the decision to invest in CIM seems fraught with risk as a result of the relative uncertainty of the benefits to be derived, the significant commitment of resources (both human and capital), and the rapid changes in technology. What results is inertia caused by indecision. Faced with competitive pressures, managers are thrust into a reactive mode that slowly drains competitive strength.

Recent studies indicate that, although the potential of CIM is great, financial justification remains the major consideration for companies interested in pursuing CIM investments. This is understandable: manufacturing companies are in business to make money, and profitability is their primary goal.

How, then, do managers resolve these conflicting forces and implement business strategies that offer growth, profitability, and long-term survival? For most manufacturers, the issue involves coordinating strategy with economic justification. Traditional justification practices encourage low-risk investments based on narrowly focused financial analysis. Today, managers are being challenged to rethink these approaches and to be creative in responding to future opportunities.

Gone is the steady state

The environment in which manufacturers operate is changing dramatically. Gone are the days of product stability, domestic markets with few competitors, predictable economic trends, and specialization. The environment is and will continue to be characterized by the following important changes:

- Product life cycles are shortening, and customers are demanding increased variety—From athletic shoes to automobiles, manufacturers are finding that quick response is a necessity for continued survival.
- Markets are global and fragmented—The economic development of many countries and advances in communications have expanded market opportunities and the ability to serve those markets. Customer demand is more specialized now. The result is a fragmentation of what used to be mass markets. Growth in the future will be in niche markets that focus on product differentiation.
- Competition, both foreign and domestic, is increasing—The Japanese are obviously a threat, but there is growing competition from Europe, South Korea, Taiwan, and Latin American countries. Domestically, the increase in corporate consolidations has

***David J. Storm** is a partner in Andersen Consulting, Arthur Andersen & Co, and head of the manufacturing practice in Andersen's Milwaukee office.*

***Steven J. Sullivan** is a manager in Andersen Consulting and a member of the firm's manufacturing industry team.*

caused more competition (rather than less, which some people expected), as financial resources have been combined.
- Technological advances are coming in waves, and, as a result, costs are falling—IBM has stated that the cost of computing is being cut in half every four-and-a-half years.

The implications of these changes for manufacturers are significant. Strategies of the past generally pursued economies of scale and low-risk new-product introductions. By following these strategies, companies have traded flexibility for rigidity and innovation for imitation. US companies were managed for stability and steady states.

Don't do it first?

A top executive for General Motors during the 1970s, faced with increased pressure to introduce front-wheel drive cars, recalled some advice he had been given many years before and that he had never forgotten: "Whatever you do, don't have GM do it first." For many industries, wait-and-see strategies have worked successfully in the past—and may still work. However, as manufacturers increasingly face pressures brought on by changes in their marketplace, strategies must also change.

Michael Porter, in his book *Competitive Strategy,* points out that the two key techniques a company has for dealing with competitive forces of tomorrow are:
- To differentiate products and become the low-cost producer.
- To attain competitive advantage through manufacturing flexibility, innovation, and the minimization of nonvalue-added activities.

A strategic response

CIM, when viewed in the appropriate context, is a strategy that can yield significant improvements in product and service value. It addresses the entire logistics pipeline—from new product idea through design to prototype and from supplier to production to customer. Companies cannot view CIM as merely engineering systems or flexible machining cells. It is a companywide program. Unless all functional areas of the organization buy into the CIM program, CIM is destined to be pigeonholed as an eccentric whim that has to be justified piecemeal through low-risk, narrow-minded financial analysis.

Ability to earn a profit

Because the goal of every business entity is to make money, all business problems inevitably relate to a company's ability to earn a profit. Profitability is ultimately a function of market share and product margins. It follows that companies providing the greatest product value will achieve the highest market share, the best margins, and the greatest profit. Product value is enhanced by reducing costs, upgrading product quality, and improving customer service. Manufacturers invest in CIM because they recognize that doing so offers, in almost all cases, solutions that directly and significantly contribute to improvements in product value. Among the solutions CIM can provide are the following:
- CIM improves a manufacturer's ability to respond to the marketplace by shortening the logistics pipeline—Electronic data communications with suppliers and customers reduces correspondence lead times. Computer-aided design (CAD) systems can shorten product development times by 30% to 60%. As a result, the cost of innovation drops, so manufacturers can more effectively differentiate their products and services in the marketplace.
- CIM improves quality—Experience shows that robots can weld to tighter tolerances and with greater consistency than humans. CIM technology reduces the process variability inherent with human participation and places the quality emphasis on prevention and correction through the use of automated process controls. In addition, by contributing to lower inventory levels, CIM exposes quality problems faster.
- CIM reduces costs by shrinking the logistics pipeline through the elimination of nonvalue-added activities—This is illustrated in Exhibit 1. Flexible manufacturing systems are tightly coupled, multifunctional cells that offer quick changeovers and, consequently, the ability to process small lots. Small lots mean less inventory and less space, both of which are nonvalue-added costs.

Maintaining lower inventories means having

Exhibit 1. *Logistics Pipeline*

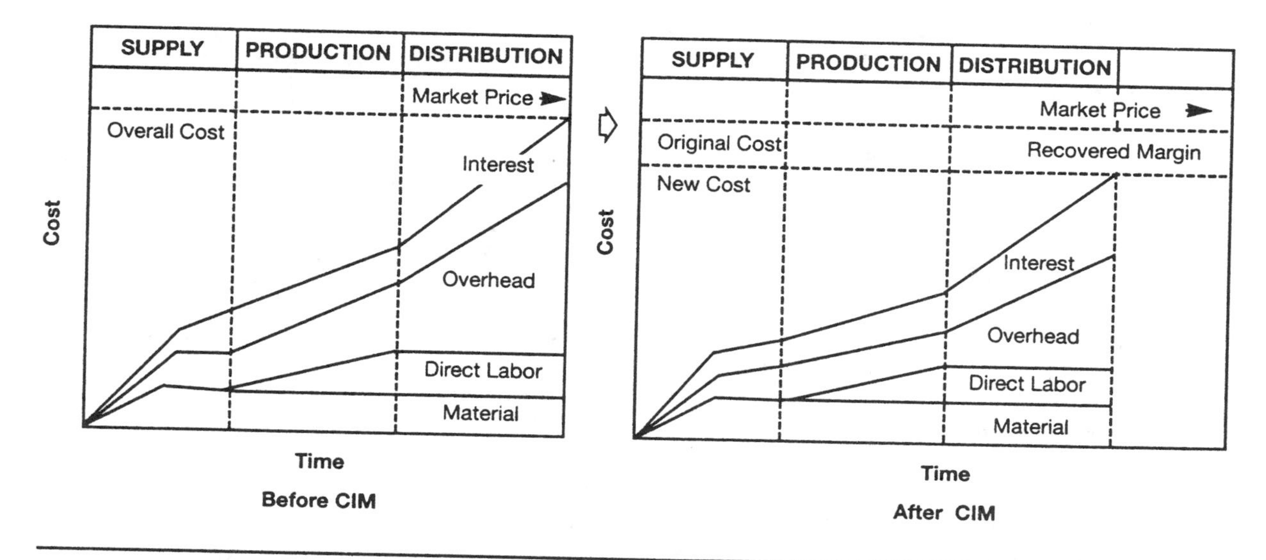

to do things right the first time. First-time quality means reductions in the nonvalue-added costs associated with scrap, rework, returns, and warranties. Finally, sharing common data among engineering, production, and the office yields recordkeeping efficiencies and improves communication. Information can be captured and reported on a timely basis because duplicate data entry and manual reconciliation are eliminated.

Not whether, but how much and when?

The crux of the issue is not so much whether to invest in CIM but rather at what rate. Eventually, all plants and equipment must be replaced, so establishing a plan for introducing and integrating new technology makes business sense; decisions, then, center on the rate and priority of investments. This is where financial decision models and management information systems play an important role, provided they overcome deficiencies of past approaches.

Traditional approaches

Integrating business strategy and economic justification historically has been a major management failing in capital budgeting. Broad strategies (which may include CIM) are established to achieve competitive advantage, but then the execution of the strategy is justified and managers' performance is measured along very narrow lines. Inevitably, strategies are compromised by inappropriate incentives or decisions made based on short-term financial criteria. Examples frequently encountered in industry include:

- The reliance on traditional cost management and performance measurement techniques that foster a piece-cost mentality—These systems, designed during the 1920s when labor was a scarce resource, encourage the capitalization of costs through high labor efficiency. Such approaches reward managers who shift costs from the profit-and-loss statement to the balance sheet by producing inventory. The hidden costs of inventory are buried in overhead accounts.
- The bias toward incremental investments and short-term results—Typical financial analyses view the introduction of new equipment in a piecemeal fashion and require paybacks based on the anticipated incremental benefits of each machine. The long-range benefits of synergies that technological integration brings (e.g., shared information and communications networks) are ignored. Further, mature industries with largely depreciated asset bases forego capital investments to meet short-range return-on-assets (ROA) targets, even though the investments may increase earnings and, ultimately, determine the company's very survival.
- The inability to simulate and analyze alter-

native business scenarios—The only vision of the future that many companies have is based on the facts of the past (which, as some have pointed out, is like driving a car using only the rearview mirror). Managers need the tools to evaluate different alternatives given different assumptions about the future.

- The failure to quantify important benefits and assess opportunity costs—Frequently overlooked are cost reductions in the areas of quality, flexibility, customer service, and space. Although these costs are not as easily identified as labor and inventory reductions, they are no less important. As Robert S. Kaplan has observed, timid accountants who give zero values to intangible benefits would rather be precisely wrong than vaguely right. In addition, managers need to know the opportunity cost of doing nothing. What, for example, is the impact of maintaining the status quo when competitors are investing?
- Excessively high hurdle rates not tied to the company's product and market strategies—Many companies maintain a single hurdle rate for all investment decisions, irrespective of product lines and their market positions. As manufacturers work toward product and process simplification, the natural tendency is to focus operations on a product-line basis. Investments tend, therefore, to be specific to product lines. Most companies have several products at different stages in their life cycles. Should investments in emerging products with high market potential be held to the same hurdle rates as investments in mature products with dwindling market potential? Exhibit 2 shows relevant guidelines used by one manufacturer.

Number-one enemies

Justifiably or not, the blame for inertia in adopting CIM typically falls on a company's financial managers. This has led some experts to dub management accountants the number-one enemy of productivity. The problem, however, pervades the entire organization and cannot be effectively resolved until attitudes change at all levels. CIM must first be viewed as a strategic commitment; then, justification can be approached based on the contribution

Exhibit 2. ***Investment Guidelines***

	Product		
	Strategic	New	Cash
Market Potential	High	Medium	Low
Life Cycle Stage	Growing	Emerging	Maturing
Investment Hurdle Rate	8%	12%	16%

CIM will make to financial and strategic objectives.

A fresh start

Most manufacturers are already investing in CIM. Investments in information systems, numerically controlled machine tools, and office automation equipment are investments in CIM.

Optimum benefits can never be reached, however, if technology is implemented in isolation or incrementally with no vision of how to integrate the various pieces. Unfortunately, this is precisely what traditional investment justification techniques encourage. A far better approach is to develop a CIM plan that establishes a strategy of action—not reaction—and unifies a company's marketing, product, manufacturing, and information plans.

CIM planning

As shown in Exhibit 3, a CIM plan should first establish some principles or key objectives that provide direction and focus for the planning effort.

As is. The existing (as-is) environment should be reviewed and opportunities assessed. Questions regarding process simplification, lead times, existing technologies, product mix, product quality, product cost structures, and customer service performance need to be investigated. Benchmarks are established for evaluating future improvements.

To be. Following documentation of the as-is environment, CIM planners must conceptualize the planned to-be environment. This must be a practical exercise that uses state-of-the-market technologies to achieve integration

Exhibit 3. *CIM Planning*

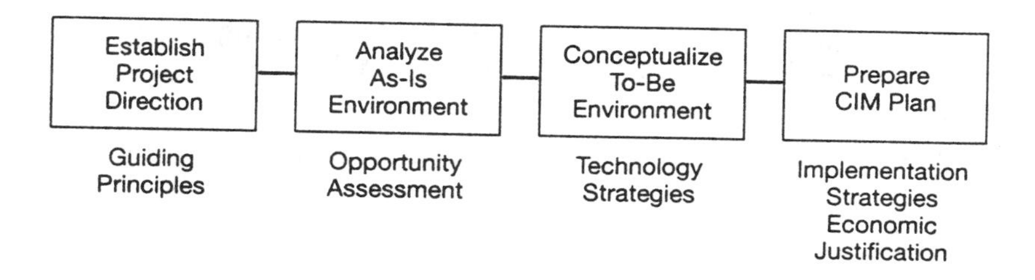

objectives. Typically, a technology strategy defines standards for hardware, cell configurations, data management, and communications. In addition, an implementation strategy is prepared that defines a logical and realistic schedule for the transition to a CIM environment.

Economic justification. The last step in the planning process is economic justification. This is, by necessity, an iterative process that identifies anticipated future cash flows. Unfortunately, this is the step that can turn a feasible strategy for competitive advantage into a pipe dream if the investment-justification analysis is performed along traditional lines.

Creativity

Companies that exercise responsible creativity in identifying potential benefits provide a clearer picture of the future than those that practice so-called fiscal stewardship. Managers must look beyond traditional savings in labor and inventory costs toward the benefits of increased revenue generation that come with higher-quality products and improved customer responsiveness. These benefits are real and must be quantified because they contribute significantly to product value, increased market share, and profitability.

To be balanced, responsible creativity must extend to the cost side of the equation as well. Given the rapid changes in technology, justification analysis must deal with technological life cycles (rather than accounting life cycles) when determining useful asset life. Frequently overlooked or underestimated is the effort necessary to accomplish CIM planning, design, and installation tasks. Special technical expertise, in the form of a systems integrator, is usually important to stimulate new ideas and facilitate progress.

Finally, the experience of many companies that have made a strategic commitment to CIM shows that the technological complexities of CIM pose a lower risk than a lack of worker acceptance of CIM. The combination of new manufacturing concepts and new technology creates an environment in which education and transition play key roles. The effort needed to accomplish this must therefore be built into the CIM plan.

The old versus the new

Traditional investment-justification methods use one set of assumptions, define the most obvious benefits and costs, and then run the figures through financial models to determine ROI and payback. These analyses are one dimensional and biased toward the short term.

The fresh-start approach

Justification constraints posed by past approaches can be largely eliminated by using a fresh-start approach to justification analysis. This means comparing a company's competitive position assuming these two scenarios:

- A 100% CIM environment.
- The status quo (or assuming only modest technology upgrades).

In both cases, the assumption is that the company starts from scratch by:

- Acquiring new plants and equipment.
- Hiring new employees with appropriate skills.
- Investing in necessary hardware and software.

Exhibit 4 illustrates these scenarios.

Since CIM improves flexibility, speed, and

quality, a situation of stable or increasing market share is assumed for the 100% CIM scenario. However, the status-quo scenario assumes that a company's competitors invest in technology and that the company itself therefore faces a declining market share.

Only by considering these two scenarios can the full impact and opportunity costs of the two baseline strategies be determined. The underlying premise of fresh start is that, if cash flows support the use of a CIM environment, assuming that the company were to start from scratch, the conversion will make sense even if traditional investment-justification analyses suggest to the contrary. Decisions then center, as they should, on strategy execution.

The fresh-start decision model takes a long-range strategic view and eliminates the inappropriate incentives of traditional methods. It allows managers to overcome the investment-justification hurdle and respond to competitive pressures with a strategy of action instead of reaction.

What if?

For fresh start to be effective, managers need tools that will help them quickly evaluate a multitude of future possibilities. Within today's framework, they must set reasonable boundaries that define tomorrow's expectations. The analysis should evaluate decisions along these three basic dimensions:

- Alternative CIM implementation strategies and rates of investment.
- Varying business assumptions regarding competition, market potential, technological lives, product lives, costs, and saving potential.
- The risks and probabilities of different outcomes.

All too often, justification assumptions are determined at a management level where the strategic implications are not fully understood. Furthermore, the assumptions are buried, rarely getting proper exposure at the decision-making level.

Top company managers must understand the impact of CIM and be involved in setting assumptions if CIM is to be viewed as a companywide strategy.

Instead of evaluating only one set of possible outcomes, simulation should assess many future possibilities. What if, for example, the product mix changes, market growth lags, margins are pinched, consolidations occur, or CIM implementation is accelerated? The following fresh-start example illustrates the process. The example was simplified to facilitate its presentation, but it does compare two different strategies and illustrates the simulation of varying market and cost parameters.

Fresh-start example

The fresh-start analysis was used to evaluate an investment decision for a new product with an anticipated life cycle of 10 to 12 years. The market, as shown in Exhibit 5, was projected to be 10,000 units in the first year and to peak at 18,000 units in years 6 and 7.

The product represented a strategic opportunity to invest in a related product line with solid growth potential. Although some existing, fully depreciated equipment was available for the manufacture of the new product, the company decided to avoid taking a short-term approach based solely on traditional financial-justification techniques. Instead, an attempt was made to assess the long-range strategic implications of the investment decision.

After market potential had been established, anticipated market share was estimated given two investment scenarios:

- A 100% CIM environment.
- A status-quo environment.

Exhibit 4. ***Fresh-Start Approach***

Current Environment
- New plant and equipment
- Current payroll costs
- New hardware and software
- Declining market share

100% CIM Environment
- New plant and equipment
- Reduced payroll costs
- New hardware and software
- CIM development costs
- Increasing market share

Exhibit 5. ***Product Life Cycle***

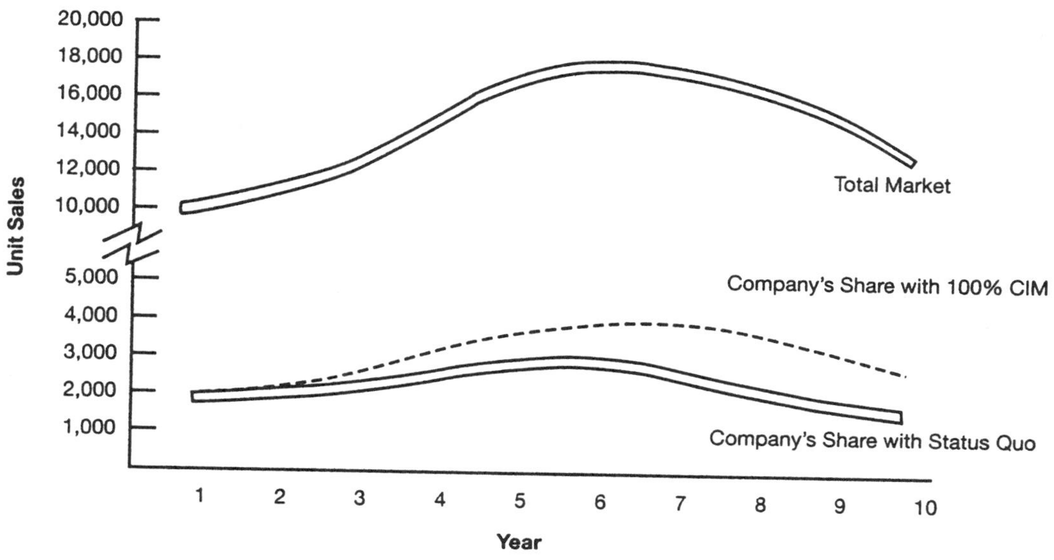

Assumptions

The assumptions were:

- If a CIM strategy were pursued, market share would increase gradually from 20% to 24% over the 10-year life cycle, and unit prices would decline from $10 to $8.50.
- If a status-quo strategy were pursued, market share would be lost because of investments in CIM by competitors, which would give the competitors greater flexibility and higher product values. It was assumed that market share would, therefore, decrease from 20% to 15% and that unit prices would fall from $10 to $8.

Other cost-of-production assumptions made were:

- Material cost would be the same for both alternatives because the design of the product would be the same and automation would not appreciably change base material costs.
- Direct labor costs would be 30% less in a CIM environment because of higher levels of automation.
- Cost of quality would decline by 60% in a CIM environment as a result of the elimination of variability that comes with a repeatable process—Specifically, internal (scrap and rework) and appraisal (inspection and testing) costs are substantially lower in a CIM environment.
- Inventory turns would improve from 3.5 to 10 in the CIM scenario, which would result in lower inventory carrying costs—This assumption is based on the flexibility that a CIM environment provides for producing toward a lot size of one and reducing work-in-process inventory.
- Occupancy costs per square foot would be 33% higher in the CIM alternative because of greater electricity consumption and special environmental needs.
- Engineering and technical support would be 50% higher for the CIM alternative because of the skill sets necessary to manage and operate the technologies employed in a cost environment.
- Plant-management and supervisory costs would be 25% less in a CIM environment because of lower head counts and improved information networks.

Results

The assumptions were then entered into a cash-flow model. Exhibit 6 shows that, while payback based on after-tax cash flows occurs a year later under the CIM alternative, total cumulative cash flows under the CIM alternative are three times those of the status-quo alternative by year 10. Moreover, if cash flows are discounted, cumulative cash flows never ex-

Exhibit 6. ***Cumulative Cash Flows***

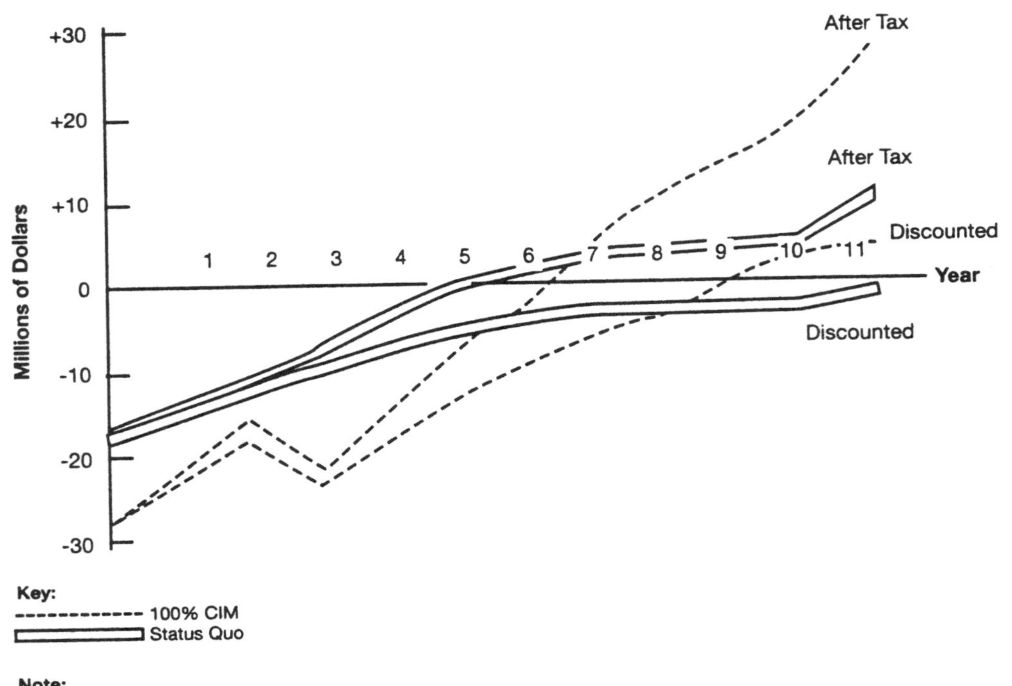

Note:
Graph assumes liquidation of property and equipment in year 2.

ceed zero for the status-quo alternative.

The long-range impact of the two strategies is even more dramatically illustrated in Exhibit 7. Under the status-quo alternative, contribution margin falls to zero in year 10 while hovering at 20% under the CIM alternative. Although unit volume is largely responsible for the difference, the CIM alternative enjoys a 17% advantage in the area of nonvalue-added costs.

In both cases, margins decline over the product life cycle due to competitive pressures. However, the market remains profitable in year 10 (and beyond) in a CIM strategy; it is far less attractive if the status-quo strategy is followed.

Initial investments

The initial investment to launch production for the status-quo alternative was determined to be $19 million. The initial investment in the CIM scenario was determined to be $28 million.

Because of better space use under CIM, a smaller plant would be needed initially. However, the total costs for equipment and tooling, computer hardware and software, and implementation were assumed to be nearly twice as much as the corresponding costs would be under the status-quo scenario. Further, because of market growth and increasing market share in the CIM scenario, increased capacity that would cost another $13 million would be incurred in year 3. No additional capital investments were assumed for the status-quo alternative.

Exhibit 8 shows how deceiving a short-range ROI approach to justification can be. In the first four years, the ROI under the status-quo scenario is higher than the ROI under the CIM alternative. However, as status-quo market share continues to drop, so does the ROI. By year 6, if the ROI is used as the yardstick, it becomes clear that the CIM scenario is the hands-down winner.

Unless manufacturers exercise creativity and strategic judgment in their investment justifications, they are doomed to make narrowly focused investment decisions that do not

Exhibit 7. *Contribution Margin*

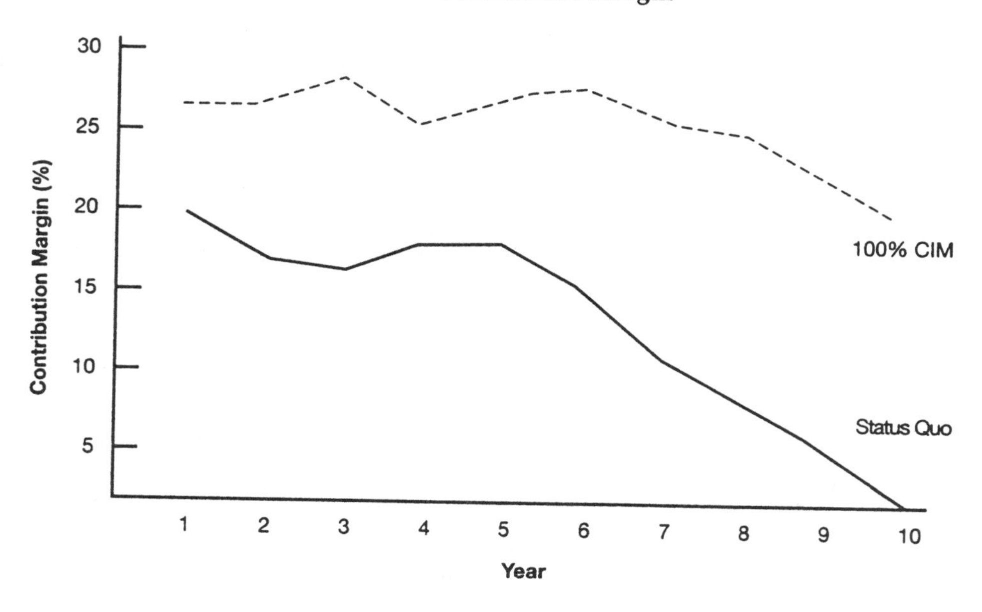

support the strategic directions that their companies should be taking. The fresh-start approach is one technique for forcing managers to step back and consider all aspects of an investment equation, including both revenue and cost components. By comparing alternatives on the proverbial level playing field, the fresh-start approach enables managers to evaluate the long-term implications of major strategic investments in technology.

Managing success

It is not enough to change the way decisions are made; the real challenge is implementing the CIM strategy and realizing the benefits. Success depends on skilled and committed

Exhibit 8. *Cash Flow Return on Investment*

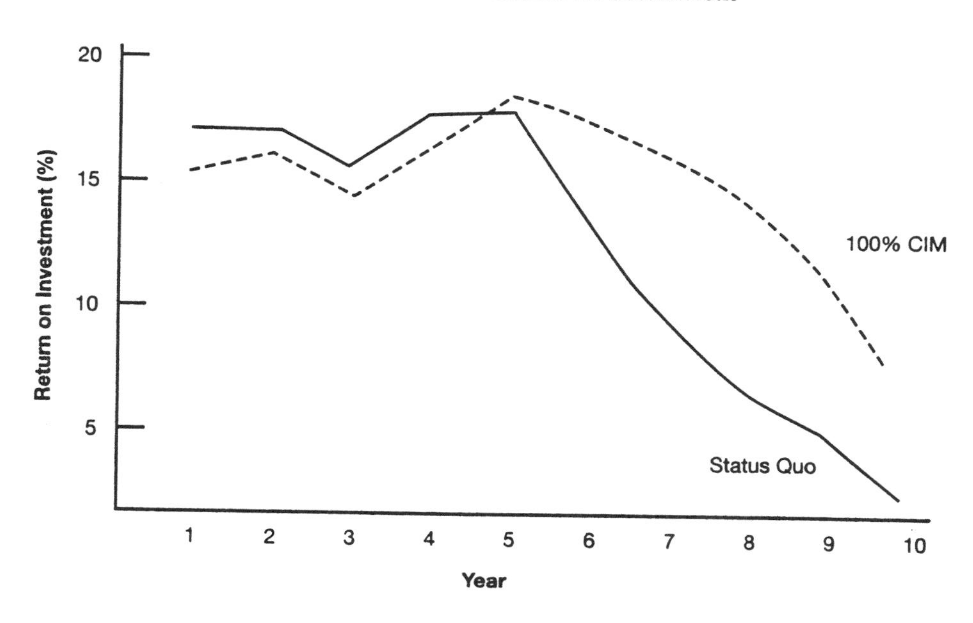

project managers who keep their eyes on the objectives while juggling daily project activities. Project managers—with support from the executive steering committee—are responsible for the timely and competent completion of tasks.

Success also depends on incentives that are consistent with the strategy. The old axiom that says people behave as they are measured is true today and will be tomorrow. As obvious as it may sound, it is ignored routinely. Performance continues to be based on labor productivity even though labor is becoming less and less important to total production costs.

Truly significant cost generators like product design complexity, process complexity, excessive lead times, quality problems, space use, and setups are also routinely ignored. By focusing on these factors as measurements, success can be planned for and attained.

Summary

The reluctance of US manufacturers to make major investments in CIM stems from:

- Inappropriate techniques for justifying the investments.
- A view of CIM as a manufacturing program instead of a companywide strategy.

The result has been low-risk, low-growth decisions based on short-term financial pressures.

In the business environment that confronts manufacturers today, low-risk strategies that maintain the status quo have caused the competitive strength of many US companies to decline to the point that their very survival is threatened. For many companies, CIM has become a competitive necessity because it contributes to flexibility, speed, and quality—three attributes that will ultimately lead to improved product value and competitive advantage.

As a result, progressive companies are looking at CIM differently. CIM is considered a vital link in a company's logistics pipeline and therefore integral to the company's overall strategy. Traditional justification approaches that encourage incremental, short-range investments must therefore be scrapped in considering CIM. A far better approach is the fresh-start approach explained in this article because it assesses the opportunity costs of not investing, identifies all benefits and costs, and enables managers to use simulation as a decision tool. Once a CIM strategy is justified, its success then depends on effective project management and ongoing performance measures that provide incentives consistent with the strategy. ▲

MANAGING DISCRETE PROJECTS IN A SYSTEMS INTEGRATION INITIATIVE

David B. Pratt
Research Engineer
School of Industrial Engineering and Management
Oklahoma State University
Stillwater, OK 74078

ABSTRACT

This paper identifies the fundamental nature of the problem associated with effectively managing and coordinating discrete projects within a systems integration initiative. An organized framework for addressing this problem is suggested. The concepts and methodologies presented are based upon the author's experiences in several manufacturing companies in which large-scale systems integration initiatives were being pursued.

INTRODUCTION

Most companies that have attempted a large-scale integration initiative have experienced numerous difficulties. The technical difficulties are formidable enough, but the most significant and most commonly encountered problem is that of effectively managing and coordinating the numerous discrete projects which collectively constitute the comprehensive systems integration plan.

Systems integration programs generally start off with a "big bang". There is an official management blessing, a feature story in the company newsletter, and an announcement that this effort "will change the way we do business". A "champion" is usually designated, along with a team of technical experts having an array of backgrounds, knowledge, and experience. In some cases (the better managed ones), appropriate charters are written for the team and for various team advisory committees.

The integration initiative is usually very broad in scope; in fact, it is typically described as "totally comprehensive". Rarely does the initiative fail to include a statement about "redefining the company".

THE PROBLEM SYMPTOMS

Perhaps this approach to launching a systems integration effort would have a reasonable chance of succeeding if it were not for the fact that so many other "total solutions" can be found within the company's current portfolio of active or proposed projects. For example, within the last five years, how many companies have not considered or pursued at least one of the following comprehensive initiatives: Manufacturing Resource Planning, Zero Defects, Quality Circles, Just In Time, Total Quality Management, etc.

As each of these comprehensive initiatives are presented to the workforce, the employees are

Reprinted from *1991 IIE International Industrial Engineering Conference Proceedings*

asked to become "fully committed" to the program. Implied is an emotional, as well as an intellectual commitment that sometimes transcends logic and, indeed, may even be counter to long-held beliefs and understandings.

Employees are, in effect, being asked to take a leap in faith regarding the appropriateness and expected outcomes of such initiatives. Subsequently, many of these initiatives either fall victim to "making shipments" or fade away due to workforce or management apathy. Still others are subsumed by other "grand initiatives". The resulting consequence is that a significant portion of such initiatives fall far short of their proclaimed benefits, with many actually being detrimental to the overall performance of the company.

The employees are finally worn down and become cynical of such initiatives. Each new effort is simply "brushed off" with the attitude, "this, too, shall pass". Others who strive steadfastly to be "good soldiers", continuously seek to discover the common threads that might run through the various initiatives. These insightful people often discover contradictory tenets between various initiatives.

One of the results on both types of employees is effectively the same: <u>a diffusion of commitment</u>. After all, how many things can one person be fully committed to at one time? Another result is a confusion over meanings and fundamentals. Where are the "first principles" upon which these initiatives are based?

Not only are there numerous "grand scheme" initiatives under way at all times, there are also many, many projects and programs on-going which pertain directly to the execution of specific operational functions. For example, the Information System group seems to always have many dozens of backlogged projects waiting to be completed. Cost reduction programs are commonly found in most organizations. Similarly, projects related to energy conservation, inventory reduction, set-up reduction, ...etc., are frequently being pursued.

It is not unusual in a low volume, high value discrete part manufacturing environment for over one half of all effort to be devoted to projects and special initiatives. With such a high expenditure of resources, it is extremely crucial that this part of the organization's efforts be rationalized and managed well. Sadly, this is rarely the case.

THE FUNDAMENTAL PROBLEM

The fundamental problem is the absence in most companies of a systematic mechanism for effectively responding to suggested initiatives and of managing the many special projects resulting from these suggestions in a unified, cohesive, rationalized way.

The chaotic situation just described is usually in "full bloom" at the time the integration initiative is launched. When the initiative is announced, it typically does not replace, supersede, or subsume any of the existing projects and other initiatives, even though it is charged with "redefining" the company. The integration initiative is simply heaped on top of the other "grand schemes", projects, and programs.

Many people within the workforce will recognize the inconsistencies in the multiple projects and either become very frustrated and cynical or, perhaps worse, originate a new "integration project". Others within the workforce will resist the integration initiative, since they see it as a threat to their "pet" project or as competition for scarce resources.

A SOLUTION FRAMEWORK

Overview

A large-scale systems integration initiative should not be pursued unless top management is willing to decree and substantiate that all existing projects and special initiatives are subject to review in light of the systems integration strategic plan. This review in all likelihood will lead to the suspension, cancellation, or modification of many existing projects. If the integration initiative is truly intended to "change the way we do business", and is to be "totally comprehensive", then all current initiatives must be subject to re-assessment

Companies that have successfully pursued a large-scale integration initiative have had to make a commitment up front to a major re-thinking and renovation of their primary performance measures, the design of the product itself, the production system, the management control systems, the information and decision support systems, vendor relations, reward structures, organizational structures, distribution and customer service strategies, purchasing practices, quality assurance, etc.

Such a comprehensive re-examination and re-structuring amounts to nothing less than "re-inventing the company". Consequently, it is futile at best and perhaps counterproductive to continue pursuing on-going, individual projects and special initiatives, without at least assessing the viability of each one.

Evolution of the Integration Initiative

SPECIAL INITIATIVES & GRAND SCHEMES
TQM
JIT
CAD
CAPP
FUNCTION SPECIFIC INITIATIVES
• DECISION TO DEVELOP INTEGRATION PLAN
• SUSPEND ON-GOING PROJECTS
DEVELOP INTEGRATION PLAN
IMPLEMENT INTEGRATION PLAN
• ASSESS PREVIOUS PROJECTS
• IMPLEMENT MULTI-PROJECT MGMT

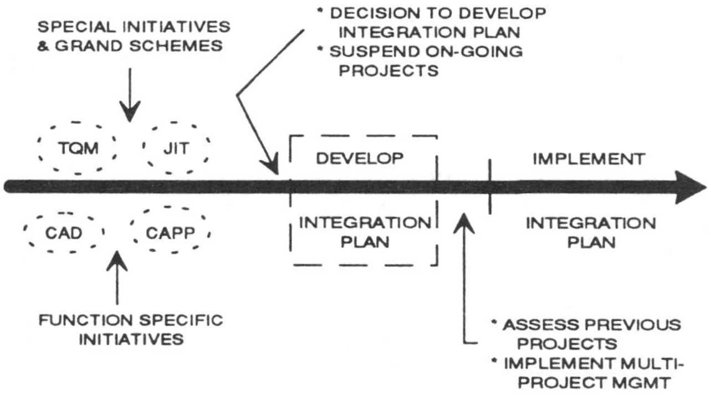

Figure 1

A systematic mechanism is needed to successfully carry out a systems integration initiative. The three major elements of such a system are (1) developing a comprehensive systems integration plan, (2) reviewing existing projects and initiatives in light of the integration initiative, and (3) managing discrete projects within the integration initiative including the continuing stream of proposed "good ideas". Figure 1 illustrates this progression from the old modus operandi to the new.

Developing the Integration Plan

Detailed procedures and methodologies for developing a comprehensive integration plan are beyond the scope of this paper. Suffice it to say that such a plan should represent a thorough rationalization of the management systems, the production systems, and the information/communication systems of the organization.

One prevailing concept is that the integration plan should be a minutely detailed road map of every action required to achieve systems integration. Companies which have attempted developing such a detailed plan have discovered that this task is virtually impossible. Long before the plan can be fully developed, changes in technology and user requirements will have rendered early portions of the plan obsolete.

A more practical approach is to consider the integration plan a high level template, including appropriate standards and a broad functional framework. The high level plan itself is subject to change, though not nearly as often as would be required for a more detailed plan.

Within this context, continuing generation of "good ideas" is not only supported, but is essential to the successful execution of the integration plan. A powerful advantage of this approach is that essentially everyone in the company can play a role in the actual realization of the integration initiative.

The integration initiative should never be considered completed. It systematically evolves into a mechanism for continuous improvement and organizational renewal. It becomes analogous to adaptive feedback control which continuously seeks improved performance.

Reviewing Existing Projects

The first essential step in determining the disposition of all on-going projects and special initiatives is to document their current status. This requires a thorough process of identification and documentation. To insure consistent reporting of data, a survey instrument such as the one in Figure 2 should be used.

Project Justification Form

JUSTIFICATION FOR CURRENT PROJECT CONTINUATION

Project Title & Description:

Responsible Manager/Organization:
Project Leader:
Project Start Date:
Original Project Completion Date:
Current Status (percent complete):
Current Estimated Completion Date:
Nature of Project:
___ Company is Contractually Obligated to Complete
Contract Number:
___ Project Required to Support Production Program
Program Name:
Brief Justification for Continuing Project:

Consequences of Discontinuing Project:

Compatibility of Project with Integration Plan:

If More Than 30% of Project Remains, Estimate:

Benefits	Optimistic	Most Likely	Pessimistic
ROI	________	________	________
Payback	________	________	________
Other	________	________	________
Costs			
Capital	________	________	________
Manpower	________	________	________
Expense	________	________	________

Figure 2

Upon completing the survey of existing projects, each project can be categorized and prioritized using an appropriate ranking procedure. This assessment procedure should include not only an evaluation of expected benefits, but also, evaluation of optimistic and pessimistic outcomes. Thus, a measure of the risk associated with the project can be established. Complete development of this procedure is beyond the scope of this paper.

Another critical aspect of the post-survey evaluation is to determine which portion(s) of the comprehensive systems integration plan each individual on-going project addresses. This assessment provides the major input to determining whether the existing project needs to be suspended, cancelled, or modified.

Managing Discrete Projects

The third and final part of the systematic mechanism is a process for managing on-going discrete projects and handling new initiatives, both "grand schemes" and "good ideas" from technical staff. Two aspects of this mechanism are critical; (1) a formal organization structure to coordinate projects and (2) a formal project management methodology.

It is important to establish a formal organizational structure to coordinate the individual projects that will be proposed and/or underway during the pursuit of a systems integration initiative. In a large multi-division manufacturing company, this structure should take on a hierarchical structure such as the one illustrated in Figure 3.

Three distinct committee levels should be implemented; the Steering Committee, the Advisory Committee, and the individual project teams. It is important to note the hierarchical nature of these committees. It is through this hierarchy that coordination and adherence to overall integration goals are maintained.

The Steering Committee is typically chaired by the Division VP of Manufacturing Operations. The primary charter of this committee is to ensure that the integration plan remains congruent with corporate strategic objectives. This committee also assists the Advisory Committee in acquiring financial and human resources to implement the integration plan.

Project Organization for Systems Integration

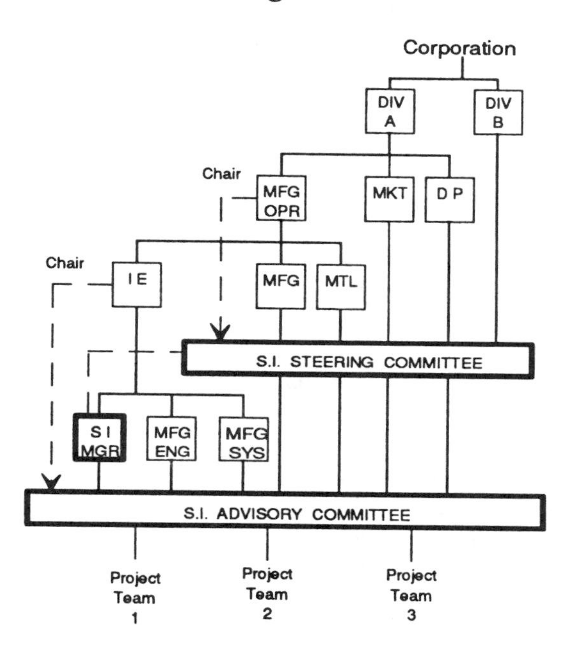

Figure 3

The Systems Integration Advisory Committee is typically chaired by the Director of Industrial Engineering. The primary charter of this committee is to ensure that the integration plan is effectively and efficiently implemented. This committee also coordinates and reviews project team and inter-functional activities.

Project teams are headed by a project leader with technical responsibility for accomplishing the project's objectives. Project team members are selected based on their ability to contribute technical and/or functional expertise to the team.

When considering the management of a discrete project, it is helpful to characterize it in terms of the several phases through which it progresses. The five phases traditionally found within project management literature are (I) concept/proposal development, (II) basic system definition, (III) detailed design and development, (IV) system implementation and (V) tracking and maintenance. Figure 4 depicts a bar chart illustrating the level of effort typically required for each project phase.

Level of Effort per Project Phase

PHASE	I CNCPT DEVL	II SYS DEFN	III DTL DSGN	IV IMPL	V MAINT
LEVEL OF EFFORT					

Figure 4

A work break down structure (WBS) is commonly used to depict the detailed steps required to accomplish the tasks within each phase of a complex project. WBS task lists of the typical steps required in each phase of a discrete project within an integration initiative are shown in Figure 5.

The major objectives of Phase I, the Concept/Proposal Development phase, are to formalize the conceptual definition of the project, estimate the project's impact on the integration plan, and determine the feasibility of pursuing the project further.

WORK BREAKDOWN STRUCTURES BY PROJECT PHASE

PHASE I - CONCEPT/PROPOSAL DEVELOPMENT

Review Present System
Conduct Literature Review
Prepare Concept/Proposal Statement
Evaluate Compatibility with Integration Plan
- Current Factory Status
- Factory of the Future Vision
- Other Active/Planned Projects

Evaluate Initial Feasibility
Develop Phase II Work Breakdown Structure
Conduct Management Review
- Perceived Value of Concept
- Compatibility with Integration Plan

Obtain Approvals
Schedule into Integration Plan

PHASE II - BASIC SYSTEM DEFINITION

Appoint Project Leader & Team
Review Phase I Results
Determine Basic User & System Requirements
- Functional Needs (New or Replacement)
- Equipment
- Data/Information
- Hardware & Software
- Control/Communication Interfaces
- Manpower & Facilities (Capacity Constraints)
- Technology (Current or Future)

Develop Basic System Definition
Establish Alternative Evaluation Criteria
Generate Potential Alternatives
Evaluate Alternatives
- Economic Benefit/Cost
- Strategic & Competitive Value
- Technological Feasibility
- Corporate Culture and other Intangibles
  - Risk Posture
  - Adaptability to Change
  - Human Resource Policies

Recommend Preferred Alternative
Develop Phase III Work Breakdown Structure
Conduct Management Review
- Compatibility with Integration Plan
- Strategic Importance/Competitive Advantage
- Feasibility
- Benefit/Cost Analysis

Obtain Approvals
Schedule into Integration Plan

PHASE III - DETAILED DESIGN & DEVELOPMENT

Review Team Membership
Review Phase II Results
Prepare Detailed System Specifications
- Functional Design Specs
- Equipment & Supply Specs
- Control Procedures
- Training & Education Plan

Conduct Technical Review
Acquire/Develop and Test Subsystems
- Physical Equipment Subsystems
- Data & Information Subsystems
- Logic & Control Subsystems

Initiate Training & Education Plan
Install Subsystems
Conduct System Tests
Develop Phase IV Work Breakdown Structure
Conduct Management Review
- Compatibility with Integration Plan
- Strategic Importance/Competitive Advantage

Obtain Approvals
Schedule into Integration Plan

PHASE IV - IMPLEMENTATION

Review Team Membership
Review Phase III Results
Prepare Detailed Operations Turnover Plan
Conduct User Acceptance Testing
Conduct Final User Training
Convert to User Control
Develop Phase V Work Breakdown Structure
Conduct Management Review

PHASE V - TRACKING AND MAINTENANCE

Review Team Membership
Review Phase IV Results
Monitor On-Going System Operation
Schedule & Conduct Periodic Benchmark Tests
Audit Anticipated Benefit/Cost Analysis
Implement Corrective Measures as Needed
Conduct Periodic Management Reviews

FIGURE 5

The major objectives of Phase II, Basic System Definition, are to develop a detailed understanding of the requirements of the system's users, evaluate potential solution alternatives, recommend a preferred solution, and determine the desirability of developing and implementing the preferred solution.

Phase III of the project, Detailed Design and Development, is typically the largest and most complex phase. This phase involves development of detailed system specifications, acquiring and testing subsystems, preparing control procedures, initiating education and training, and conducting system tests.

The major objectives of Phase IV, the Implementation phase, are to conduct final user training and convert the system to user control.

Phase V is the post-implementation Tracking and Maintenance phase. The major objective of this phase is to audit the on-going performance of the system, ensuring that it provides the benefits upon which it was justified. In the event that benefits are not as planned, corrective measures may be required. It is important to note that this is a continuous and on-going process. There is no definitive end point to this feedback control loop.

The project management structure described above may appear to be unnecessarily complex and constraining. Perhaps this would be true for a set of projects that are relatively small. If, however, an individual project may cost $10 million dollars or more, it is important to provide a mechanism that forces a logical and financial justification at several checkpoints within the life of the project. At the completion of each phase, more and more will be known about the nature of the project, its contribution to corporate objectives, its compatibility with and support of the comprehensive integration plan.

SUCCESSFUL IMPLEMENTATION

Keys to Success

There are three keys to successful implementation of this multi-project management methodology. They are (1) recognizing that the integration plan is dynamic not static, (2) reviewing existing projects within the integration plan when a new project proposal is evaluated, and (3) reviewing the viability of on-going projects against the plan at various points within the project's life cycle.

Integration initiatives are by their very nature long-term activities. Over time, as technology and our ability to use it changes, so too must our vision of the integrated enterprise. For this reason the systems integration plan must be seen as a dynamic plan rather than a static one.

To ensure that the integration initiative remains viable, two measures must be implemented. First, on a periodic basis the entire integration plan should be reviewed and updated. The timing of these reviews is a function of the particular industry within which the company operates and the pace of technological change. Second, when the integration plan is updated, all on-going projects must be reviewed for consistency. The Project Justification Form (presented previously in Figure 2) is an appropriate document for this purpose.

The second key to success involves reviewing existing projects within the integration plan whenever a new project is initiated. In the project management methodology outlined above, the Phase I task 'Evaluate Compatibility with Integration Plan' explicitly considers the impact a proposed project has on other on-going projects.

The third and final key to success is the monitoring of continued compatibility with the integration plan during the design and development of a discrete project. This process is explicitly considered during Phases II and III of the on-going project within the 'Management Review' WBS task.

The Unifying Concept

The three keys to success all unify in a single concept. Simply stated, this unifying concept is that a formal mechanism must be in place to prevent the initiation and/or pursuit of multiple projects which are redundant, overlapping, inconsistent, or worst of all, contradictory within the framework of the systems integration initiative.

Each key to success addresses a unique aspect of this unifying concept. Periodic reviews and use of the Project Justification Form protect against incompatibilities caused by changes in the integration initiative itself. The Phase I 'Compatibility Evaluation' prevents incompatibilities caused by the introduction of new projects which have ramifications on existing projects. Finally, the Phase II and III 'Management Reviews' prevent incompatibilities caused by increased insights and understanding which evolve within the project development cycle.

The Approval Process

To ensure that this unifying concept is managed effectively, a formal approval process is required during the Management Review tasks on the WBS. Figure 6 illustrates an overall approval process which is consistent with the organization structure of Figure 3 and the Phased WBS of Figure 5.

Of particular emphasis in Figure 6 is that, depending upon the action being taken, a particular person's authority may be approval, recommendation, review, request, or appeal. The intent of this structure is to place project decision making at the most appropriate level while still allowing a sufficient review and approval process. Here again the emphasis is on coordination and compatibility with overall integration goals.

Management Approval Process

Project Phase	S. I. Adv. Comm.	S. I. Steer. Comm.	Project Leader	Dept. Funct. Mgr
I	App	Apl	Req	Rec
II	Rev/Rec	App	Rev/Rec	Rev/Rec
III	App	Apl	Rev/Rec	Rev/Rec
IV	Apl	Apl	Rev/Rec	App
V	Apl	Apl	Rev/Rec	App

Legend:
Apl - Appeal
App - Approve
Rec - Recommend
Rev - Review
Req - Request

FIGURE 6

SUMMARY

Enormous inefficiencies exist in most manufacturing firms in terms of how they manage (or fail to manage) discrete projects within an overall integration initiative. Without a disciplined approach and a systematic mechanism, the integration initiative is likely to fail in a maze of overlapping, inconsistent, and contradictory projects.

This paper outlines a framework for effectively managing discrete projects within a multi-project systems integration initiative. It is a mechanism that evolved over several years and within several companies with which the author has been associated. The key success factors for implementation as well as the unifying concept which they support, have been identified. Finally, an approval mechanism which ensures implementation of the success factors and unifying concept is presented.

ACKNOWLEDGEMENT

The author thanks Dr. Joe H. Mize for his significant contributions and insightful comments which have considerably improved this paper.

REFERENCES

Kerzner, Harold, Project Management, Van Nostrand Reinhold, New York, 1984.

Mize, J. H., D. J. Seifert, and G. Berry, "Strategic Planning for Factory Modernization: A Case Study", National Productivity Review, Winter, 1984- 85.

Mize, J. H., D. J. Seifert, and F. S. Settles, "CIM From a Corporate View- The Garrett Experience," Industrial Engineering, November 1985, Vol. 17, No. 11.

Mize, J. H., "Success Factors for Advanced Manufacturing Systems", Proceedings, 1987 IIE Spring Conference.

Mize, J. H., "Prerequisites for CIE", Proceedings, CAM-I Computer Integrated Enterprise Interest Group Meeting, October 15-16, 1987.

Seifert, L. C., "Product Realization Processes at AT&T", 1987 ASEE Annual Conference Proceedings, June, 1987.

Seifert, L. C., "Design and Analysis of Integrated Electronics Manufacturing Systems", Design and Analysis of Integrated Manufacturing Systems, National Academy Press, 1988.

Seifert, Lawrence C. and Alfred D. Zeisler, National Manufacturing Policy: An Industry Perspective, Commissioned paper appearing in THE CHALLENGE TO MANUFACTURING: A PROPOSAL FOR A NATIONAL FORUM, National Academy of Engineering, Washington, DC, 1988.

Sibbald, G. W., "Roadblocks to CIM Success", CIM Review, Vol. 4, No. 3, Spring, 1988.

Solberg, J. J., "Integrated Manufacturing Systems: An Overview", Design and Analysis of Integrated Manufacturing Systems, National Academy Press, 1988.

AUTHOR

David B. Pratt is a Research Engineer in the Center for Computer Integrated Manufacturing within the School of Industrial Engineering and Management at Oklahoma State University. He holds BS and MS degrees in Industrial Engineering. Following twelve years experience in the petroleum, aerospace, and pulp & paper industries, he returned to pursue his Ph.D. at Oklahoma State in 1989 with completion expected in 1991. His research interests include manufacturing systems modeling, applied operations research, and the strategic implications of CIM. He is a registered Professional Engineer, an APICS Certified Fellow in Production and Inventory Management, and an ASQC Certified Quality Engineer. He is a member of IIE, NSPE, APICS, TIMS, and ASQC.

ACHIEVING SYSTEMS INTEGRATION THROUGH PROJECT MANAGEMENT TECHNIQUES

Adedeji B. Badiru, Ph.D., P.E.
School Of Industrial Engineering
University of Oklahoma
Norman, OK 73019

ABSTRACT

This paper discusses the use of project management approaches in systems integration efforts. The paper covers the basic characteristics of a system, system integration procedure, potential role of project management, and the expected functional characteristics of an integrated system. Guidelines are provided for enhancing systems communication, systems coordination, and conflict resolution in integrated systems.

INTRODUCTION

Systems integration is now a major objective in many organizations. Even though systems integration problems have been recognized for a long time, no specific methodologies and procedures for achieving the desired integration have been developed. This paper explores the role that project management techniques can play in the effort to successfully integrate the various components of a system. Full system integration involves integrating and coordinating all aspects of a corporate enterprise including engineering, design, sales, planning, management, finance, accounting, and manufacturing.

This paper recommends the integration of the various subsystems in an organization with respect to a given set of objectives by using proven project management techniques [4, 6]. Many unintegrated subsystems exist in a typical organization. These subsystems may include management subsystem, manufacturing subsystem, design and engineering subsystem, customer service subsystem, financial subsystem, marketing subsystem, inventory subsystem, personnel information subsystem, product quality information subsystem, and production subsystem [3]. In an unintegrated system, some of these subsystems may possess different, or even conflicting, priorities.

In order to achieve the benefits of integration, management must be willing to change its prevailing organizational culture. Organizational objectives must be prioritized, integrated, and applied uniformly throughout the organization with a global systems view. With systems integration, the benefits of both centralized and decentralized subsystems can be pooled in a synergistic fashion.

SYSTEMS DEFINITION

An operational definition of a system must be established before the task of integration can begin. The classical definition refers to a *system* as *a collection of interrelated elements brought together to pursue a common objective*. One can argue that the lack of a uniform definition of what each person (or department) perceives as a system is a major stumbling block to systems integration in practice. Many individual departments within an organization perceive themselves as individual self-sufficient systems; and they tend to operate as such, without regard for the essential interactions with other departments. The classical characteristics of a system include interaction with the environment, possession of an objective, ability to self-regulate, and ability to self-adjust.

For example, in a manufacturing system [5], interaction with the environment may be defined in terms of what the market environment (the

Reprinted from *1990 IIE Integrated Systems Conference Proceedings*

customer) wants. The objective of the system may be to achieve an acceptable level of product quality. The self-regulation characteristic may relate to the system's ability to maintain the stipulated quality level once it is achieved. The self-adjustment characteristic may relate to the system's ability to make amendment should the quality level deviate significantly from the required level. An appreciation of how one system affects another in the pursuit of the common objective is a significant component of systems integration.

The various elements of an integrated system act simultaneous in a separate but interrelated fashion to achieve a common goal. This synergism helps to expedite decision and operating processes and to enhance the collective effectiveness of an organization. The supporting commitments from other subsystems of the organization serve to counterbalance the weaknesses of a given subsystem. Thus, the overall effectiveness of the system is greater than the sum of the individual results from the subsystems. Systems integration should address the integration of people, hardware, software, and other tools in an organization. For example, software and hardware integration has been proven [9] to be useful in creating faster design and better communication in several manufacturing operations

SYSTEMS PROJECT

Project management can play a multitude of roles in systems integration [1]. A conceptual model for the potential role of project management in systems integration is presented in Figure 1.

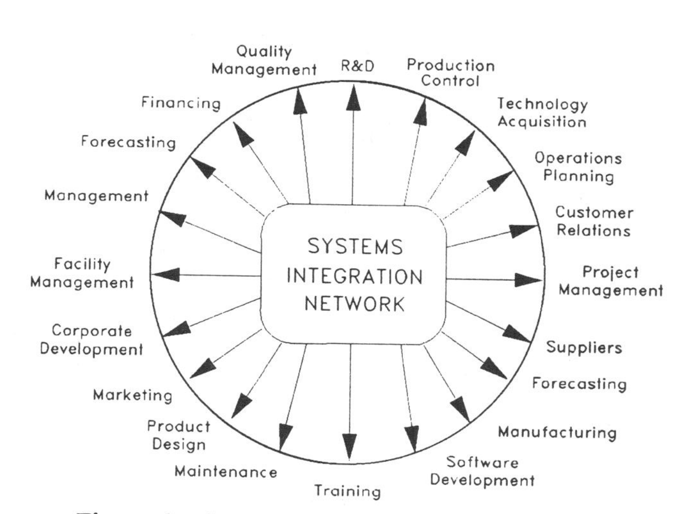

Figure 1. Systems Integration Network

The global view provided by a systems approach [8] ensures that all the factors that can influence organizational productivity are considered. The objectives of systems integration can be achieved through basic project management approaches as discussed below:

Systems Project Management: Project management can provide guidelines for managing the operations of an integrated system just like any conventional project. The delineation of the business objectives and organizational goals can be addressed along project management functional lines.

Systems planning: Plans guide actions. The success of systems integration depends on how well plans are developed and executed across an organization. Project management can provide planning models for the systems integration effort.

Systems simulation: The network approach of project management can facilitate the creation of simulation models that permit a comprehensive study of an integrated system prior to actual implementation. Parallel systems operations and functional coordination can be studied in the safe environment of simulation.

Systems staffing: Project management techniques provide guidelines for staffing across systems requirements. Project organization models can address personnel interfaces, managerial hierarchy, and control structure required in an integrated system.

Team organization: The management of an integrated system involves the management of teams that were previously disjoint. The team organization and coordination approaches of project management can be useful in integrating the functions of cooperative system teams.

Task planning and networking: The planning and coordination of tasks across several functional groups in an integrated system can be handled by CPM and PERT analysis. Operations planning and control in an integrated environment can be addressed very well by project coordination techniques.

Work breakdown structure: The work breakdown structure approach of project management can facilitate the identification of the interdependencies of activities in an integrated system. The identification can then help in establishing proper operating guidelines for the system.

Resource allocation: Proven resource allocation techniques used in project management can be employed in the assignment of resources to components of an integrated system. Resource sharing and intra-subsystem or inter-subsystem resource interfaces can be defined in terms of project resource loading and leveling. Project-oriented resource allocation can facilitate the sharing of resources in an integrated system.

Integrated scheduling: The components of an integrated system may be viewed as elements of a system consisting of multiple projects. The operations of the components can be scheduled in an integrated fashion that emphasizes the interrelationships of functions within the integrated system.

Systems monitoring and control: Activity networking, Gantt charts, resource loading graphs, and cost charts can be utilized in monitoring the operations of an integrated system and prescribing corrective actions for system deficiencies.

Systems implementation: Project management procedures can provide guidelines for implementing systems integration. Implementation strategies designed to fit systems goals can be developed through the techniques of project implementation models.

Systems justification: Both the strategic and economic justification of integrated systems can be evaluated by employing standard models used in project management [2].

Performance appraisal: The appraisal of an integrated system can be established in terms of productivity, cost, efficiency, responsiveness, and output based on project management models.

SYSTEMS INTEGRATION PROCEDURE

Systems integration involves the linking of components to form subsystems and the linking of subsystems to form composite systems. It can be implemented within a single department or across departments in an organization. It facilitates the coordination of diverse technical and managerial efforts to enhance organizational functions, reduce cost, save energy, improve productivity, and increase the utilization of resources. Systems integration emphasizes the identification and coordination of the interface requirements between the components in a system. The components and subsystems operate synergistically to enhance the performance of the total system. Systems integration ensures that all performance goals are satisfied with a minimum of expenditure of time and resources. One of the reasons that industrial engineering has been flourishing is the fact that the profession emphasizes the need for systems integration. Systems integration, as stressed by Meckler [7], is based upon integrating compatible functions to achieve the following benefits:

1. Dual-use integration: This involves the use of a single component by separate subsystems to reduce both the initial cost and the operating cost during a project life cycle.

2. Dynamic resource integration: This involves integrating the resource flows of two normally separate subsystems so that the coordinated flow of resources from one subsystem to another minimizes the total resource requirements in the system.

3. Restructuring of functions: This involves the restructuring of functions and re-integration of subsystems to optimize costs when a new subsystem is introduced into an integrated system.

Systems integration is particularly important when introducing new technology into an existing system. It involves coordinating new operations to coexist with existing operations. It may require the adjustment of functions to permit sharing of resources, development of new policies to accommodate product integration, or realignment of managerial responsibilities. Presented below are important questions relevant for systems integration:

* What are the unique characteristics of each component in the system to be integrated?

* How do the characteristics complement one another?

* What physical interfaces exist between the components?

* What data/information interfaces exist between the components?

* What ideological differences exist between the components?

* What are the data flow requirements for the components?

* Are there similar integrated systems operating elsewhere?

* What are the reporting requirements in the integrated system?

* Are there any hierarchical restrictions on the operations of the components of the integrated system?

* What are the internal and external factors expected to influence the integrated system?

* How can the performance of the integrated system be measured?

* What benefit/cost documentations are required for the integrated system?

* What is the cost of designing and implementing the integrated system?

* What are the relative priorities assigned to each component of the integrated system?

* What are the strengths of the integrated system?

* What are the weaknesses of the integrated system?

* What resources are needed to keep the integrated system operating satisfactorily?

* Which section of the organization will have primary responsibility for the operation of the integrated system?

* What are the quality specifications and requirements for the integrated system?

If implemented properly, an integrated system should exhibit the following characteristics:

1. Systematic solution of organizational problems
2. Interaction with subsystem environments
3. Specification of the interrelationships of subsystems
4. Dynamic integration of activities into an effective total system
5. Uniformity of objective
6. Cooperative regulation of efforts
7. Coordinated adjustment of functions to solve problems

The Triple C model for communication, cooperation, and coordination [1] can be effective in achieving the objectives of integrated systems. An implementation of the model for systems integration is discussed below.

SYSTEMS COMMUNICATION

Communication is a valuable resource that can contribute significantly to productivity improvement in an integrated system. The communication function involves making all system components aware of the requirements of integration. Those that will be affected by the integration process directly or indirectly should be informed as appropriate regarding the following:

* The scope of the integrated system
* The need for integration
* The person or group in charge of the integration effort
* Alternatives available
* Expected cost of the integration
* Disadvantages associated with lack of integration
* Persons that will be affected by the failure of the integration effort
* Potential weaknesses of the proposed integrated system
* Potential direct and indirect benefits of the integrated system
* Resource requirements versus resource availability
* Prevailing sources of support for the integration effort
* Personnel contribution needed for the integration to succeed
* Schedule for launching the integration project
* Organization of the teams involved in the integration project

The communication channel must be kept open throughout the integration effort. In addition to in-house communication, external sources should also be consulted as appropriate. The communication effort should succeed if the integration teams follow the strategies below:

* Demonstrate action-oriented commitment to the integration effort
* Institute flexible communication hierarchies
* Develop communication responsibility matrix

* Create multi-channel communication networks

* Identify internal and external communication needs

* Endorse both formal and informal communication links

SYSTEMS COOPERATION

The cooperation of the members of the integration team must be explicitly sought. Merely saying "YES" does not constitute enough assurance of full cooperation. The potential supporters of integration must be convinced of the merits of the effort. Some of the factors that influence cooperation include manpower requirements, work overload, resource requirements, budget limitations, past experiences, destructive competition, conflicting priorities, too much departmental individualism, and lack of uniform organizational support. A structured approach to seeking cooperation should provide details for the following:

* The rewards of cooperation

* The cooperative efforts required

* The implication of a lack of cooperation

* The criticality of cooperation for the integration effort

* The time-frame for implementing the proposed integration

* The organizational impact of integration

SYSTEMS COORDINATION

The activities of the integrated system must be coordinated after proper communication and cooperation have been secured. Coordination facilitates harmonious organization of system functions. The development of a responsibility chart can be very helpful at this stage. A responsibility chart is a matrix consisting of columns of individual or functional departments and rows of required actions. Cells within the matrix contain relationship codes that indicate who is responsible for what. The responsibility chart helps to avoid overlooking critical communication requirements, personnel interactions, and functional responsibilities. It can help resolve questions such as:

1. Who is to do what?

2. Who is responsible for which results?

3. What personnel interfaces are involved?

4. Who is to inform whom of what?

5. Whose approval is needed for what?

6. What support is needed from whom for what functions?

CONFLICT RESOLUTION

When implemented as an integrated process, the Triple C approach can help avoid conflicts in an integrated system. When conflicts do develop, it can help in resolving the conflicts. Conflict resolution through Triple C can be achieved by observing the following rules:

1. Admit that conflicts exist and make efforts to identify the underlying causes.

2. Use cooperation and negotiation as mechanisms for resolving system conflicts.

3. Distinguish between proactive, inactive, and reactive subsystem behaviors in a conflict situation.

4. Use communication to defuse internal conflicts and competition within an integrated system.

5. Recognize that short-term system compromise can lead to long-term system gains.

6. Use coordination to work towards a unified system goal.

CONCLUSION

With the increasing shortages of resources, more emphasis should be placed on the sharing of resources. No organization or component of an organization should operate as a self-sufficient island of functional specifications. Resource sharing, which may involve physical equipment, knowledge, personnel, information, policies, and procedures, provide a common basis for achieving systems integration.

This paper has presented the potential use of project management approaches in pursuing and achieving systems integration. The paper covers the basic characteristics of a system, system integration procedure, specific project management approaches, and the characteristics of an integrated system. Guidelines for implementing systems integration are also presented. The communication and coordination requirements and conflict resolution strategies in an integrated system are also outlined.

REFERENCES

[1] Badiru, Adedeji B., *Project Management In Manufacturing And High Technology Operations*, John Wiley & Sons, New York, 1988.

[2] Badiru, Adedeji B., "A Management Guide to Automation Cost Justification," *Industrial Engineering*, Vol. 22, No. 2, Feb. 1990a, pp. 26-30.

[3] Badiru, Adedeji B., "A Systems Approach to Total Quality Management," *Industrial Engineering*, Vol. 22, No. 3, March 1990b, pp. 33-36.

[4] Cleland, David I., *Project Management: Strategic Design and Implementation*, TAB Professional & Reference Books, New York, 1990.

[5] Gessner, Robert A., *Manufacturing Information Systems: Implementation Planning*, John Wiley & Sons, New York, 1984.

[6] Kerzner, Harold *Project Management: A Systems Approach to Planning, Scheduling, and Controlling*, 3ed, Van Nostrand, New York, 1989.

[7] Meckler, Gershon, "Systems Integration - A State-of-the-Art Report," *Consulting-Specifying Engineer*, August 1987, pp. 44-51.

[8] Roman, Daniel D., *Managing Projects: A Systems Approach*, Elsevier Science Publishing Co., Inc., New York, 1986.

[9] *Industrial Engineering*, "Systems Integration Creates Faster Design, Better Communications," *Industrial Engineering*, Vol. 22, No. 7, July 1990, p. 65.

ABOUT THE AUTHOR

Dr. Adedeji B. Badiru, P.E. is a member of the industrial engineering faculty at the University of Oklahoma. He is a senior member of IIE. He is a past president of the Oklahoma City Chapter of the Institute. He is a registered professional engineer in the state of Oklahoma. He holds BS and MS in Industrial Engineering and MS in Mathematics from Tennessee Technological University. He holds a Ph.D. in Industrial Engineering from the University of Central Florida. Dr. Badiru has published numerous papers on project management, expert systems, and microcomputer applications. He is the author of *Project Management in Manufacturing and High Technology Operations*, John Wiley & Sons, 1988 and coauthor of *Computer Tools, Models, and Techniques for Project Management*, TAB Professional Reference Books, 1990. He is a member of IIE, SME, PMI, AAAI, ORSA, and TIMS.

STRATEGICALLY PLANNING PROJECTS: THE PROJECT MANAGEMENT REQUIREMENTS ANALYSIS

Deborah S. Kezsbom, Ph.D.
President
MRA Management Resources, Inc.
560 Sylvan Avenue
Englewood Cliffs, NJ 07632
(201) 871-1640

ABSTRACT

Project Management strategies must be coupled to the project's business objectives. This paper presents a methodology for achieving company-specific project management strategies and training.

INTRODUCTION

Technology, manufacturing processes, market demands, management philosophies, government regulations, and organizational structure are a few of the factors that determine the appropriateness of project management strategies within today's competitive corporate environment. Developing professionals with strong project planning, scheduling, controlling and team building skills for ensuring project realization has become a necessity. Project management tools and procedures must address the surrounding culture or typical manner of doing business, so that they can be readily accepted and implemented. Project planning, scheduling, controlling and team building strategies, moreover, must be tightly coupled to the project's business objectives. The dilemma, however, concerns **how** to assess the needs of each specialized project effort and do it as **quickly** and **efficiently** as possible.

A major issue associated with determining project needs is that of outlining a specific methodology that is orderly and detailed. The primary purpose of a **Project Management Requirements Analysis** is (1) to assess and validate present strategies used in accomplishing project objectives, (2) to determine specific project requirements contingent upon scope, timeliness, performance, regulatory requirements, and organizational structure, and (3) to pinpoint strategies that may be improved upon, eliminated, or added to increase the likelihood of project success, given the particular characteristics and demographics of the project.

MEETING THE DEMANDS OF PROJECT WORK

The distinguishing characteristics that render a project environment unresponsive to traditional (functional) policies and procedures have tremendous implications for determining planning, scheduling, and controlling strategies that **facilitate**, rather than hinder, the accomplishment of project objectives.

Projects place great demands on the organization and the people who are responsible for achieving project objectives. Projects typically are dynamic, unique undertakings that have a specific beginning, and end with a well-defined accomplishment or goal. Projects exist for finite periods of time and as they move through a sequence of activities and events, require constant surveillance and adjustment of people, materials, time and money. The fluid, multi-disciplined, cross-functional nature

Reprinted from *1990 IIE Integrated Systems Conference Proceedings*

horizontal coordination of talents and materials that may be quite unique to the organization from where the project initiated. This horizontal effort encourages a variety of opinions, methodologies, and procedures that also create a tremendous potential for conflict.

Project management strategies must be structured to **encourage** and facilitate the accomplishment of project goals within this demanding and challenging environment. This involves developing planning procedures, reward structures, control mechanisms, etc. that parallel the nature of the project work and direct human interactions towards goal accomplishment. Project management strategies must therefore:

- provide a structure in which people are encouraged to accomplish a particular task within the framework of the total project;
- encourage collaborative efforts across the organizations; and
- create a system that minimizes effort and maximizes cost efficiency.

CREATING THE "RIGHT" MATCH

Project planning, scheduling and monitoring strategies must be closely coupled to environmental and organizational factors which impact project processes and procedures. Adopting specific project management approaches requires a long hard examination of, for instance, current planning, scheduling and controlling techniques and the managerial philosophies associated with project units.

No one set of project management approaches will ever address all project needs. However, there are certain factors to consider when contemplating the "right match." These factors include:

- The scope, length and/or technological magnitude of project efforts.
- Market scheduling, cost issues and demands.
- Geographic dispersion of project team members.
- The nature of government regulations and/or of requirements.
- Manufacturing and Quality processes and/or philosophies.
- Organizational Structure and Reporting Relationships.
- Managerial Philosophies.

Strategies that are effective for managing projects must be sensitive and responsive not only to the nature and challenges of the specific technology on which the product is based, but to the social and professional relationship(s) it creates.

Long-term projects, for instance, will demand regularly scheduled reporting activities, that encourage the team to focus on milestones and foster good relationships. Short-term projects, on the other hand, may require more frequent, but brief team meetings that pinpoint problems and identify immediate solutions.

Project management strategies should be established, therefore, to provide sufficient channels of communication, result in properly integrated project work, address the market and scheduling demands, and meet the social and professional needs of its contributing specialists.

THE PROJECT MANAGEMENT REQUIREMENTS ANALYSIS

While implementing good sound project management, methodology often appears to be the logical starting point for high project performance, it is often much more important to make certain that the strategies selected are the **right** ones, and that their implementation will lead to project success.

The **Project Management Requirements Analysis** is an interpretative approach to developing sound project management strategies and subsequent training. It provides direction through identifying, documenting and selecting **appropriate** process methodology. It is based upon certain key assumptions that include the following:

- The project environment consists of a dynamic flow of information. We can make sense of this information by placing it in meaningful "chunks" or project "end-result areas."
- Relationships exist between the information obtained from the project environment and developing a framework to explain **how** the data should be organized.

An effective **Project Management Requirements Analysis** is the process of connecting the fragments of information and activities of project life into a coherent and efficient pattern, with project participants coming to some agreement on the interpretation and strategies.

A PROJECT MANAGEMENT REQUIREMENTS ASSESSMENT MODEL

A **Project Management Requirements Analysis** provides the direction for problem identification and solving, and the development of project management methodology through identifying, documenting and selecting appropriate strategies. Project leaders and facilitators improve the effectiveness and efficiency of their project organization, by identifying certain critical project issues and/or needs before rushing off to apply project management scheduling or tracking methodology.

The following model presents six (6) recommended steps to the project management requirements which involve a variety of tools and techniques.

Step 1.0 - Information Gathering/Preliminary Interface Meeting with Sponsor Organization

The objective of the **Preliminary Interface Meeting** with the sponsor project management team is to establish a clearer focus or vision of the team's perceived planning, scheduling, tracking needs. The data generated by this preliminary meeting becomes the basis of the Requirements Plan. It is a **blueprint** for action that outlines customer needs with regard to, for instance, project management software, planning and scheduling assistance, and on-going tracking and maintenance. The preliminary data generated becomes the input for Step 2.0, **The Requirement Analysis Interview Document**.

Step 2.0 - Develop a Project Requirements Analysis Interview Document

Interviewing represents a **subjective process** to gather pertinent information in order to develop what should be as objective a strategy as possible. One of the difficulties one wrestles with in subjective situations is outlining an interpretive methodology that is orderly, detailed and straight forward. Through the development of a structured interview document, we attempt to develop an interpretive process with "objectivist" language.

This document, therefore, is intended to:

- pinpoint present project management strategies
- determine subsequent planning, tracking, cost estimating, and directing strategies
- determine specific software needs (eg. reporting, budgeting, planning, updating operating systems requirements).

As illustrated in Table 1, the four major end-result areas of the requirements analysis document, therefore, parallel the four functional responsibilities of the typical matrix project management system.

Table 1

TOPICAL AREAS FOR PROJECT MANAGEMENT REQUIREMENTS ANALYSIS

- **PLANNING AND SCHEDULING TECHNIQUES**
  - WBS
  - Work Packages
  - Resource Constraints
  - Scheduling Constraints
  - Change Process
- **CONTROLLING AND TRACKING**
  - Change COntrol Strategies
  - Progress Against Schedules
  - Reporting Mechanisms
- **ORGANIZING AND ORGANIZATIONAL TECHNIQUES**
  - Responsibility Assignments
- **INTERFACE MANAGEMENT**
  - Communication Strategies
  - Team Building
  - Typical Conflict Areas

Step 3.0 - Identifying Project Subject-Matter-Experts/Requirements Assessment Partners

A Project Management Requirements Analysis may at times, pose a threat to present project participants. It may, unfortunately, create a misconception that something is "wrong" with the present strategies, or create a desire for finger pointing at contributing project functions. Successful project requirements assessments, therefore, depend on choosing the "correct" partners (i.e. subject-matter-experts) to guide the process and to eventually own it when it is completed. An otherwise successful analysis may fail simply because uninvolved or unrepresented people may not see that the change or new strategies might benefit them. Project plans are more readily accepted by having affected people act as partners or be represented in its creation.

The planning partners/subject-matter-experts interviewed during the data collection process should, therefore, include:
(a) management and select project specialists who are directly affected by the results
(b) those key players that will implement the plan.

The exact partners or "SME's" depend on the projects' technological and functional mix, but should include a balance of hands-on project specialists who are really involved in what's going on and first and second level management who are responsible for the acceptance and shifting of priorities; that is the **implementers** and their immediate supervisors! Final selection should be in the hands of the project manager or project management team with assistance from the Interviewer/Expert.

Step 4.0 - Data Collection

To objectify a somewhat subjective procedure, data collection should be a team process. Face-to-face meetings are arranged and a combination of interviewing and questionnaire strategies are used. The questions are broad-based enough to cover an array of possible needs, without burdening respondents with pre-meeting research or being or overly time-consuming length. Any "hard" data (i.e. status reports, milestones charts, etc.) that respondents bring to the interview is certainly welcome, but not necessary. When such data is offered, however, review the validity and reliability of this data and use it only when it supplies useful information concerning project needs and performance.

It is important for the interviewing team to gain acceptance of themselves and the requirements analysis process. Share with the subject-matter-experts the intent and the basic concepts of the assessment process. Answer any questions they have about the interview document and address their concern(s) (if any) of anonymity.

Step 5.0 - Compilation and Interpretation of Data

The data generated by the Requirements Assessment interviews is independently analyzed by a team of project management experts/Interviewers. The goal of this synthesis is to determine the strategies presently used that are effective and should be maintained and to identify and fill the gaps between ideal situations and the client organization's needs. It is critical to realize the importance of having two or more individuals at the data gathering and interpretive data analysis steps. The interviewer is in an position to determine realities. By choosing to exclude, include, ignore or deflect certain pieces of information, the interviewer can influence how the situation is defined. True needs of the project organization must be separated from hearsay or even perceived end results. The interviewer's, therefore, must always bear in mind that they are helping to define the situation both by the questions asked and the end-results created. Recommendations for maintenance and improvements are compiled in an **Executive Summary/Feedback Report**.

Step 6 - Management/Specialist Feedback Session

Before setting priorities, and deriving specific strategies or interventions, a feedback session outlining and presenting the compilation of information is held. All assessment participants, first and second-level management, and interested parties in an authority or decision-making position are invited. Identified and

documented needs are presented and any frequencies or hard data associated with the derivation of the information explained.
Based on the previous steps and all resulting data, provide requirements assessments with partners an actual list of the gaps in results. It is important to further illustrate how the information derived from the requirements analysis define the extent to which the project team can move towards a more productive and satisfying planning, scheduling,tracking and interface system for all.

CONCLUSION

Creating project management strategies that serve in the long run to actually modify the organization, means more than merely changing the reporting structure or regrouping functional tasks assignments. People enter into projects and organizations with a variety of expectations. These typically include the need for professional respect, technical growth, social contracts or affiliation, consistency and even some sense of security. Project management strategies must include adapting systems and procedures to the social requirements of its primary resource, people.

Dr. Deborah S. Kezsbom, president of MRA Management Resources, Inc. advised technical and management teams on Project Management analysis, software and implementation strategies. She is an expert in project team building approaches and developing project management systems. Her clients include, among others, AT&T, Compugraphic Corporation, 3M, General Instrument, Motorola and various agencies, within the Department of Defense. Co-author of Dynamic Project Management: A Practical Guide for Managers and Engineers, John Wiley 1989, Dr. Kezsbom was Associate Editor of the IEEE Communications Society and responsible for a regular column on engineering management issues, entitled "Communicating." A Phi Beta Kappa Graduate, Dr. Kezsbom is an active member the Project Management Institute, IEEE, ASEM, and the IIE.

TRAINING EMPLOYEES IN INDUSTRIAL ENGINEERING PROJECT TEAM CONCEPTS

Charles S. Elliott
Associate Professor of Engineering and
Director-Center for Professional Development
College of Engineering and Applied Sciences
Arizona State University
Tempe, AZ 85287-7506
(602) 965-1740

Participative management and especially the use of project teams is now widely practiced in much of our country's industry and increasingly in the service sector also. Training is an absolute necessity for high quality results and long term sustinance of this most important management system. The purpose of this paper is to outline some suggested basic topics, methods and procedures which need to be considered and to report on some successful examples which have been utilized in this area.

WHY PROJECT TEAMS?

Whether they are called Quality Circles, Employee Involvement Teams, Productivity Improvement Teams, Union-Management Participation Teams, or otherwise similarly labeled, the basic concept of radically altering ones traditional approach to managing business, industry, governmental or service firms has gained wide, and increasing usage throughout the United States. [1] While often stimulated by our learning about the great success of Japanese management [2] and our requirement to become much more competitive internationally, the empowerment of workers, reduction in levels of management, and seeking wide participation by the entire work force has gained such broad based (but far from unanimous) support and acceptance. Although often hard to quantify the value (ROI, etc.) of such efforts, we have forged ahead and consistently expanded the use of teams. Now headed for what many consider perhaps the ultimate team concept - self managed work teams, [3] we need to reflect on what has worked (and not worked) [4] and why. Training stands out as a very key element. Of course we have found ourselves much more successful with teams when we have the strong support of top management, have successfully coped with the somewhat natural "resistance" of middle and first line management, have gotten the union on board if one is present in the firm, have established good measurement systems, have maximized our reward systems, and have taken a long term perspective rather than our usual short term results mentality to any such efforts. The key to success is that we now recognize that this is a whole new way of managing and not just another "program" to get more out of the workforce.

While all of the above listed factors are important, the central role of training of employees will be addressed here - with particular reference to the special roles that Industrial Engineers can (and should) play in this process.

Reprinted from *1990 IIE Integrated Systems Conference Proceedings*

WHAT KIND OF TRAINING FOR TEAMS?

You answer the first part of this question with another question -what are you expecting teams to do? Several companies have been encountered which have not really addressed this issue. Without a clear understanding of the objectives, it is very difficult to design team training activities.

Scope of Team Activities

Are the teams to be highly involved in statistical process control (SPC)? Involved with a broad range of problem solving? Do extensive data gathering? Develop work simplification and process improvements? Perform productivity measurement and improvement? Deal with the total quality of work life in the facility? Address union-management concerns? (very few do) Deal with overall management systems and improvements? Will they ultimately be involved in team member selection? Interview new hires? Provide training to their co-workers? Make formal presentations to others? Are teams a part of a comprehensive Total Quality Management (TQM) program? In short, training needs to be addressed thoroughly and somewhat differently in each area.

Starting Where They Are

Next, we need to consider the target training group(s) and assess where the individuals are in knowledge and skills as to what is expected. Two very common mistakes occur here - we start out with unrealistically high expectations (we only hire the cream of the crop so employees don't need that low level stuff) or we set our expectations too low (no body knows anything about this so everybody has to be trained on everything). Either error turns off a lot of otherwise highly motivated participants in the very beginning of Project Team efforts.

Basic Skills. Many instances have been noted of employees needing skills training in reading, writing and basic computational skills. [5,6]. None of us seems to want to admit this and we spend too much time blaming the educational system for its obvious failures in this area. Motorola has established a basic skills level for communication and computational skills for their employees to be the seventh grade level and wants to increase this to eighth and ninth. Many workers did not meet this so they spent $60 million over four years in basic skills education which led to the founding of their Motorola University to provide a wide variety of internal education and training. [7]

Fundamental Skills. Only after being sure that employees are ready for the "real" training in the fundamentals of team project skills should you proceed. Too many firms have naively assumed they could start here and be successful and some can, but most cannot.

What to include here? While again dependent upon team objectives, the following fundamental team skills may need to be provided - (if assessment shows they are lacking!):

- Data identification
- Data gathering techniques
- Basic statistics
- Control charts
- Pareto diagrams
- Cause and effect problem analysis
- Process capability
- Measurement systems
- Computer applications - particularly data input
- and many phases of Design of Experiments

Communication Skills. How to make presentations, listening, interpersonal communications, team record keeping, etc. are very important team project skills which need to be included in training programs.

Group Process Skills. Employees will especially need training in the "people skills" aspects of teams. This can include:

- group dynamics
- dealing with conflict
- leading meetings
- motivation
- positive feedback
- and many other topics

While IE's often are consulted or even directly involved in the technical skills training identified above, they should also be considered in the communication and group process areas - if they are adequately prepared to do so. Many of our IE consulting skills and processes readily lend themselves to being effective leaders in this area. Of course, being an IE does not automatically qualify you to be an effective trainer in these skills and processes. Colleges and universities should pay more attention to helping IE's gain some of these skills in their undergraduate and graduate curriculums.

Another important aspect of consideration is the type of training. A minimum of lecturing and a maximum of involvement - hands on activities, case studies, problem solving, role playing, simulation, practice, constant feedback, and related approaches must be included. Effective learning is not a spectator sport!

WHO DOES THE TRAINING?

Having addressed the many issues above as best that the firm can (while recognizing that the real power of teams often arises well into their activities and their value has frequently gone far beyond even the highest expectations of their champions and initiators) so that it now has a relatively clear vision of what training it needs (at least to start), the question of who best to do the training needs to be considered. Of course, the numbers of people to be trained, the time available to train them, and available budgets must be factored in also. There are many possible sources for the training.

In-House Trainers

Many firms, especially large ones, may have this internal expertise or could quickly obtain it. They have the advantage of knowing the firm, its culture, past efforts, special needs, and related knowledge base not readily available to outsiders. Unless previously experienced in group processes, team building, statistics, and other defined needs however, they may not be the best source - particularly in the beginning phase of team efforts. Long term however, internal training will be a must for sustaining the program, dealing with small numbers of newcomers to the firm, and teams themselves will assume some of the training tasks the firm needs. Corporate headquarters (rather than the division or site) may have central training services to draw upon or in-house consultants may be available. Industrial engineers are often involved with training and frequently are assigned the task to head up the effort and may do much of the first phases. While often very competent in statistics, design of experiments, work methods and processes, etc. all IE's have not necessarily been adequately prepared to conduct extensive training efforts. Although usually better prepared in "people skills" issues than most other engineers, IE's may well need special assistance in performing the high quality training that is going to be needed.

Outside Consultants

As any one who attends national conferences or reads relevant trade journals probably knows, there are multitudes of outside consultants available. It often seems that every early retiree has gone into this business! Their background, experience, past successes (check references very carefully), time availability and of course fees needs careful consideration. While some "off the shelf" training efforts can be quite adequate, internal tailoring to the particular firm is usually needed for best results.

Local Institutions

Community colleges, universities, government agencies, and other local institutions may offer training resources with considerable experience and at very reasonable costs. If formal centers or departments are not present, individual faculty members may be available. Their ability to deal with large numbers, tailor efforts to special firm needs, etc. needs to be considered. They are too often taken for granted and not considered adequately. Many are well prepared to assist firms in many phases of training that team projects will need. Of course, others may not be available to help. They should be checked out however.

Packaged Programs

There are also many commercially available print, audio, and video based training packages available today. They can be very cost effective when large numbers of people are going to need the same basic training. When combined with local mentors or facilitators, they can be especially effective. As total, stand alone approaches however, their value can be quite questionable.

Mixtures of internal and external sources usually are required for the total training effort to be adequately provided. A major factor of success is to make sure that someone with a good advisory group (steering committee, etc.) is in charge of the total training efforts and that they are given the resources - time, funds, physical facilities, etc. to adequately do the job.

WHEN TO DO THE TRAINING

A carefully thought out, well planned, training program should grow out of the above analysis. The time scale to consider will vary with the size of the firm, number of locations and employees, work schedule requirements, etc. The key factor is to provide the training immediately before it is needed and certain parts of it -continuations or refinement as it is actually being used on the job. If trained too early in the cycle, much is forgotten. Training after the first useage (usually because we forgot or overlooked the "need"), is perhaps better than not at all but it may also require "unlearning" bad processes. It is several orders of magnitude more difficult to "correct" poor group process skills than it is to train individuals properly at the start.

HOW MUCH WILL THE TRAINING COST

When asked this (and top managers and accountants always do), there is a great temptation to answer "nothing" - it is an investment and not a cost factor! Unfortunately this valid concept is still not yet widely enough accepted. There are far too many variables to provide meaningful figures here but Karp [8] back in 1983 estimated $45,000 to $60,000 for the first year of a new effort involving six quality circle teams including the cost of time off the job. Cost figures (what they count varies widely!) reviewed at 16 different organizations over the last three years show a bewildering range - $85,000 to $6+ million! Each organization has to address this using its own methods and procedures. A general rule of thumb seems to be that it amounts to at least twice what you originally estimated and still is only half of what you need!

SOME PROBLEMS/CHALLENGES TO FACE

In our rush to adopt team approaches - because so many good results from others are noted [9,10], it certainly fits our democratic political structure (though not our long standing industrial management structure!), and it "seems" like at least worth trying ("because not much else is working around here!"), our enthusiam often overwhelms our normal, logical thought processes. But we must be careful not to "over engineer" this either. Teams are not easy! Just considering the training aspects discussed in this paper should cause some pause for reflection, not to mention the many other factors involved which require consideration.

Gregory Huszco [11] presents a good outline of ten common pitfalls of team-training approaches - including counting on training alone to develop effective teams. Karp [8] provides three major causes for failures with quality circles which the writer fully concurs with from his own experiences:

1. Management impatience
2. Lack of management support
3. Inadequate implementation and planning

With the added experience many now have with teams, the subject of rewards and results needs to be added to the list. If teams work and directly impact the bottom line results of the firm, top management cannot take all the financial rewards. They will have to, over the long term, share them with the affected employees - or at least communicate accurately and thoroughly that such "gains" are keeping the organization (and their jobs) alive!

Managers have to <u>really</u> believe in teams and that this approach will work. Unfortunately we have not yet developed a true "litmus" test to easily measure this but wise change agents will make every effort to assess this as thoroughly as possible within the organization. The organization often is not yet <u>ready</u> for team project usage.

Lastly, an observation on expectations is in order. The powerful impact of management expectations has been well presented by Livingston [12]. We usually are much too low in our expectations of workers [13], especially as they apply to team projects. Managers must personally face this challenge and do everything they can to show that they expect high level results and positively demonstrate this in a sincere manner when they see teams delivering them (or find out what happened and redirect if this does not occur). The number of managers who have been very positively (often overwhelmingly) surprised with the good results from their teams is legion. The few who have been otherwise stricken may very well have only themselves to blame.

REFERENCES

[1] Lawler, E.E., <u>High-Involvement Management: Participative Strategies for Improving Organizational Performance.</u> Jossey - Bass, San Francisco, 1986.

[2] Harper, S.C., "Now That the Dust has Settled: Learning from Japanese Management," <u>Business Horizons</u>, July - August 1988, p. 43+.

[3] Dumaine,B., "Who Needs a Boss?", <u>Fortune</u>, May 7, 1990, p 53+.

[4] Wolff, M.F., "Building Teams - What Works (Sometimes"), Research - Technology Management, November - December, 1989, p. 9+.

[5] Szabo, J.C., "Learning at Work", <u>Nation's Business</u>, February, 1990, p.27+.

[6] Dreyfuss, J., "Thre Three R's on the Shop Floor", <u>Fortune</u>, Education 1990 Special Issue, p. 86+.

[7] Wigginhorn, W., "Motorola U: When Training Becomes an Education", Harvard Business Review, July - August, 1990, p. 71+.

[8] Karp, H.B., "A Look at Quality Circles", 1983 Annual for Facilitators, Trainers and Consultants, University Associates, 1983, p. 157+.

[9] See any Annual Proceedings for the Association for Quality and Participation.

[10] Lee, C., "Beyond Teamwork",Training, June, 1990, p. 25+.

[11] Huszczo, G.E., "Training for Team Building", Training and Development Journal, February, 1990, P. 37+.

[12] Livingston, J.S., "Pygmalion in Management", Harvard Business Review, July - August, 1969, p. 81.

[13] Shimko, B.W., "The McPygmalion Effect", Training and Development & Journal, June, 1990, p. 64+.

BIOGRAPHICAL SKETCH

Dr. Charles S. Elliott. Associate Professor of Engineering, Industrial and Management Systems Engineering Department and Director, Center for Professional Development, Arizona State University. He is a past chapter president of the Central Indiana Chapter of IIE and currently active with the Phoenix, Arizona Chapter and is a senior member of IIE. He is actively involved as a consultant in continuing education, training, engineering management, and team activities. He is also a member of ASEE, AQP, SME, IEEE, and ASEM.

Using the Nominal Group Technique Effectively

The Nominal Group Technique helps groups generate ideas and reach consensus through a five-stage structured process.

D. Scott Sink

Since 1969, when the Nominal Group Technique was first tested, interest in and applications of this structured group process have grown exponentially. What began as a technique to enhance the effectiveness and efficiency of program planning in health services has rapidly expanded into the areas of productivity measurement systems development, strategic planning and strategy implementation, participative problem solving, and many others.

This article will describe what the Nominal Group Technique is, briefly outline how to use it, and describe case studies of actual applications.

Background of the Nominal Group Technique

Developed by André P. Delbecq and Andrew H. Van de Ven in 1968, the Nominal Group Technique (NGT) is a special-purpose behavioral science technique that is useful in situations where individual ideas and judgments need to be tapped but where a group consensus is the desired outcome. The NGT is essentially a very structured and therefore very effective and efficient mechanism for idea generation and group consensus seeking. It is useful when a specific task or question already identified requires a group's ideas and judgment.

The NGT has come along at a particularly appropriate time in the evolution of management thought, technique, and practices. Most American managers are reexamining basic philosophies and practices, and for many this has involved giving increased attention to group processes and techniques. Improved commitment, understanding, communication, coordination, and cooperation are viewed as valuable outcomes. Quality circles, productivity action teams, quality-of-work-life programs, team building, and productivity gain-sharing plans such as Improshare, Scanlon, and Rucker all place important emphasis on group processes and behavior.

Experience is proving that the quality, effectiveness, and efficiency of specific group processes play a

The Nominal Group Technique is a social-science breakthrough that managers can apply relatively easily.

---

significant role in the overall success of these programs and techniques. And since managers often spend as much as 80 percent of their time in meetings, the quality of the group processes utilized in these meetings has a great impact on managerial productivity.

Unfortunately, group processes too often leave participants exhausted and discouraged because of the seemingly endless meanderings into unfruitful byways in what has been called "reactive search"—a search focusing on initial responses rather than a continuing creative flow; the focus effect that occurs when a group is unable to extricate itself from one channel of thought; or the mixing of solutions with problems and problems with solutions. What group has not rushed to "solutions" before the problems were clear? Felt frustrated by overbearing extroverts who dominate the sessions? Suppressed disparate or conflicting ideas because of differences in authority, prestige, age, race, sex, or levels of professionalization? What group has not experienced the general lack of creativity and absence of a sense of closure or accomplishment that leaves participants feeling impotent, bored, and frustrated?

In a management era where participative decision making and problem solving are increasingly common, techniques like the Nominal Group Technique have been quite welcome. For while the NGT is not itself a program but a participative data collection and consensus-forming device, it can be an important component of participative, group-oriented programs.

The NGT has proved to be extremely effective in that it is (1) easy to learn, (2) applicable to a wide variety of areas and situations, (3) easy to integrate into programs and projects of larger scope, (4) highly satisfying to participants, and (5) quite successful at inspiring a commitment to action, follow-through, and follow-up.

In short, the Nominal Group Technique is a real and very timely social-science breakthrough that managers can relatively easily and successfully apply.

How to execute the Nominal Group Technique

The NGT is a structured group meeting that proceeds along the following basic format:

1. Individual "silent generation," Delbecq's and Van de Ven's term for the writing of ideas by the group members;
2. Individual round-robin feedback from group members of their ideas, which are recorded in a succinct form on a flip chart;
3. Group clarification of each recorded idea;
4. Individual voting and ranking on priority ideas; and
5. Discussion of group consensus results and focus on potential next steps.

Silent generation

Imagine a typical meeting room in which five to twelve individuals are seated around a U-shaped conference table. A facilitator/leader addresses the group, stating the purpose of the meeting, the desired outcomes, and the general character of the NGT. A written task statement is passed out to each participant, and the individuals are asked to silently respond to the task statement.

Round-robin feedback

After from five to fifteen minutes, a structured sharing or presentation of ideas takes place. Each individual, in round-robin fashion, presents one idea in a succinct, three- or four-word phrase from his or her list. An assistant, recorder, or the facilitator writes that idea on a flip chart in full view of the other participants. Each idea is given a sequential number. There is still no discussion at this stage in the NGT session.

Round-robin listing of ideas continues until all of them have been recorded on the flip charts. As the charts are filled, usually with from three to four ideas per chart page, they are taped up on walls in full view of the participants. The round-robin phase typically will take from fifteen to twenty-five minutes, depending upon the number of participants, producing a list of from fifteen to sixty ideas or responses to the task statement. One can typically expect from two to three ideas per blue-collar participant and from four to five ideas per professional and/or managerial participant. Figure 1 depicts a sample flip chart page.

Hierarchical consistency between ideas on the list is often a problem.

Figure 1
Sample Flip Chart Page

1. Improve planning

(leave space between ideas for clarification)

2. Reduce absenteeism

(leave a column on right to record votes)

3. Improve quality

Group clarification of ideas

The next stage of the NGT incorporates a structured group discussion of all the ideas. The purpose is for the facilitator to ensure that the list is collectively exhaustive and nonrepetitive. This can be done by asking the participants to scan the list to ensure that each item is clear to them, items do not overlap, items have not been left off, and there are no items that need to be combined or even deleted. The clarification stage is typically the part of the technique where the group is most likely to get bogged down. Hierarchical consistency between ideas on the list is often a problem. That is, one idea may be broader or more general than another and questions will arise as to which should be combined with which and how specific or narrow an idea should be. The facilitator should let the group decide on hierarchical consistency but be alert to the danger of prolonged and disruptive argumentation. A steady pace must be maintained. This stage takes between twenty and thirty minutes.

A convenient way to overcome the difficulties that arise during the clarification period is to allow participants to depict overlap or hierarchical dependencies on the voting cards to be discussed in the next section. In this way, each participant can depict his or her perception of interdependencies between ideas without forcing the group to agree with the logic in the session itself.

For example, a number of ideas posted during the round-robin stage may overlap or be similar in terms of focus, intent, and/or context. Often, no clear consensus with respect to the combining of like ideas may emerge during the clarification stage. In these situations, the facilitator may allow individuals to identify what they perceive as overlap between ideas and to note this during the next stage. At that time they can vote for the idea within the overlapping group that most represents the point they feel is important and simply record overlapping ideas on the back side of that particular voting card. When the votes are tabulated and recorded on the flip charts, ideas that members perceived as overlapping may be identified accordingly.

Individual voting on ideas

The fourth stage of the NGT provides an opportunity for individual voting on the ideas. Each participant is provided with from five to nine 3x5 index cards. (If there are fifteen ideas on the list, use five cards; if there are from twenty to thirty, use seven cards; and if there are more than thirty, use nine cards.) Figure 2 depicts such a card.

Each participant, privately, selects his or her n (from five to nine) most important or highest priority idea subset from the larger list of ideas. One idea is written in the center of each card. The sequential idea number from the flip chart pages is recorded in the upper left-hand corner of the card for each of the n selected ideas. Again, if members feel there is overlap among ideas that did not get resolved during the last

If the ranking process fails for any reason, the entire NGT meeting is essentially wasted.

Figure 2
Typical Preprinted Card for Voting and Ranking

IDEA #________

IDEA ________________________________

____

____

____

RANK ________

stage, they may, on the back side of the voting card, note ideas that they perceive have commonality. The sequential number of the overlapping idea is sufficient notation. Next, each participant privately identifies priorities by ranking the ideas.

A very reliable and valid method for structuring this ranking process is to ask the participants, once they have finished identifying their *n* most important ideas, to spread the cards out in front of them. The participants are then each asked to choose the most important ideas of the ones selected and assign the highest number (from 5 to 9, depending on how many ideas were selected in the priority subset) to that idea. This value is recorded in the lower right-hand corner of the card. The participants are then asked to set that ranked card aside and to select the least important idea of the remaining ideas. This idea is assigned a value of 1. The participants are then asked to identify the most important idea of the remaining ideas and to assign it the second highest number. This outside-in ranking process continues until all ideas have been assigned a number and have thereby been arranged in priority order.

It is important for the facilitator to maintain pace during this stage. If the ranking process fails for any reason, the entire NGT meeting is essentially wasted, since the primary goal is to achieve a group consensus.

Discussion of results

Figure 3 depicts a typical flip chart sheet after this stage. The discussion of results typically focuses on participants' perceptions of consensus priority ideas. The facilitator should point out that there are two types of consensus: strength of vote score and number of votes. For example, in a group of ten persons, three members could give one idea an 8 while the other seven members might not select it at all. On the other hand, all ten members might have given another idea a 1. The former idea has a score of 24 while the latter has a score of 10. On the basis of strength of score, the first idea is "more important." But on the basis of number of votes the second idea is clearly "more important"—more people have selected it as belonging among the *n* (from five to nine) most important ideas although its ranking among those ideas may be low.

Discussion then typically focuses on what the group might do next if the results are to be used as a component of a larger effort, such as a quality-circle program, a productivity improvement effort, a planning activity, a productivity measurement process, etc. The NGT, as mentioned, is primarily a data collection and a consensus-shaping methodology. Therefore, the session produces a prioritized list of responses to a particular task statement or question (e.g., identify and list ideas for productivity improvement). (For summary format for NGT results, see Figures 5, 6, 7.)

But the primary value of the NGT lies in its capacity for developing commitment and other behavioral outcomes that spur action on behalf of a larger effort (productivity improvement, quality circles, etc.). Such post-process outcomes are generated by the nature of the NGT activity itself.

The final discussion stage is crucial for the effective utilization of the NGT results—both the ideas, or process output, and the post-process behavioral outcomes. For this stage to achieve its purpose, two conditions are crucial. First, the program designers and related management personnel must have thought through the short- and long-term goals of the larger effort and how the NGT results can be used in achieving them. Then, during the discussion stage of the NGT, an action plan focusing on use of the NGT results can be presented, evaluated, and developed further. For example, if a particular NGT session focused on developing

Figure 3
Typical Flip Chart Sheet
at End of NGT Session

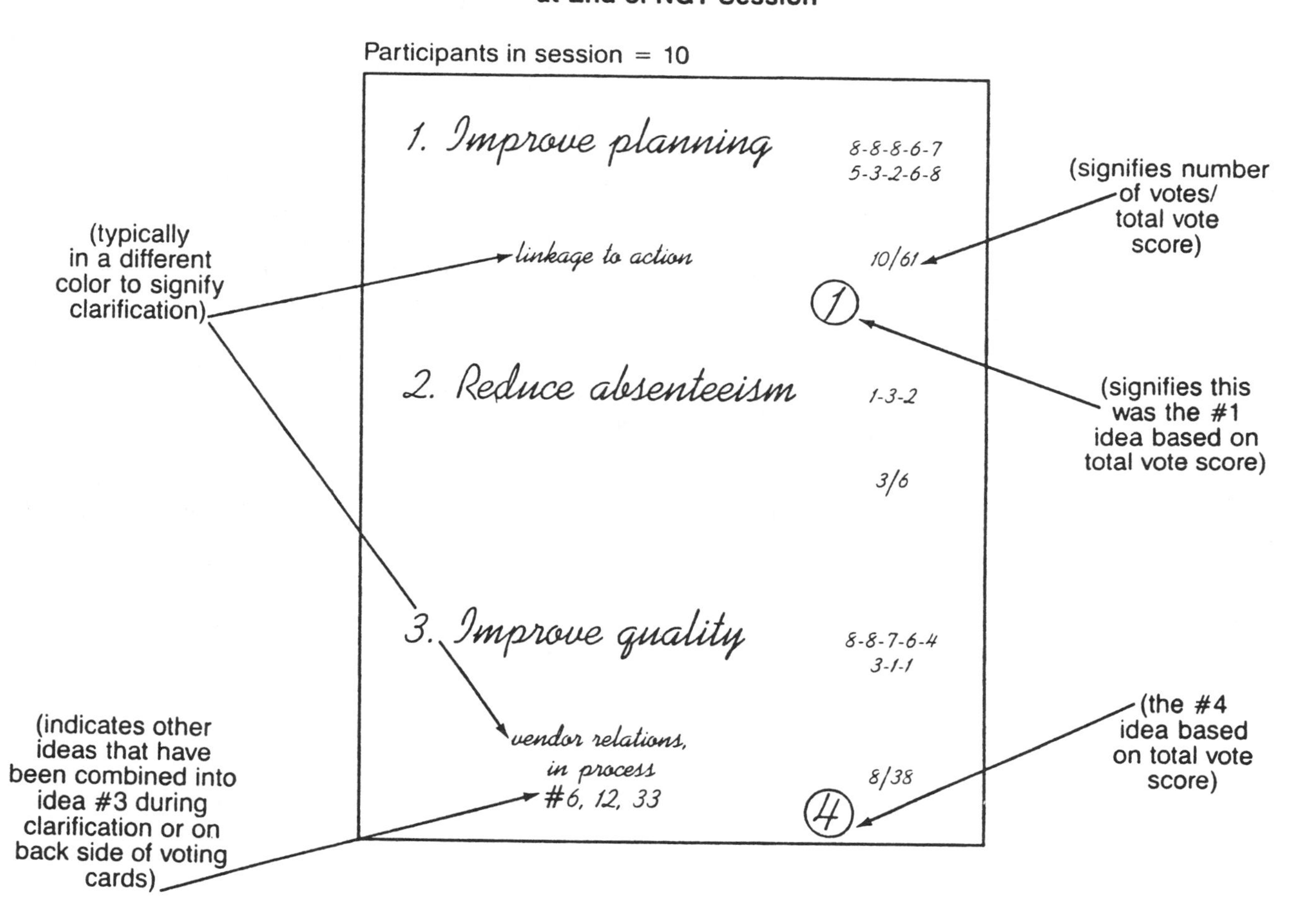

a prioritized list of problems within a particular work group, an action plan for tackling the top priority problems can be developed. Such an action plan would necessarily deal with a list of activities, sequencing for those activities, costs and benefits, the persons responsible and accountable for the overall plan as well as specific activities, alternative approaches to solving the problems, and target dates for obtaining the expected results.

Second, considerable leadership judgment is required during the discussion stage of the NGT. The NGT process itself imposes structure and requires specific facilitator behaviors in the first four stages but not in the final discussion stage. In that stage, the level of task-directed leadership that is appropriate seems to vary significantly from group to group. The facilitator needs to be able to perceive and respond appropriately to a particular group's need for structuring of the ensuing steps. Without sufficient leadership, participants will leave the meeting with a commitment but without a mechanism for follow-through. In some cases leadership emerges from the group itself, and this should be encouraged. However, in other cases a significant degree of direction and perhaps even task assignment will be required from the facilitator.

NGT results and participation programs

One way to conceptualize the linkage of results from an NGT session to the goals of a larger program, such as a quality-circle effort, a productivity measurement process, or a planning activity, can be depicted in Figure 4. The Figure suggests that in those cases where programs involving participation are desirable and appropriate, the NGT can play a central role.

Programs listed in box A have been growing in popularity and, probably, in appropriateness in the U.S. All of these programs entail a relatively high degree of participation and therefore are prime candidates for application of the NGT. In fact, the NGT has been utilized successfully in all of the types of programs listed. It has been found to be a very effective and efficient mechanism for structuring early stages of participation. It has also been found to be highly reliable for producing desirable process output (box B). That is, the probability is quite high that a correctly managed NGT session will create a high quality list of prioritized ideas. In the past six years I have personally conducted over fifty successful NGT sessions that generated such lists; only one session ended in failure.

The generation of certain post-process behavioral outcomes (box C) is also highly likely. Experience has shown that satisfaction with the session itself, willingness to follow up, actual follow-through, commitment to future involvement and future steps, and general employee development are all reliable post-process outcomes. In a sense, the NGT process sets up certain critical psychological states deemed necessary for the achievement of many of the desired outcomes (box E) from the larger program effort. Critical psychological states reflect job-related attitudes such as a felt responsibility for outcomes associated with the job, experienced meaningfulness of the particular job, and perceived significance of the task. Job design and job enrichment theory and research suggest that any process that can positively affect these and other critical psychological states has a high probability of causing one or more of the desired outcomes to occur.

Regardless of the success of the NGT in terms of creating process output and post-process outcomes, it cannot be utilized in larger-scoped programs without management leadership and support. Note that box D, leadership structure for action planning and integration of NGT results into the larger program effort, reflects a critical component in the process of utilizing NGT results. It influences, and in some cases even directly

Figure 4
Role of NGT in Employee Participation Programs

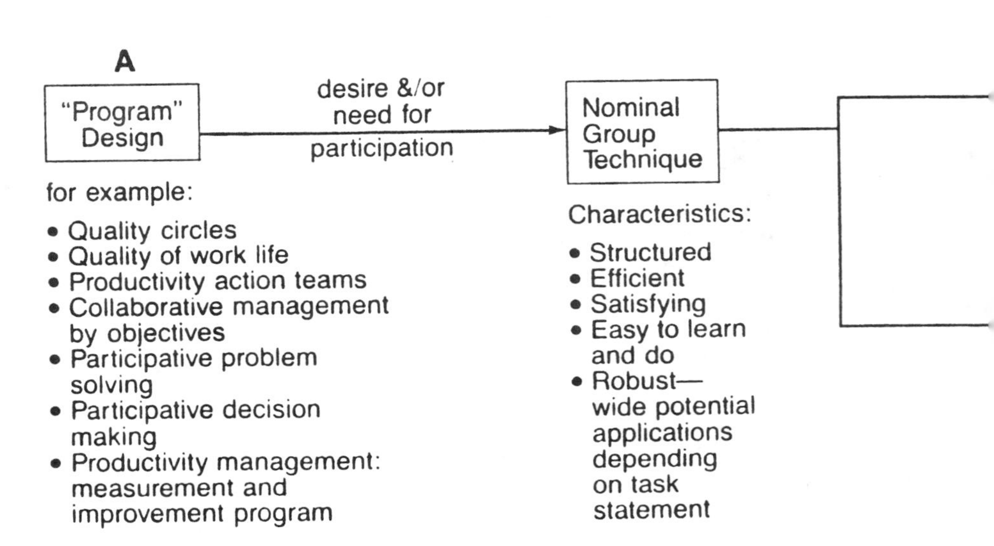

The range of NGT applications has increased sharply in 13 years.

---

determines, the extent to which desired outcomes are actually achieved. Experience suggests that without leadership intervention, the desired outcomes from programs such as those listed under box A are less likely to take place.

In the past six years, the author has been involved in numerous applications of the NGT in a variety of organizations. In each case, management utilized the NGT as an early, sometimes even initial component of a program designed to achieve one or more of the desired outcomes listed under box E. Three specific case applications of the NGT are briefly described in the following section. Their purpose is to assist the reader in thinking through applications that might be appropriate in his or her organization. These case studies do not represent empirical research on the effectiveness of NGT applications. They do represent specific systematic yet pragmatic attempts to improve the effectiveness and efficiency of employee involvement activities in organizations.

Case studies

In the roughly thirteen years that the NGT has been available to managers, educators, and researchers, the range of its applications has increased significantly. But perhaps the most important areas for application of the NGT in U.S. organizations have been the following:

1. Planning processes—NGT is used to increase the amount and quality of participation in and thus, it is hoped, commitment to the planning process, to link strategic planning to strategic as well as tactical and operational actions;
2. Productivity measurement system design and development—NGT is used to generate "normative" productivity measurement systems composed of measures, ratios, and/or indexes;
3. Participative problem solving—NGT is used in programs like quality circles and productivity action teams as a mechanism for increasing the effectiveness and efficiency of identification and selection of priority opportunities, problems, causes, alternatives, etc.

Brief descriptions of an actual case study for each of these three applications will now be presented.

NGT applications in the planning process

An academic department at a major university

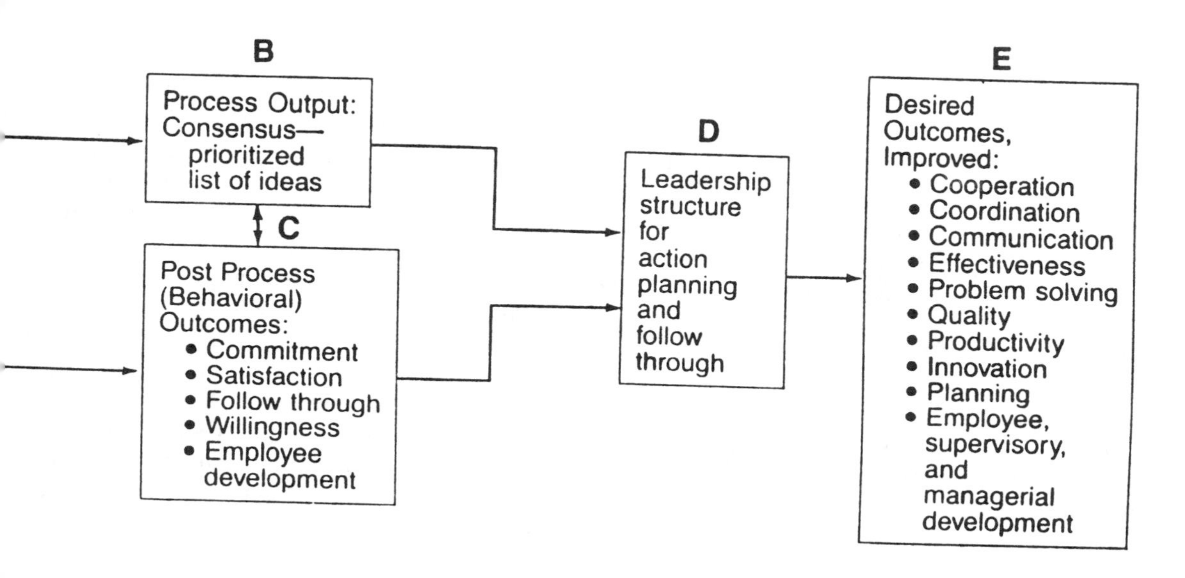

A university department faculty using the NGT in short-range planning tends to "buy into" the objectives.

initiated a planning process several years ago. The basis for the planning process came from work done by Peter Drucker in 1954.

Phase I of the planning process implementation focused on getting the faculty acquainted with planning concepts and comfortable with thinking in longer range terms. The stress was on developing department goals that were congruent with overall university goals and were acceptable to the faculty. Several years were spent on this early phase.

When the department head concluded that the faculty was ready to move to Phase II of the planning process, attention began to focus on linking strategic planning to strategic, tactical, and operational actions.

In the first year of Phase II, the stress was almost entirely on developing objectives, targets, and responsibilities for each major goal. At an annual planning retreat during the second year of Phase II, the department head decided to incorporate the NGT as a component in the planning process. Specifically, the NGT was used to generate one-year objectives for the department. The faculty received a brief explanation of the process and the technique and were then asked to respond to the following task statement: "Please list below specific objectives that our department should be working on during the next year." Each faculty member had received information on the university's as well as the department's long-range goals. Therefore, short-range planning incorporated thinking about strategic actions. Results of this NGT session are shown in Figure 5.

Evaluation

This particular academic department has now been utilizing the NGT as an integral part of the planning process for four years. Initially there was some skepticism concerning the technique, particularly regarding its highly structured nature. But subsequently, acceptance of the technique has been so high that over a third of the department's faculty has used it in other applications outside the department itself.

Prior to inclusion of the NGT to set one-year objectives for the department, the planning process was significantly more top-down. Goals and objectives for the department were generated by its head and then cursorily discussed by faculty. And the general perception was that substantive follow-through on objectives was minimal.

Since the inclusion of the NGT in the planning process, setting one-year objectives for the department is significantly more bottom-up and participative in character. General perceptions are that faculty are significantly more involved in the short-range planning process and as such have tended to "buy into" the objectives themselves. In the four years this process has been used, the department of fourteen faculty members accomplishes an average of between ten and twenty significant department-related objectives per year. Note that these objectives are group objectives and are in addition to individual faculty position responsibilities.

The department for a number of years has earned a reputation as one of the better in the country from an administrative and academic standpoint. There is considerable agreement among the faculty that a primary reason for this success is the quality of the planning process. The NGT has been a significant component in this process.

Design and development of productivity measurement systems

A supervisor of a department of twelve engineers in an equipment service division of a major U.S. corporation returned from a three-day course on productivity measurement and improvement. She was particularly impressed by her exposure to a multifactor, firm level productivity measurement model and what had been called a "normative" productivity measurement approach. The latter approach, she had been told, was appropriate at the group level and, in particular, was very useful for hard-to-measure applications.

Using the NGT, this approach assigned members of the organization the task of developing the components of a productivity measurement system. The approach appeared to have been quite successful in other organizations at linking productivity measurement to productivity improvement actions. She was very excited about implementing it, as her greatest difficulties in the past had been precisely the linking of planning for improvement to specific individual engi-

Figure 5
Academic Department Application of NGT for Setting One-Year Objectives

Group: Industrial Engineering Department

Facilitator Name: Scott Sink

Date: January 27, 1978

Number of participants = 14
n (number of objectives/action programs asked to vote for) = 8

Task Statement: "Please identify specific action programs or objectives our department should be working on and accomplishing during the next year."

Key: O—Programs which are aggregates as result of clarification
S—Programs for which subprograms were identified in subsequent sessions
D—Programs which became topics of discussion during open session
I—Programs for which implementation was discussed

	Action Program/Objectives	Votes Received 8 = Most Impt., 1 = Least	Total (number of votes/total vote score)
ISO	1. Faculty Development	8-8-8-8-8-8-8-7-6-5-4-4-1	13/83
O	2. Graduate Recruitment	8-7-7-7-5-5-5-4-3	9/51
DSO	3. Industrial Liaison	8-8-7-6-6-4-4-3-2-2	10/50
DO	4. Faculty Support	7-7-7-6-6-3-3	7/39
SO	5. Intra, Communication	7-6-5-2-2-2-1	7/25
	6. Working Papers	8-6-4-3-1	5/22
	7. Minors Ph.D.	6-4-4-2-1-1	6/18
O	8. Course Content Coordination	4-3-3-3-2-1	6/16
	9. National Conference	5-5-5-1	4/16
D	10. Coop Program	8-3-2-2	4/15
	11. Interdisciplinary Research	7-6-1	3/14
	12. Graduate Support	7-4-3	3/14
	13. Timeliness of Info on Funding Opportunities/RF	6-5-3	3/14
O	14. Begin-End Courses Evaluation	6-5-3	3/14
	15. Manufacturing Institute	5-4-2	3/11
O	16. Indices	7-3	2/10
	17. Material Handling/Plant Layout	6-4	2/10

(Note: There were 41 objectives; 33 received votes; the 17 top vote getters are listed in the figure.)

neer behaviors. The risks and costs, she had been told, were that the quality of the measurement system initially might not be as good as she would like and that her engineers would have to spend some time participating in the development of the measurement system. She was reluctant to involve her engineers in activities that were not directly productive but decided it was an investment that could have significant direct as well as indirect benefits.

Based on what she had learned at the course and on further suggested readings, she developed the following plan of action:

1. Hold productivity basics seminar with engineers (one day, outside consultant);
2. Make informal assessment of response to the subject with individual engineers;
3. Present proposal to engineers for develop-

ment of productivity measurement system;
4. If accepted, run NGT session to develop consensus list of productivity measures, ratios, and/or indexes;
5. Review results and discuss next steps with group; and
6. Integrate and operationalize results with current control system.

The plan was well received by the engineers, and an NGT session was run to develop productivity measures, ratios, and indexes. Results of the NGT session are presented in Figure 6.

Evaluation

The engineers were highly pleased with the NGT session and its results but were unclear as to what came next. They voiced concern over how the measurement system would integrate with the existing individual performance appraisal and merit evaluation system.

In response to their concern, the supervisor asked them if they would like to go a step further and become involved in using the NGT-generated results to design the department's productivity system. Their unanimous response was that they wished to be involved so long as she felt they could contribute and so long as their participation did not interfere with their projects and duties. The result has been that as she has developed the productivity measurement system—guided by the NGT results—she has periodically submitted it to the engineers for review.

The group has since used the NGT to generate consensus ideas for productivity improvement. Action teams are set up and subgroups of engineers work on specific priority projects identified by the department itself.

The supervisor and the engineers are pleased with the role the NGT has played in structuring group

Figure 6
Results from NGT Session to Develop Productivity Measures, Ratios, and/or Indexes

Group: Engineers
Facilitator: Scott Sink
Date: 6/25/82

Number of participants = 12
n (number of measures asked to vote for) = 8

Task Statement: "Please list below measures, ratios, and/or indexes of productivity that you feel this engineering group should use to monitor, evaluate, and control productivity performance for the group."

Measure, Ratio, Indexes	Votes Received 8 = Most Impt., 1 = Least	Total (number of votes/total vote score)
Number of times our clients ask for our help	8-8-6-4-8-3-1	7/38
Measurable output / Resources utilized	8-8-8-8-8-8 8-7-4-8-5-1	12/81
Projects completed on time and within budget	7-7-6-7-5-3-8-7-5	9/55
Customer satisfaction	1-3-7-6-6-6	6/29
Percent of group objectives accomplished on time	7-7-7-6-6-7-4-3-5-4-5	11/61
Percent of successfully implemented projects	7-6-6-6-5-4-3-1	8/38

activity in both the area of productivity measurement and improvement. They sense substantial progress within the department regarding group goals and objectives. The engineers seem pleased that they have been able to contribute to the design of a productivity measurement system. They are particularly pleased with the opportunity to work together as a group on common problems and opportunities.

Participative problem-solving programs

A plant manager for an oil-industry-related manufacturing and fabricating company with 250 employees implemented an employee involvement program with the help of a local state university. The program began with a pilot-study group of eleven welders plus their supervisor. The NGT was an integral component of the program.

During the first meeting of the group, the NGT was used to identify and set priorities for ideas on productivity improvement. The results of this session are shown in Figure 7.

After the first session, a management review committee screened the list to verify the quality of the ideas. The group of welders was then broken down into smaller action teams of three to five workers and assigned a specific idea to develop. Once taught how to develop alternatives for carrying out the ideas, each action team presented a proposal to a management review committee for approval. The committee then decided how to proceed.

Evaluation

The plant manager has noted progress in all of the four top priority areas targeted by the welding group for productivity improvement. Perhaps even more important is the improvement in worker attitudes that he has observed. In fact, the plant manager has commented that the major gain achieved through the par-

Figure 7
Results from NGT Session to Develop Ideas for Productivity Improvement

Group: Welders on shop floor
Facilitator: Scott Sink
Date: 6/17/82

Number of participants = 11
n (number of ideas asked to vote for) = 8

Task Statement: "Please list below ideas for improving productivity, effectiveness, and/or efficiency of your work group."

Idea	**Votes** 8 = Most Impt., 1 = Least	**Total** (number of votes/total vote score)
Interdepartment and intershift cooperation and coordination	8-8-8-8-8-8-7-7-6-4-3	11/75
Labor-management relations	8-8-8-8-8-7-7-7-7	9/68
Unnecessary non-conformances, improve quality control procedures	7-6-5-4-4-3-2-2-2-2-2	11/39
Improve scheduling, reduce set-up time	7-6-4-3-3-2-2	7/27

(Note: The welders identified 22 ideas for productivity improvement. The NGT helped them reach the level of consensus shown above in an hour and a half.)

ticipative problem-solving program has been the consensus and commitment generated by the NGT at the outset of the program itself. This consensus and commitment has created a willingness to identify and resolve long-standing productivity problems, which no one had ever taken the time to tackle.

Conclusion

The Nominal Group Technique can assist management in operationalizing participative management philosophies, concepts, and approaches. A structured group process, each of its five stages was developed to fulfill specific design specifications (i.e., equality of participation, functional search behaviors, closure to the decision process, etc.) When viewed as a component that can be designed into participative activities, it will assist in improving the effectiveness and efficiency of group decision-making processes.

The 1980s and 1990s will present tremendous challenges to management. If management develops the skills necessary to take full advantage of the NGT, its ability to regulate group behavior and hence organizational behavior will play a key role in helping it to meet these challenges.

SELECTED BIBLIOGRAPHY

1. Bradford, L.P., ed., *Group Development*. 2nd Edition. La Jolla, California: University Associates, 1978.

2. Delbecq, A.L.; Van de Ven, A.H.; and Gustafson, D.H. *Group Techniques for Program Planning: a guide to nominal group and Delphi processes*. Glenview, Illinois: Scott, Foresman and Company, 1975.

3. Sink, D.S. "Productivity Action Teams: An Alternative Involvement Strategy to Quality Circles," Annual Industrial Engineering Conference *Proceedings*. Atlanta, Georgia: Institute of Industrial Engineers, 1981.

4. Sink, D.S. and Mize, J.H. "Role of Planning and Its Linkage to Action in Productivity Management," 1981 Spring Annual Conference *Proceedings*. Atlanta, Georgia: Institute of Industrial Engineers, 1981.

5. Van de Ven, A.H. and Delbecq, A.L. "Nominal versus Interacting Group Process for Committee Decision-Making Effectiveness." *Academy of Management Journal* 14(2):203–212, June 1971.

D. Scott Sink is an associate professor in the School of Industrial Engineering and Management at Oklahoma State University and director of the Oklahoma Productivity Center at OSU. His areas of interest include productivity management and measurement, work measurement and improvement, and organizational behavior.

Additional Readings on Tools/Techniques for the Systems Integrator

Baudin, Michel. 1990. *Manufacturing Systems Analysis*. Englewood Cliffs, NJ: Yourdon Press.

Boznak, Rudolph G. 1988. Achieving a competitive manufacturing advantage through effective multi-project management. *Proceedings 1988 IIE International Industrial Engineering Conference*. Norcross, GA: Institute of Industrial Engineers. pp. 285-290.

Burstein, M.C. and P. Graham. 1990. Strategic justification of CIM: A systematic market-based approach for plant determination of distinctive manufacturing capabilities. *Proceedings, Manufacturing International 1990*. American Society of Mechanical Engineers. pp. 21-25.

Canada, John R. and William G. Sullivan. 1989. *Economic and Multiattribute Evaluation of Advanced Manufacturing Systems*. Englewood Cliffs, NJ: Prentice Hall, Inc.

Dao, Ky-Lan, James Comer, and David Rodjak. 1991. CIM flows from data flow diagrams. *Manufacturing Systems*. July, pp. 67-72.

Delberg, Vandeven and Gustafson. 1975. *Group Techniques for Program Planning: A Guide to Nominal Group and Delphi Processes*. Glenview, IL: Scott, Foresman and Co.

Miltenberg, G. John and Itzhak Krinsky. 1987. Evaluating flexible manufacturing systems. *IIE Transactions*, Vol. 19, no. 2, June, pp. 222-233.

Noble, Jean L. 1990. Strategic benefits of CIM in cost justification. *CIM Review*. Summer, pp. 66-70.

Putrus, Robert. 1990. Accounting for intangibles in CIM justification. *CIM Review*. Winter, pp. 23-29.

IX. SYSTEMS INTEGRATION CASE STUDIES

Most people benefit from seeing an example of whatever they are attempting to do. Consequently, we have included three case studies which give a flavor of system integration initiatives.

"There is no generic manufacturing system, so there can be no generic CIM" is a phrase often quoted. The case studies are not intended as a step-by-step cookbook for pursuing systems integration. Rather, they will demonstrate basic principles and suggest some of the general steps that need to be followed.

Reprinted with permission from *National Productivity Review*, V4N1, Winter 1984-85.

Strategic Planning for Factory Modernization: A Case Study

One company's experiences with the complex but essential process of manufacturing modernization are presented.

Joe H. Mize, Deborah J. Seifert, and Gayle Berry

Introduction

Manufacturing firms in the U.S. are under increasing pressure to modernize their production equipment and their management systems in order to remain competitive and profitable. Since about 1980, essentially all business-oriented publications have run numerous articles on automation, robotics, CAD/CAM, computerized processes, etc. The term "second industrial revolution" has been used to characterize the changes that are occurring and that are predicted to occur by the beginning of the twenty-first century.

This movement comes at a time when many U.S. manufacturers are operating forty-to-sixty-year old plants, filled with twenty-to-thirty-year old equipment. The grave problems faced in recent years by the U.S. auto industry due to this situation should be a serious lesson to other U.S. manufacturers who do not yet believe that modernization is essential.

This article describes the initial efforts of one high-tech company to develop a long-range strategic plan for factory modernization. Its purpose is *not* to present a step-by-step procedure for factory modernization that will work for every organization. Rather, the authors hope to be of assistance to others engaged in efforts of this type by describing our general approach and by sharing some of the important lessons we have learned.

It is difficult for many U.S. manufacturers to meet the new customer expectations regarding cost, quality, etc.

Company profile

The Garrett Turbine Engine Company (GTEC) has been manufacturing turbine engines in Phoenix, Arizona since 1951. GTEC is an operating company of the Garrett Corporation, which in turn is one of the Signal Companies. GTEC enjoys a reputation for being a cost-competitive manufacturer of high quality turbine engines, personal and business jet aircraft, and auxiliary power systems. A majority of the company's market is civilian, and the remainder military. The global market for small-to-medium thrust jet engines is expected to grow substantially during the next fifteen years. To capitalize on increased defense spending by the federal government, the company would like to increase substantially its participation in the military market.

The manufacture of turbine engines involves very complex and sophisticated manufacturing processes. Exotic materials must be machined into complex contours within very small tolerances. Very high levels of skill and knowledge are required of the engineers who design turbine engines and the machine operators who actually operate the complex processes.

GTEC's manufacturing facilities consist of a mix of state-of-the-art equipment (CNC machine cells, FMS systems, robots), relatively modern equipment (five-axis grinding machines, electrochemical discharge machining), and older equipment (1950 vintage grinders, etc.).

GTEC's management control systems have many modern, innovative features but still incorporate work order splitting, expediting, and occasional crisis management.

The driving forces for modernization

The manufacturing world is being revolutionized by technological advances in microelectronics, material science, processing technologies, automation technologies, robotics, and computer-assisted processes. Rapid advances are also being made in management systems for modern manufacturing facilities. Of particular significance is the emerging science of networking the many components of a factory under a centralized hierarchical control discipline.

These advances are hastening the day when a discrete part manufacturing facility can be operated in a near optimal state, with proper consideration given to overall system performance measurements. The concept of a "central control room" for managing a factory, analogous to the heavily instrumented control rooms in continuous flow industries, is on the verge of being implemented within a few discrete part manufacturing companies.

A particularly strong driving force for modernization comes from the marketplace itself. Customer demands and expectations are increasing rapidly in terms of product performance, product quality, cost, serviceability, and responsiveness to scheduled delivery requirements. Domestic and foreign competition are feeding the more demanding customer expectations. It is difficult for many U.S. manufacturers to meet the new expectations regarding cost, quality, etc., using the aging equipment and management systems currently in place.

The work force is also changing rapidly in terms of expectations regarding employment security, meaningfulness of work, work environment, and desire to participate in decisions that affect the workers' jobs. Perhaps the single greatest failing of U.S. management has been its inability or unwillingness to change a system that uses workers' hands but not their brains.

Another important driving force is the fact that the U.S. Department of Defense is putting great pressure on defense contractors to improve their performance regarding cost, quality, and schedule. The Air Force is pushing its Technology Modernization (Tech Mod) program, while the Army and Navy programs are called Manufacturing Technology (Man Tech). All three programs have the common and innovative feature of enticing contractors to hold down costs through a shared savings scheme. Through this mechanism, everybody wins when savings are realized.

Finally, the global marketplace is increasingly competitive. Both domestic and foreign manufacturers are improving their capabilities to satisfy customer demands for higher quality products at prices that are extremely competitive.

Collectively, these driving forces are bringing about extremely rapid change in discrete part manufacturing in general and turbine engine manufacturing in particular. This change is so rapid and so pervasive that it can only be characterized as revolutionary. Thus, it is not an exaggeration to say that manufacturing is now entering a phase that can be called "the second industrial revolution."

GTEC's decision to modernize

Garrett Turbine Engine Company made the decision to embark upon an extensive modernization program in the summer of 1983. After tracking technology advances over several years, and considering all the above mentioned driving forces at work, GTEC manufacturing management concluded that the time was ripe for an intensive effort to upgrade its manufacturing capabilities. This actually amounted to a decision to accelerate its factory modernization efforts, since the company has always attempted to maintain a reasonably modern manufacturing capability.

A key factor in GTEC's coming to this decision was the realization that factory modernization is something the company should be doing anyway, independently of the Department of Defense's encouragement of modernization by enticing defense contractors with financial incentive programs. Even though the Air Force's Tech Mod program was a contributing factor, GTEC would not have made the decision if it had not been a sound management decision in its own right.

An early decision was made to rely primarily on internal talent to conduct the in-depth analysis required and to formulate a long-range strategic plan for factory modernization. GTEC is fortunate in having an unusually large number of highly qualified professionals on its staff who have both the conceptual abilities and the practical knowledge of GTEC manufacturing systems necessary to conduct the modernization study. The internal project team was augmented with two university faculty consultants who brought their own specialized talents to the team. But perhaps the major contribution of the consultants was the objectivity they brought to the project.

Another early decision was to name the program Fac Mod (Factory Modernization) rather than Tech Mod after the Technology Modernization program of the Air Force. The reason was that Fac Mod at GTEC is comprehensive, covering all aspects of the company, not just technology. GTEC's Tech Mod effort is a subset of its comprehensive Fac Mod program.

Early activities

The following subsections describe some of the major activities engaged in by the project team. Only the highlights are covered, since the details of how GTEC accomplished these activities are not of general interest to other organizations facing different circumstances The important thing is to recognize the existence of this basic set of activities.

Project team composition

Once the fundamental decision had been made to embark upon a comprehensive factory modernization effort, the first task was to form the Fac Mod study team. This turned out to be a difficult task. The team needed the best talent available in the company, but these were the very people most valuable in their home departments. Their managers were understandably reluctant to release some of their most valuable professionals. This dilemma forced GTEC management to adopt the guiding principle that Fac Mod is the most important thing GTEC will do in the next five to ten years. Thus, the needed talents were assigned to the Fac Mod team. The array of talents assigned to the team included people with current knowledge of equipment and processes, others with detailed knowledge of the manufacturing management systems, and others with knowledge of the costing and budgeting processes.

Organizational placement of the Fac Mod team

The next critical issue concerned the proper place in the organizational structure to place the team. The team needed to be placed low enough in the organization to relate directly with the operating units, but high enough so as to have an overview of the important

Developing a specific project plan proved to be very time-consuming and difficult to achieve.

---

components of the organization that would be heavily involved in the study.

Members of the Fac Mod team were assigned full time to the project, so that their complete attention and efforts could be devoted to Fac Mod activities. A unique accounting code was assigned for reporting Fac Mod team efforts.

The decision was made to have the Fac Mod manager report directly to the manager of industrial manufacturing engineering, who in turn reports to the vice-president for manufacturing operations. Since the vice-president reports to the president of GTEC, the Fac Mod manager is only two levels removed from the top of the organization.

Defining the Fac Mod mission

The Fac Mod management team experienced greater difficulty than expected in crisply defining the mission of the Fac Mod effort. Although everyone had a general concept of what the Fac Mod effort would entail, not everyone's concept was the same.

After considerable thought and extensive interaction with all management levels, the following mission statement shown was adopted:

> Develop, with comprehensive manufacturing participation, a strategic plan for factory modernization that has the highest potential for providing significant long-term profitability, schedule, and quality benefits to GTEC.

This mission statement recognizes the three most important long-term performance measures of the Garrett Turbine Engine Company: product quality, delivery schedule, and company profitability.

Defining the desired outputs, creating a project plan

Once the basic mission of the Fac Mod team had been agreed upon, defining the desired outputs was a relatively straightforward matter. It was decided to follow the basic three-phase format used by the Air Force's Tech Mod program to specify the outputs to be achieved. Figure 1 presents the major components of each of the three phases: Factory Analysis/Project Identification, Project Design/Development, and Implementation.

In order to achieve the desirable outputs just described, it was necessary to develop a specific project plan. This proved to be very time-consuming and difficult to achieve. The general flow chart in Figure 2 illustrates the major elements of the project plan that eventually evolved. Note that this project plan pertains only to Phase I, the Factory Analysis phase of Fac Mod.

The concept of "Experts"

It was recognized from the beginning that a truly comprehensive factory modernization effort must incorporate the inputs of many, many people at various levels within the organization. The Fac Mod team was to serve the role of integrating their inputs into a comprehensive plan.

The "Executive Experts" consisted of appropriate GTEC vice-presidents and their direct reports. Input from this group tended to focus on policy issues and overall factory management needs.

The "Management Experts" consisted of management levels below the "Executive Expert" group and from within the organizational groups of manufacturing, purchasing, material, assembly, industrial/manufacturing engineering, plant engineering, quality assurance, and data processing. The inputs from this group tended to focus on short-range operational problems, such as shop floor scheduling, inventory issues, quality issues, and meeting month-end shipment quotas.

The "Technical Experts" consisted of line supervisors, people who live in the trenches. The inputs from this group tended to focus on technical issues, such as equipment performance, the difficulties of meeting engineering specifications, material handling problems, etc.

The concept of "Experts" worked extremely well. Not only was it an excellent source of inputs and suggestions, it was also an excellent way to begin the

Figure 1
The Three-Phase Format for Specifying Fac Mod Outputs

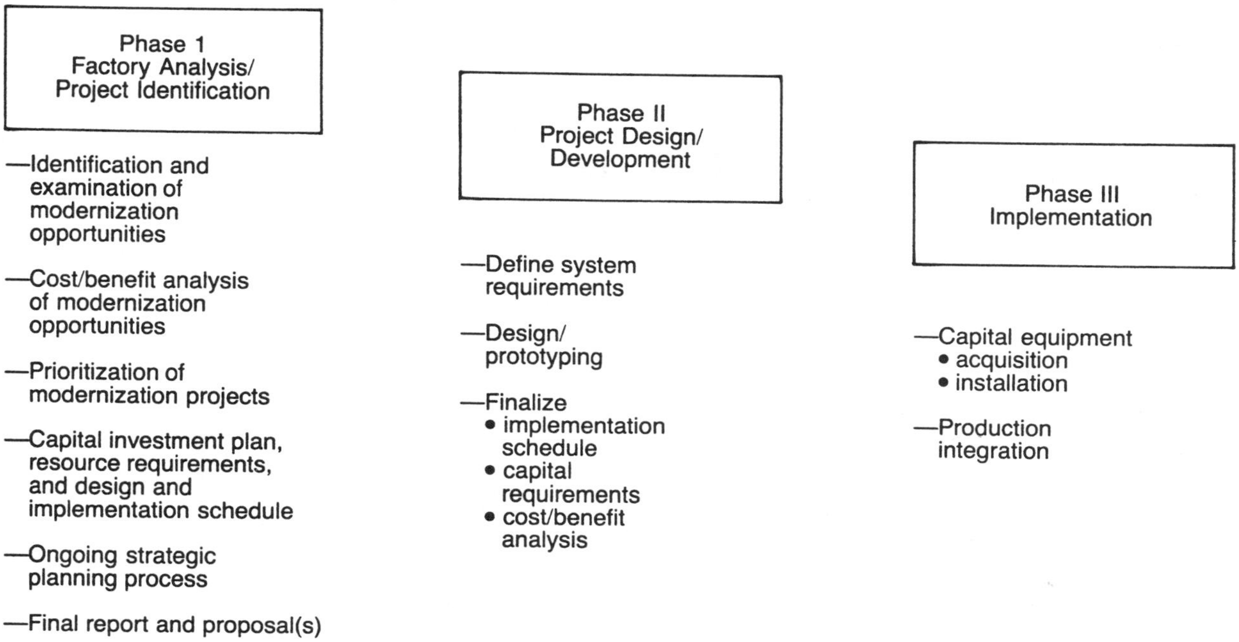

Phase 1 Factory Analysis/ Project Identification	Phase II Project Design/ Development	Phase III Implementation
—Identification and examination of modernization opportunities —Cost/benefit analysis of modernization opportunities —Prioritization of modernization projects —Capital investment plan, resource requirements, and design and implementation schedule —Ongoing strategic planning process —Final report and proposal(s)	—Define system requirements —Design/ prototyping —Finalize • implementation schedule • capital requirements • cost/benefit analysis	—Capital equipment • acquisition • installation —Production integration

plantside education effort regarding factory modernization. By communicating the long-term goals of Fac Mod at an early stage, many potential fears and apprehensions were avoided.

Phase I: Factory Analysis/Project Identification

A series of meetings was scheduled to initiate the project. The "Executive Expert" meeting was the first to be conducted and consisted of approximately fifteen people, excluding members of the Fac Mod team. The project was explained, and the Nominal Group Technique was utilized to gain their inputs regarding "what we do well," "what we do poorly," and "potential solutions."

Completely unplanned and unforeseen by the project team was the high degree of consensus that emerged among the Executive Experts regarding the potential opportunities for improving GTEC's manufacturing capabilities. Thus, the meeting served to facilitate better communications among the executive experts in the conduct of their daily interactions.

Similar meetings were conducted with the "Management Experts" and "Technical Experts." These meetings, in total, generated a very large number of potential projects to be considered for inclusion in the Fac Mod strategic plan.

The mid-project "downer"

GTEC's Fac Mod program began with a lot of excitement and fanfare. Support for the program was high at all levels of the company. The project team was extremely busy during the first three months conducting meetings, collecting data, listening to suggestions, etc.

As happens in all such programs, the euphoria slowly subsided, especially when the team had collected more imput than it knew how to effectively

Figure 2
Project Plan for Phase I

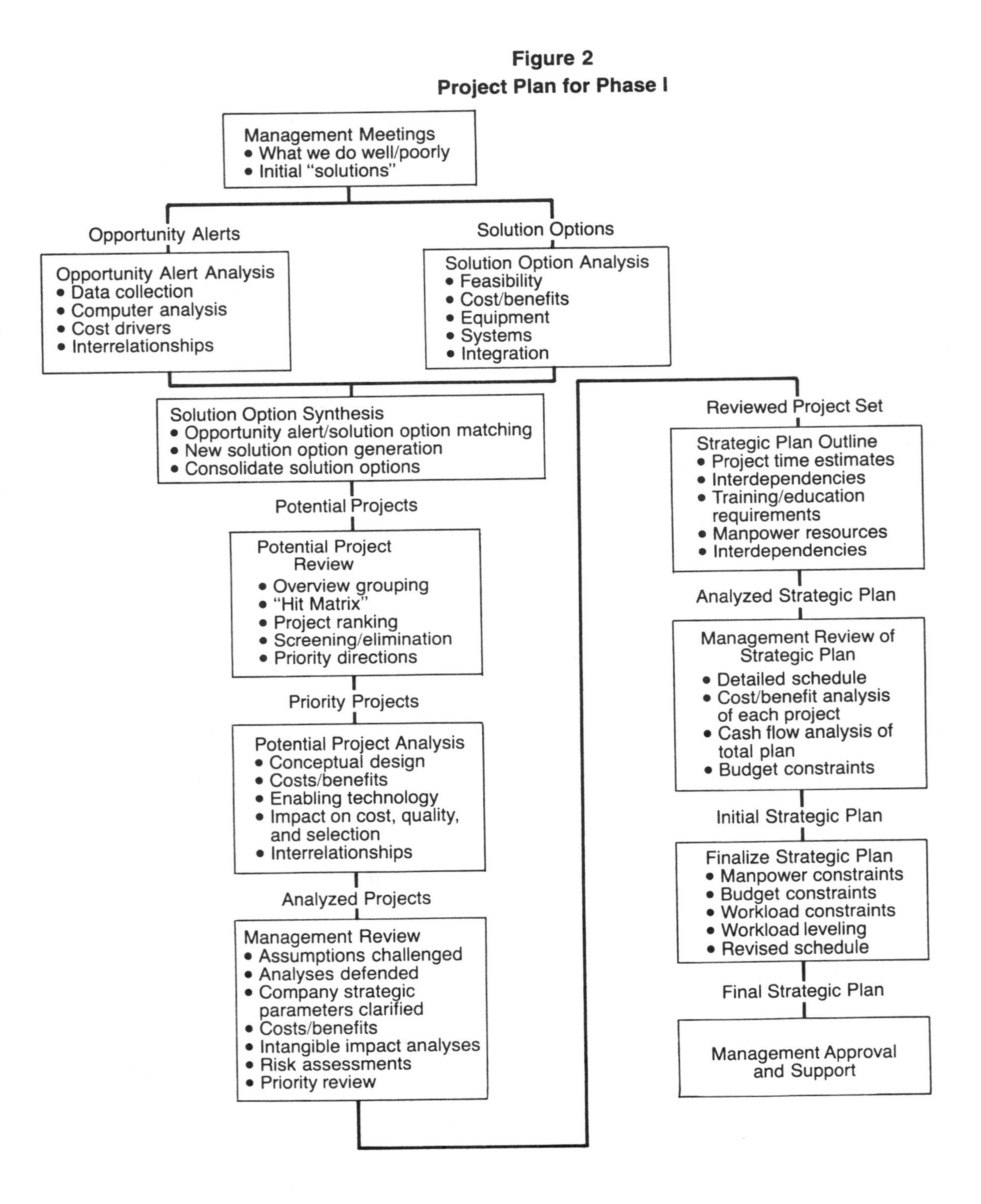

The project emphasis initially had been on **analysis,** *but later the project team had to face the more difficult job of* **synthesis.**

utilize. The project emphasis initially had been on *analysis*; but after the analysis had been done, the project team then had to face the more difficult job of *synthesis*. Based upon the analyses performed, a plan of attack must be *created*. Synthesis is inherently more difficult, more frustrating, than analysis.

The project team conducted a second series of meetings with the three groups of experts. The team presented its analyses to the experts and attempted to obtain guidance on priorities. The experts failed to respond, instead asking the project team to do the prioritization. The euphoria that characterized the beginning of the program was replaced with self-questioning and concern.

The Fac Mod program leaders analyzed the situation and determined that there was a missing ingredient in the project approach. There was no general framework that outlined the overall constraints and parameters of GTEC's "Factory of the Future." Without such a framework, there was no way to objectively assess the relative "worth" of an individual modernization project, or of the total set of projects.

The need for a vision

The mid-project "downer" prompted the Fac Mod leaders to begin conceptualizing GTEC's manufacturing environment in 1995 and beyond. A vision of GTEC's Factory of the Future began to emerge.

It was important to develop this conceptualization within the context of major trends and forces at work throughout industry and society. This effort benefitted from the fact that one of the Fac Mod leaders had spent several years analyzing major trends for the Institute of Industrial Engineers. This analysis provided a starting point for the creation of GTEC's vision of the future.

This effort turned out to be a major step in creating a framework for factory design and also for communicating the program plans to all levels within the organization. The "vision" stimulated the Fac Mod team to overcome the barriers that were encountered during the mid-project downer.

The need for a "Champion"

While the comprehensive vision was being created, it became increasingly clear that there was a strong need for a "Fac Mod Champion"—a person who is not only a true believer in factory modernization, but also one who is willing to give top priority to the overall effort to sell the program to management, acquire needed resources, and effect the organizational and company cultural changes necessary for successful implementation of the long-range strategic manufacturing plan.

The Fac Mod Champion should be someone other than the manager of the Fac Mod study team. The Champion must be a manager within the permanent organizational structure. This person should either have direct responsibility for long-range factory performance, or report directly to the person who has this responsibility.

In GTEC's case, the Fac Mod Champion turned out to be the manager of industrial manufacturing engineering, who reports directly to the vice-president for manufacturing operations.

Each company must assess its own situation and determine the appropriate person to be the focal point. The Fac Mod Champion should not be a "Johnny or Joanie come lately" to the factory modernization movement, but one who knows the company, understands the manufacturing technologies involved, and is committed to a successful modernization effort, no matter what obstacles may arise.

The Fac Mod Champion should be a "practical visionary." This is to say, he or she must be able to think conceptually and visualize what the factory of 1995 will be like but must also be able to distinguish between pipe dreams and realistic visions. Such people are rare and extremely valuable.

The Fac Mod strategic plan

As a result of the analyses of inputs and suggestions, and within the framework of the vision of GTEC's

The second major phase is to generate detailed designs of the various components of GTEC's Factory of the Future.

Factory of the Future, the Fac Mod team formulated a strategic plan for review, approval, and implementation.

The strategic plan consists of a set of explicitly defined projects that have been placed within a five-year time frame. Each project spelled out in detail in terms of specific tasks, capital equipment acquisitions, facility modifications, personnel requirements, and training/education programs.

The interdependencies between various tasks of the several projects were identified and observed when the overall schedule was generated. A resource leveling exercise resulted in shifting certain projects to later time periods, so that the total set of projects remained within the resource constraints of personnel, capital, space, and technical expertise.

The projects were eventually grouped into three categories:

Equipment Projects. Projects involving new or upgraded manufacturing equipment. These projects range from a single machine upgrade to a complete Flexible Manufacturing System (FMS).

Factory Management Projects. Projects involving new or improved factory management systems. Examples are computer-aided process planning, MRP II, and capacity requirements planning.

Personnel/Organizational Projects. These projects recognize the critical importance of personnel and organizational issues in factory modernization. Extensive retraining and orientation/education programs will be required at various points in the implementation schedule. Different modes of organizational communications will be required. These must be considered explicitly in developing the Fac Mod strategic plan.

It is not feasible to describe the specific projects included in GTEC's strategic plan for factory modernization. To give the reader an appreciation for the magnitude of this effort, the plan for the next five years involves approximately forty major equipment projects and twenty major projects aimed at upgrading GTEC's factory management capabilities.

Phase II: Project Design/Development

As of late summer 1984, when this article was being written, the Fac Mod study team had completed its initial efforts to create a long-range strategic plan for factory modernization. The next major phase is to generate detailed designs of the various components of GTEC's Factory of the Future. This essentially amounts to creating detailed scaled drawings, specifications, documentation, and operating procedures for each of the large number of individual projects included in the strategic plan. In so far as possible, specific equipment/process devices must be specified, complete with speeds/feeds, operating characteristics, interface parameters, etc.

This factory design phase must be conducted along a carefully charted "migration path" established by the strategic plan. The creation of an achievable migration path is one of the most difficult tasks to accomplish in the entire modernization effort. No large company is going to move from its current factory design to its "ultimate" factory design in one step. Not only would the capital investment requirements be prohibitive, but the total amount of pervasive change involved in such a move could not be accommodated.

The migration path should be constructed as a set of carefully integrated development stages, beginning with a complete understanding and documentation of the factory as it exists today. Using the vision of the Factory of the Future as a general long-term framework, a series of development stages is then constructed. Knowing the general characteristics of the "ultimate" factory design, it is a relatively straightforward matter to construct the first one or two major development stages. Many of the components of these early stages will consist of equipment/process upgrades that were on the "wanted" list of factory management before the Fac Mod effort was formally initiated. Now that a comprehensive study has been conducted, and a vision of the ultimate factory has been conceived, much better decisions can be made regarding immediate requests for specific items of equipment. Management can better determine whether these components will contribute to the achievement of the ultimate factory design.

Only the first few stages of the migration path can be developed in complete detail at this point. Subsequent stages are progressively less specific. As the company implements the early stages, it will have a sounder basis for becoming more specific about the next set of stages. This suggests that the Fac Mod planning effort be a continuing activity. This is discussed in greater detail below.

It is important to recognize that the factory management system undergirding the physical factory must evolve in conjunction with the evolutionary equipment/process advances being made in the factory. Also, education and training must be provided at critical, strategic points within each project and for the modernization program as a whole.

Ongoing Fac Mod organization

One of the most difficult decisions confronted in the Fac Mod effort was the question of how to organize for the management of the long-range strategic plan for factory modernization. The basic question was, "Now that the strategic plan has been developed, why not do away with a formal Fac Mod team, and assign to permanent units in the organization the responsibility for accomplishing various components of the plan?"

Existing organizational units at GTEC have the basic responsibility for many of the several types of projects included in the strategic plan. Even so, it was eventually decided to continue Fac Mod as a distinct organizational entity. The Fac Mod office will serve as a central focal point for visibility, project coordination and integration, resource leveling, and performance tracking.

It was decided to initiate a matrix organization structure as the mechanism to implement the Fac Mod strategic plan. A specific project is assigned to the particular line organization that would normally have responsibility for such a project. A project leader is selected from that line organization, and a team is formulated according to the specific project needs. The Fac Mod office assists in identifying the appropriate project leader, in putting together the project team (usually including individuals from *other* line organizations, in obtaining budget approvals, in establishing a well-defined project plan, and in monitoring actual project accomplishments.

Figure 3 shows a diagram of the manner in which the Fac Mod matrix management system operates. The Fac Mod office consists of project engineers who have the responsibility for coordinating several related projects in the manner described above.

Ongoing strategic planning

The initial Fac Mod strategic plan should not be viewed as a static document, set in concrete forever. Rather, it should be viewed as a living, changing document that reflects the dynamic nature of an ongoing strategic planning process.

The early projects in the strategic plan will be described in considerable detail because these projects will involve the implementation of equipment or systems that are currently obtainable in the marketplace. Relatively little risk is associated with these projects, since their underlying technologies are relatively known quantities.

But the farther into the future a project falls, the less we are able to know the specific nature of the project. It will likely entail evolving technologies, thereby increasing the risk factor. As time goes on, and as the evolving technologies emerge, we will be able to become more specific about the project.

The formulation of the initial Fac Mod strategic plan was based on a set of assumptions regarding the global market for turbine engines, GTEC's competitors, and a wide array of economic factors. The uncertainty of these assumptions, along with the rapidly evolving technologies described earlier, make it imperative that strategic planning continue as an ongoing activity at GTEC.

The type of strategic planning being suggested here is of the same basic type that a few well-known companies have practiced for many years. General Electric, IBM, and Texas Instruments are typical of the companies that do a good job in strategic planning.

It is emphasized, however, that we are speaking of *strategic planning for factory modernization*. All

Figure 3
Matrix Organization for Fac Mod Implementation

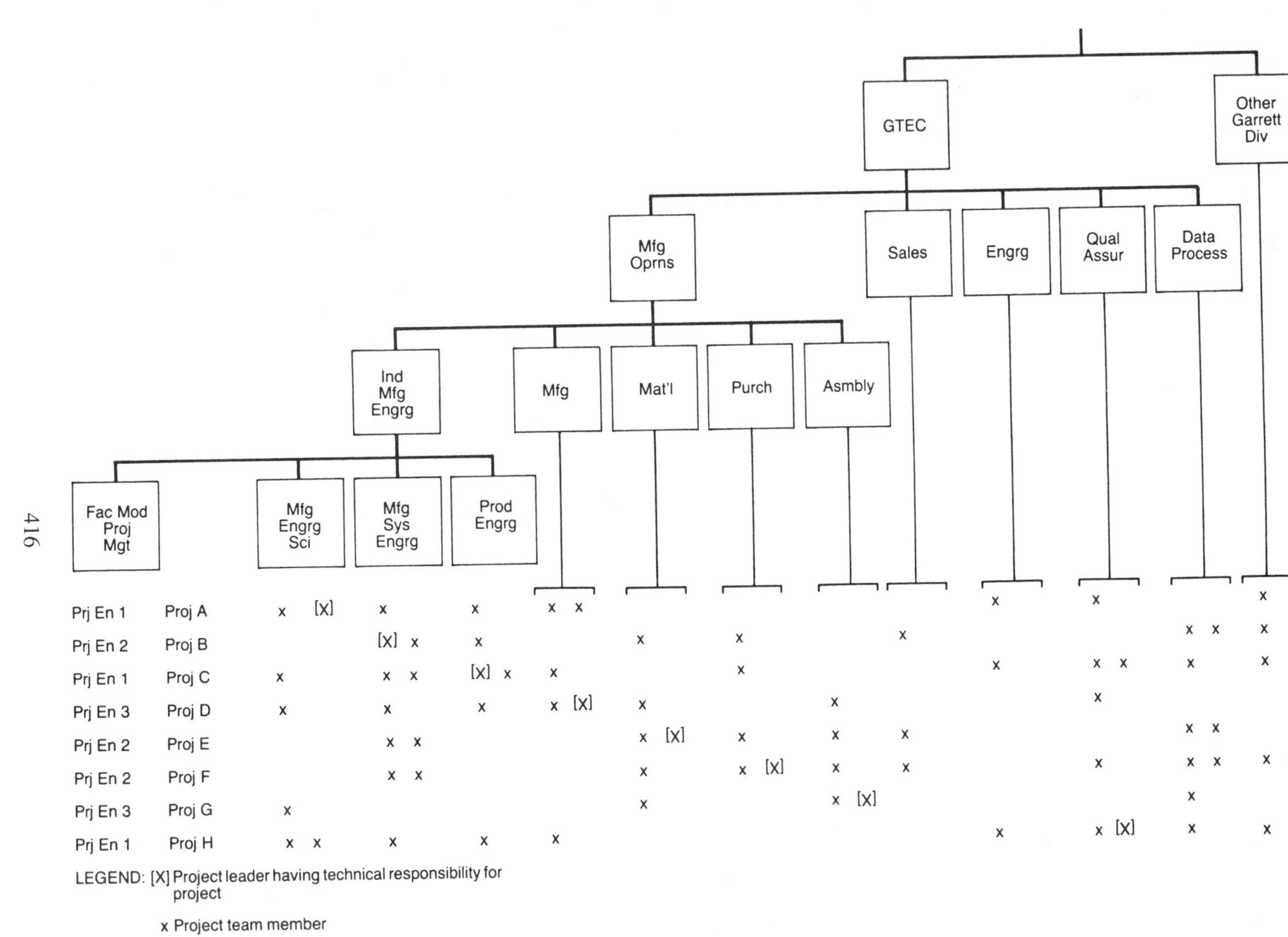

All U.S. companies need to increase the amount of attention paid to strategic planning for manufacturing.

U.S. companies, including those that have long practiced strategic planning, need to increase the amount of attention paid to strategic planning for manufacturing.

Conclusion: lessons learned

Managing the development of a strategic plan for factory modernization is somewhat like having a baby—you really can't tell others what it is like; they must experience it themselves. Nevertheless, the authors have filtered from their experiences the following things they wish they had known before embarking on this endeavor:

1. There is no "cookbook" for developing a strategic plan for factory modernization. The dearth of useful literature on factory planning methodology should be viewed as a golden opportunity by those who are capable of contributing to this field.

2. The manufacturing world is changing rapidly and dramatically, driven by pervasive forces in three categories:
   - Technological advances in microelectronics, material science, processing technologies, automation, robotics, and computer-assisted processes.
   - Customer demands and expectations regarding product quality, product cost, and responsiveness to scheduled delivery requirements.
   - Social/societal changes regarding employment security, meaningfulness of work, and desire to participate in decision processes.

3. Companies must take the initial steps toward modernization, even in the face of great uncertainties. They must be willing to take risks, recognizing that mistakes *will* be made. This will require a management style that combines risk taking with mental toughness, disciplined analytical skills, and a sense of knowing when the company must proceed with a project just because it is the right thing to do.

4. *Whether* to modernize is no longer the question; the only questions are how far, how fast, and in what directions to modernize.

5. Factory modernization must be a decision that makes sense for the company independently of whether or not there is an external incentive, such as the Air Force Tech Mod program.

6. While external consultants may be extremely valuable in the overall Fac Mod effort, there is no substitute for developing and maintaining in-house capability for planning and executing the entire effort. This is no small task, since there is a severe shortage of people who have the background and experience necessary to develop a long-range strategic plan for manufacturing. Few people in the manufacturing organization are trained to think in terms of integrated systems.

7. Factory modernization must have an internal champion (a "true believer"), high enough in the organization to effect change and willing to give top priority to "the cause."

8. Every project of this type can expect to experience at least one "mid-project downer," a period of frustration, fumbling, and self-questioning. If projects like this were easy, effortless, and always smooth sailing, someone would already have done it.

9. The creation of a vision for the organization's factory of the future is an essential, crucial step in the strategic planning process. Without it, there is no framework, no reference, no basis for effective communication.

10. A carefully designed program of communication to all levels is an essential component. Everyone must feel that this is their program. A comprehensive education and training program must be included as explicit components of the Fac Mod strategic plan.

11. Traditional cost/benefit analyses are not sufficient for justifying a comprehensive factory modernization program, since they fail to explicitly consider the synergistic benefits to be derived from the integration of system components.

12. A Fac Mod strategic plan must demonstrate a migration path from the current factory status to an appropriate level of data-driven automation, through time. The "appropriate level" is itself a moving target, necessitating a periodic update of the migration path and associated cost/benefit analysis.

13. Beyond the analysis phase, the ongoing effort requires a unique organizational entity for coordination, visibility, progress tracking, work-load leveling among interdependent projects, and reporting. A matrix type organizational control is required.

Traditional cost/benefit analyses fail to explicitly consider the synergistic benefits of the integration of system components.

---

14. Factory modernization should not be considered a one-time activity, but an ongoing process that never ends. This implies that corporate strategic planning efforts must explicitly include manufacturing strategic planning inputs.

15. Changes of the magnitude involved in Fac Mod will be extremely traumatic for any company. Not only is the capital investment requirement (and associated risks) extremely high, but the very culture of the company, its very personality, must undergo significant changes.

16. The limiting factors in how rapidly an organization can modernize its manufacturing facilities are not a lack of capital or advanced technology, but (1) a severe shortage of engineers and systems analysts who know how to use the available technology, and (2) the difficulties in managing the rate of organizational change that will be required. The simple fact is that few companies know how to cope with change, let alone manage it. Perhaps the Fac Mod team of the future should include professional change agents.

Joe H. Mize is Regents Professor of industrial engineering and management at Oklahoma State University, and a consultant to industry in strategic planning for factory modernization. The author of five engineering texts, Dr. Mize's teaching and research interests are in the areas of simulation, manufacturing systems, automation, CAD/CAM, and CIM. He is past president of the Institute of Industrial Engineers.

Deborah J. Seifert is manager of manufacturing systems engineering at Garrett Turbine Engine Company, Phoenix, Arizona, where she is responsible for managing short- and long-range capacity and factory planning, production system development, CAM activities, and financial analysis support. Dr. Seifert has numerous publications in the areas of manufacturing systems, man-machine modeling, and computer simulation. She is currently director-elect of the Operations Research Division of the Institute of Industrial Engineers.

Gayle Berry is manager of operational planning at Garrett Turbine Engine Company, Phoenix, Arizona. Dr. Berry has experience in computer-integrated manufacturing and has published numerous technical papers on shop floor information systems, computer-aided process planning, and human factors in manufacturing.

APPLYING THE LESSONS LEARNED IN IMPROVING MANUFACTURING OPERATIONS

Dan C. Krupka
AT&T Bell Laboratories
Murray Hill, NJ 07974
(908) 582-4102

ABSTRACT

During the past several years, AT&T has made major improvements to its manufacturing operations. In a number of cases, engineers from Bell Laboratories, AT&T's R&D organization, have played an important role. According to the original plans, R&D was to develop concepts, tools, and systems to support these programs. Over time, the R&D engineers have become much more intimately involved. Along the way, they have learned many lessons about applying and transferring technology. Most of these are universal; they apply to anyone attempting to introduce novel quantitative ideas into manufacture. This paper interweaves some concepts recently developed by AT&T Bell Laboratories with lessons learned in reducing them to practice. The themes emphasized are simplicity and a regard for cultural issues.

INTRODUCTION

Since the mid 1980s, AT&T's factories have been making significant improvements in their operations. To assist in some of these activities, they have drawn upon Bell Laboratories whose more traditional linkage with manufacturing has been through product and process development. In assuming this new role, engineers from Bell Laboratories have been learning many lessons. Many of these are similar to those to be learned by any engineer who attempts to apply "academic" ideas to the "real" world of manufacturing.

This paper is a report of the lessons learned. The first lesson deals with the order for attacking improvement programs and suggests that hard automation be the last step. Simplicity is the theme of the second: Much can be accomplished without resorting to highly sophisticated methods. The third lesson stresses the need to combine the flow-oriented thinking of industrial or manufacturing systems engineers with total quality management. In the fourth lesson, the need to recognize that developing an algorithm or system is but the first step - albeit a critical one - in improving an operation. Obtaining operational results requires much persistence. Technology transfer is the theme of the fifth lesson which teaches that a proponent who moves from factory to factory may be the most effective way to transfer technology. The final lesson - and the most important one - stresses the need to recognize that many operational improvements require cultural change.

In most cases, the lessons are illustrated by concepts and results developed during the past few years in AT&T.

LESSON 1

In applying resources to improve manufacturing operations, the recommended sequence is the following:

1. *Improve existing processes by modifying product flows, applying quality techniques, simplifying procedures, etc. All these steps are relatively inexpensive and offer opportunities to experiment.*
2. *The process improvements may require a supporting software system to operate consistently. This is the stage where it makes sense to specify it and install or develop it.*
3. *As a final step, consider the installation of physical automation - robots, conveyors, AGVs.*

Reprinted from *1991 IIE International Industrial Engineering Conference Proceedings*

The above sequence cannot be considered as dogma because some situations clearly require a different sequence. It has been our experience, however, that it can be perilous to proceed otherwise.

A large potential for wasting resources arises when one attempts to develop software systems to support processes that have not been improved and simplified. These systems inevitably are complex and expensive, reflecting the complexity of the underlying process. Moreover, they will potentially institutionalize mediocrity. In some cases, a software system may not even be required. One of our factories had installed an inventory tracking system on one of its lines.

This was necessary when the manufacturing interval was long. After it had been reduced by an order of magnitude, the system was eliminated. We have also learned that robots are not a panacea, and we have become much more selective in their use.

The foregoing prescription is now generally accepted, but such was not the case five years ago when we embarked in our journey. In fact, the approach is now known as "frugal manufacturing" a term popularized by Schonberger.[1]

Following the proper sequence in an improvement program does not guarantee success, and it certainly does not ensure rapid progress. We have found that we have been most successful when we have been judicious in our use of tools. This brings us to our second lesson.

LESSON 2

Use the simplest methods that will accomplish the task. Focus on the goals, not the means.

Several years ago we became experts at building simulation models for integrated circuit (IC) fabrication lines. These allowed us to validate some of the rules of thumb that had been used by more experienced but less technically sophisticated engineers. The models also allowed us to estimate the effects of "hot" lots on a line's performance and the consequences of non-uniform start rates.[2] Most importantly, they enabled us to develop an intuitive feel for the performance of IC lines. Nonetheless, it became our habit to launch operations improvement programs on these lines by building simulation models. Our habit was finally broken when a line was shut down before our model was completed.

Experiences such as this have led us to adopt the simplest methods first. The simplest of these is taking data and showing it to operators. Figure 1 shows the average time per wafer to pass through a photolithography step. The considerable improvement shown during the first four months of this program was achieved merely by posting the data. Subsequent improvements used only a slightly more sophisticated approach.

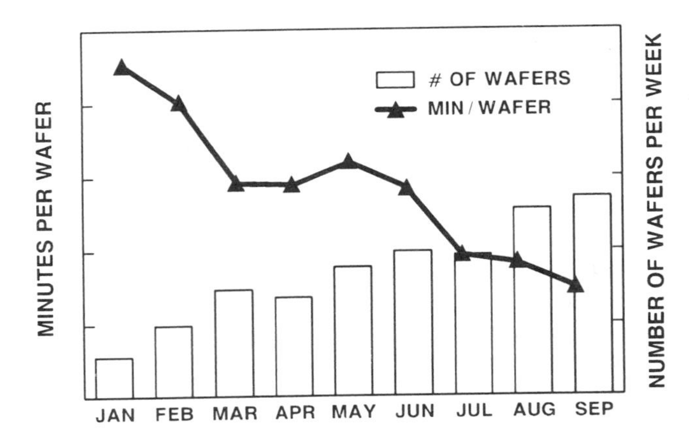

Figure 1 - Average time required to expose a wafer as a function of calendar time. Note that production is ramping up.

In many of our projects, we are asked to assist in reducing a line's manufacturing interval in the face of rising demand, without the prospect of adding more facilities. Apart from the standard approaches, we have relied on simple queueing theory such as shown in Figure 2. The curves show average sojourn times for a single-server queue with generalized arrival and service rates. We have used such a figure to explain that a project aimed at reducing all sources of variability and service times can result in a shorter interval at a higher production rate. This intuitive approach has allowed us to help our customers much more rapidly than when we relied exclusively on simulations models.

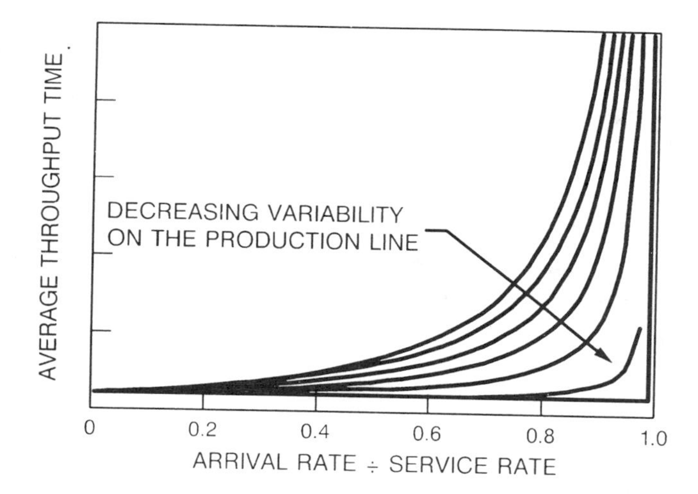

Figure 2 - Throughput time for a single server queue as a function of traffic intensity.

When we began our heavy involvement with AT&T's manufacturing facilities, we restricted our efforts to managing product flows. Our tools consisted primarily of simulation programs and queueing analyzers such as QNA.[3] We learned, however, that they alone are not sufficient. This leads to the next lesson.

LESSON 3

The best results are obtained by using an approach that combines manufacturing systems engineering and total quality management.

This marriage of approaches is illustrated by the results of a circuit pack line at our Denver Works. Figure 3 shows the distribution of the manufacturing interval over roughly a two-year span. In mid 1986, the line had an average interval of nearly 20 days and a very broad distribution. One year later, the average was reduced to four days. By April 1988 the interval was approximately two days and the distribution was dramatically shrunk. These results were achieved as the result of a combination of programs.[4] These included simulating the line to estimate transfer lot size, the introduction of a rigorous materials management program, improvements in storeroom operations, the application of statistical process control, and the establishment of a total quality control team. The performance has continued to improve.

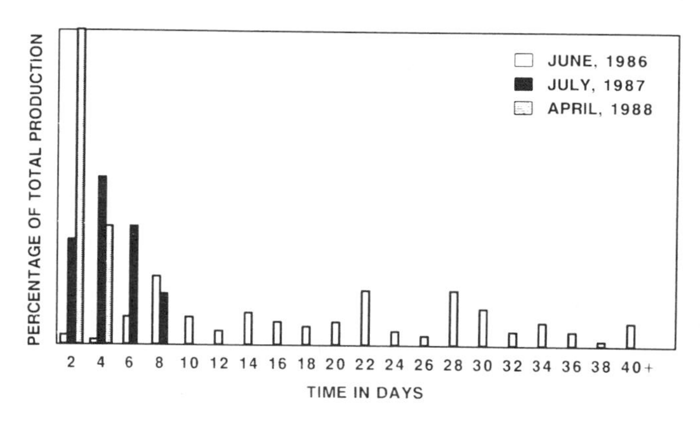

Figure 3 - Distributions of the manufacturing interval for a circuit pack assembly line over approximately a two- year interval.

The experience in Denver and at other AT&T manufacturing locations suggest that improvements in operations are accelerated by applying a multi-disciplinary approach. As more is learned about improving operations, and as time becomes more widely recognized as a source of competitive advantage,[5] we may see the birth of "industrial operations engineering" practiced by engineers trained in industrial engineering, total quality management, materials management, and computer science. The task of these engineers would be to design and manage operations in such a way as to minimize the (manufacturing) interval and its variance.

All of the foregoing discussion has been centered on the application of established or traditional tools and systems. We're also learning how to introduce new systems.

LESSON 4

When planning to develop and implement a new software system, bear in mind that the creation of the algorithm and the writing of the software are a small part of the total effort.

This lesson is illustrated by the introduction of a system to schedule part of a cable shop at our Atlanta plant. Manufacture of a cable, comprised of a large number of twisted pairs of wire, begins with drawing copper rod to the desired gauge, insulating it with colored plastic and then twisting pairs of wires of different colors. Subsequent steps are depicted in Figure 4. Twisted pairs are combined into 50- or 100-pair strands and then the strands are further

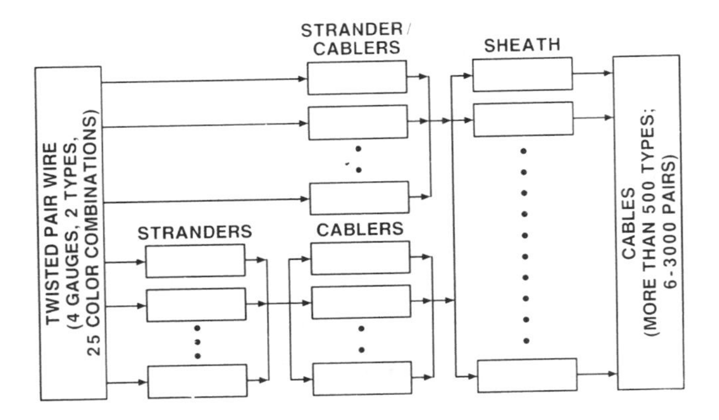

Figure 4 - Stranding, cabling, and sheathing processes in one of AT&T's cable shops.

combined into cables comprised of as many as 3000 twisted pairs. In the final step, the cable is wrapped in a protective layer, often corrugated aluminum, and then given a protective jacket. Complexity in the shop arises from the very large number of final products that follow many different routings, long setup times, the need to run certain products through specific sheathing lines, and some double runs through sheathing lines.

For many years the stranders and sheathing lines were scheduled by experienced people who were capable of keeping track of the complexity. It was evident, however, that the shop could operate much more productively if a more sophisticated, computer-generated schedule were developed. Help was sought from R&D and a scheduler was duly developed. Unfortunately, the time that elapsed from the first attempt to use the scheduler to successful implementation was much longer than anyone involved would have desired.

The scheduler is now being used and the shop's lead time has been significantly reduced. The system takes as its input the weekly load for the shop, routings, priorities, and equipment descriptions. It produces shift-by-shift schedules for the stranders, cablers, and sheathing lines using a backward scheduling algorithm that seeks to minimize setups in sheathing. The result is synchronization of all areas and shifts in the shop.

Success was finally achieved by overcoming a number of obstacles. They are listed below, along with descriptions of the countermeasures that were used.

Goal Statement

The goal, as initially stated, was to implement the scheduler. Progress was accelerated when a more meaningful goal - to use the scheduler to shorten the interval to a specified target - was enunciated.

Shop Ownership

The shop felt a loss of control and was skeptical regarding the validity of the schedulers. By asking the shop to evaluate schedulers, it became evident that the system did not adequately represent shop operations. As the proposed schedules, amended with input from the shop, began to make more sense, the shop developed greater confidence in the validity of the system.

Metrics

The published schedules conflicted with the traditional metric for the shop - weekly production measured in total length of wire produced. The new metric is adherence to schedule.

Date Integrity

Some schedules failed to make the desired products. One major cause of this problem was traced to incorrect shop routings. Previously, these were maintained, in large part, in the heads of the shop's management. Successful implementation of a computer-based system required the establishment of processes to insure the accuracy of routings and all other data used as inputs to the system.

Communications

The computer printout generated by the scheduler was difficult to understand. By creating a wall chart that translates the printout into a more readily understandable format, acceptance of the new schedule was enhanced.

The foregoing discussion demonstrates that the technically challenging aspects of system development form but a small portion of the total effort that leads to a successful implementation. Regrettably, reward systems in technical communities, whether in R&D or manufacturing, have tended to favor the creation of technically elegant or sophisticated solutions over their implementation.

LESSON 5

The spread of concepts, tools, and systems from factory to factory is greatly accelerated by migrating advocates.

Factories are understandably reluctant to tamper with their operations. Meeting production targets and shipping schedules are of primary importance. Even relatively simple applications of just-in-time methods have not occurred overnight in the U.S. It is not surprising, therefore, that more sophisticated concepts, tools, or systems spread even more slowly. The rate of adoption can be accelerated, however, when someone familiar with the method travels from factory to factory to facilitate introduction.

This lesson is well illustrated by the history of the Final Assembly Sequencer. In the following, a description of the algorithm is followed by an account of its spread among several AT&T factories.

The development of this sequencer was motivated by the desire to ensure a smooth product flow in a facility that produces make-to-order highly optioned telecommunications systems. In 1986, when the need was first identified, a typical system included about 40 circuit packs of approximately 25 different types. Since the systems were all different, the number of circuit packs of any one type varied considerably from system to system. To ensure the smooth operation of feeder lines that assembled the circuit packs, an algorithm was developed[6] to sequence a week's worth of production of the systems that were master-scheduled in weekly buckets.

The concept is further illustrated in Figure 5 which shows how the usage of parts varies from product to product. Since the end product may be completed at any time within the schedule horizon, typically equal to one week, an efficient search heuristic was developed to select a nearly optimal sequence from many feasible alternatives.

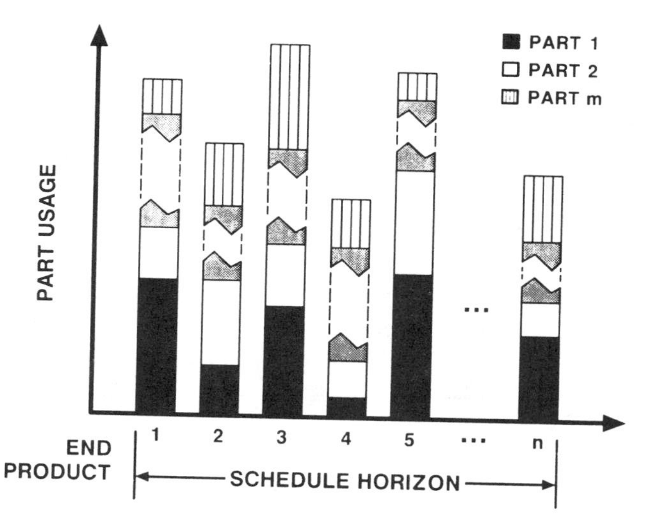

Figure 5 - Part usage for a complex make-to-order end product over the schedule horizon. By appropriate resequencing of the end products within the schedule horizon, the usage variability of parts can be reduced, thereby imposing a smoothed demand on the feeder shops.

Since its initial and successful use in Denver, the sequencer has been applied in at least three other AT&T factories. In each case, it was introduced by one of the engineers who first identified a need for it in Denver, who contributed to its development, and who participated in its successful implementation. (The initial implementation was rapid and successful because all of the potential obstacles listed in the previous section were appropriately addressed or did not exist).

Figure 6 shows how the introduction of the sequencer contributed to the performance of a shop in the AT&T factory (now closed) in Winston-Salem. A particularly gratifying outcome of its introduction into this facility was the satisfaction expressed by the operators. The line performed as never before, they claimed.

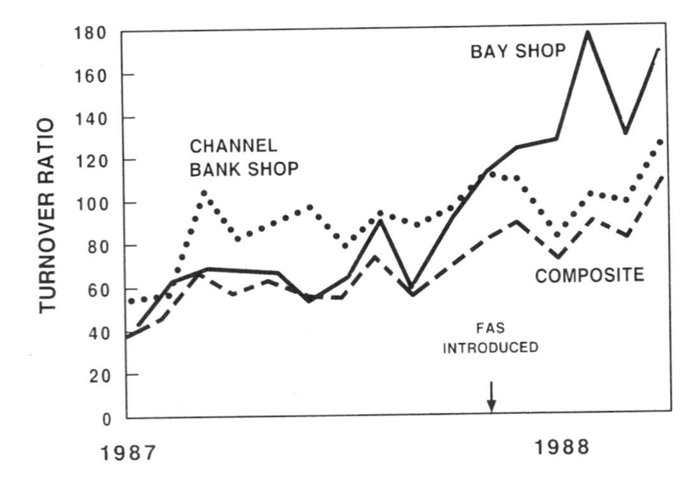

Figure 6 - Effect of applying the final assembly sequencer in one of AT&T's manufacturing lines.

More recently, the sequencer was introduced into a factory that makes the cabinets that house remote terminals. There, in conjunction with an extensive pull system that links parts fabrication, painting, and final assembly, the sequencer has been used to level load the shops and to synchronize their operations. As a result, the inventory has dropped and the synchronization has facilitated a major ramp-up.

Most of the foregoing lessons possess cultural themes. These are so important, however, that they merit a lesson of their own.

LESSON 6

People make the difference: The most significant operational improvements have been accompanied by cultural change.

As engineers, we are trained to believe in calculations. But calculations alone do not improve operations. As long as there are people in our factories, we cannot expect to have a major operational impact unless we're prepared to tackle cultural issues. Yet this is an area where few of us have been adequately prepared.

Our goal over the past several years has been to improve AT&T's manufacturing operations through the application of quantitative methods. We have learned that these do not suffice. The cultural issues that we have faced include the concern of operators that streamlined high-quality operations will lead to the loss of jobs; anxiety of shop supervision over potential loss of power; the slowing down of a project caused by the loss of an essential or charismatic manager; and the reluctance of R&D engineers to enmesh themselves in the details of manufacturing operations. We continue to learn, and recording our experiences is part of that learning.

ACKNOWLEDGMENTS

The lessons described in this paper represent the shared learning of many of my colleagues in AT&T. I am particularly grateful to C. R. Bourquin for his help in illustrating Lesson 4.

REFERENCES

1. Schonberger, R. September/October 1987 Frugal Manufacturing, Harvard Business Review p. 95.
2. Burman, D.Y., Gurrola-Gal, F.J., Nozari, A., Sathaye, S., and Sitarik, J.P. July/August 1986. Performance Analysis Techniques for IC Manufacturing Lines, AT&T Technical Journal, 65, Issue 4, p. 46.
3. Whitt, W. November 1983. The Queueing Network Analyzer. Bell System Technical Journal, 62, No. 9, Part 1, p. 2779.
4. Albano, R.E., et al. July/August 1990. Manufacturing Execution: Circuit Packs. AT&T Technical Journal, 69, Number 4, p. 64.
5. Stalk, Jr., George and Hout, T. M. 1990. Competing Against Time. The Free Press.
6. Luss, H., Rosenwein, M.B., and Wahls, E.T. 1990. Integration of Planning and Execution Operations: Final Assembly Sequencing. AT&T Technical Journal, 69, Number 4, p. 99.

BIOGRAPHICAL SKETCH

Dan C. Krupka is Head of the Manufacturing Systems Engineering Department at AT&T Bell Laboratories. His organization works with AT&T's factories to improve their manufacturing operations. He has a B. Eng., in Engineering Physics from McGill University, Montreal, Canada, a Ph.D. in Experimental Physics from Cornell University, and an Advanced Professional Certificate in Economics from New York University. He has been with AT&T Bell Laboratories since 1967.

CIM AT APPLE COMPUTER, INC.
A CASE STUDY

Michael Kaskowitz
Apple Computer, Inc.

ABSTRACT

From the original Macintosh Factory in Fremont California, to facitlities around the globe, Apple Computer has applied its own technology on the factory floor to enhance its manufacturing capabilities. In the process, we have gained a great deal of experience in CIM. This paper will highlight Apple's experiences in this area, using its Fremont FMS as a case study. We will discuss various aspects of the CIM discipline at Apple; its organization, focus, architecture and challenges. In addition, we will discuss how the Macintosh has been successfully applied in our manufacturing facilities.

INTRODUCTION

In early 1980's, Apple Computer revolutionized the computer industry by demonstrating that computing could be done by the individual on small computers, rather than by large groups on mainframes. With the introduction of the Apple II, the personal computing industry was born.

The early growth of Apple was tremendous. In order to meet the growing demand for our product, manufacturing facilities were opened domestically, as well as in Ireland and Singapore.

In January of 1984, Apple Computer introduced Macintosh, a revolutionary platform in personal computing. To build Macintosh, it was decided that a highly automated factory be built in Fremont, California. In addition, this factory would be controlled by the Macintosh!

Initial applications of Macintosh technology on the Fremont factory floor took advantage of the Macintosh's early strengths. These applications utilized the Macintosh's powerful user interface for database "front-end" collectors, its AppleTalk interface to support local networks, and its serial I/O to support limited device interfaces to barcode printers and scanners.

SITE SPECIALIZATION

With the early successes of the Fremont factory, our facilities around the world were soon automated. Interestingly, due to regional differences in our markets, the sites tended to specialize in different aspects of automation.

The Fremont site built all of the Macintosh's for Apple. Given the high domestic demand for the Macintosh, Fremont developed high speed lines dedicated to the manufacture of the Macintosh. This allowed Fremont to build high volumes, but limited the configurations that could be built. In addition, due to its proximity to Apple's design centers in Cupertino, Fremont also served as the lead site in manufacturing R&D.

The Cork, Ireland facility built Apple II product for the European Region. Europe provided a diverse customer base, with varying regional issues. As a result, the Cork facility specialized in building several configurations at lesser volumes on highly flexible manufacturing lines.

Reprinted from *1991 IIE International Industrial Engineering Conference Proceedings*

The Singapore facility was the counterpart to the Fremont factory. It built the Apple II product line for domestic usage, as well as the Pacific Region. This led Singapore to develop high speed dedicated processes.

DEGREES OF MANUFACTURING FLEXIBILITY

As our business grew, so did our product diversity. To be more responsive to our customer base, all products (Apple II and Macintosh) were being built in all three sites. In addition, the volumes of each configuration increased.

With these changes in volume and product mix, the degree of automation varied (refer to Diagram #1 below). To keep up with our changing business requirements, Apple found it necessary to target two areas of automation, namely: 1) High Volume, Dedicated Serial Processes, and 2) Medium Volume, Flexible Manufacturing Systems (FMS).

In order to provide more flexibility in product deployment, it was decided that all sites have both capabilities.

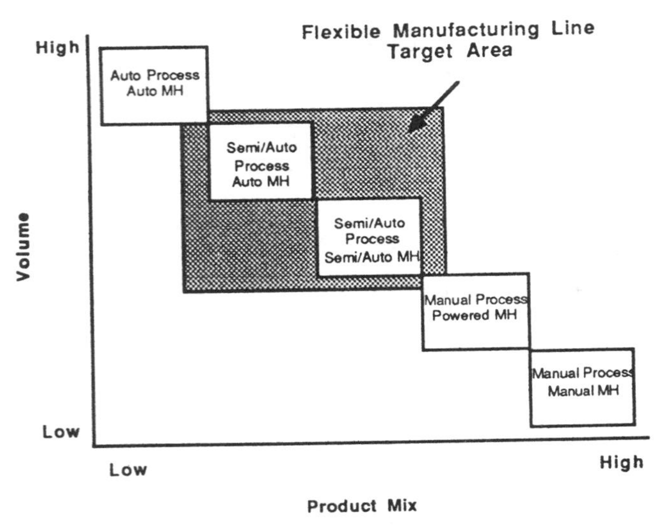

Diagram #1- Degrees of Automation

High Speed Dedicated Lines

To meet the needs of high-volume manufacturing, each site developed several lines which are dedicated to a single product family. Each line is serial in nature, and product is "pushed" through in batches.

Material delivery is achieved through the use of KAN-BAN carts. We use carts to manually deliver parts to the lines. When a cart is consumed, it is brought back to recieving where new parts are loaded based on its KAN-BAN card. If we wish to change the product being built, the KAN-BAN card is replaced with a different one, and the new material is brought to the line. While manual, this method reduces WIP, expedites changeovers and provides high throughput, with very low inventory levels.

Flexible Manufacturing System (FMS)

In addition to our dedicated processes, each site has a Flexible Manufacturing System, the most recently of which was completed in Fremont. Due to the highly automated nature of the our FMS's, they vary with each site, yet have many common similarities. For the purpose of this paper, I have decided to highlight the Fremont FMS.

The primary goal of the FMS is to build multiple products, simultaneously. By integrating a flexible assembly system into our production facilities, we ensure our ability to handle production changes due to changing market situations, resource changes, and process changes, much more cost effectively.

The FMS system is capable of supporting multiple product builds concurrently. In addition, the FMS system also supports multiple configuration builds within each product build schedule, rather than conventional "batch" scheduling. This flexibility places emphasis on several new processes, such as coordinating mixed production schedules, investing in shared operator functions, and verifying product configurations.

Material cost has historically represented a significant component of product cost, as well as a contributor to induced material defects. Therefore, a reduction in the amount of unnecessary material handling was also a major objective of the FMS system. The logical evolution of this involved an automated delivery of materials to their "point of use".

Another important discipline is that our FMS is implemented as a "pull" system, rather than a conventional "push" system. In-line queuing of both assemblies and material is therefore substantially reduced, resulting in several immediate advantages.

In-process material and assembly inventories are substantially reduced, thereby streamlining individual process flows and balancing production operations more efficiently. Fluctuations in production yields may also be tracked more closely, increasing product quality by improving engineering reactance.

Problems of Gaining Acceptance - We learned a great deal in the implementation of our FMS, namely that the major obstacles were not technical, rather they were in gaining site acceptance.

To achieve these advantages of the FMS requires a certain degree of discipline. For example, inventory buffers in a "push" system afford slack which can hide inefficiencies in process flow, leading to poor product cycle times and higher work-in-process costs. Since station-to-station buffering is substantially reduced in a "pull" system, operator and process cycle time consistency is critical. Variability in individual operations is more noticable, so line balancing becomes more important.

To reduce material inventory levels, conventional storage systems are replaced by minimized staging systems. This requires greater discipline in guaranteeing availability and timely delivery of material. The skills learned in practicing Just-In-Time concepts must be extended to manage reduced material levels without risk of "material starvation".

Another difficulty encountered in the FMS was in coordinating the efforts of dedicated operational teams. Most of the philosophy on which the FMS is based centers around a "process-oriented" focus. This is in contrast to the "product-specific" serial lines on which all Apple manufacturing has been based, since the introduction of the Macintosh Factory in 1983.

Much of the organizational infrastructure in Fremont is structured around product lines. When specific products were selected to manufacture on the FMS system, their product-specific organizations accompanied them. Since each product team was measured solely on the quality and schedule accomplishments specific to their product, integration of product functions and coordination of shared tasks were avoided. Instead, competition for common resources detracted from the overall efficiency and acceptance of the system.

Once this was recognized by the project team and upper management, the metrics by which the operations teams were measured were altered to promote cooperative involvement. Operator cross-training was encouraged for process-oriented functions, production scheduling across products was better coordinated, and process and quality engineers were reorganized to concentrate across a broader base of products.

While the importance of following these disciplines is recognized by upper management, it is often difficult to convey these advantages to line operators, who recognize only the difficulty in meeting stricter constraints. Operator training and continuous feedback are critical to improving operator awareness and commitment to these practices.

In addition, engineering and supervisory training continues to be a critical area of focus. Selecting the right product combinations, defining well-coordinated process routings, and intelligently scheduling configuration builds are all key to optimizing line utilization and product changeovers.

SOFTWARE CONTROLS ARCHITECTURE

Over the years, as more automation has been added to the sites, the need to develop a comprehensive controls architecture became apparent. This new platform needed to reduce manufacturing cycle times, control inventory and support mixed build schedules. To meet these needs, an advanced architecture utilizing new philosophies was developed.

Apple's current software controls architecture consists of five(5) levels - Device, Workstation, Cell, Factory, and Site (refer to Diagram #2 below). This controls hierarchy is implemented primarily using Macintosh and Tandem computers and programmable controllers.

This tightly-integrated architecture manages the shop-floor schedule, automatically routes both assemblies and material, provides workstation control and data collection, tracks product build detail and pallet content electronically, and tracks in-process inventory levels. The following sections describe briefly the architecture and responsibilities of each layer of this hierarchy.

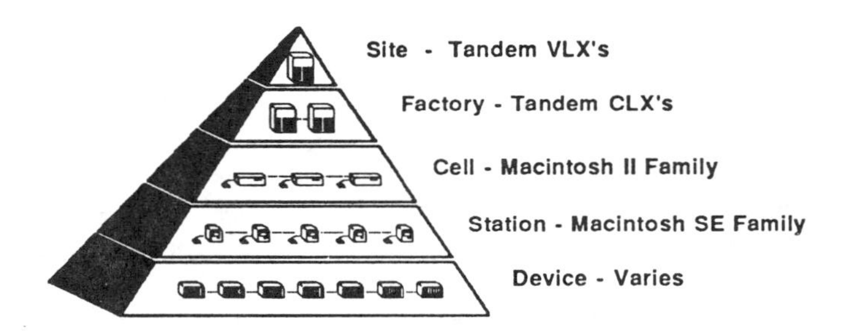

Diagram #2 - Control Architecture

Site Layer

The Plant layer of the hierarchy is composed of a network of Tandem VLX fault-tolerent computers. This network is responsible for managing all site business functions, including production control and scheduling, materials procurement, and inventory management.

Initially, functional integration with our Business System has been confined to reporting bulk material movement and assembly completion transactions. Current design efforts will address the addition of shop-floor scheduling and inventory control interfaces within a subsequent phase of development.

Factory Layer

The Factory Host primarily serves as a shop-floor interface to the Plant Host. It captures and communicates material movement and assembly completion transactions in real-time, and maintains shop-floor data for Production Build and Pack schedules, product configurations, part numbers, and abbreviated build lists, and bulk material attributes.

In addition, this layer acts as the server for all production quality and efficiency metrics. This shop-floor database collects failure and repair data, production downtime, station and product line yields, and product history tracking. Cycle time counts and line utilization are also monitored.

It has been our experience that the preferred hardware platform varied in different part of the world. The primary reason for this is that different standards exist and varying degrees of hardware support are available.

While our hardware needs were different, we realized that our automation and data collection needs were the same. As a result, we have strived for a common software layer amongst differing hardware platforms.

Cell Control Layer

The Cell Control Layer manages the shop-floor build and pack schedules, downloads line configuration and routing setup information, maintains material and assembly pallet records and electronic travellers, manages the Material Staging and Automated Run-In systems, satisfies material replenishment and assembly movement transactions, and communicates material movement and assembly completion transactions to the Factory layer.

The Macintosh II family is used extensively at this layer, due to its expandability and higher computation speeds. In addition, the color capabilities of the Macintosh II lends itself to fault annunciation and monitoring.

Workstation Control Layer

The Macintosh SE is used extensively at the Workstation Layer. These Macintoshes range in application from Assembly, Test, Inspection, Rework, Barcode Generation, and Material Induction stations. They communicate serially with the PLC layer to coordinate workstation pallet flow, and communicate via AppleTalk to the Cell layer. They also provide the workstation user interface, which manages local data collection and product configuration build prompting to the operator.

The workstation Macintoshes also serve as collectors for the Automated Information System. These collectors report assembly failure symptoms and associated repairs, material defects, workstation cycle times, downtime statistics, product history tracking, and product content.

The Macintosh SE was chosen for its modularity and lower cost. The sophisticated user interface provides a consistent and easy method for the operators to interact with our data collection and control systems.

Note that only the Cell and Device levels of the controls hierarchy are critical to production operation. This is reflected in the design decision to route critical PLC-to-Cell transactions through a direct interface, rather than through the workstation controller layer. Loss of components such as workstation controller or Host interface do not affect production flow. If a workstation Macintosh goes down, the PLC will recognize loss of active communications, and will continue normal operation in a degraded mode. When the Macintosh resumes communication, normal operation is automatically resumes. Only a minor loss of data is experienced.

Device Control Layer

Programmable Logic Controllers (PLC's) comprise the lowest level of the hierarchy. They are responsible for all conveyor control, barcode network management, and physical pallet tracking.

All PLC's are networked over their own LAN, for peer-to-peer communication, as well as having direct serial connections with the Workstation Layer. In addition, a serial driver for the Macintosh was developed to allow the Cell Controller to communicate as a peer device on the PLC local area network. This allows unsolicited messaging between the Master PLC and Cell Controller to occur for several key transactions. Routing information is downloaded via this interface to setup workstation function, material queue assignments, and material and assembly flows.

CIM MODEL

During the early years of Apple manufacturing, a great deal of the efforts of the site CIM departments was dedicated towards shopfloor control systems. As our business needs have expanded, a greater emphasis has been placed on data collection and integration.

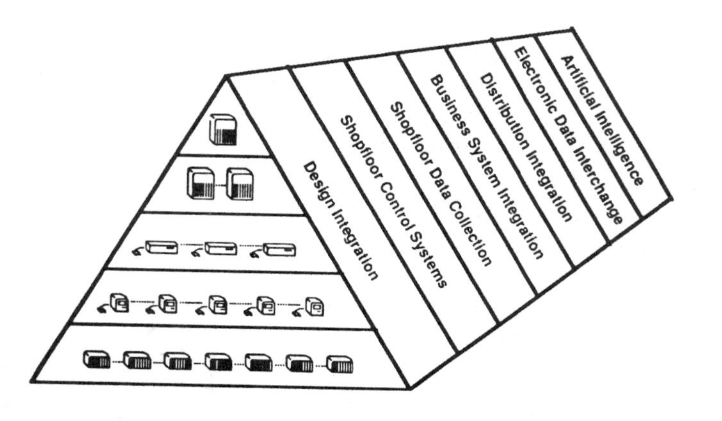

Diagram #3 - Areas of Concentration

To address these needs, we have developed a model of seven areas that we need to concentrate on over the next few years (refer to Diagram #3). These areas of concentration are detailed in the following section.

Design Integration

Apple has a long history of being involved early in the design process. Each site has an New Products Organization (NPO) which is responsible to work with design engineering and assist in the product design and manufacturing introduction. A representative from the NPO is assigned to each new product team, and serves as a liason between the design community and manufacturing engineering.

The manufacturing sites are responsible for developing new processes, and provide guidelines to design engineering. Through this "Design for Manufacturing" process, we have been extremely successful in reducing our product costs and our manufacturing cycle times.

Our CIM goals for design integration, are to improve the automatic interfaces between Engineering CAD/CAM and our shopfloor machines. This will allow us to reduce the time required to introduce new products.

Shopfloor Control Systems

Shopfloor Control Systems will continue to be an important focus for the CIM groups. In addition to conveyor control, we extensively interface to AS/RS systems and AGV's. Apple also utilizes robotics and "smart" machine controllers in our Printed Circuit Board operations.

Shopfloor Data Collection

By continuing to strive for improved product quality, the demand for reliable and timely data from the shopfloor becomes critical. We are utilizing data collection to control inventories, monitor yields, and provide statistical process control.

Business Systems Integration

Although the shopfloor data collection systems give real-time information as described above, the need still remains to provide our business system with completions and other information which can augment our MRP systems.

Distribution Integration
A major strategy for the future is to shorten the time required for product leaving manufacturing to arrive in our customers hands. By managing this supply chain, we reduce inventory, while increasing our responsiveness and flexibility.

To accomplish this, we need tighter integration between our business systems and distribution systems. We want to expand the visibility of the DRP systems to the manufacturing shopfloor, so that distribution can better plan what is ready to leave the factory.

Electronic Data Interchange
Another component of supply chain management, is managing our suppliers. To this end, we are involved extensively in the committees working on EDI. EDI will allow us to handle purchase orders, reciepts and sales via electronic links with our supply base and our dealer base. This in turn will drastically reduce costs and increase responsiveness.

Artificial Intelligence
We believe that in the years ahead, AI will play an increasing role in the manufacturing environment. As our systems and processes become more complex, the application of expert systems can improve system uptime. In addition, we are investigating heuristics as a method of implementing sophisticated scheduling rules, allowing us to potentially schedule around material shortages and "down" equipment.

CIM ORGANIZATION

As each site developed its own automated systems, it also developed its own CIM organization. As the development was driven by the field, there was limited interaction between the CIM organizations. In short, they were decentralized.

Being decentralized had the advantage of allowing the CIM organizations to be highly responsive, yet led to a lack of coordination in Systems Architecture. As a result, we tried to manage CIM in a centralized manner, with Corporate taking the lead role. While the centralized organization allowed for more co-ordination between the sites, it was very inefficient.

Having tried both decentralized and centralized methods of managing CIM, and not being happy with the results of either, we have embarked on a new approach. The new model is what we call Distributed CIM. Simply put, it implies that research is co-ordinated in a shared manner, but implemented independently by the sites (in a decentralized manner).

All projects are managed by the sites, with one site taking on a leadership role. This site will then be assigned resources (possibly from the other sites) to implement the project. When the project is completed, it will be utilized by all manufacturing sites. It has been proposed that a new role be created, that of a CIM Liason. The responsibility of this person is to assure that the sites are working in a unified manner, and to serve as a liason with Corporate.

SUMMARY

Since 1983, Apple has successfully applied the Macintosh to its manufacturing environment. We have taken advantage of its strengths in user interface and networking, and have designed a systems strategy around it. In the process, we have gained some valuable insights...

Introducing any new manufacturing system to Production is a challenging task. The amount of effort required to debug and integrate our software controls architecture, responsible for automatically routing and tracking all materials and assemblies within the system, is significant.

Involvement in training production and support resources on new equipment, concepts and technologies also consumed a substantial portion of our CIM groups efforts. Since little support infrastructure was in place at production startup, much of the burden of supporting the system fell upon our project teams.

The lessons learned over the years highlight the importance of continued involvement of production when the introduction of new disciplines are involved. While top-level management buy-in is critical to project justification and continued support, low-level commitment and buy-in is equally as important to guarantee the system's success.

When new manufacturing concepts and technologies are introduced, adequate training, necessary to promote an awareness of associated benefits, is crucial to ensuring an efficient use of the system. Once the advantages are understood, its users are usually quick to champion implementation efforts.

Within any flexible manufacturing system, process focus is every bit as important as individual product focus. Therefore, it is important to promote new measures to judge the system efficiency and operator productivity. In addition, early involvement of support engineering groups guarantees that proper processes and procedures will be in place when the system is introduced to production.

Recognizing similar symptoms within any flexible manufacturing system will prevent problems that may plague system operation. By anticipating such problems with adequate training and early participation of key support groups, process inefficiencies and unnecessary production downtime can be avoided, and expected gains may be realized.

Over the years, we have learned that the key to good systems integration is to remain flexible. Not only do our manufacturing systems need to remain flexible, but our organizations need to change as required to meet our growing needs. We believe that a distributed model for CIM will allow us to become consistent on company-wide metrics, while remaining responsive to the individual needs of the manufacturing sites.

BIOGRAPHICAL DATA

Michael Kaskowitz, Apple Computer, Inc.
48105 Warm Springs Blvd. Fremont CA, 94539.
Mr. Kaskowitz has held numerous positions in Apple Computer over the past ten years, including Fremont CIM manager, FMS project manager, and most recently working to establish the Global CIM model. He currently devotes much of his time to lecturing on Macintoshes in Manufacturing, and is involved with DeAnza College and sits on its CACT advisory committee. He is a member of SME, and IEEE.

Additional Readings on Systems Integration Case Studies

Emrich, Mary. 1991. Dissecting a systems integration project, part 1. *Manufacturing Systems*. May, pp. 18-22.

Emrich, Mary. 1991. Dissecting a systems integration project, part 2. *Manufacturing Review*. June, pp. 46-50.

ABOUT THE EDITOR

Joe H. Mize is Regents Professor in the School of Industrial Engineering and Management at Oklahoma State University. He is also Director of the Center for Computer Integrated Manufacturing at OSU. He is a Senior Associate of CIM Systems, Inc. of Dallas where he is a leading consultant to manufacturing firms in developing long-range strategic plans for CIM implementation. Mize holds a BSIE from Texas Tech and a MSIE and Ph.D. from Purdue University. Dr. Mize is a Past-President and Fellow of the Institute of Industrial Engineers and the 1990 winner of the Frank and Lillian Gilbreth Industrial Engineering Award. He is also a member of the National Academy of Engineering.